SMALL ARMS, ARTILLERY AND SPECIAL WEAPONS OF THE THIRD REICH

SMALL ARMS, ARTILLERY AND SPECIAL WEAPONS OF THE THIRD REICH

Terry Gander and
Peter Chamberlain

MACDONALD AND JANE'S · LONDON

First published in 1978 by
Macdonald and Jane's Publishers Limited
Paulton House, 8 Shepherdess Walk,
London N1 7LW

ISBN 0 354 01108 1

Designed by Judy Tuke

Printed in Great Britain by
Tonbridge Printers Ltd,
Tonbridge, Kent.

Contents

Errata

p10 Datenblätter für Heeres Waffen, Nuremberg 1944 should read 1974.

p39 8.5 mm Gewehr 241 (f). Picture transposed with 8 mm Karabiner 551 (f) on p 48.

p88 sMG 08 with AA adapter should read Yugoslav Mitralez 7.9 mm M 8 M.

p88 sMG 08 as harbour AA defence weapon should read sMG 08 with AA adapter.

p302 12 cm Granatwerfer 42. Caption should read 12 cm GrW 42 on the Eastern Front.

p337 schweres Wurfgerät 41. Captions of bottom two pictures transposed.

p352 Blendkörper IH and Nebelkerze 39. Transpose data entries.

Introduction

Although World War II has been over for more than thirty years, the social and political results of that conflict are with us still, and will be for many years yet. One of the major instigators of those changes was the armed might of the German nation, a might embodied in the German Army, Navy and Air Force along with the numerous German quasi-military organisations. Over the years since 1945 it has become an accepted truth that the German fighting machine was a supreme example of military efficiency, superbly trained and equipped, and utilising revolutionary tactics that swept all before it. While all this was certainly true of the front-line German forces of 1939 it was most certainly not applicable to the vast bulk of the German forces in 1945. But the years 1939–1945 were spent in a gradual harnessing of the efforts of three of the world's major nations into a concerted campaign to crush what was one single European nation. Such a diversion of effort can only be explained by the unique military tradition of Germany and its founder states, a tradition that reaches back many hundreds of years to produce one of the most autocratic and military-based nations the world has ever known.

The emergence of the German military state can be traced in modern terms to the gradual emergence of Prussia as a major European nation during the early 19th century. Prussia pursued a major re-unification of the German nation by the brutal expedient of armed expansion, and this re-unification became an accepted political fact in 1870 when the new Germany beat and humbled the traditional German foe, France. Thereafter Germany grew richer and more powerful and her expansion of trade and influence was only finally curbed by World War I. The end result of that war was a major disaster for Germany. Her Army and Navy were reduced to mere shadows of their former selves, but in typical German fashion, even before the victorious Allies had drawn up the terms of the Treaty of Versailles, the German General Staff and other military planners were busy preparing for the future.

One of their first moves was to carefully stockpile and conceal as much military equipment as was feasible. In this way much valuable material and many weapons were hidden away from the gaze of the various inspecting Treaty commissions, ready for any possible future application. After 1919 the strength of the German Army was restricted to 100,000 men but with such a small force the General Staff were able to place the accent on quality training and preparation for the future, a future that was further assured by the use of a short-service scheme that soon built up a large Reserve Army. Further preparation for the future was ensured by close tactical analysis of the lessons of World War I with an eye to any possible future trends of weapon design and tactical developments.

Although the 1920s were years of economic stringency in Germany, as they were for most of Europe, German military expenditure was kept at a relatively high level although little of this expenditure was subject to public scrutiny. The terms of the Versailles Treaty severely restricted defence and other military spending but despite this the German War Office quietly funded a great deal of commercial and military research into a wide range of military topics. Little of this work was carried out in Germany itself during the 1920s for the Treaty commissions were very active during those early years. Most of the major German armament concerns made commercial arrangements with foreign industrial companies or quite simply gained control or took over existing firms in nearby countries and converted them to their own use. For example, Krupp formed a fruitful association with the Swedish firm of Bofors while Rheinmetall took over a Swiss watch factory at Solothurn and converted it into a research centre for automatic weapons. Other German companies took over concerns in Denmark and Holland. These 'foreign' shadow factories had the approving paternalism of the German War Office which often funded the work carried out in them and fed guidance and design requirements to the development teams that spent the 1920s and early 1930s working to perfect new weapons for the future. By the late 1920s War Office support had spread to the diplomatic front, for in 1927 the German-funded research and development facility at Kazan, deep inside the Soviet Union, was secured for vehicle and weapon testing.

Many of the early development projects in the 1920s were devoted to pure research into all facets of modern technology. Chemistry, metallurgy, optics, mechanics, materials, all were investigated for what they might offer to future weapon application. Also investigated were new weapon concepts such as the general-purpose machine gun, the anti-tank gun, armour-piercing ammunition, rockets and an array of other weapons of modern warfare.

By the early 1930s the German forces were in a state of gradual and unobserved expansion to be ready for any future conflict. By 1932 the results of the decade of research were appearing in the workshops of German factories ready for future mass production, and in 1933 the NDSAP came to power. All remaining Treaty restrictions were cast aside and a frantic period of growth in all arms of the German forces began. Manpower flocked to the forces, but as the numbers grew there commenced a serious shortfall in equipment levels that was never to be fully reconciled, not even before 1945. The new armies were issued with, and trained with, the stockpiled equipment from 1918. Many of these old weapons were still in use after 1939 as new equipment was slow in reaching the ever-expanding forces. The truth was simply that, while Germany had the will and manpower resources to expand the Army, Navy and Air Force, German industry was not geared or organised to produce weapons on the scale required, and only a major reorganisation carried out under the stress of war enabled the needs of the forces to be met. Even as late as 1942 some sectors of German industry were still busy producing perambulators and garden furniture.

By the late 1930s the fruits of the years of research in the form of superb modern weapons and equipment were at last reaching the troops. Most of this new material was issued to a relative handful of elite formations to form the spearhead of the new German forces. Many of these elite units had a difficult period of growth during the mid-1930s, for in 1935 conscription was introduced in Germany and most trained formations were often broken up to form the cadres of an ever-growing number of new units. Training was thus fragmented and rushed but the political upheavals of mid-Europe often provided practical experience that was more than a substitute. A typical example of this was the German occupation of the Rheinland in 1936, and further valuable experience was gained in the Spanish Civil War from 1936 to 1938. The Spanish experience was especially valuable in weapon testing, for any modifications could be easily incorporated on the German assembly lines at a time when mass production was only just commencing.

The German General Staff viewed the years 1937–1939 with trepidation, for their war plans were based on the premise that Germany was unlikely to be involved in a major war before 1942 or 1943. Consequently the Army strengths and equipment levels were well below what was deemed necessary when Hitler's political opportunism precipitated the invasion of Poland in September 1939 and World War II began. At the time the situation was considerably helped by the results of the final annexation of Czechoslovakia in March 1939, which placed the entire armoury of that country at the disposal of the German forces, along with the capacity of the important Skoda Works at Pilsen. Thus when the Germans invaded Poland, and later France, whole divisions were equipped with Czech material and this material was kept in

German use, very often until 1945.

Although the Czech windfall provided a useful numerical boost for the Germans in 1939 and 1940 the expanding German forces made increasing demands on German industry for more and more war material. While German industry strove to meet these needs it was only a dramatic reorganisation that began in early 1941 which finally placed the German nation on a fully-committed war footing. The man responsible for the upheaval was Reichsminister Speer, who devoted his energies to organising the German war effort into building up one of the most powerful military nations the world has ever encountered. Duplication and waste were gradually eliminated and the power of Speer's ministry grew to such an extent that by 1944 every back-street workshop was under his control. Industry mushroomed, and to serve its needs labour was coerced, conscripted or forced from all the German-occupied territories in attempts to bolster the overstretched German manpower reserves.

The conquests of 1939, 1940 and 1941 placed huge tracts of territory under the domination of the Reich, but at the same time they brought their own military problems. First and foremost, occupation units had to be found and equipped. Industry was only just able to meet the front-line demands so the new garrison and occupation divisions had to be issued with whatever was left over. As the stocks of old German equipment were limited, recourse had to be made to captured weapons and equipment. Some captured equipment was considered good enough for front-line issue and very often captured armament facilities were kept in being to provide weapons for the German forces. But the bulk of the war booty was often of limited value due mainly to age or lack of spares and ammunition back-up. Nevertheless, it was considered suitable for second-line service, and as a result the bulk of the German forces, those not actually in the front-line fighting, were issued with a great variety of weapons and equipment that soon grew into a quartermaster's nightmare. The weapons were gathered from every corner of occupied Europe and their variety was increased after the German invasion of the Soviet Union produced vast stockpiles of all kinds of Soviet weapons and equipment. Thus the German forces gradually adopted a two-tier scale of weapon issue – the front-line troops were issued with the best that German designers and industry could provide, while in the rear areas and in the occupied territories and training camps virtually anything that came to hand was in service.

Throughout the war the fighting German troops were superbly equipped and led. The quality of their weapons was recognised by all who encountered them but as the war continued not only did the quantitative demands on German industry increase but an ever-growing number of requests for improved and new weapons flowed from the battlefields. Meeting these demands was no easy matter for the German war machine as major upheavals in production totals were costly and inevitable, but until 1945 the German soldier generally retained his equipment initiative. This was largely due to the overall weapon development carried out under the control of three departments – *Oberkommando des Heeres* (OKH), *Oberkommando der Luftwaffe* (OKL) and *Oberkommando der Marine* (OKM). Each of these departments was responsible for the weapon needs of its own arm of the service, but often liaison between the three was poor or non-existent, a state of affairs which often led to duplication of effort. The three departments were responsible also for analysing future weapon needs and the subsequent issuing of specifications and production orders as well as the testing and inspection of new designs. Unfortunately, their task was often made more difficult by the machinations of the SS, who often used their considerable influence to interfere with the smooth running of procurement programmes for their own political ends. The SS even had its own separate supply organisation, which often led to yet more duplication of effort and non-standardisation of equipment.

Unfortunately for the German war effort the unique political structure of the Third Reich produced a whole host of problems that proved detrimental to all aspects of the military organisation. All power flowed from Hitler himself, and thus around him grew a relatively small clique of staff officers, party functionaries, industrialists and hangers-on who continually bickered and squabbled amongst themselves for the political power and rewards that were available. Consequently inter-service rivalry grew from its normal healthy competition into counter-productive scheming and plotting for advantages in raw materials and facilities. The influence of the SS in this internecine warfare has already been mentioned, but very often their sinister burrowing went unnoticed until they emerged in a position of influence or power in important weapon projects. Very often other such attempts to steer or influence important weapon projects were made more brazenly, for to be seen in control of such important weapon projects as artillery production or V-weapon development was a good method of attracting attention, followers and political influence. Other interested parties would therefore often spend time and effort to block or disrupt programmes to divert some of the kudos away from their rivals, and in this blatant manner many important development programmes were continually disrupted or disorganised to the detriment of the German war effort. Another aspect of this empire-building was that frequently odd and quack weapon ideas were pushed ahead with great diversion of effort and facilities away from more important projects. These odd weapon systems were sponsored for little other reason than that their possible success might add to the personal standing of their sponsors with a subsequent rise in their political influence. This covert and overt struggle became more acute and the whole situation more chaotic during the last year of hostilities, when the German research and development organisations had to battle hopelessly with such bizarre contrivances as wind guns, vortice projectors, sound reflectors and the like.

The fact that such weird and useless projects ever got under way in the Third Reich was mainly due to the lack of an overall body to assess and co-ordinate research and development of weapons. Each of the services carried out its own research programmes and very often did so in complete isolation from any other work of a similar nature. Apart from the services, industrial and academic research also worked without any form of overall guidance. The results were often chaotic and wasteful. Another aspect of this lack of an overall watchdog was that many weapons that can only be described as gimmicks or technical novelties were lavished with a disproportionate amount of attention and some even reached the soldiers in the field. It would seem that the German mechanical genius was often bemused by technical innovation to the detriment of a viable tactical application, so the German soldier sometimes found himself issued with, and spent valuable time in getting trained to use, all manner of useless or odd weapons. Typical of these oddities was the flare pistol converted to fire miniature (and practically useless) anti-tank grenades, the rifle that fired round corners, the fantastically expensive 21 cm K(E) long-range gun and the needlessly complex gun sights used on many light anti-aircraft guns. There were many other examples that could be quoted.

Despite all the difficulties outlined above, the German services and industry were able to provide weapons that had no peer anywhere, and which were produced in ever-increasing numbers, despite Allied bombing raids. The output of 1944 and early 1945 was greater than at any time previous, but by then the main problem was that the weapons being produced were not the ones needed in the front lines. After about mid-1943 the technical race to keep abreast of current demands was gradually being lost by Germany and the needs to produce sufficient numbers of modern weapons were met less and less. A typical example of this was the '43' group of artillery pieces, intended to give the German artillery arm a supremacy over their opponents. The specification was issued in 1943 but by 1945 only single prototypes or mock-ups were all that industry could show, and many German gun-

ners saw the end of the war with the same artillery weapons they had been using in 1939.

Until late 1944 German weapon design and development were directed towards producing first-class weapons that, despite the gradual introduction of advanced mass production techniques of sheet-metal stampings and the like, were often of high-grade quality and performance – the *MG 42* was a case in point. But by late 1944 the Reich itself was in danger of attack over its borders and air raids were already a major disruptive element. Under the circumstances lower standards had to be accepted, and the accent was switched from quality to quantity with the *'Primitiv-Waffen-Programm'* of October 1944. Very much a last-ditch effort, this programme was primarily intended to produce infantry weapons for the Volkssturm divisions formed in the same month but the ultimate aim was to simplify weapon manufacture for all arms of the German services. But the whole *'Primitiv-Waffen-Programm'* practically came to nothing and only a handful of prototypes had been completed by the end of World War II in Europe. The reasons for the failure are not hard to discover. The programme had been launched with over-optimistic expectations and unrealistic goals at a time when the very fabric of the Third Reich was crumbling under land and air attack. By that time damage to road and rail transport was already almost impossible to repair, fast-dwindling fuel reserves progressively lamed most weapons of modern warfare, and conditions in all German-controlled areas were becoming increasingly chaotic. Inside the Reich itself the huge foreign slave labour pool suffered untold miseries under the German yoke but such unwilling workers were no substitute for skilled and well-organised labour, and production schedules were delayed and disorganised. Raw material supply dwindled to a trickle and by May 1945 the conditions inside Germany were almost medieval. In the various front lines German troops fought to the bitter end but behind them the German military machine collapsed and the once powerful armed might of the German nation was no more.

Fremden Gerät Numbers

Prior to the outbreak of the 1939–45 war, the Heereswaffenamt (Army Ordnance Department) inaugurated a system for listing data on all known foreign weapons and equipment. This information, with illustrations where possible, was published in a series of loose-leaf books known as the *Kennblätter Fremden Geräts* (D.50 series) – Recognition Books for Foreign Equipments. These books were periodically amended after 1939 as captured weapons and equipment fell into German hands.

The D.50 series consisted of fourteen volumes of which the following were concerned with weapons:

D.50/1	*Handwaffen*	Handweapons – pistols, rifles and sub-machine guns
D.50/2	*Maschinengewehre*	Machine guns – light and heavy
D.50/3	*Werfer*	Mortars
D.50/4	*Leichte Geschütze*	Light artillery
D.50/5	*Schwere Geschütze*	Medium artillery
D.50/6	*Schwerste Geschütze*	Heavy and super-heavy artillery
D.50/12	*Kraftfahrzeuge*	Motor vehicles – armoured cars, tanks, self-propelled guns, armoured tractors, artillery tractors, half-tracked vehicles
D.50/14	*Pioniergerät*	Engineer equipment – mines, grenades, fuses, demolition charges

The *Fremden Gerät* designation system consisted of the German nomenclature for the weapon or equipment involved, and a group number for that particular weapon, both of which were followed by the initial letter of the country of origin. From the very start of World War II the Germans made the maximum use of almost every weapon that came into their possession. Normally the weapons were used with their own captured ammunition but in a few cases the Germans went to the extent of manufacturing more ammunition for them and in some cases re-bored some weapon types to conform to German calibres. Examples of the latter were the Soviet 7.62 cm and 8.5 cm anti-aircraft guns bored out to 8.8 cm as stocks of ammunition for this calibre were more readily available.

Weapons taken into service with the German armed forces were identified by their *Fremden Gerät* designation which was also applied to the German-published workshop manuals and spare parts relevant to each weapon. There were exceptions to the system, one of which applied to the many Czech weapons used by the German forces which had been impressed before the *Fremden Gerät* system was officially approved for service. Another anomaly arose when troops in the field often referred to their charges by their year of origin or model number – one example is the French 7.5 cm FK 231(f) which was often known as the 7.5 cm FK 97(f). However, such colloquial designations were unofficial, even if they did appear in reports and Ordnance returns.

Acknowledgments

Photographic credits: Imperial War Museum (London), Bundesarchiv (Koblenz), Etablissement Cinématographique des Armes (Paris), C. Yust, I. Hogg, K. Pawlas, US Official, R.H. Mayne, Chamberlain Collection and Gander Collection.
The authors would also like to acknowledge most gratefully the assistance and guidance given by Mr Alex Vanags-Baginskis, who edited this work.

List of sources

The information and data contained in this book have been culled from a myriad of sources. In the course of researches that began as a mild hobby and grew to become an obsession numerous works, published and unpublished, have been used to glean facts and figures which have been combined in this book. Many of these sources have not been included in the list below as they consisted of magazine articles, odd pages removed from unremembered works, unpublished notes, paragraphs from intelligence bulletins, scraps of conversation and the odd fleeting glimpse of a weapon in a newsreel. Such items cannot be usefully recorded in a book of this nature, but the listing below does include the bulk of the sources used to compile this book. Not included are the actual service manuals for the weapons themselves, but many of these were consulted in the Imperial War Museum Library and the Library of the Royal Artillery Institution at Woolwich. More data has come from the various editions of *Waffen Revue*, edited by Karl Pawlas at Nuremberg, a publication that acts as valuable archive material for weapons of many kinds. Reliance on these manuals has produced some anomalies in the data tables which in some cases are at variance from previously published material. Where such anomalies have occurred we have placed our reliance on the original German material.

A Pocket History of Artillery. Kosar. Light Field Guns. London 1974; Mittlere Feldgeschütze. Munich 1974

Aircraft Rockets and their Installations. Halstead Exploiting Centre translation. Undated

Armi della Fanteria Italiano. Pignato. Parma 1971

Armi Portali, Artiglierie e Semoventi del Regio Esercito Italiano 1900–1943. Benussi. Milan 1975

Artillerie im Küstenkampf. von Harnier. Munich. Undated

Artilleristische Rundschau. Various volumes. Munich 1933–1943

Artiglierie e Automezzo dell'Esercito Italiano. Pignato. Parma 1972

Artiller of the World. Foss. London 1974 and 1976

Artyleria Ladowa. Pataj. Poland 1975

Artyleria i Rakiety. Warsaw 1972

BAOR Technical Intelligence Report No 79. The Development of the Heavy AA Gun in Germany from 1918 to 1945. 1946

Barbarossa. Clark. London 1965

BIOS Report. Development of Panzerfaust. London 1945

BIOS Report No 2/26. Interrogation of the Personnel of the Ballistics Section of Rheinmetall-Borsig. London. Undated

BIOS Report 182. German Coast Artillery Equipment employed in the Defence of the West Coast of Denmark. London. Undated

Book of Pistols and Revolvers. Smith. Harrisburg 1968

Brassey's Artillery of the World. London 1977

Canadian Military HQ Summaries:
- Artillery Equipments
- Pistols, Rifles and Machine Carbines
- Mortars and Projectors
- Rocket Projectors
- London 1945

Captured Stores Equipment Catalogue. Alexandria 1943

Catalogue of Enemy Ordnance Materiel. Vol 1 – German. Washington 1945

Ceskoslovenske Delostrelecke Zbrane. Karlicky. Prague 1975

CIOS Reports:
- Artillery Carriage and Gun Development by the Rheinmetall-Borsig AG
- Artillery Experimental Range, Hillersleben
- Development of Weapons by Rheinmetall-Borsig
- Investigation of Artillery Design and Development performed by Rheinmetall-Borsig AG
- Krupp AG and Bochumer Verein
- Recoilless Gun Development of Rheinmetall-Borsig
- Reich Ministry of Armaments and War Production. Part 2 – Summary of Technical Information
- Skoda Works, Pilsen
- All from SHAEF 1945

D 97/1 (Ordnance listing). Berlin 1944

D 1864/1. Panzerschreck (8.8 cm RPzB 54 mit 8.8 cm RPzBGr 4322). Berlin 1944

Data concerning AA Weapons. Halstead Exploiting Centre. HEC 5309. Undated

Datenblätter für Heeres Waffen, Fahrzeuge, Gerät. Pawlas reprint of OKH document dated August 1944. Nuremberg 1944

Dati Tecnici sulle Artiglierie in Servizio. Rome 1938

Design and Development of Weapons. Postan, Hay and Scott. HMSO London 1964

Deutsche Artillerie 1934–1945. Engelmann und Scheibert. Limburg/Lahn 1974

Deutsche Raketenwerfer. Engelmann. Friedberg 1977

Dictionary of Explosives, Ammunition and Weapons (German Section). Picatinny Arsenal Technical Report No 2510. 1958

Die Deutsche Geschütze 1939–1945. Munich 1959

Die Entwicklung der Maschinenwaffen, insbesondere bei Rheinmetall-Borsig bis zum Kriegsschluss 1945. Unterlüss 1946

Die Walther Leuchtpistole. Zella-Mehlis 1939

Enemy Equipment. German Mines and Traps. War Office London 1943

Enemy War Materials Inventory List. Section 2 – Armaments. SHAEF 1945

Equipment 'Hammer'. Halstead Exploiting Centre HEC/10 831. 1945

50 Jahre Rheinmetall Düsseldorf 1889–1939. Düsseldorf 1939

Field Rocket Equipment of the German Army 1939–1945. Gander. London 1972

Fire and Movement. RAC Tank Museum 1975

88 – Flak and Pak. Chamberlain and Gander. Windsor 1976

Flak-Gerät des Auslandes. Berlin 1938

Flugzeug-Bewaffnung. Schliephake. Stuttgart 1977

German Aircraft of the Second World War. Smith and Kay. London 1972

German Army in Pictures. War Office London 1944

German and Japanese Solid-fuel Rocket Weapons. MID War Dept. Washington 1945

German Artillery in the Channel Islands 1941–1945. Ginns (original unpublished form). Jersey 1975

German Artillery of World War 2. Hogg. London 1975

German Explosive Ordnance Vol 1 (OP 1666). Forest Grove reprint 1969

German Fortifications in Jersey. Ginns and Bryant. Jersey 1974

German Infantry Weapons of World War 2. Barker. London 1972

German Infantry Weapons Vol 1. Washington. Reprint 1966

German Machine Guns. Musgrave and Oliver. Washington 1971

German Order of Battle 1944. Reprinted London 1975

German Railway Guns in Action. Engelmann. English translation 1976

German Research in World War 2. Simon. New York. Undated

German Weapons Illustrated. War Office London 1943–1945

German Weapons, Uniforms, Insignia 1841–1918. Hicks. La Canada, California 1937

Geschichte des zweiten Weltkrieges. Berlin 1960
Glossary of Captured German Material 1945. War Office London 1945
Gunners at War. Bidwell. London 1970

Handbook of Enemy Ammunition. Pamphlet No 15. German Ammunition Markings and Nomenclature. War Office London 1945
Handbook of German Anti-aircraft Artillery (Flak). War Office London 1946
Handbook of German Military Forces (TM E 30–151). Washington 1945
Handbuch: die Tschechoslowakische Wehrmacht. Berlin 1938
Handfeuerwaffen Band I und II. Lugs. East Berlin 1962. Translated from Czech. Prague 1956
Hitler's Atlantic Wall. Partridge. Guernsey 1976
Hitler's Fortress Islands. Toms. London 1967

Illustrated Record of German Army Equipment 1939–1945:
Vol 2. Artillery. Parts 1 and 2
Vol 5. Mines, Mine Detectors and Demolition Equipment. War Office London 1948
In der Festung Guernsey. Guernsey 1945
Intelligence Bulletins Vol 1–5. War Department. Washington 1942–1945
Intelligence Bulletin. 'Guided Missiles . . . the weapon of the Future.' Washington 1946

Jane's Infantry Weapons 1975. London 1975

Kennblätter fremden Geräts (D.50 Series):
D.50/1 Handwaffen
D.50/2 Maschinengewehre
D.50/3 Werfer
D.50/4 Leichte Geschütze
D.50/5 Schwere Geschütze
D.50/6 Schwerste Geschütze
Berlin 1941–1945
Kriegssoll (Heer) an Vorschriften (KaV) Heft 8. Panzertruppe-Feldheer. OKH 1943
Kystartillerist i Norge. Norsk Militaert Tidsskrift. Hefte 12. Oslo 1974

Liste der Vorschriften über Beutegerät. Berlin 1944

Mauser Aktennotiz Nr 3679. MK 103 Ortfest für Flugabwehr. Oberndorf 1944
Mauser Aktennotiz Nr 3761. Vorführung von Flugzeugbordwaffen in Lafetten für Erdeinsatz in Kummersdorf am 10.10.44. Oberndorf 1944
Mauser Rifles and Pistols. Smith. Harrisburg 1946
Military Pistols and Revolvers. Hogg. New York 1970
Military Smallarms of the Twentieth Century. Hogg and Weeks. 1st edition. London 1973

Ordnance Target Report No 41. Recoilless Gun Development of Rheinmetall-Borsig. ETOUSA
Oruzhiye Pobiedi 1941–1945. Molodaya Gvardiya. Moscow 1975

Panzernahkampfmittel. OKH 1943
Panzernahkampfwaffen (Panzerfaust). OKH 1944
Pictorial History of the Machine Gun. Hobart. London 1971
Pictorial History of the Sub-machine Gun. Hobart. London 1973
Pocket Book of the German Army. War Office London 1943
Polska Bron. Kobielski. Warsaw 1975
Prezentuj Bron Orez Zolnierza Polskiego 1939–1970. Magnuski. Warsaw 1971
Proiottili, Cannoni, Semovente Controcarro e Trattori dell'Esercito Tedesco 1936–1945. Pirella. Milan 1976

Rail Gun. Batchelor and Hogg. Poole 1973
Rakiety Bojowe 1900–1970. Burakowski and Sala. Poland 1973
Recoilless Guns, development and data. Halstead Exploiting Centre HEC/5351. 1945
Rheinmetall-Borsig AKH Division. Artillery Development 1922–1940. 1945 translation
Rodowod Katiuszy. Galkowski. Warsaw 1972
Rozwoj artylerii przeciwlotniczej. Przeniczny. Warsaw 1973

Small Arms in Profile. Various parts. Windsor
Small Arms of the World. Smith & Smith. 10th edition. Harrisburg 1973
Soviet Projectile Identification Guide. Ordnance Technical Intelligence. US Army 1952
Special Series. Various. MID War Dept. Washington 1944
Stand der Entwicklung und neue Gesichtspunkte für die Küstenverteidigung. Unterlüss 1947
Summary of Research and Development being conducted in Germany when the War ended. War Office London 1945

Table of German Armament Equipment (Approved and Experimental). Parts I–X. Unterlüss 1947
Taschenbuch Jugoslawisches Heer. Berlin 1941
Technika Wojskowa LWP XXX LAT Rozwoju 1943–1973. Warsaw 1973
Textbook of Automatic Pistols. Wilson/Hogg. Reprint London 1975
The Big Guns at War. Cleeve. London 1945–1946 (unpublished)
The Book of Rifles. Smith. New York 1948
The German Occupation of the Channel Islands. Cruickshank. Guernsey 1975
The German 'Sturmgewehr'. von Lossniter. Unterlüss 1947
The Guns of World War 2. Hogg. London 1976
The Machine Gun. Chinn. Washington 1951
The Mare's Nest. Irving. London 1964
The Revolver 1889–1914. Taylerson. London 1970
The World's Submachine Guns Vol 1. Nelson & Lockhoven. Cologne 1963
Typy Broni I Uzbrojenia:
7. Katiusza
45. Armata przecipancerna wz.36. Poland

Umlcene Zbrane. Prague 1966

Waffen des Auslandes, Frankreich. Berlin 1938
Waffen Revue. Various editions. Pawlas. Nuremberg
Walther Pistols and Rifles. Smith. Harrisburg 1962
Weapons of the German Infantry during World War 2. Harms and Fiest. California 1968
Weapon Mounts for Secondary Armaments. Noville and Assoc. Detroit USOC 1957
Wehrmacht Truppen-Transportvorschrift. OKW 1943
Wojsko Polskie 4,7,10. Warsaw 1970–1975
World War 2 Fact Files. Chamberlain and Gander:
Anti-tank Weapons
Machine Guns
Mortars and Rockets
Anti-aircraft Guns
Light and Medium Field Artillery
Heavy Artillery
Infantry, Mountain and Airborne Guns
Axis Pistols, Rifles and Grenades
Allied Pistols, Rifles and Grenades
Sub-machine guns and Automatic Rifles
London 1974–1976

Zusätze für den artl. Ausbau der Kanal- und Atlantikküste. Berlin 1942

Pistols

To the German soldier the term 'Pistole' meant the automatic or self-loading pistol, as from just before 1900 the entire output of German pistol designers and manufacturers had been devoted to this type of handgun. Conversely, from about the same time the revolver was almost entirely neglected as a military weapon. For some reason, peculiar to the German mechanical genius, the self-loading pistol attracted a great deal of attention with the result that, by 1939, German pistols had reached a high degree of sophistication in design, finish and reliability. In fact, German-designed pistols used during WW II were second to none when compared to similar weapons evolved elsewhere.

The role of the self-loading pistol with the German forces was much the same as the role of the pistol had always been, in that it was a light convenient weapon carried on the person in such a manner that it allowed the normal use of both hands and arms for other tasks. That was, and still is, the textbook definition – but the pistol has always been a bit more than that to the average soldier. The pistol, when worn or flourished, imparts an air of authority on the user to a remarkable degree, and this air of authority can also be manifested in the form of improving the self-confidence of the user. Thus, although the pistol was intended for the use of officers, aircrews, tank troops, signallers, drivers and the like the use of the pistol spread until, in the German forces, it virtually became second nature to wear a pistol whatever their duties. In addition to front-line troops the pistol also became part of the uniform of officers both in Germany and abroad and, in accordance with regulations, was worn by all NCOs and many junior ranks when on duty in occupied territories. The result was an ever-increasing demand for pistols which the German industry was unable to meet and large numbers of captured weapons had to be impressed into service. In addition to that, certain foreign designs were kept in full production for the German forces to supply the necessary new and replacement handguns, but even these measures never quite satisfied the needs.

The two most important German types were the *Pistole 08 (P 08)* and the *Pistole 38 (P 38)*. The most well known was the *P 08*, known as Luger or Parabellum in other countries, which has by now become so firmly identified with the popular concept of the German soldier or officer that no Hollywood wartime epic seems complete without every 'German' brandishing or wearing one. The *P 08* entered extensive German military use during WW I but it had been part of the German armoury since 1908 when it was adopted by the German Army and Navy. Thereafter its military use spread and a bewildering variety of models and calibres appeared. Early models were made usually in 7.65 or 9 mm, but by 1939 the 9 mm Parabellum version was the accepted military calibre. The basic military model with a barrel length of 102 mm had been standardised already before 1939, but early models did remain in service.

For all its popularity the *P 08* was far from being an ideal service pistol. Its upward-operating toggle breech mechanism was open to the elements and the entry of dirt and grit, and the pistol was not really suitable for large-scale production under wartime conditions.

Trials to find a replacement for the *P 08* were carried out from about 1935 onwards, and in 1938 the successful design was accepted for service as the *P 38*. This new pistol, developed and produced by the Waffenfabrik Carl Walther of Zella-Mehlis, Thuringia, was of very advanced construction that included such features as a double-action trigger mechanism and a protruding button indicating an empty magazine. Apart from that, the *P 38* was designed from the outset for large-scale production. At an average cost of RM 32, – it was also cheaper than the *P 08* (production cost RM 35,–) but despite these advantages it never did supplant the *P 08* before 1945 and the two types continued to provide the bulk of pistols used by the German troops until the end of WW II. Well over 1 million *P 38*s were made before hostilities ended. In 1957 the same basic pistol was adopted as the standard sidearm of the new West German Army, and production continued in slightly modified form as the *P 1*.

Far behind the *P 08* and *P 38* in quantity came another German design, the Mauser *C 96*. This rather obsolete handgun was still very much in use as a service pistol until 1945, but its bulk, relative complexity and decreasing numbers gradually relegated it to second-line duties. Even so, it was carried by many German soldiers on active service because, of all the various self-loading pistol designs ever produced, the Mauser *C 96* had an aura and appearance that put it in a class of its own. The *C 96* was built in many different models and calibres but by 1939 only the basic 7.63 and 9 mm and the *'Schnellfeuerpistole'* were in service. Evolved in 1931, the *'Schnellfeuerpistole'* was a selective fire handgun fitted with a 20-round magazine, used with the standard wooden holster as a shoulder stock. Despite the fact that when fired fully automatically this weapon was almost uncontrollable, it was favoured by Waffen-SS troops and often used in anti-partisan campaigns.

Apart from these better-known German designs there were many others in military use; practically every kind of handgun was taken over by the services after 1939. Many pistols produced for the commercial market became military issue, typical of these being the *Walther PP* and *PPK* originally designed for police use. Many thousands were manufactured for the German forces and issued to officers, aircraft crews, the military and political police, and Party officials. Other designs that, because of their relatively small size, fell into the same category were the Mauser *HSc*, Sauer *Modell 30* and *38*, Stock, Beholla, and the Ortgies pistols. Numerous commercial 6.35 mm models also found their way into use as ornamental sidearms for high-ranking officers but in that realm individuals often provided their own weapons and consequently their pistols were often non-service issue.

When Czechoslovakia was completely occupied by the Germans in 1939 the various Czech armament facilities came under direct German control, along with a huge quantity of Czech handguns and other weapons. Among these were the handy vz.24 pistols, most of which were taken over by the Luftwaffe, while two other types of pistols were kept in series production in the Czech factories for the German forces. These were the vz.27 and vz.38, which were adopted as the *7.65 mm P27(t)* and *9 mm P 39(t)*. Another foreign design adopted by the Germans in 1939 was the Polish 'ViS' wz.35. This pistol was kept in production at the Radom works until 1944, with almost the complete output going directly to the German Army and the Waffen-SS.

In 1940 the German invasion of the Low Countries brought another important arms facility under German control, the world-famous Belgian Fabrique Nationale d'Armes de Guerre (FN) at Herstal, Liège, which at that time was producing large quantities of self-loading pistols. Most of these were superlative weapons based on original designs evolved by the American John M. Browning. Large numbers of mle 1900, 1903, 1910 and 1922 pistols fell into German hands and were reissued to various Army and Luftwaffe formations, but for the Germans one of the chief prizes were the production lines for the Browning 'Grande Puissance' (High Power) pistol. This superb military handgun was kept in large-scale production and became an important part of the German front-line armoury as the *9 mm P 640(b)*.

By arrangement, Hungary also produced pistols for the German forces – the Pisztoly 37M, manufactured as *7.65 mm P 37(u)* at the Femaru-Fegyver es Gepyar, Budapest.

Thus by 1941 German troops were being issued with new pistols produced not only in Germany but also in Poland, Czechoslovakia, Belgium and Hungary, but even these increased

deliveries could not meet the needs of the expanding German forces. As a result, purchases were negotiated with a neutral country, Spain, and in 1941 an order for 6,000 Astra Mod.400 pistols was made. Known as the *9 mm P(Astra)* these weapons supplemented similar pistols captured from the French troops but their use was limited owing to the different cartridge. A much more useful Spanish pistol was the Astra Mod.600 designed for the 9 mm Parabellum cartridge but only 10,450 out of over 38,000 ordered could be delivered to the German authorities across the French-Spanish border before the German retreat in 1944.

Yet despite strenuous efforts to increase the output of handguns at home and in German-occupied countries there were never enough new pistols to cover the seemingly endless demands made by all branches of the German armed forces. The only solution was to issue the vast number of captured pistols to the various occupation, garrison, local auxiliary police and other second-line formations. The scope of types involved was enormous and a comprehensive list of all these captured and requisitioned weapons is beyond even a book of this nature. German pistol inventory lists were expanded with handguns from Denmark, Norway, France, Belgium, Holland, Greece, Yugoslavia, the Soviet Union and Italy, as well as British and American weapons; the logistics problems with ammunition supply must have been complex indeed.

By 1944 the number of pistols issued to German forces must have run into millions, but still the demand for more pistols came from the field as they were lost, damaged, captured or simply worn out. By that time the German war production machine was under increasing strain and of necessity concentrating on other weapons, with less capacity allocated to the manufacture of pistols. However, subsequent late attempts to develop a new, easily-produced 'war emergency' pistol resulted in some interesting designs. No final standard model was put into full production before the end of hostilities but a number of prototypes and trial weapons were made. Principal of these was the so-called *'Volkspistole'*, a crudely finished design with an unusual blow-back delay mechanism that retarded the slide until the bullet had left the muzzle. It was designed from the outset for mass production using unskilled workers and simple machines but only a limited number were ever completed.

A Waffen-SS trooper in action with a Schnellfeuerpistole *using its wooden holster as a shoulder stock*

9 mm Pistole 08

German designation 9 mm P 08
Calibre/cartridge 9 mm Parabellum
Magazine capacity 8 rounds
Length 222 mm
Length of barrel 102 mm
Weight unloaded 0.87 kg
Muzzle velocity 320 m/sec
Max effective range 50 m
Original manufacturer DWM (Deutsche Waffen- und Munitionsfabrik) 1898
Other manufacturers Simson & Co., Suhl; Mauser-Werke; Krieghoff; Erma, Erfurt; and others

Remarks: One of the best known pistols, generally described as 'Luger'. Originally produced in 1899 in 7.65 mm calibre, but in 1908 changed to an exceptional 9 mm round (Parabellum) which became world's most widely used pistol and sub-machine gun cartridge. Official sidearm for German NCOs and special troops. Production phased out in 1943. Over 2 million made in many different versions.

◀ German reconnaissance patrol armed with the Pistole 08 on the Western Front, early 1940

▼ Naval version Modell 1908/14 with a 140 mm barrel

---

9 mm Pistole 38

German designation 9 mm P 38
Calibre/cartridge 9 mm Parabellum
Magazine capacity 8 rounds
Length 215 mm
Length of barrel 125 mm
Weight unloaded 0.94 kg
Muzzle velocity 340 m/sec
Max effective range 50 m
Original manufacturer Carl Walther Waffenfabrik AG, Zella-Mehlis
Other manufacturers Mauser-Werke AG, Oberndorf-am-Neckar; Spreewerke, Grottau; Walther-Werke, Strakonitz; Brünn (Brno), occ. Czechoslovakia; Herstal, Belgium; and others

Remarks: An advanced and reliable double-action handgun evolved from prototype Walther Heerespistole (HP). Designed as replacement for P 08 but eventually produced in parallel; over 1 million made 1939–1945. Widely acknowledged as the best military pistol of WW II.

Walther Modell HP

Walther Armeepistole

German designation Walther AP
Calibre/cartridge 9 mm Parabellum
Magazine capacity 8 rounds
Length 217 mm
Length of barrel 124.5 mm
Weight unloaded 0.793 kg (steel frame); 0.644 kg (dural frame)
Muzzle velocity 340 m/sec approx
Original manufacturer Waffenfabrik Carl Walther AG, Zella-Mehlis

Remarks: Hammerless predecessor of Walther HP(P 38). Not accepted for service but some used for troop trials.

Walther Armeepistole

Mauser C 96

German designation C 96; also Schnellfeuer-Selbstladepistole M 30; Zwanziglader 1931 (7.63 mm)
Calibre/cartridge 7.63 mm Mauser or 9 mm Parabellum
Magazine capacity 10 to 20 rounds
Length without butt 299 mm
Length with butt 647 mm
Length of barrel 139 mm
Weight without butt 1.33 kg
Weight with butt 1.78 kg
Muzzle velocity (7.63 mm):480 m/sec; (9 mm):430 m/sec
Max effective range 50–300 m (with shoulder stock)
Original manufacturer Waffenfabrik Mauser AG, Oberndorf-am-Neckar

Remarks: Oldest and longest-ranging service pistol, heavy and clumsy but effective. Over the years made in some 30 versions in Germany and elsewhere. 9 mm Parabellum variant introduced in German Army in 1917. In German service during WW II often carried by anti-partisan troops and military transport drivers behind Eastern Front.

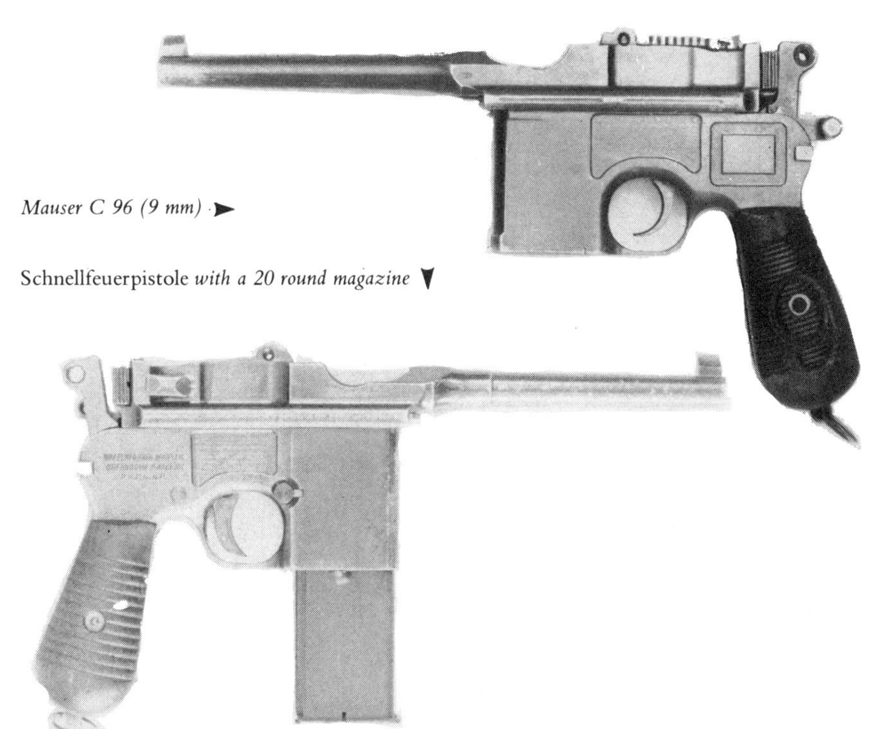

Mauser C 96 (9 mm) ➤

Schnellfeuerpistole *with a 20 round magazine* ▼

Volkspistole

Calibre/cartridge 9 mm Parabellum
Magazine capacity 8 rounds
Length 286 mm
Length of barrel 130 mm
Weight unloaded 0.96 kg
Muzzle velocity 381 m/sec approx
Max effective range 30 m
Original manufacturer Mauser-Werke AG, Oberndorf-am-Neckar

Remarks: Emergency pistol design intended for assembly from simple pressings and a minimum of machined parts. Only prototypes produced before end of WW II, all of which used an unusual locking system directing propellant gases forward to retard the barrel until bullet had left the muzzle. Prototypes also made by Walther and Gustloff-Werke using slightly different actions.

Walther PP

German designation Walther PP (Polizei-Pistole)
Calibre/cartridge 7.65 mm (.32 ACP); also 9 mm Short (limited numbers only)
Magazine capacity (7.65 mm) 8 rounds
Length 162 mm
Length of barrel 85 mm
Weight unloaded 0.708 kg
Muzzle velocity (7.65 mm) 289 m/sec
Max effective range (7.65 mm) 40 m
Original manufacturer Waffenfabrik Carl Walther AG, Zella-Mehlis

Remarks: A very advanced pistol for its time, and most reliable. Commercially introduced in 1929, becoming the most widely used police sidearm in Europe before WW II. Design copied in many other countries.

Walther PPK

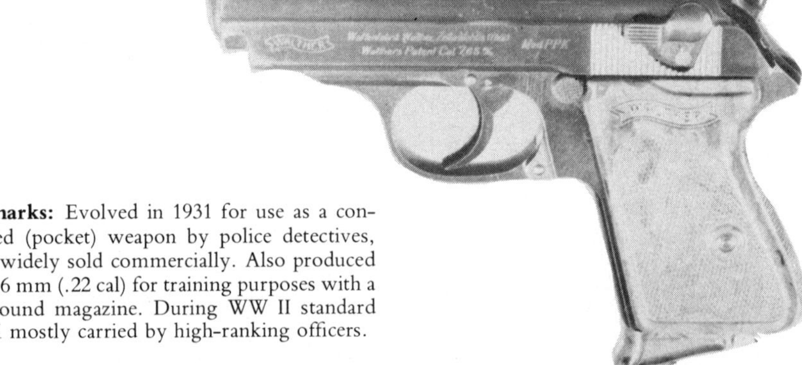

German designation Walther PPK (Polizei-Pistole Kriminal = Police Detective Pistol)
Calibre/cartridge 7.65 mm (.32 ACP); also 9 mm Short (limited numbers only)
Magazine capacity 7 rounds 7.65 mm; 6 rounds 9 mm Short
Length 148 mm
Length of barrel 80 mm
Weight unloaded 0.568 kg
Muzzle velocity (7.65 mm) 280 m/sec
Max effective range (7.65 mm) 40 m
Original manufacturer Waffenfabrik Carl Walther AG, Zella-Mehlis

Remarks: Evolved in 1931 for use as a concealed (pocket) weapon by police detectives, and widely sold commercially. Also produced in 5.6 mm (.22 cal) for training purposes with a 10 round magazine. During WW II standard PPK mostly carried by high-ranking officers.

Mauser Modell HSc

German designation Mauser Pistole neuer Art (MnA)
Calibre/cartridge 7.65 mm (.32 ACP)
Magazine capacity 8 rounds
Length 165 mm
Length of barrel 86 mm
Weight unloaded 0.596 kg
Muzzle velocity 290 m/sec
Max effective range 40 m
Original manufacturer Mauser-Werke AG, Oberndorf-am-Neckar

Remarks: Very advanced design when introduced in 1938. During WW II widely used by Luftwaffe aircrews and Navy personnel. A 9 mm Parabellum cartridge version designated HSv was proposed as standard German service sidearm but not accepted.

Sauer Modell 30

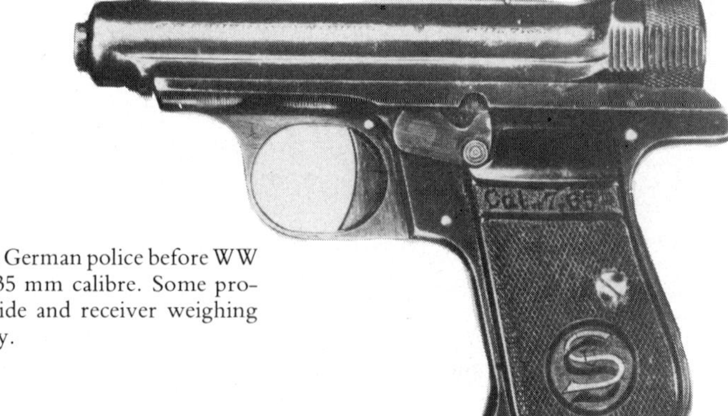

German designation Sauer Behördenmodell ('for officials'); also S&S BM
Calibre/cartridge 7.65 mm (.32 ACP)
Magazine capacity 7 rounds
Length 146 mm
Length of barrel 77 mm
Weight unloaded 0.625 kg
Muzzle velocity 274 m/sec
Max effective range 40 m
Original manufacturer J. P. Sauer & Sohn, Suhl

Remarks: Issued to German police before WW II. Also made in 6.35 mm calibre. Some produced with dural slide and receiver weighing only 0.411 kg empty.

Mauser Modell 34

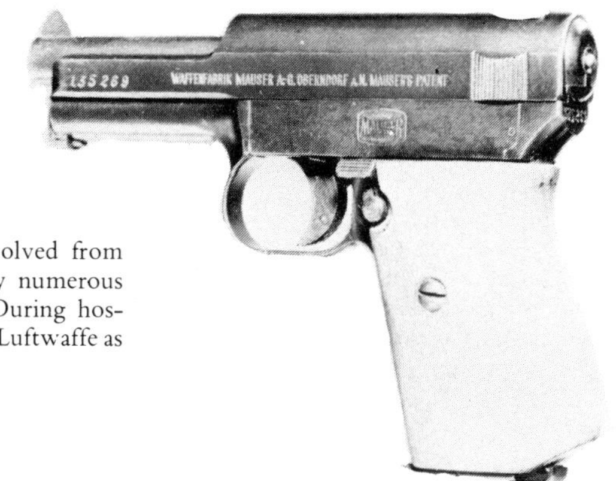

German designation Mauser Selbstlade-pistole 1934; Mauser-Pistole MaA
Calibre/cartridge 7.65 mm (.32 ACP)
Magazine capacity 8 rounds
Length 159 mm
Length of barrel 87 mm
Weight unloaded 0.6 kg
Muzzle velocity 297 m/sec
Max effective range 40 m
Original manufacturer Mauser-Werke AG, Oberndorf-am-Neckar

Remarks: Commercial model evolved from original 1910 design. Adopted by numerous police services prior to WW II. During hostilities issued to German Navy and Luftwaffe as substitute sidearm.

Sauer Modell 38(H)

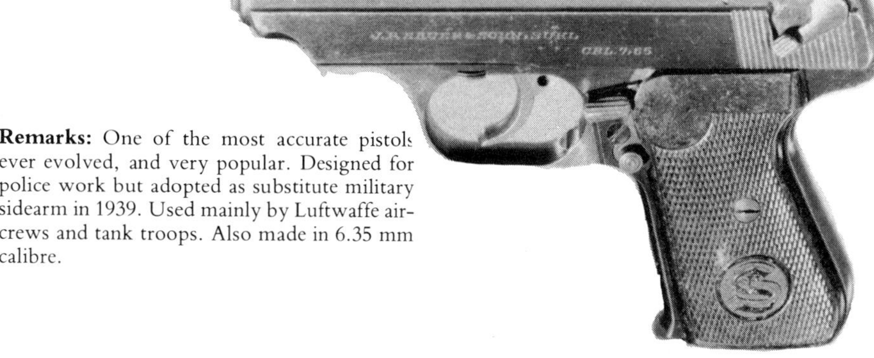

German designation Sauer-Selbstlade-pistole Modell 38(H) oder 38H
Calibre/cartridge 7.65 mm (.32 ACP); also 9 mm Parabellum
Magazine capacity 8 rounds
Length 171 mm
Length of barrel 83 mm
Weight unloaded 0.72 kg
Muzzle velocity (7.65 mm) 280 m/sec
Max effective range (7.65 mm) 30 m
Original manufacturer J. P. Sauer & Sohn, Suhl

Remarks: One of the most accurate pistols ever evolved, and very popular. Designed for police work but adopted as substitute military sidearm in 1939. Used mainly by Luftwaffe air-crews and tank troops. Also made in 6.35 mm calibre.

Ortgies Pistols

German designation Ortgies
Calibre/cartridge 7.65 mm (.32 ACP) and 9 mm Short
Magazine capacity 8 rounds (7.65); 7 rounds (9 mm)
Length 165 mm
Length of barrel 87 mm
Weight unloaded 0.6 kg
Muzzle velocity (7.65 mm) 280 m/sec (approx)
Max effective range 40 m
Original manufacturer Deutsche Werke AG, Erfurt and Kiel

Remarks: Basically commercial models, impressed as substitute military sidearms. Also made in 6.35 mm calibre.

Ortgies 7.65 mm ►

Ortgies 9 mm ▼

Rheinmetall Pistole

German designation Not known
Calibre/cartridge 7.65 mm (.32 ACP)
Magazine capacity 8 rounds
Length 164 mm
Length of barrel 92.2 mm
Weight unloaded 0.67 kg
Muzzle velocity 280 m/sec approx
Max effective range est 30 m
Original manufacturer Rheinische Metallwaren-und Maschinenfabrik, Sömmerda

Remarks: Based on Browning designs. Relatively few in service, mostly with German naval personnel.

Beholla

German designation Beholla Kal. 7.65 mm
Calibre/cartridge 7.65 mm (.32 ACP)
Magazine capacity 7 rounds
Length 140 mm
Length of barrel 73 mm
Weight unloaded 0.64 kg
Muzzle velocity 280 m/sec approx
Max effective range 30 m
Original manufacturer Becker & Hollander, Suhl
Other manufacturers Stendawerke GmbH Waffenbau, Suhl; August Menz, Suhl; and others

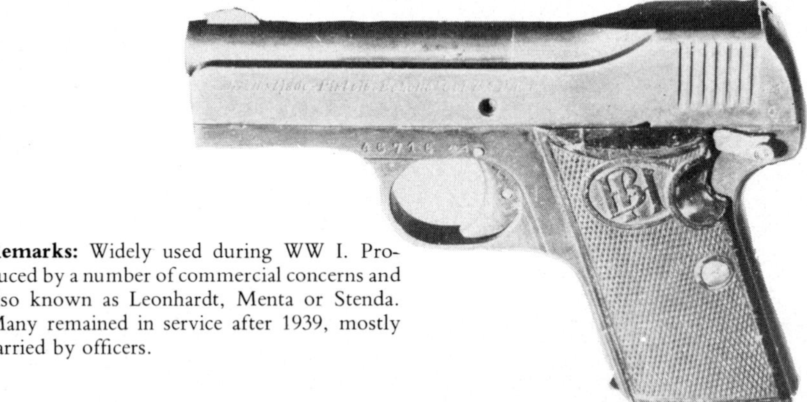

Remarks: Widely used during WW I. Produced by a number of commercial concerns and also known as Leonhardt, Menta or Stenda. Many remained in service after 1939, mostly carried by officers.

Stock

German designation Stock Kal. 6.35 oder 7.65 mm
Calibre/cartridge 6.35 mm (.25 ACP) or 7.65 mm (.32 ACP)

	6.35 mm	**7.65 mm**
Magazine capacity	7 rounds	8 rounds
Length	121 mm	173 mm
Length of barrel	63 mm	92 mm
Weight unloaded	0.35 kg	0.67 kg
Muzzle velocity (approx)	210 m/sec	280 m/sec
Max effective range	25 m	40 m

Original manufacturer Franz Stock, Berlin

Remarks: Well-made commercial model adopted for personal use by many German officers.

Simsonpistole Modell 1922 and 1927

German designation Simsonpistole Kal. 6.35 mm
Calibre/cartridge 6.35 mm (.25 ACP)
Magazine capacity 6 rounds
Length 114 mm
Length of barrel 56 mm
Weight unloaded 0.37 kg
Muzzle velocity 210 m/sec approx
Max effective range 25 m
Original manufacturer Simson & Co, Suhl (later part of Gustloff-Werke)

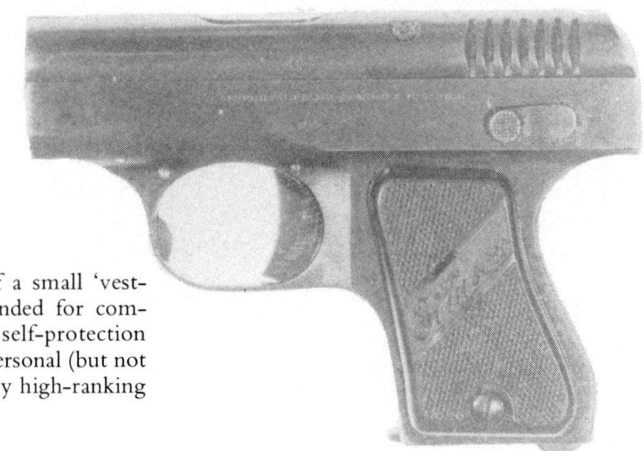

Remarks: Typical example of a small 'vest-pocket' pistol. Originally intended for commercial sales as concealed self-protection weapon; after 1939 carried as personal (but not 'official issue') sidearm by many high-ranking staff officers.

9 mm Pistole 12(ö)

German designation 9 mm P 12(ö)
Original designation 9 mm Repetierpistole M.12
Calibre/cartridge 9 mm Steyr or 9 mm Parabellum
Magazine capacity 8 rounds
Length 216 mm
Length of barrel 128 mm
Weight unloaded 1.02 kg
Muzzle velocity 340 m/sec
Max effective range 50 m
Original manufacturer Österreichische Waffenfabrik, Steyr

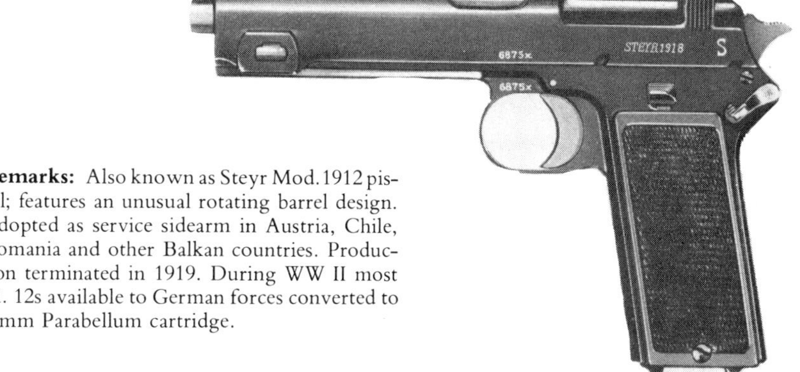

Remarks: Also known as Steyr Mod.1912 pistol; features an unusual rotating barrel design. Adopted as service sidearm in Austria, Chile, Romania and other Balkan countries. Production terminated in 1919. During WW II most M. 12s available to German forces converted to 9 mm Parabellum cartridge.

9 mm Pistole 24(t)

German designation 9 mm P 24(t)
Original designation Automaticky Pistole ČZ vz.24
Calibre/cartridge 9 mm Short; 9 mm M.22 (German)
Magazine capacity 8 rounds
Length 152 mm
Length of barrel 91 mm
Weight unloaded 0.7 kg
Muzzle velocity 295 m/sec
Max effective range 40 m
Original manufacturer Česká Zbrojovka, Prague

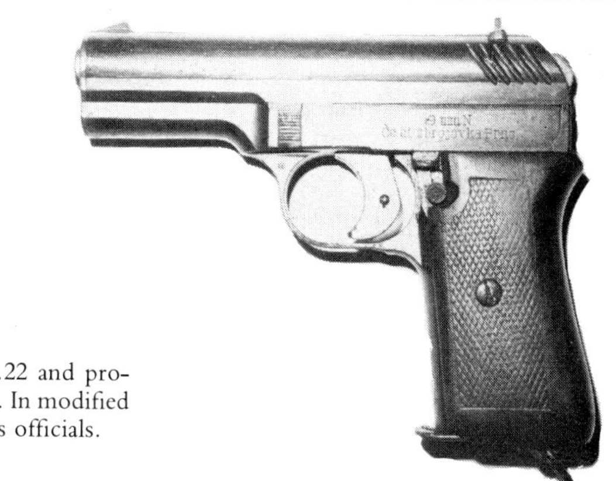

Remarks: Evolved from ČZ vz.22 and produced for various 9 mm cartridges. In modified form also used by Polish Customs officials.

7.65 mm Pistole 27(t)

German designation 7.65 mm P 27(t)
Original designation Automaticky Pistole ČZ vz.27
Calibre/cartridge 7.65 mm; also 9 mm Short (.380 ACP)
Magazine capacity 8 rounds
Length 160 mm
Length of barrel 99 mm
Weight unloaded 0.7 kg
Muzzle velocity 280 m/sec
Max effective range 40 m
Original manufacturer Česká Zbrojovka, Prague. Under German occupation: Böhmische Waffenfabrik AG, Prag

Remarks: Produced in large numbers before WW II. In 1939 adopted as substitute sidearm by German forces and manufactured in slightly modified form. Special 205 mm long silencer for this pistol was also evolved during WW II.

7.65 or 9 mm Pistole 37(u)

German designation P 37(u) Kal. 7.65 oder Kal. 9 mm
Original designation Pisztoly 37M; PT 37M
Calibre/cartridge 7.65 mm or 9 mm Short
Magazine capacity 8 rounds 7.65; 7 rounds 9 mm
Length 182 mm
Length of barrel 110 mm
Weight unloaded 0.77 kg
Muzzle velocity (7.65 mm): 280–290 m/sec (9 mm): 300 m/sec
Max effective range 40 m
Original manufacturer Femaru-Fegyver es Gepyar, Budapest

Remarks: Originally made for commercial sales. Accepted by Hungarian Army in 1937. During WW II adopted as substitute sidearm by German forces and produced for their use in slightly modified form.

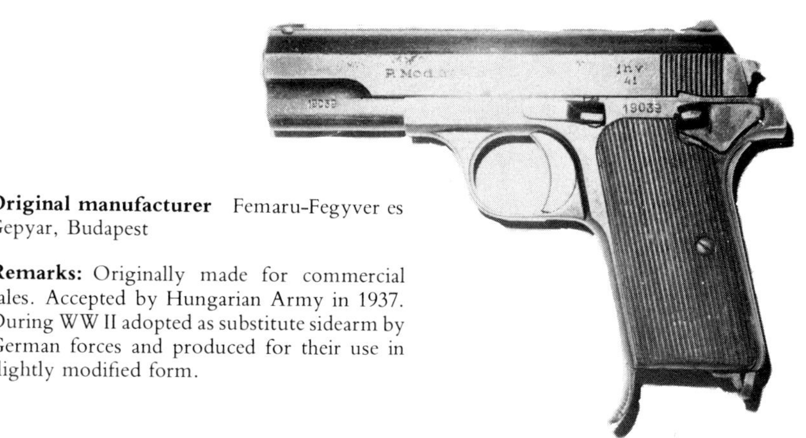

9 mm Pistole 39(t)

German designation P 39(t) Kal. 9 mm
Original designation Automaticky Pistole ČZ vz.38/39
Calibre/cartridge 9 mm Browning Short (.380 ACP)
Magazine capacity 8 rounds
Length 198 mm
Length of barrel 119–119.5 mm
Weight unloaded 0.909 kg
Muzzle velocity 299 m/sec
Max effective range 50 m
Original manufacturer Česká Zbrojovka, Prague

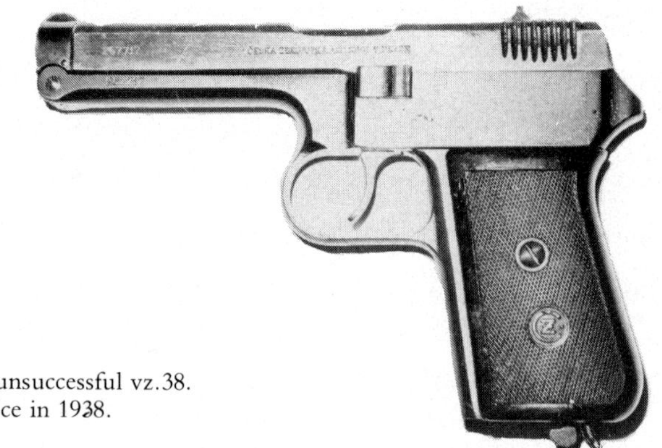

Remarks: Evolved from unsuccessful vz.38. Introduced into Czech service in 1938.

7.62 mm Pistole 615(r)

German designation 7.62 mm P 615(r)
Original designation *Samozaryadny Pistolet Tula-Tokareva obr. 1930 g*; TT-30
Calibre/cartridge 7.62 mm Type P
Magazine capacity 8 rounds
Length 195 mm
Length of barrel 117 mm
Weight unloaded 0.854 kg
Muzzle velocity 420 m/sec
Max effective range 50 m
Original manufacturer Soviet State Arsenal, Tula, and several other State plants

Remarks: *Pistolet obr. 1933 g* (TT-33) essentially similar and was given same German designation. Manufactured and used in very large numbers before and during WW II.

7.65 mm Pistole 620(b)

German designation 7.65 mm P 620(b)
Original designation Pistole Automatique Browning, mle 1900; Pistolet 1900
Calibre/cartridge 7.65 mm (.32 ACP)
Magazine capacity 7 rounds
Length 162.5 mm
Length of barrel 102 mm
Weight unloaded 0.615 kg
Muzzle velocity 290 m/sec
Max effective range 30 m
Original manufacturer Fabrique Nationale d'Armes de Guerre (FN), Herstal, Liège

Remarks: One of commercially most successful handguns of pre-WW I years. Substantial numbers taken over by German forces in 1940.

7.65 mm Pistole 621(b)

German designation 7.65 mm P 621(b)
Original designation Pistole Automatique Browning mle 1910; Pistolet 1910
Calibre/cartridge 7.65 mm (.32 ACP)
Magazine capacity 7 rounds
Length 154 mm
Length of barrel 88.5 mm
Weight unloaded 0.57 kg
Muzzle velocity 299 m/sec
Max effective range 40 m
Original manufacturer Fabrique Nationale d'Armes de Guerre (FN), Herstal, Liège

Remarks: Some were also produced modified for 9 mm Short (.380 ACP) cartridges. Sold commercially from 1912 until today (USA).

9 mm Pistole 622(b)

German designation 9 mm P 622(b)
Original designation Pistole Automatique Browning, mle 1903; Pistolet 1903
Calibre/cartridge 9 mm Browning Long
Magazine capacity 7 rounds
Length 203 mm
Length of barrel 127 mm
Weight unloaded 0.91 kg
Muzzle velocity 320 m/sec
Max effective range 40 m
Original manufacturer Fabrique Nationale d'Armes de Guerre (FN), Herstal, Liège

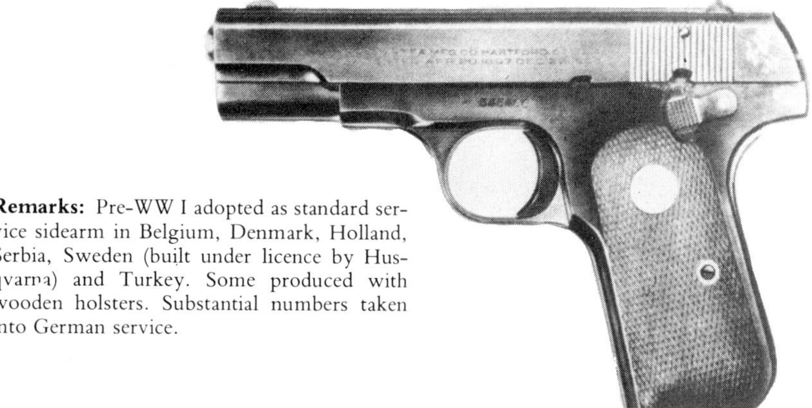

Remarks: Pre-WW I adopted as standard service sidearm in Belgium, Denmark, Holland, Serbia, Sweden (built under licence by Husqvarna) and Turkey. Some produced with wooden holsters. Substantial numbers taken into German service.

7.65 mm Pistole 623(f)

German designation 7.65 mm P 623(f)
Original designation Pistolet automatique type Star
Calibre/cartridge 7.65 mm (.32 ACP)
Magazine capacity 9 rounds
Length 190 mm
Length of barrel 120 mm
Weight unloaded 0.668 kg
Muzzle velocity 280 m/sec
Max effective range 50 m
Original manufacturer Bonifacio Echeverria y Compania SA, Elgoibar, Spain; also other Spanish armament factories

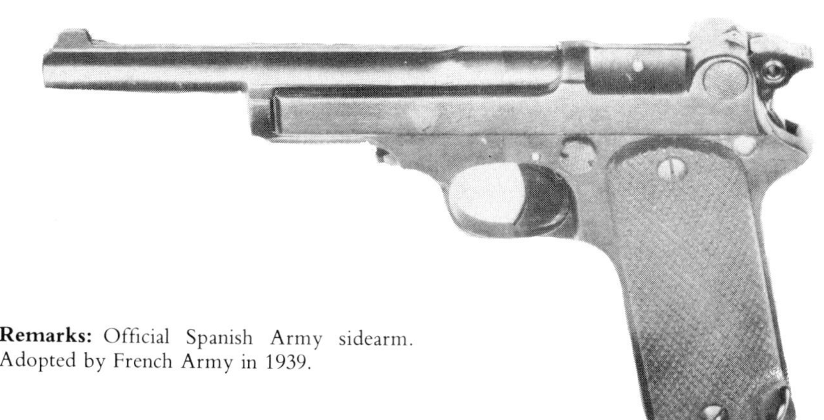

Remarks: Official Spanish Army sidearm. Adopted by French Army in 1939.

7.65 mm Pistole 624(f)

German designation 7.65 mm P 624(f)
Original designation Pistolet automatique type Ruby
Calibre/cartridge 7.65 mm (.32 ACP)
Magazine capacity 9 rounds
Length 150 mm
Length of barrel 82 mm
Weight unloaded 0.81 kg
Muzzle velocity 280 m/sec approx
Max effective range 50 m
Original manufacturer Gabilondo y Compania SA, Elgoibar, Spain

Remarks: Adopted by French Army and in service until 1939 when partly replaced by mle 1935A and 1935S pistols.

7.65 mm Pistole 626(b),(d) and (h)

German designations 7.65 mm P 626(b), (d) oder (h)
Original designations (b): Pistole Automatique Browning, mle 1922: Pistolet Browning; (d): 'Browning'; (h): Pistool M 25 No 1
Calibre/cartridge 7.65 mm (.32 ACP)
Magazine capacity 9 rounds
Length 154 mm
Length of barrel 88.5 mm
Weight unloaded 0.57 kg
Muzzle velocity 299 m/sec
Max effective range 40 m
Original manufacturer Fabrique Nationale d'Armes de Guerre (FN), Herstal, Liège

Remarks: Enlarged version of mle 1910. Used as service sidearm in Denmark, France, Greece, Holland, Turkey and Yugoslavia. After German occupation limited numbers issued to Luftwaffe aircrews. *See* also 9 mm Pistole 641(b).

7.65 mm Pistole 625(f)

German designation 7.65 mm P 625(f)
Original designation Pistole automatique mle 1935A et 1935S; also MAS 1935, etc. (see below)
Calibre/cartridge 7.65 mm Longue
Magazine capacity 8 rounds

	mod. 1935A	mod. 1935S
Length	193 mm	188 mm
Length of barrel	109 mm	104 mm
Weight	0.735 kg	0.795 kg
Muzzle velocity		345 m/sec
Max effective range		40 m

Original manufacturer Société Alsacienne de Construction Mécanique (SACM). Also made by Chatellerault (MAC), Tulle (MAT), Société d'Applications Générales Electriques et Mécaniques (SAGEM) and at St. Etienne (MAS)

Remarks: Based on US Colt/Browning M1911A1, differing only in minor details. Adopted as official French service sidearm in 1936 but large-scale production not started until 1938.

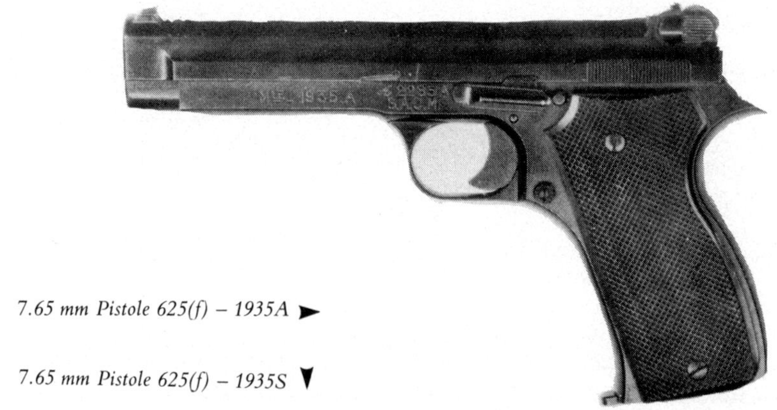

7.65 mm Pistole 625(f) – 1935A ►

7.65 mm Pistole 625(f) – 1935S ▼

9 mm Pistole 640(b)

German designation 9 mm P 640(b)
Original designation Pistolet Automatique Browning, mle à Grand Puissance; Pistolet G.P.
Calibre/cartridge 9 mm Parabellum
Magazine capacity 13 rounds
Length 197 mm
Length of barrel 118 mm
Weight unloaded 0.89 kg
Muzzle velocity 354 m/sec
Max effective range 45 m
Original manufacturer Fabrique Nationale d'Armes de Guerre (FN), Herstal, Liège

Remarks: Generally known as Browning High-Power. Before WW II adopted as service sidearm in Belgium, Denmark, Holland, Lithuania and Romania. Production continued after German occupation in 1940, mainly for Waffen-SS troops.

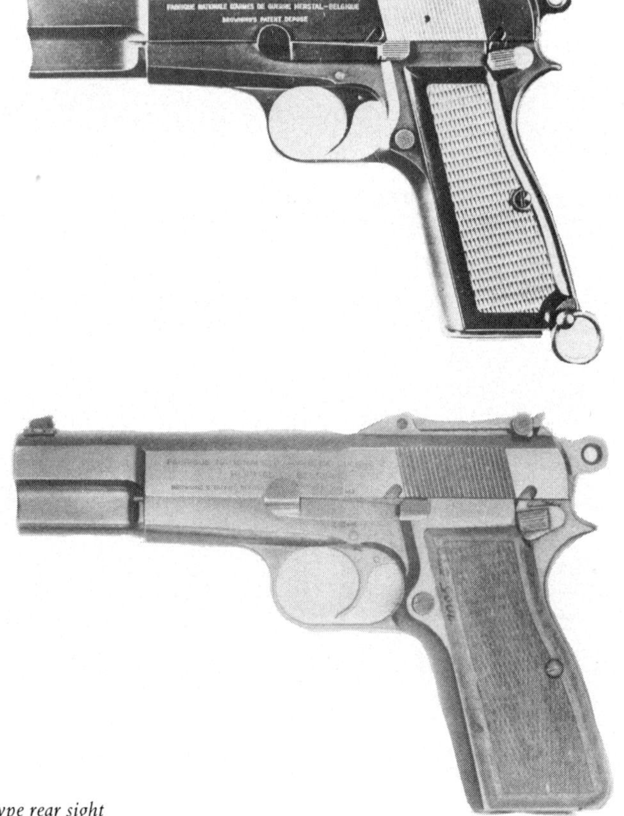

9 mm Pistole 640(b) with a tangent type rear sight

9 mm Pistole 641(b),(h) and (j)

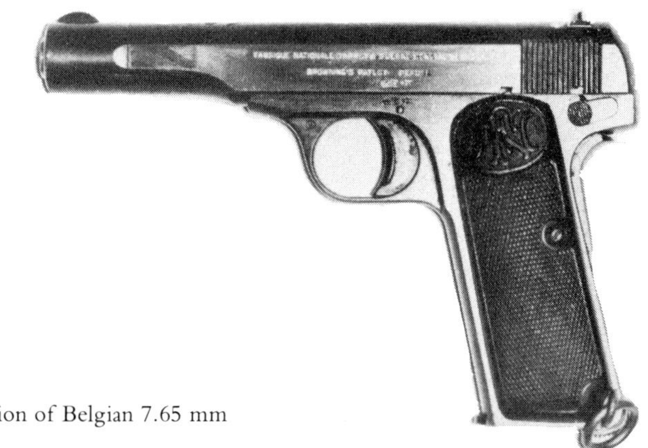

German designations 9 mm P 641(b), (h) oder (j)
Original designations (b): Pistolet Automatique Browning mle 1922: Pistolet Browning cal. 9 mm; (h): Pistool M 25 No 2; (j): M 10/22 Browning
Calibre/cartridge 9 mm Short (.380 ACP)
Magazine capacity 8 rounds
Length 180 mm
Length of barrel 113 mm
Weight unloaded 0.7 kg
Muzzle velocity 266 m/sec
Max effective range 40 m
Original manufacturer Fabrique Nationale d'Armes de Guerre (FN), Herstal, Liège

Remarks: 9 mm version of Belgian 7.65 mm Browning mle 1922.

---

9 mm Pistole (Astra) and 9 mm P 642(f)

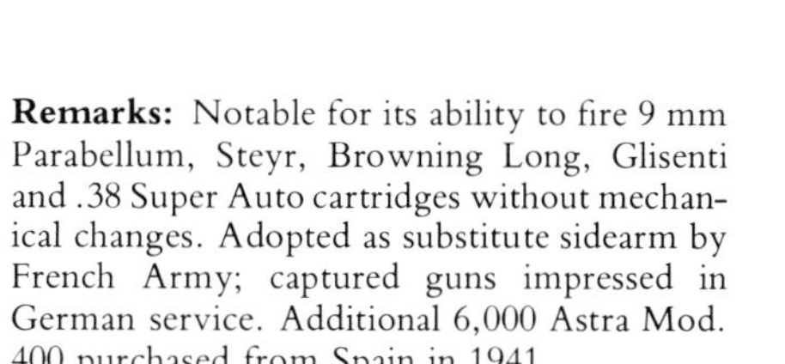

German designations 9 mm P(Astra); 9 mm P 642(f)
Original designations (Spain): Pistola Automatica Astra modello 400; (f): Pistolet automatique 'Astra' mle 1921
Calibre/cartridge 9 mm Long, or others (see below)
Magazine capacity 8 rounds
Length 221 mm
Length of barrel 150 mm
Weight unloaded 1.15 kg
Muzzle velocity 343 m/sec
Max effective range 50 m
Original manufacturer Unceta y Compania SA, Guernica, Spain

Remarks: Notable for its ability to fire 9 mm Parabellum, Steyr, Browning Long, Glisenti and .38 Super Auto cartridges without mechanical changes. Adopted as substitute sidearm by French Army; captured guns impressed in German service. Additional 6,000 Astra Mod. 400 purchased from Spain in 1941.

---

9 mm Pistole (Astra) 43

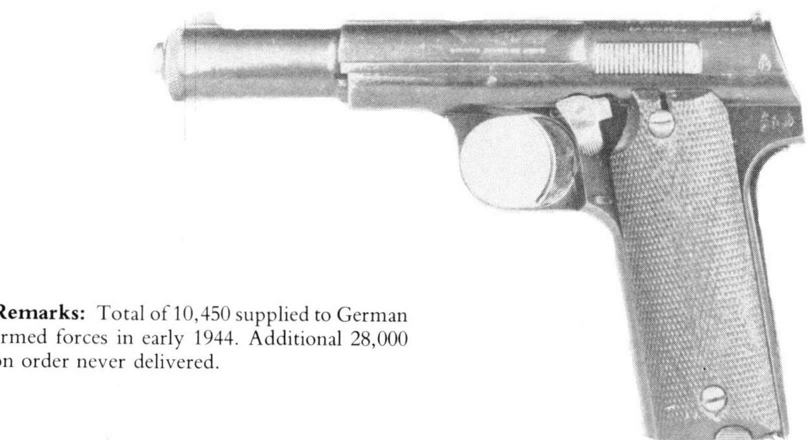

German designations 9 mm P(Astra) 43; 9 mm P(Astra) 600/43
Original designation Pistola Automatica Astra modello 600/43
Calibre/cartridge 9 mm Parabellum
Magazine capacity 8 rounds
Length 205 mm
Length of barrel 134 mm
Weight unloaded 0.99 kg
Muzzle velocity 340 m/sec
Max effective range 50 m
Original manufacturer Unceta y Compania SA, Guernica, Spain

Remarks: Total of 10,450 supplied to German armed forces in early 1944. Additional 28,000 on order never delivered.

---

9 mm Pistole 644(d)

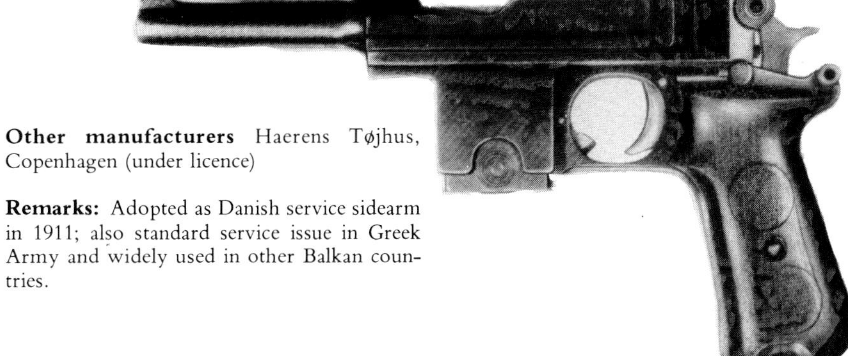

German designation 9 mm P 644(d)
Original designation 'Bergman' M 1910/21
Calibre/cartridge 9 mm Bergmann-Bayard
Magazine capacity 6, 8 or 10 rounds
Length 254 mm
Length of barrel 101 mm
Weight unloaded 1.022 kg
Muzzle velocity 340 m/sec
Max effective range 40 m
Original manufacturer Anciens Etablissements Pieper, Herstal, Liège

Other manufacturers Haerens Tøjhus, Copenhagen (under licence)

Remarks: Adopted as Danish service sidearm in 1911; also standard service issue in Greek Army and widely used in other Balkan countries.

9 mm Pistole 35(p) and P 645(p)

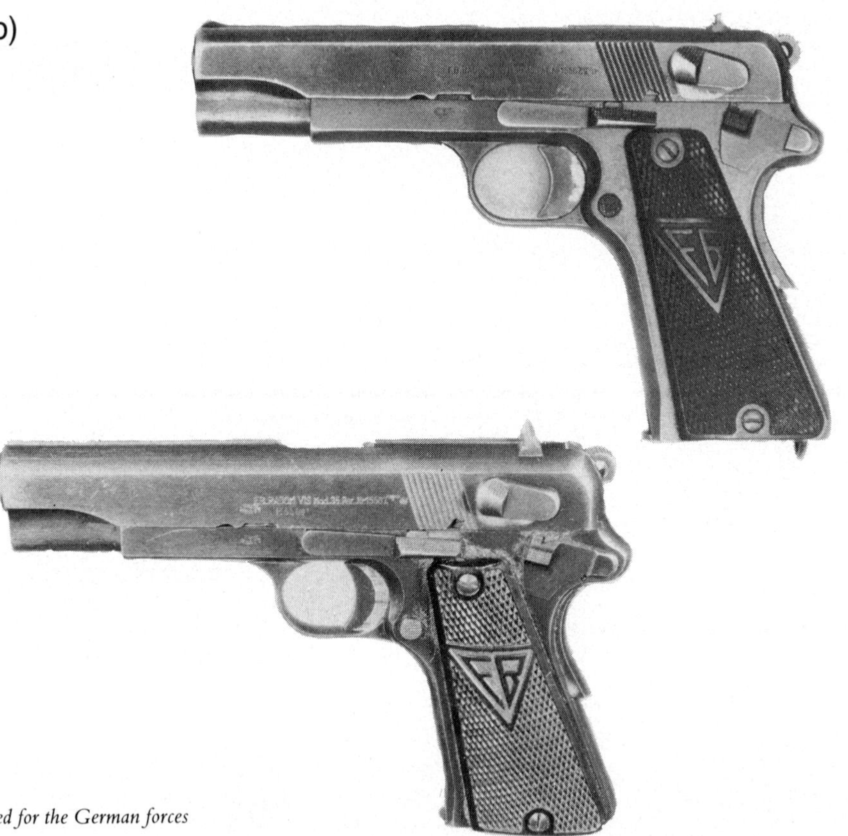

German designation (official): 9 mm P 645(p); (usual): 9 mm P 35(p)
Original designation Pistolet 'ViS' wz.35
Calibre/cartridge 9 mm Parabellum
Magazine capacity 8 rounds
Length 200 mm
Length of barrel 120 mm
Weight unloaded 1.022 kg
Muzzle velocity 350 m/sec
Max effective range 50 m
Original manufacturer Fabryka Broni, Radom, Poland

Remarks: Also known as ViS or, more generally, 'Radom'. Standard Polish service sidearm from 1935. Produced for German forces after 1939 with gradually declining standard of finish and eventual omission of grip safety and other details. Production ceased in 1944.

9 mm Pistole 35(p) produced for the German forces

11.25 mm Pistole 657(n) and 11.43 mm Pistole 660(a)

German designations P 657(n); P 660(a)
Original designations (n): 11.25 mm automatisk pistol modell 1914: Pistol m/14; (a): US Pistol, Automatic, Caliber .45, M1911 and M1911A1
Calibre/cartridge 11.43 mm (.45 M1911)
Magazine capacity 7 rounds
Length 217 mm
Length of barrel 127 mm
Weight unloaded 1.1 kg
Muzzle velocity 262 m/sec
Max effective range 50 m
Original manufacturer (USA) Colt's Patent Firearms Manufacturing Co., Hartford, Conn.
Other manufacturers Norwegian State Arsenal, Kongsberg (M1911 under licence)

Remarks: Standard sidearm of US armed forces since WW I. Mod. 1911 adopted in Norwegian service in 1914. Only limited numbers in German use owing to non-standard ammunition.

---

9 mm Pistole 671(i)

German designation 9 mm P 671(i)
Original designation Pistola Automatica Beretta modello 1934 and 1938
Calibre/cartridge 9 mm Short (.380 ACP)
Magazine capacity 7 rounds
Length 155 mm
Length of barrel 86 mm
Weight unloaded 0.617 kg
Muzzle velocity 245 m/sec
Max effective range 40 m
Original manufacturer Pietro Beretta SpA, Gordone-val-Trompia and Brescia

Remarks: Evolved from M1915 via M1921 and M1923. Also manufactured in 7.65 mm calibre. Well-liked and handy defensive weapon.

Revolvers

Following the introduction of *Pistole 08* the German military gradually lost all interest in revolvers and only a small number remained in service with various garrison units and the police after WW I. However, such was the demand for pistols after 1940 that stocks of captured revolvers had to be issued to the growing number of second-line, garrison and auxiliary police formations. Such substitute sidearms distributed to German occupation units included the French Pistole-Revolveur mle 1892, various Spanish Astra revolvers (some bought direct from Spain), British Webleys, Russian Nagant obr. 1895g, Smith & Wesson and US Colt service models. Revolvers captured in the field mostly passed directly into use with German front-line troops, but those stockpiled were eventually issued to Volkssturm formations from late 1944 onwards. In many cases the available ammunition was strictly limited and these handguns had little more than morale and last-ditch defence value.

7.62 mm Revolver 612(r), (g) and (p)

German designation (single action): 7.62 mm R 612/2 01(r) oder (p); 7.62 mm R 612(g); (double action): 7.62 mm R 612/1 01(r) oder (p)
Original designation (r): *Revolver Nagant obr. 1895g*; (g): Nagant 95; (p): Rewolwer 'Nagan'
Calibre/cartridge 7.62 mm 'Nagant'
Cylinder capacity 7 rounds
Length 230 (235) mm
Length of barrel 110 (114) mm
Weight unloaded 0.795 kg
Muzzle velocity 272 m/sec
Max effective range 40 m
Original manufacturer (r), (p): Tulsky Oruzheniya Zavod, Tula, and other State armament factories; (g): Manufacture d'Armes Nagant Frères, Liège, Belgium

Remarks: Of Belgian origin. Adopted as standard Russian service sidearm in 1895 and still produced during WW II. Single and double-action versions were issued to enlisted men and officers respectively. German use of captured guns was restricted by need for special Nagant ammunition. Greek Army also used Nagant mod. 1912 manufactured in the USA.

8 mm Revolver 637(f)

German designation 8 mm R 637(f)
Original designation Pistole Revolveur mle 1892; mle d'Ordonnance 1892
Calibre/cartridge 8 mm mle 1892
Cylinder capacity 6 rounds
Length 235 mm
Length of barrel 118.5 mm
Weight unloaded 0.792 kg
Muzzle velocity 225 m/sec
Max effective range 30 m
Original manufacturer Mre. d'Armes St. Etienne; other French arsenals

Remarks: First European swing-out cylinder revolver; was commonly known as 'Lebel'. Obsolete and of limited military value in WW II.

11.55 mm Revolver 665(e)

German designation 11.55 mm R 665(e)
Original designation Pistol Revolver .455 No 1 Mk 6
Calibre/cartridge .455 Webley; actual calibre: 11.2 mm
Cylinder capacity 6 rounds
Length 286 mm
Length of barrel 152 mm
Weight unloaded 1.09 kg
Muzzle velocisty 189 m/sec
Max effective range 50 m
Original manufacturer Webley & Scott, Birmingham
Other manufacturers Royal Small Arms Factory, Enfield Lock

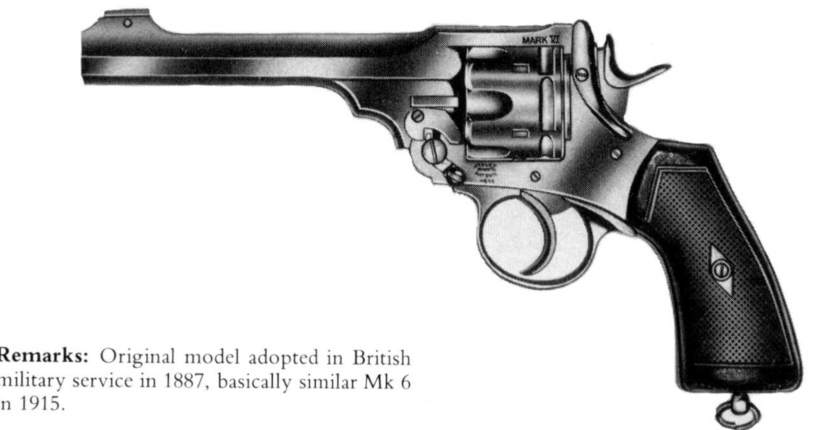

Remarks: Original model adopted in British military service in 1887, basically similar Mk 6 in 1915.

Flare pistols

The standard German flare pistol was the *27 mm Leuchtpistole* developed and produced by Carl Walther at Zella-Mehlis. It was accepted for service in 1928, and the early examples were made of steel. Later, increasing demands made on the steel production dictated the use of aluminium and finally zinc alloy. The *Leuchtpistole* fired a wide range of coloured flares including the whistling *'Pfeifpatrone'*, the universal warning of a gas attack. Most of these flare pistols went to the Army, but a varying percentage were delivered to the Luftwaffe.

In 1942, growing demands from the Eastern Front for close-support weapons led to various emergency solutions. One of these was the adaptation of the *Leuchtpistole* to fire small HE grenades of special design. This involved the fitting of a rifled barrel and a simple sight, the converted flare launchers being known as *'Kampfpistole'*. Later still, the concept was taken a stage further with the *'Sturmpistole'* when it was adapted to fire small hollow-charge anti-tank grenades. The *Sturmpistole* had a folding metal shoulder stock and a complex little bubble sight. A steel liner could be removed from the barrel to fire the normal range of 27 mm flares and grenades. Of the various unusual weapons evolved by the German small-arms designers the *Kampfpistole* and *Sturmpistole* must have been the least useful, for the small grenades they fired contained too little explosive to have any real effect. Both were of very limited combat value, quite out of proportion to the effort and cost of their development and production.

Carl Walther also produced the *Sternsignalpistole*, made in both single and double-barrel versions. These were used mainly to fire illuminated flares and most were delivered to the German Navy.

Various other German manufacturers produced flare pistols for the armed forces, but many of these were commercial models diverted to service use. Most were simple single-barrel pistols designed to take the standard 27 mm flares, but one odd development was the four-barrel *Vierläufige Leuchtpistole* produced by G. Erdmann & Co. of Berlin in 1939.

Nearly all these flare pistols were issued with a variety of clamping devices that enabled them to be fired remotely while fixed to trees, walls or the ground.

By early 1944 the supply of signal pistols was becoming difficult and an improvised flare launcher was evolved and issued. Its designation is unknown, but it was made up by fitting a short barrel to the top of a standard stick hand grenade handle. The barrel was hinged for loading, and the signal flare fired by releasing a spring-loaded bolt contained inside the handle.

2.7 cm Leuchtpistole

German designation 2.7 cm LeuP
Calibre 26.65 mm
Length 245 mm
Length of barrel 155 mm
Weight unloaded (steel): 1.325 kg; (light metal): 0.73 kg
Manufacturer Waffenfabrik Carl Walther AG, Zella-Mehlis

Remarks: Standard German flare pistol, widely used by all services.

A signal cartridge being inserted into the Leuchtpistole

The Walther 2.7 mm cm Leuchtpistole *was a smooth-bore weapon used to fire some 40 different signal cartridges and two types of grenades. Of the latter, one was similar in design to the signal cartridge and loaded into the breech of the gun; the other, known as* Wurfkörper 361, *was fired from the barrel of the pistol fitted with a smooth-bore liner*

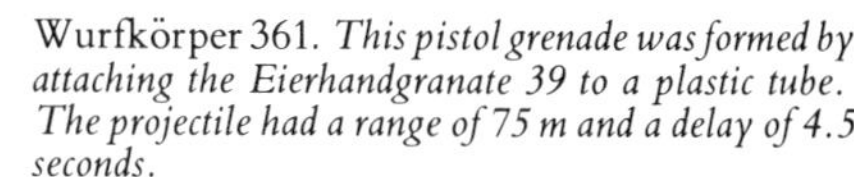

Wurfkörper 361. *This pistol grenade was formed by attaching the Eierhandgranate 39 to a plastic tube. The projectile had a range of 75 m and a delay of 4.5 seconds.*

2.7 cm Leuchtpistole 42 and 2.7 cm Kampfpistole

German designation 2.7 cm LeuP 42, 2.7 cm KmP
Calibre without liner 26.65 mm
Calibre with liner 23 mm
Length of barrel 155 mm
Weight (approx) 1.4 kg
Weight of HE grenade 0.14 kg
Approx range of HE grenade 90 m
Manufacturer Waffenfabrik Carl Walther AG, Zella-Mehlis

Remarks: Could also fire the 2.6 cm Wurfgranate Patrone 326 LeuP anti-tank charge.

2.7 cm Kampfpistole, *the standard* Leuchtpistole *rifled to fire HE, smoke, indicator and illuminating star flare on parachute. A liquid sight was attached to the left side of the pistol. The HE nose-fused grenade, shown in the picture, weighed 140 grammes.*

2.7 cm Sturmpistole

German designation 2.7 cm StP
Calibre with rifled liner 23 mm
Length with butt extended 584 mm
Length with butt folded 305 mm
Length of barrel 180 mm
Weight 2.5 kg
Weight of HE grenade 0.14 kg
Weight of HE filling 0.007 kg
Approx range of grenade 90 m

Remarks: Very limited combat value.

◄ *2.7 cm* Sturmpistole *folded. This weapon was also adopted by the German paratroops*

2.7 cm Sturmpistole *with butt extended*

2.7 cm Sternsignalpistole

German designation Sternsignalpistole Walther Mod. SL
Calibre 26.65 mm
Length 340 mm
Length of barrel 230 mm
Weight 1.81 kg
Manufacturer Waffenfabrik Carl Walther AG, Zella-Mehlis

Remarks: Single-barrel version.

2.7 cm Sternsignalpistole Modell SL, *the naval version with a wooden forestock*

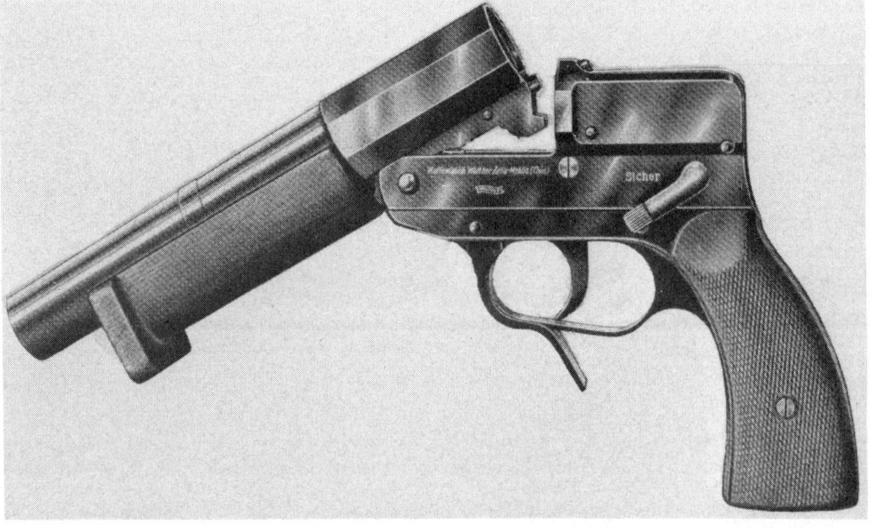

2.7 cm Doppelschuss

German designation Sternsignalpistole Walther Mod. SLd
Calibre 26.65 mm
Length 340 mm
Length of barrels 230 mm
Weight 2.65 kg
Manufacturer Waffenfabrik Carl Walther AG, Zella-Mehlis

Remarks: Double-barrel version.

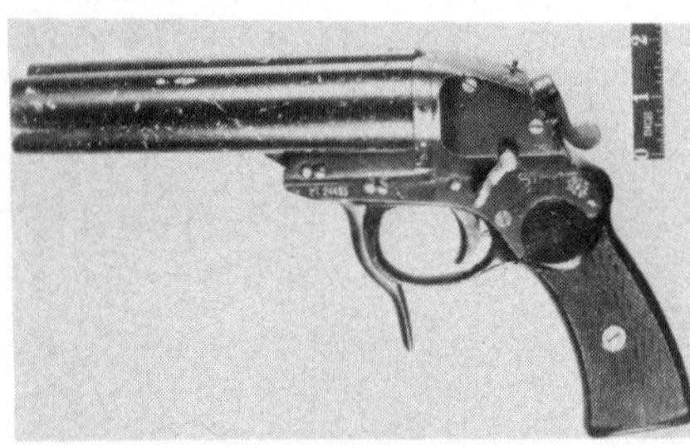

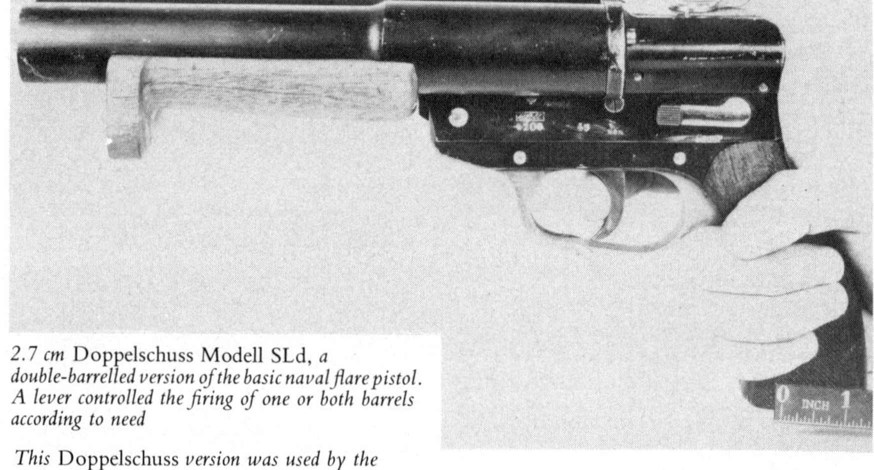

2.7 cm Doppelschuss Modell SLd, *a double-barrelled version of the basic naval flare pistol. A lever controlled the firing of one or both barrels according to need*

◄ *This* Doppelschuss *version was used by the Luftwaffe air and ground troops*

Vierläufige Leuchtpistole

Calibre 26.65 mm
Length overall 214 mm
Length of barrels 90 mm
Weight 2.35 kg
Weight of magazine 0.68 kg
Manufacturer G. Erdmann & Co. GmbH, Berlin

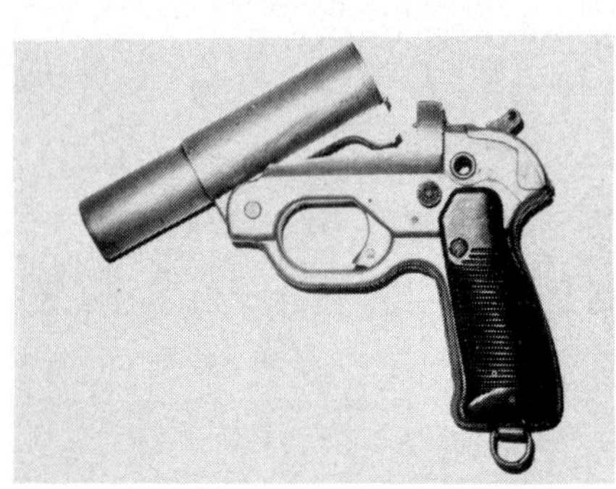

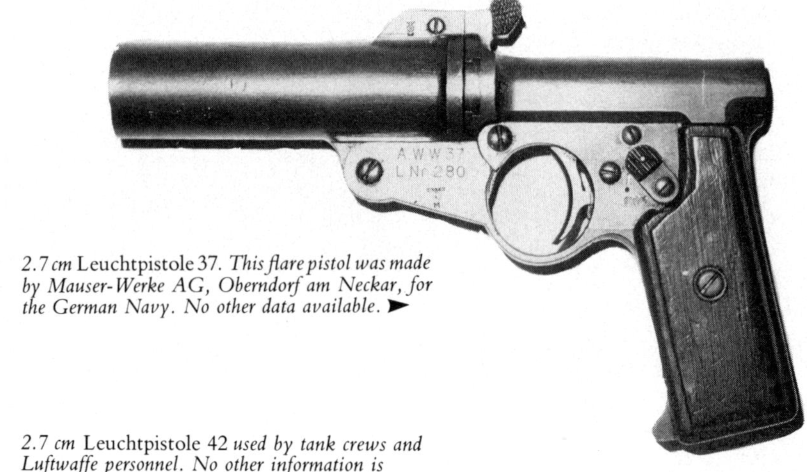

2.7 cm Leuchtpistole 37. *This flare pistol was made by Mauser-Werke AG, Oberndorf am Neckar, for the German Navy. No other data available.* ►

◄ *2.7 cm* Leuchtpistole 42 *used by tank crews and Luftwaffe personnel. No other information is available*

Bolt-action rifles

No matter what army a soldier belongs to, or any other arm of the services for that matter, the new recruit is always trained in the use of that army's standard rifle. Thus for the serviceman in the field, the use of the rifle is one of his most well-learned and basic skills and the rifle itself should be so well-known that its handling, care and use should be second nature to that serviceman. For the front-line soldier in the German Army this was an accepted fact, but the same could not be said for many other members of the German forces, for they were issued with any one of a bewildering variety of bolt-action rifles from all the corners of Europe.

As with other weapons, the basic problem during WW II was insufficient production capacity to cover the increasing demands of the German armed forces, but before that stage we should look at the years immediately after WW I. Almost as soon as the Treaty of Versailles had reduced the German Army to an attenuated shadow of its former self, the German General Staff had initiated an intensive study of future tactical requirements based on the lessons drawn from WW I. One of the main points to be learned from this study was that the average soldier was unable to realise the full potential of the rifle issued to him due to insufficient training and the fact that most infantry engagements took place at ranges well under those which the service rifle was designed for, ie. well over 1,000 metres. At that time the logical lesson from this situation was not drawn – that had to wait until the much later concept and original design of the less powerful *'kurz'* round. Instead the official report called for a rifle based on the existing well-tried *7.92 mm Gewehr 98* but slightly shorter and 'handier'.

The *Gew 98* had a fairly long history based throughout on one of the most reliable and efficient rifle bolt actions ever made, the Mauser system. The originator of this action was Peter Paul Mauser, an inventor of considerable genius who perfected his bolt-action rifle as the result of a series of rifles that started as far back as 1871. With the *Gew 98* the design became virtually 'frozen' and after that date most changes made were to the rest of the rifle rather than the bolt action and magazine. The old German Army took the *Gew 98* to war with them in 1914, and in 1918 it was still being produced in large quantities. After 1918 the 100,000-man German Army was thus left with an extensive armoury of good rifles and, despite the depredations of the various Treaty commissions that ordered the destruction or sale of much of the German war equipment, they managed to keep large stocks in storage.

The lessons of the Staff studies started to see fruition with the modification of many old *Gew 98* rifles to take new ammunition, new sights to suit the ammunition change, and small changes to the sling swivels and bolt-handle angle. The altered rifle was designated *7.92 mm Karabiner 98b (Kar 98b)* and so many were converted that they remained the 'standard' service rifle with

Luftwaffe ground troops armed with Soviet obr.91/31 *rifles*

many formations until 1945. At about the same time, the mid-1920s, numbers of old *Gew 98* rifles were also shortened to become true carbines for the cavalry and artillery troops. These were designated *Kar 98a*.

The years after WW I were not noted for drastic innovations to German Army equipment due to the financial and social restraints as Germany struggled to repair the ravages of the war, but the German armament manufacturers nevertheless quietly worked away to regain their former markets. Needless to say, the Mauser Werke AG of Oberndorf were among the firms involved in this activity, and by 1924 had evolved a new rifle based on their vast experience in producing the *Gew 98*. The new model was a shorter version of the *Gew 98* which almost exactly met the German Staff requirements, but in 1924 the Treaty commissions were still active with their restrictions on German arms production.

The result was that Mauser Werke resorted to their old contacts in other countries that had produced Mauser rifles under licence in the past, and so the Mauser design – known as the 'Standard' – went into large-scale production in Czechoslovakia and Belgium, not only to equip the armies of those nations but also to supply large export orders to such distant places as China and South America.

The Czech manufacturers were Českaslovenske Zbrojovka Brno (ZB) who produced the new rifle as the Puska vz.24, promptly adopted by the newly-formed Czech armed forces. The Belgian rifles were made by FN at Liège who also produced a wide variety of export models until the type was adopted by the Belgian Army in 1935 as the Fusil 35. In 1929 production of the Mauser 'Standard' was started in Poland at the arsenals at Warsaw and Radom, but this version, the karabinek wz.29, was manufactured under an arrangement with the Czech licensees.

When the National Socialist Party came to power in German in 1933 the new Mauser rifle was thus already in large-scale production in Europe, except Germany itself. This situation was rectified in 1935 when the 'Standard' was put into production as the *7.92 mm Kar 98k*, the first assembly lines being set up at Oberndorf followed soon afterwards by others all over Germany. By 1943 the production cost of *Kar 98k* had been reduced to RM 70, – each, but this low average was achieved only by adding the output of all German-controlled plants. These included the Czech and Polish arsenals as well as the modern FN complex in Belgium.

The German military planners had not counted on war starting before 1942 or so and, as the German armed forces expanded rapidly to fight the campaigns beginning with 1939, it became painfully obvious that they would soon face a drastic shortage of that most basic commodity, the bolt-action rifle. But here the excellence of the basic Mauser action again showed its worth: it had been adopted by many European gunmakers for a wide range of commercial sporting rifles. These commercial lines were soon converted to military purposes, a typical example of this being the rifle produced by the Austrian Steyr-Werke at Oberdonau which had been manufacturing a version of the Mauser 'Standard' for export since 1929. As this was almost identical with the *Kar 98k* it was adopted for German use and produced as the *7.92 mm Gew 29/40*, with most of the output going to the Luftwaffe. In Czechoslovakia, a carbine version of the basic

Mauser design, the karabina vz.16/33, had been produced not only for the Czech armed forces but also for export. This variant too was kept in production under German control and adopted as the *Gew 33/40*; most of these guns went to the German mountain troops and airborne formations.

The only rifle adopted as standard issue that did not use the Mauser action was the *Gew 98/40*, which was of Hungarian origin and used a Mannlicher bolt system. Due to the urgent need for rifles to supply the German armed forces it was ordered from the Hungarian factories, altered to suit German requirements and issued in some numbers.

But the saving source of supply that enabled the Germans to form and equip new divisions was the impression of captured weapons. The fall of the Low Countries and France in 1940 resulted in huge numbers of serviceable rifles which were later issued to various garrison units in the occupied territories. The same had already happened in Denmark and Norway, but there the numbers were small compared with the quantities captured from Belgian, French and British troops. The subsequent Balkan campaign also added its share to the spoils, but by far the largest amount of arms fell into German hands during the first year of 'Operation Barbarossa', the invasion of the Soviet Union. The advances of 1941 and 1942 yielded vast numbers of Mosin-Nagant rifles and carbines most of which were stockpiled, although a proportion were kept in use by Russian and other local troops fighting on the German side. Due to their awkward length the Mosin-Nagant rifles were not very popular with German troops, but large numbers of the handier carbines were issued to various second-line formations. With a few exceptions – such as the Mosin-Nagant sniper rifles with telescopic sights – captured bolt-action guns were not used by front-line formations in the interest of standardisation, but behind the lines a variety of rifles from every corner of Europe were in regular use by communication and engineering units, drivers, military police, garrison and guard troops (including some divisions along the Atlantic coast), the personnel of home-defence anti-aircraft detachments, auxiliary local militia formations and various training depots.

Apart from everything else the availability of these huge stocks of captured equipment enabled the Germans to release nearly all newly-produced *Kar 98k* rifles to their front-line troops.

By 1944 the Reich itself was in imminent danger of invasion and the facilities to produce more rifles were being rapidly reduced by Allied bombing and the loss of occupied territories. As the last-ditch gesture the Volkssturm was formed, but to provide the necessary weapons was not easy. As a result they were given anything that could be fired, including old German 11 mm rifles dating back to 1871. The armouries were combed for remaining captured weapons and these were distributed together with handfuls of ammunition and orders to 'stand fast'.

Among the many different types of guns issued to the Volkssturm were large numbers of Italian Mannlicher-Carcano rifles. When Italy surrendered to the Allies in September 1943 most of the Italian forces in the areas under German control were disarmed, resulting in huge numbers of rifles in a large variety of models being added to the German stockpiles. Unfortunately for the Germans the Italian ammunition supply situation was in a state of chaos. After WW I the standard Italian calibre had been 6.5 mm, but during the 1930s a change to a new 7.35 mm cartridge was made and rifles produced to take the new round. But with war imminent and large stocks of the old 6.5 mm ammunition still being held it was decided to revert to the old calibre. The result was a quartermaster's nightmare, for some types of rifle could then be found in either calibre. The situation was never resolved and when the Germans inherited the problem they tried to partially solve it by re-boring many Italian rifles to 7.92 mm. Some of these were issued to troops based in Italy, but the bulk of Italian rifles went to the Volkssturm, many still in their original Italian calibres.

By late 1944 all the remaining stocks of captured guns had been used up, but the demand for more guns was as acute as ever. It was at this late stage that the German industry came up with a series of crude emergency rifles, the *VG 1* and *VK 98*. The *VG 1* was a very simple weapon that could be turned out using ordinary machine tools, while the *VK 98* used old Mauser actions married to a crudely-finished barrel and stock. The planned large-scale production never materialised, which was just as well, because both rifles were likely to be as dangerous to the user as to the target.

The *Kar 98k* remained in production right up to the end of WW II despite the increasing use of automatic handguns and the introduction of the revolutionary short cartridge assault rifle. It was an excellent service rifle but it did have its faults, not the least of which was that it was rather difficult and time-consuming to manufacture. This was one of the reasons why, despite large-scale production, the *Kar 98k* never did entirely replace the old *Gew 98* in service. In the field it was reliable, accurate and easy to use but, like all other bolt-action rifles, of increasingly limited value under modern combat conditions.

A Volkssturm unit armed with the Italian Fucile Modello 91 rifles

7.92 mm Gewehr 98, 221(j), 293(j) and 299(p) or 98(p)

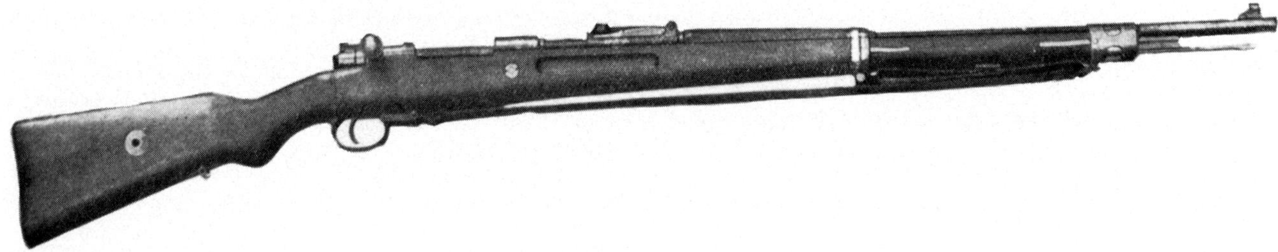

German designations 7.92 mm Gew 98, 221(j), 293(j) 299 oder 98(p)
Original designations 221(j): 7.9 mm M 10C; 293(j): 7.9 mm M 98; 98(p): Karabin 98a
Calibre/cartridge 7.92 mm × 57
Magazine capacity 5 rounds
Length 1250 mm
Length of barrel 740 mm
Weight unloaded 4.2 kg
Muzzle velocity 640 m/sec
Original manufacturer Mauser-Werke AG (various locations)

Remarks: Standard German service rifle of WW I. After 1918 many thousands stockpiled in Germany for possible future use which came after 1933. Above-listed 'foreign' versions represented German rifles delivered as war reparations, or sold post-war. By 1939 used only by second-line and garrison units, but many still in service in 1945.

7.92 mm Karabiner 98a, 492(j) and 493(p) or 98(p)

German designations 7.92 Kar 98a, 492(j), 493(p) oder 98(p)
Original designations (j): Karabini 7.9 mm M 98; (p): Karabinek 1898
Calibre/cartridge 7.92 mm × 57
Magazine capacity 5 rounds
Length 1100 mm
Length of barrel 600 mm
Weight unloaded 3.63 kg
Muzzle velocity 870 m/sec
Original manufacturer Mauser-Werke AG (various locations)

Remarks: Cut-down version of Gew 98, intended for cavalry units. After 1939 service use confined to second-line units and police.

7.92 mm Karabiner 98b

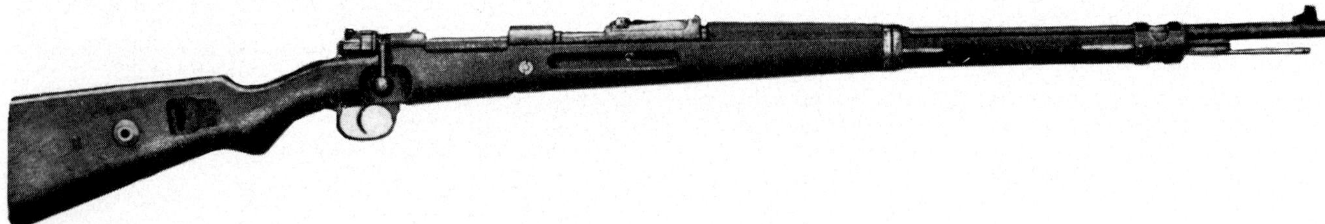

German designation 7.92 mm Kar 98b
Calibre/cartridge 7.92 mm × 57
Magazine capacity 5 rounds
Length 1250 mm
Length of barrel 740 mm
Weight unloaded 4.01 kg
Muzzle velocity 785 m/sec

Remarks: Updated conversions of Gew 98 completed during 1920s. Modifications included altered sights to suit new ammunition and changed sling swivels. Although described as carbine remained in rifle length as before. Most also appear to have retained their Gew 98 markings.

7.92 mm Karabiner 98k, 7.9 mm Gewehr 24(t), 7.92 mm Gew 29/40 or 29/40(ö), 7.65 mm Gew 262(b), 7.9 mm Gew 289(p) or 29(p), 7.9 mm Gew 290(j) and 7.9 mm 298(j)

7.92 mm Kar 98k, the standard German service rifle of WW II years. There were many variations in length, weight and furniture

German designations	Original designations
7.92 mm Kar 98k	
7.9 mm Gew 24(t)	Puska vz.24
7.92 mm Gew 29/40 oder 29/40(ö)	Steyr Modell 31
7.65 mm Gew 262(b)	Fusil 35
7.9 mm Gew 289(p) oder 29(p)	Karabin 29
7.9 mm Gew 290(j)	Puska 7,9 mm M 24
7.9 mm Gew 298(j)	Puska 7,9 mm M 29

Slight variations appeared from model to model but the basic data was as follows:

Calibre/cartridge 7.92 mm × 57 or (Belgian) 7.65 Mauser
Magazine capacity 5 rounds
Length 1107.5 mm
Length of barrel 739 mm
Weight unloaded 3.9 kg
Muzzle velocity 755 m/sec
Original manufacturers Mauser-Werke AG, Oberndorf-am-Neckar; Licence (t) and 290 (j): Česká Zbrojovka, Brno – later Waffenwerke Brünn; (p) and 298 (j): Czech licence – Polish State Arsenals Warsaw and Radom; (ö): Steyr-Werke AG, Steyr; (b): Fabrique Nationale d'Armes de Guerre (FN), Herstal, Liège
Centres of production 1939–45 Mauser-Werke AG, Oberndorf-am-Neckar. Sauer-Gruppe, Suhl. Gustloff-Werke, Weimar. Steyr-Daimler-Puch AG, St Valentin. Waffenwerke Brünn (formerly ČZ, Brno).

Remarks: A shortened version evolved from Gew 98 with side-sling. Adopted as standard German service rifle in 1935, ten years after similar export/licence versions. Late production series made without bayonet mounting bar and featured pressed bands. Captured examples (except Belgian 7.65 mm) also issued to various first-line formations.

Kar 98k ZF, a standard rifle fitted with the Zielfernrohr 39 *(ZF 39), a commercial telescopic sight adopted by the Army*

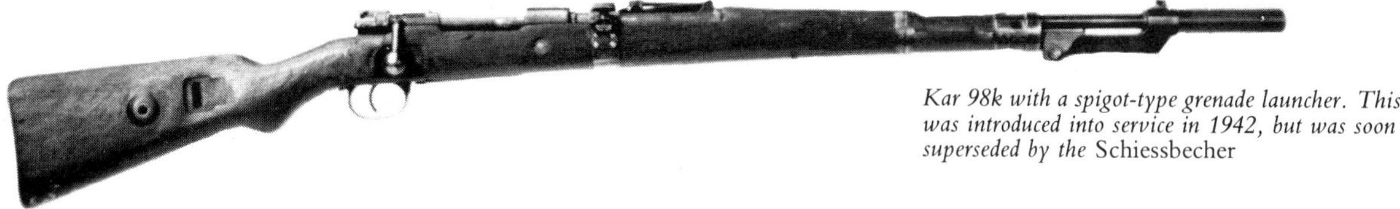

Kar 98k with a spigot-type grenade launcher. This was introduced into service in 1942, but was soon superseded by the Schiessbecher

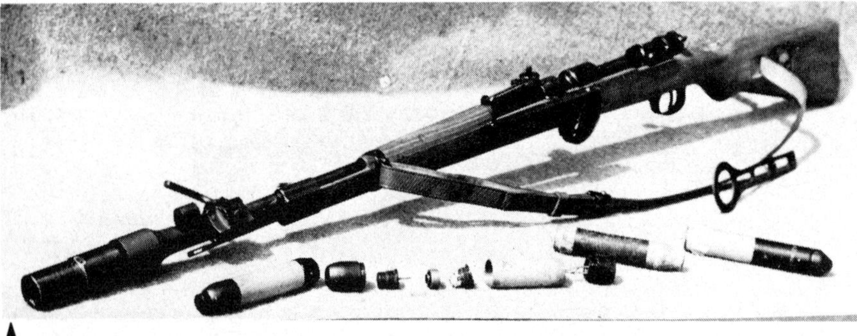

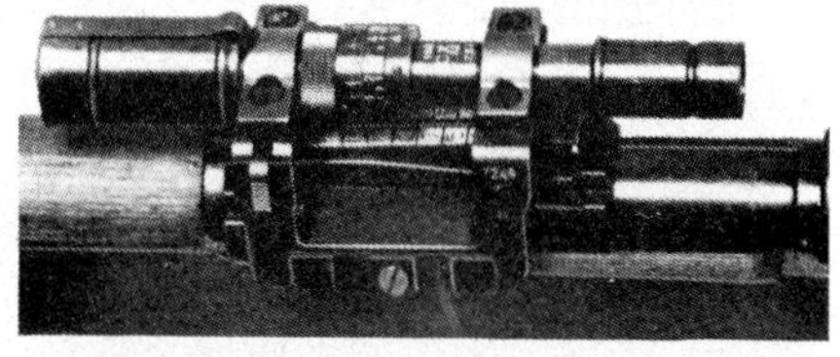

Kar 98k with attached Schiessbecher *and ammunition. The* Gewehr-Sprenggranate *(left) HE grenade could also be used as a handgrenade by unscrewing its base and pulling the exposed cord. Fired from the discharger cup it had a range of some 250 m. The* Gewehr-Panzergranate *(right) anti-tank grenade was effective only against lightly armoured targets. The grenade launcher could also be used with the MP 43/1 and 44*

The Schiessbecher *was a rifled 3 cm calibre discharger cup that could be fitted to most types of German rifles and fired spin-stabilized HE and anti-tank rifle grenades. A special sight was clamped behind the rear rifle sight for aiming purposes. The illustration shows a* Gewehr-Sprenggranate *anti-personnel grenade being loaded into the* Schiessbecher. *Note the Knights Cross decoration worn by the senior NCO demonstrator*

In an attempt to provide some form of standard rifle suitable for the parachute troops Mauser-Werke produced a version of the Kar 98k that could be broken into two separate parts for stowage and airborne use. The designation 7.92 mm Kar 98/42 has been applied to this weapon but cannot be confirmed. Only a small number were produced for troop trials

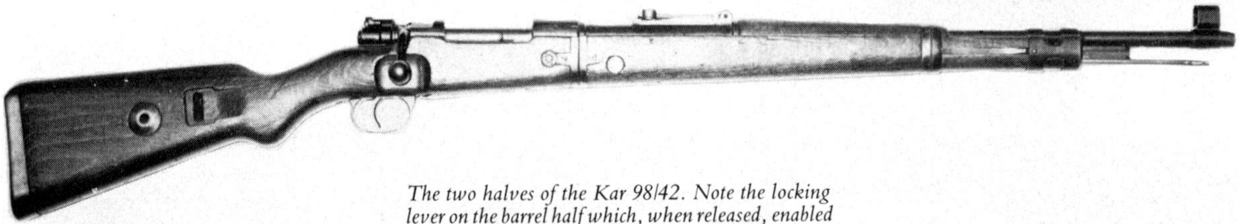

The two halves of the Kar 98/42. Note the locking lever on the barrel half which, when released, enabled the two halves to be rotated about an interrupted thread. This thread housed the chamber

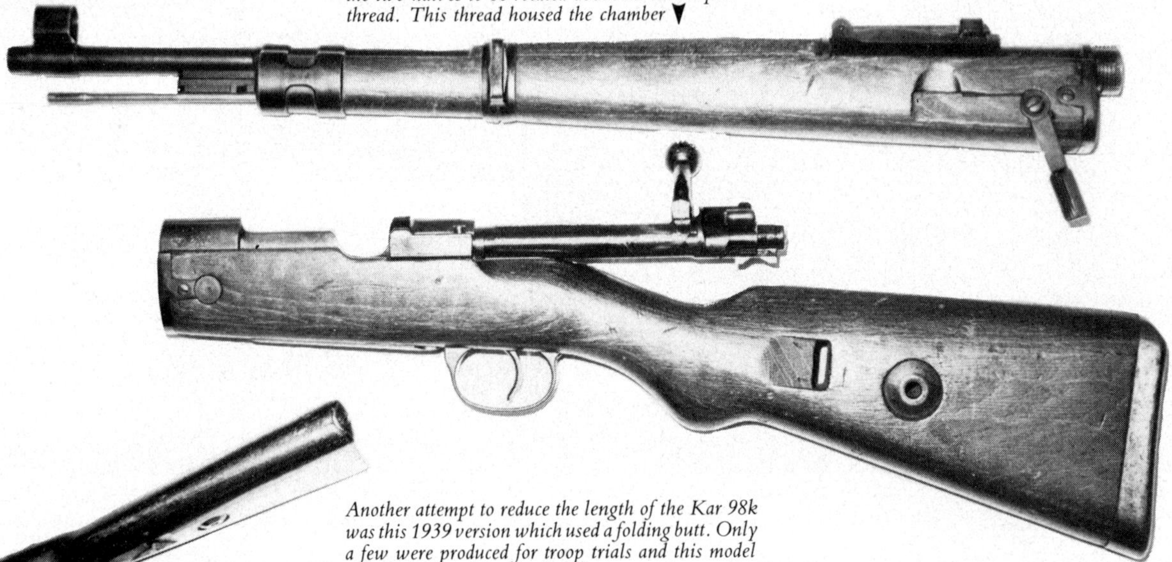

Another attempt to reduce the length of the Kar 98k was this 1939 version which used a folding butt. Only a few were produced for troop trials and this model was not adopted for service

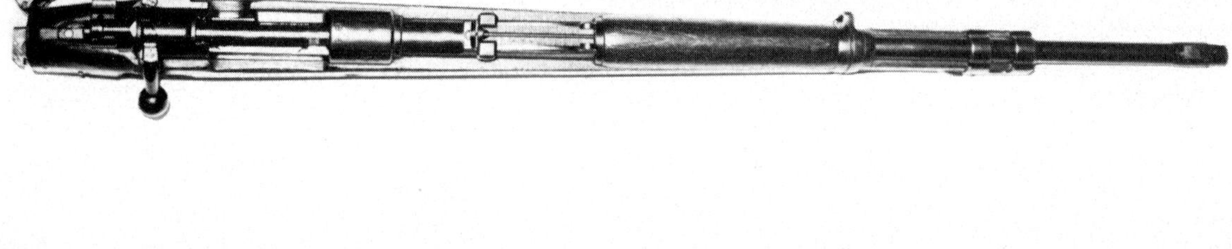

7.92 mm Gewehr 98/40

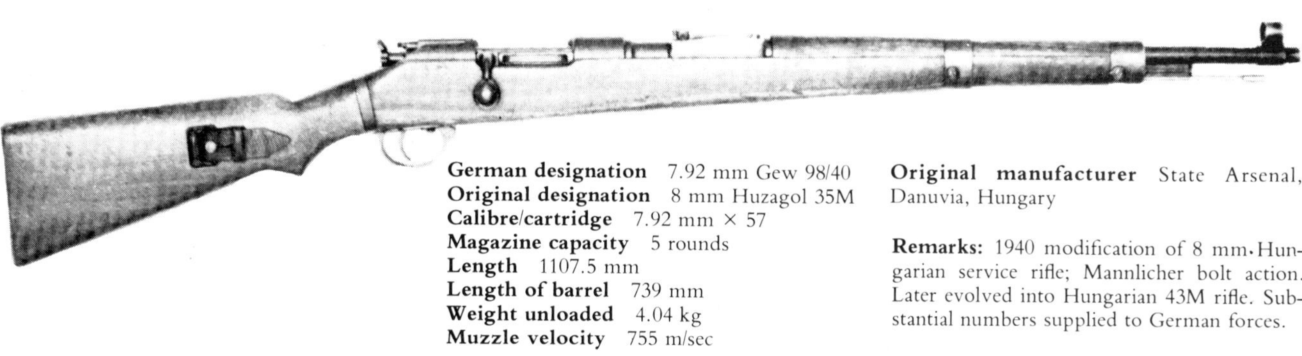

German designation 7.92 mm Gew 98/40
Original designation 8 mm Huzagol 35M
Calibre/cartridge 7.92 mm × 57
Magazine capacity 5 rounds
Length 1107.5 mm
Length of barrel 739 mm
Weight unloaded 4.04 kg
Muzzle velocity 755 m/sec
Original manufacturer State Arsenal, Danuvia, Hungary

Remarks: 1940 modification of 8 mm Hungarian service rifle; Mannlicher bolt action. Later evolved into Hungarian 43M rifle. Substantial numbers supplied to German forces.

7.92 mm Gewehr 33/40

German designation 7.92 mm Gew 33/40
Calibre/cartridge 7.92 mm × 57
Magazine capacity 5 rounds
Length 993 mm
Length of barrel 490 mm
Weight unloaded 3.58 kg
Muzzle velocity 715 m/sec
Original manufacturer Waffenwerke Brünn (formerly ČZ Brno)
Other manufacturers

Remarks: Modification of Musketon vz.16/33. Issued to German mountain troops. Limited number also made with folding stock for airborne formations.

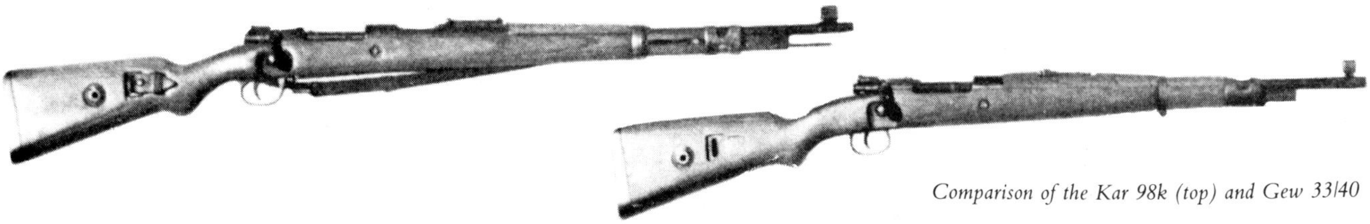

Comparison of the Kar 98k (top) and Gew 33/40

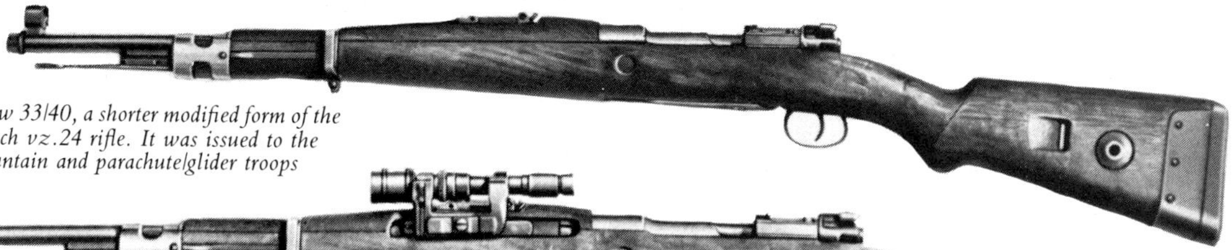

7.92 mm Gew 33/40, a shorter modified form of the original Czech vz.24 rifle. It was issued to the German mountain and parachute/glider troops

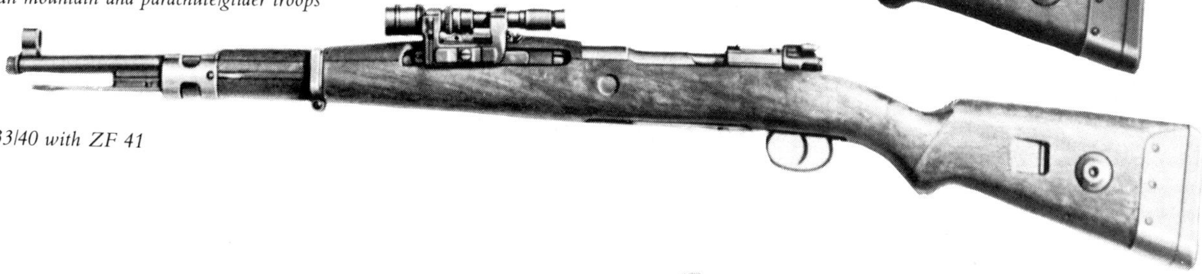

Gew 33/40 with ZF 41

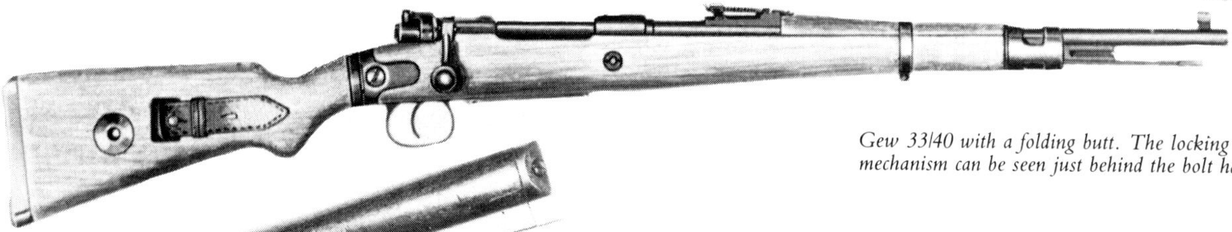

Gew 33/40 with a folding butt. The locking mechanism can be seen just behind the bolt handle

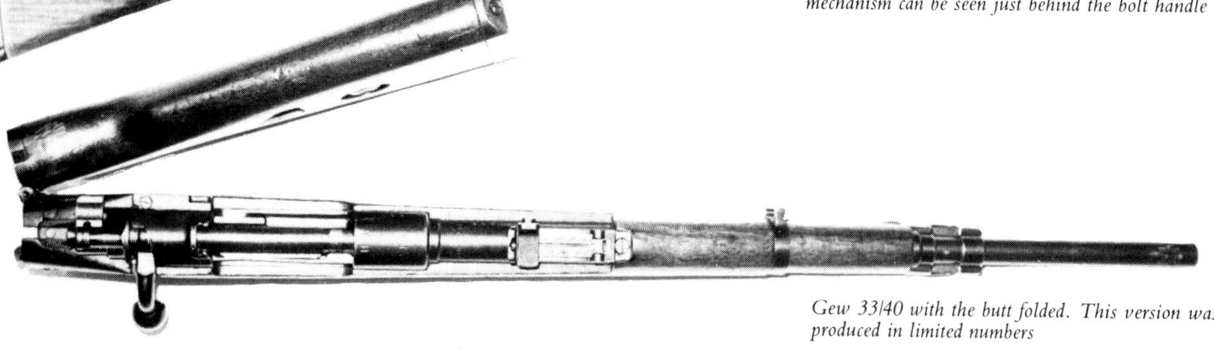

Gew 33/40 with the butt folded. This version was produced in limited numbers

7.92 mm Volkssturmgewehr 1

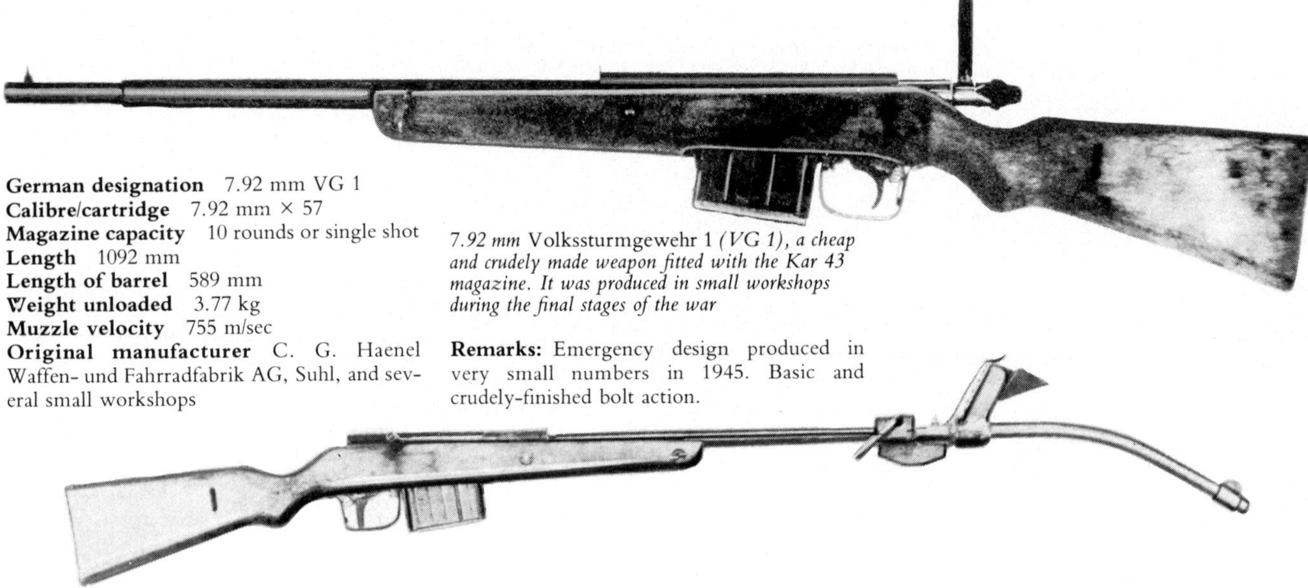

7.92 mm Volkssturmgewehr 1 *(VG 1), a cheap and crudely made weapon fitted with the Kar 43 magazine. It was produced in small workshops during the final stages of the war*

German designation 7.92 mm VG 1
Calibre/cartridge 7.92 mm × 57
Magazine capacity 10 rounds or single shot
Length 1092 mm
Length of barrel 589 mm
Weight unloaded 3.77 kg
Muzzle velocity 755 m/sec
Original manufacturer C. G. Haenel Waffen- und Fahrradfabrik AG, Suhl, and several small workshops

Remarks: Emergency design produced in very small numbers in 1945. Basic and crudely-finished bolt action.

VG 1 with a curved barrel extension for firing around corners

---

7.92 mm Volkssturmkarabiner 98

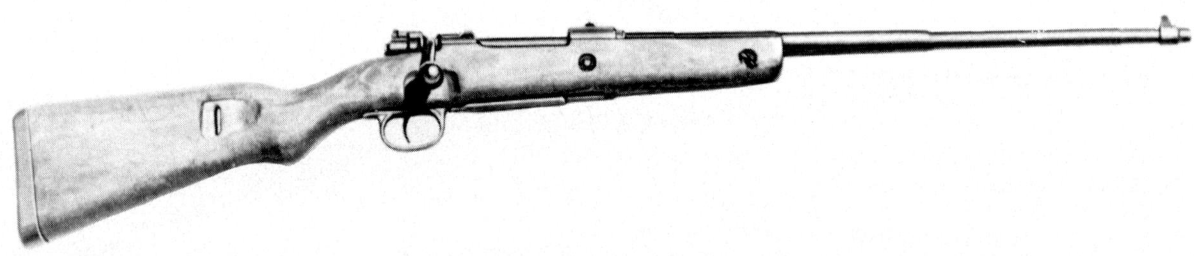

7.92 mm Volkssturmkarabiner 98, *another emergency weapon produced to arm the Volkssturm during the last few months of the war. This single-shot carbine was assembled from old Mauser rifle receivers and barrels fitted with crudely finished wooden stocks*

German designation 7.92 mm VK 98
Calibre/cartridge 7.92 mm × 57
Magazine capacity Single shot
Length 1031 mm
Length of barrel 528 mm
Weight unloaded 3.13 kg
Muzzle velocity 731 m/sec
Original manufacturer Mauser-Werke AG, Oberndorf-am-Neckar

Remarks: Simple 1945 emergency rifle utilising Gew 98 bolt action combined with various available German and foreign-made Mauser barrels; crudely finished stock and fittings. Very few issued for service.

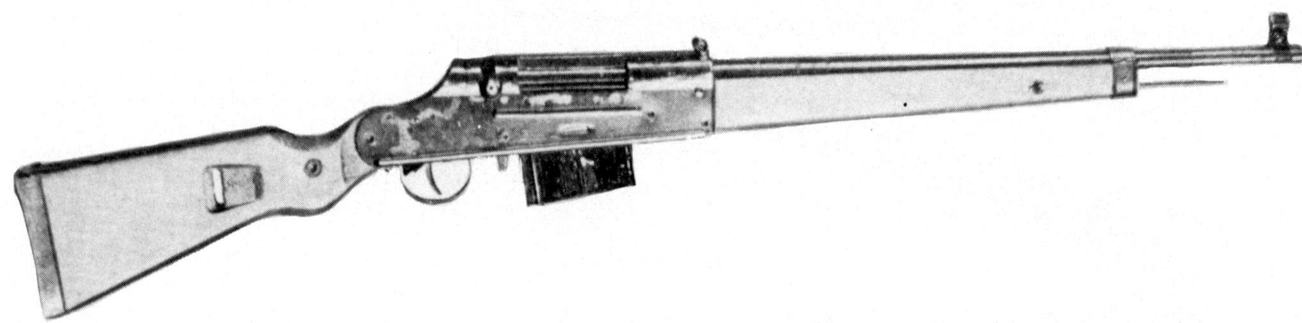

7.92 mm VG 2 differed in having a cut-down MG 13 barrel and a sheet metal housing for the bolt mechanism. It was fitted with the 10 round Kar 43 magazine

7.92 mm Volkssturmgewehr

Designation unknown

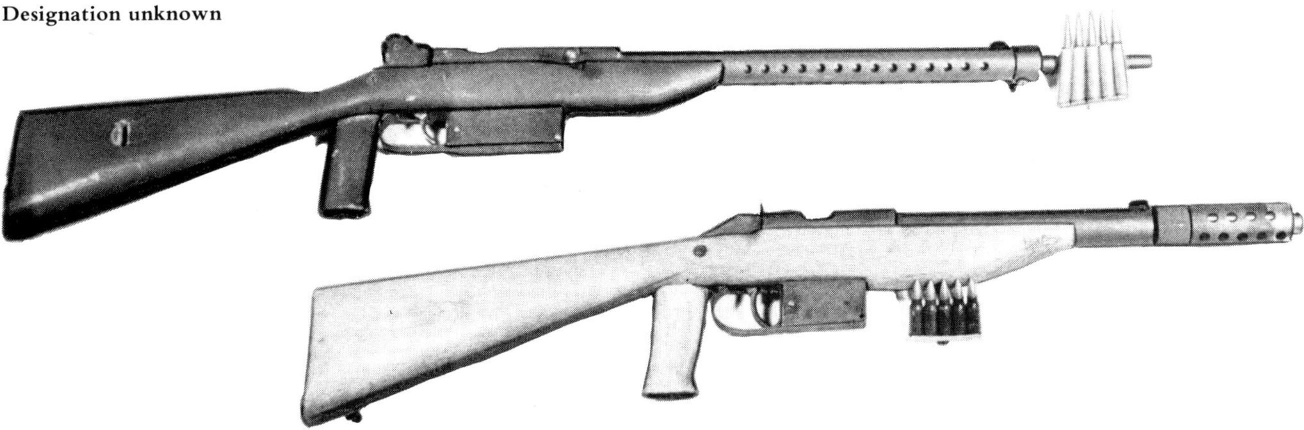

Remarks: An unusual type of infantry weapon intended for mass production was made by the Hessische Industrie-Werke at Wetzlar in early 1945. The rifle version (top) fired standard 7.92 mm × 57 ammunition and had an unusual bolt-action which moved the barrel forward when operated. The magazine had a capacity of five rounds. No other details are known.

The carbine version was a semi-automatic weapon which fired 7.92 mm × 33 '*kurz*' ammunition. It too used a blow-forward mechanism which pushed the barrel forward when fired and another unusual feature was the use of a double-action trigger mechanism. The magazine capacity was five rounds and the barrel length was approximately 410 mm. Apart from that no other details are known. Neither weapon got past the prototype stage.

7.92 mm Gewehr 33(t)

German designation 7.92 mm Gew 33(t)
Original designation Musketon vz.16/33
Calibre/cartridge 7.92 mm × 57
Magazine capacity 5 rounds
Length 995 mm
Length of barrel 490 mm
Weight unloaded 3.5 kg
Muzzle velocity 731 m/sec
Manufacturer Česká Zbrojovka, Brno

Remarks: Actually a carbine, despite its designation. Kept in production after German occupation of Czechoslovakia as Gew 33/40. Widespread German service use.

7.92 mm Gew 33(t). The standard Czech Army carbine, with modifications it was adopted by the German Army as Gew 33/40

6.5 mm Gewehr 209(i)

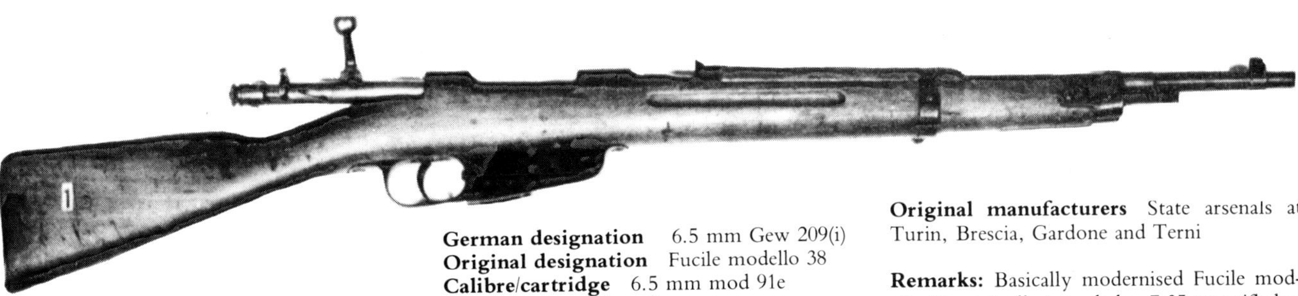

German designation 6.5 mm Gew 209(i)
Original designation Fucile modello 38
Calibre/cartridge 6.5 mm mod 91e
Magazine capacity 6 rounds
Length 1020 mm
Length of barrel 536 mm
Weight unloaded 3.45 kg
Muzzle velocity 707 m/sec

Original manufacturers State arsenals at Turin, Brescia, Gardone and Terni

Remarks: Basically modernised Fucile modello 91 originally intended as 7.35 mm rifle but calibre reverted to 6.5 mm soon after start of manufacture. After 1943 limited numbers rebored by Germans to 7.92 mm calibre and issued to local second-line units.

6.5 mm Gewehr 210(i)

German designation 6.5 mm Gew 210(i)
Original designation Fucile modello 41
Calibre/cartridge 6.5 mm mod 91e
Magazine capacity 6 rounds
Length 1170 mm
Length of barrel 690 mm
Weight unloaded 3.72 kg
Muzzle velocity 720 m/sec approx
Original manufacturer State arsenals at Turin, Brescia, Gardone and Terni

Remarks: Fucile modello 91 modernised for production during WW II. After 1943 some rebored by Germans to 7.92 mm calibre; issued mostly to second-line and garrison units.

6.5 mm Gewehr 211(h)

German designation 6.5 mm Gew 211(h)
Original designation Geweer M95
Calibre/cartridge 6.5 mm Mannlicher rimmed
Magazine capacity 5 rounds
Length 1290 mm
Length of barrel 790 mm
Weight unloaded 4.35 kg
Muzzle velocity 730 m/sec approx
Original manufacturer Hembrug Arsenal

Remarks: Mannlicher licence. 1893. Used only by German units stationed in Holland.

6.5 mm Gewehr 211(n)

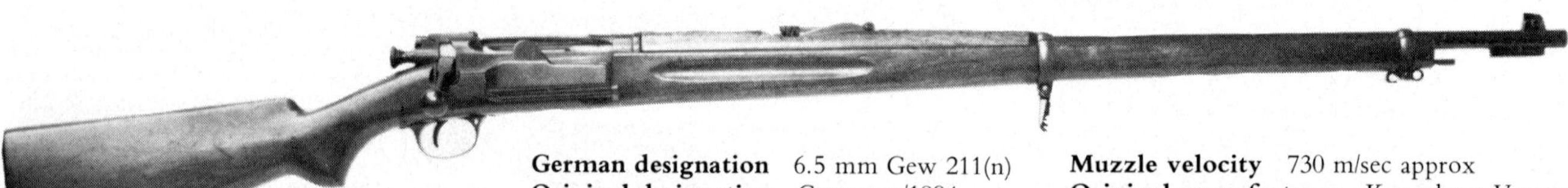

German designation 6.5 mm Gew 211(n)
Original designation Gevaer m/1894
Calibre/cartridge 6.5 mm × 55 Mauser rimless
Magazine capacity 5 rounds
Length 1263 mm
Length of barrel 763 mm
Weight unloaded 4.05 kg
Muzzle velocity 730 m/sec approx
Original manufacturer Kongsberg Vapenfabrik; limited number also by Steyr-Werke, Austria

Remarks: Known as Krag-Jorgensen rifle. Used mainly by German occupation units in Norway and some training depots.

6.5 mm Gewehr 214(i) and 214(j)

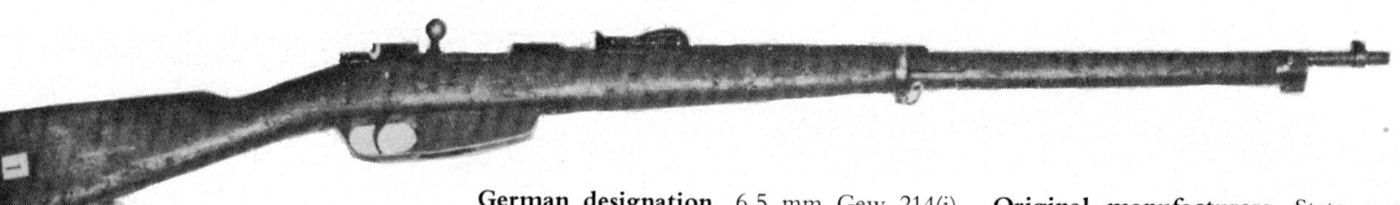

German designation 6.5 mm Gew 214(i) oder (j)
Original designation (i): Fucile modello 91; (j): Puska 6.5 mm M 91 i
Calibre/cartridge 6.5 mm Mod. 91e
Magazine capacity 6 rounds
Length 1280 mm
Length of barrel 780 mm
Weight unloaded 3.9 kg
Muzzle velocity 630 m/sec
Original manufacturers State arsenals at Turin, Brescia, Gardone and Terni

Remarks: Mannlicher-Carcano bolt action. Standard Italian Army service rifle of WW I, many still in use in 1940. Substantial numbers captured by German forces in 1943 but found of limited value; only small numbers distributed to Volkssturm formations.

6.5 mm Gewehr 215(g)

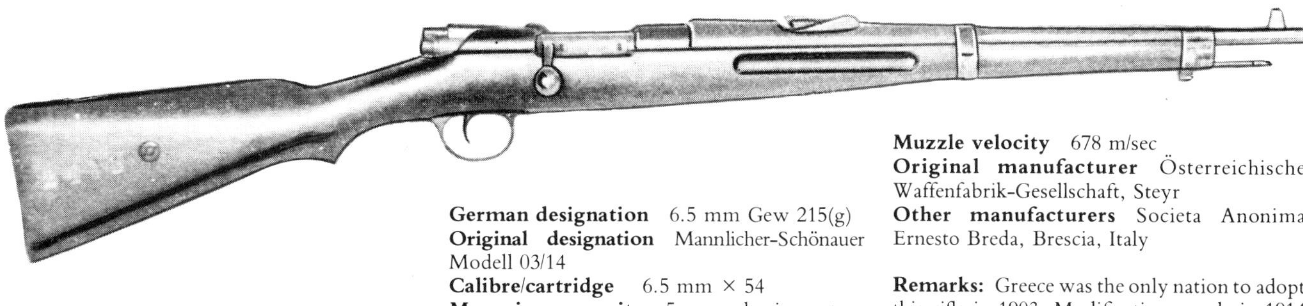

German designation 6.5 mm Gew 215(g)
Original designation Mannlicher-Schönauer Modell 03/14
Calibre/cartridge 6.5 mm × 54
Magazine capacity 5 rounds in rotary magazine
Length 1228 mm
Length of barrel 725 mm
Weight unloaded 3.9 kg
Muzzle velocity 678 m/sec
Original manufacturer Österreichische Waffenfabrik-Gesellschaft, Steyr
Other manufacturers Societa Anonima Ernesto Breda, Brescia, Italy

Remarks: Greece was the only nation to adopt this rifle in 1903. Modifications made in 1914 and in 1927. Additional rifles were produced in Breda to replace war losses. Features an unusual rotary magazine. German service use limited to garrison troops in Balkan area.

7.35 mm Gewehr 231(i)

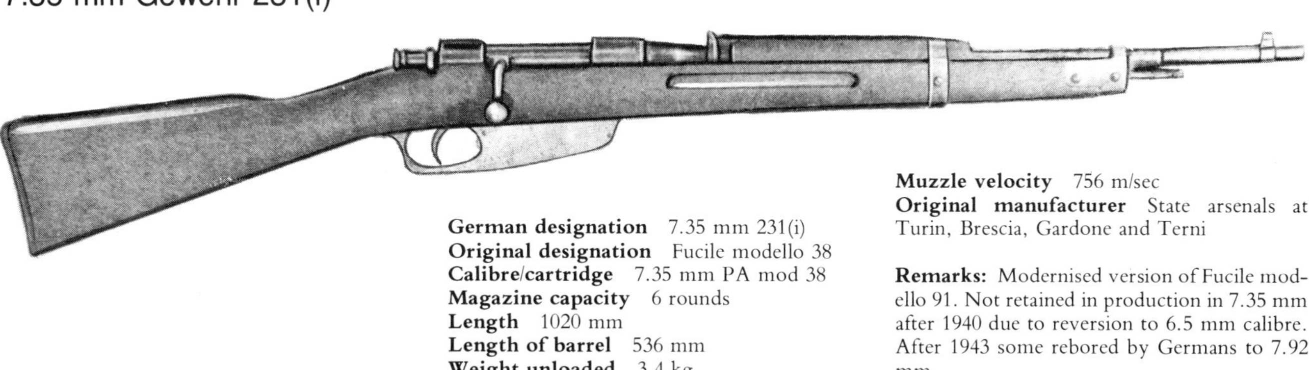

German designation 7.35 mm 231(i)
Original designation Fucile modello 38
Calibre/cartridge 7.35 mm PA mod 38
Magazine capacity 6 rounds
Length 1020 mm
Length of barrel 536 mm
Weight unloaded 3.4 kg
Muzzle velocity 756 m/sec
Original manufacturer State arsenals at Turin, Brescia, Gardone and Terni

Remarks: Modernised version of Fucile modello 91. Not retained in production in 7.35 mm after 1940 due to reversion to 6.5 mm calibre. After 1943 some rebored by Germans to 7.92 mm.

7.5 mm Gewehr 241(f)

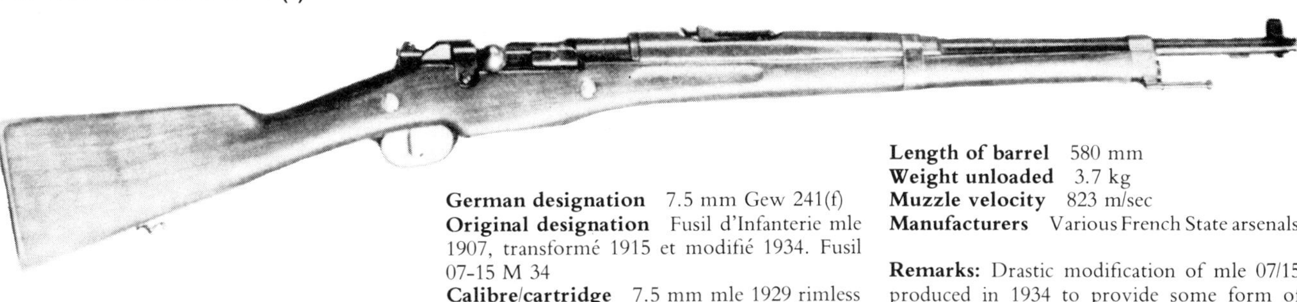

German designation 7.5 mm Gew 241(f)
Original designation Fusil d'Infanterie mle 1907, transformé 1915 et modifié 1934. Fusil 07-15 M 34
Calibre/cartridge 7.5 mm mle 1929 rimless
Magazine capacity 5 rounds
Length 1080 mm
Length of barrel 580 mm
Weight unloaded 3.7 kg
Muzzle velocity 823 m/sec
Manufacturers Various French State arsenals

Remarks: Drastic modification of mle 07/15 produced in 1934 to provide some form of 'modern' rifle for French Army. Used mainly by German occupation units after 1940.

7.5 mm Gewehr 242(f)

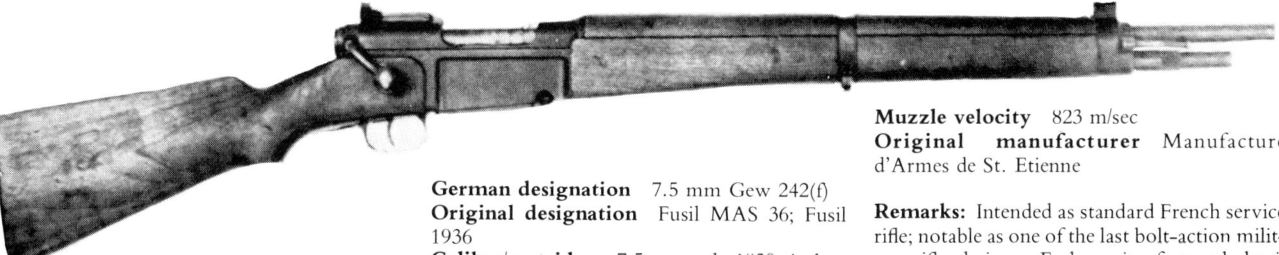

German designation 7.5 mm Gew 242(f)
Original designation Fusil MAS 36; Fusil 1936
Calibre/cartridge 7.5 mm mle 1929 rimless
Magazine capacity 5 rounds
Length 1020 mm
Length of barrel 574 mm
Weight unloaded 3.72 kg
Muzzle velocity 823 m/sec
Original manufacturer Manufacture d'Armes de St. Etienne

Remarks: Intended as standard French service rifle; notable as one of the last bolt-action military rifle designs. Early series featured detail differences and were built for various services; later production standardised. Large numbers captured by German forces and subsequently issued to various second-line and garrison units.

7.62 mm Gewehr 249(a)

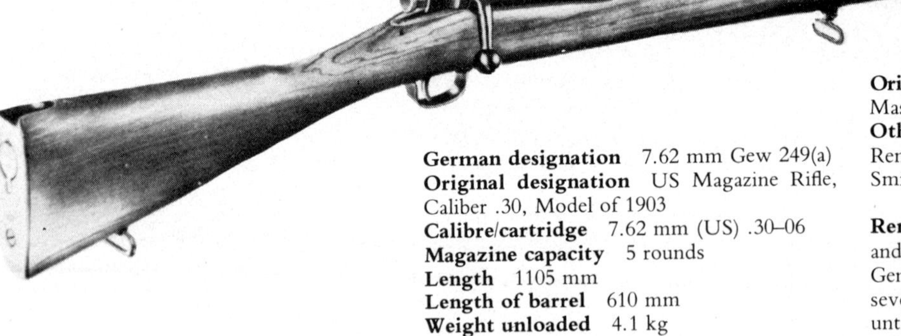

German designation 7.62 mm Gew 249(a)
Original designation US Magazine Rifle, Caliber .30, Model of 1903
Calibre/cartridge 7.62 mm (US) .30–06
Magazine capacity 5 rounds
Length 1105 mm
Length of barrel 610 mm
Weight unloaded 4.1 kg
Muzzle velocity 855 m/sec
Original manufacturer Springfield Armory, Mass., USA
Other manufacturers (WW II) Rock Island; Remington Arms Union, Eddystone, Pa.; L. C. Smith Corona Typewriter Co.

Remarks: Evolved around Mauser bolt action and adopted as standard US service rifle in 1903. Generally known as 'Springfield'. Produced in several versions, including a sniper variant, until 1943. German use of captured rifles limited to localised service. Some issued to Volkssturm.

7.62 mm Gewehr 252(r) and (j)

German designation 7.62 mm Gew 252(r) oder (j)
Original designations (r): *Vintovka obr. 1891*; (j): Puska 7,62 mm M 91R
Calibre/cartridge 7.62 mm × 54R
Magazine capacity 5 rounds
Length 1305 mm
Length of barrel 802 mm
Weight unloaded 4.37 kg
Muzzle velocity 810 m/sec
Original manufacturer Various Russian State arsenals

Remarks: Belgian-Russian design. Standard service rifle of old Tsarist armies, known as 'three-liner'. Substantial numbers still in service with Red Army in 1941; Yugoslav service rifles practically alike. Captured examples mostly stockpiled and later issued to various second-line service; also Volkssturm.

7.62 mm Gewehr 254(r) and 7.62 mm Zielfernrohrgewehr 256(r)

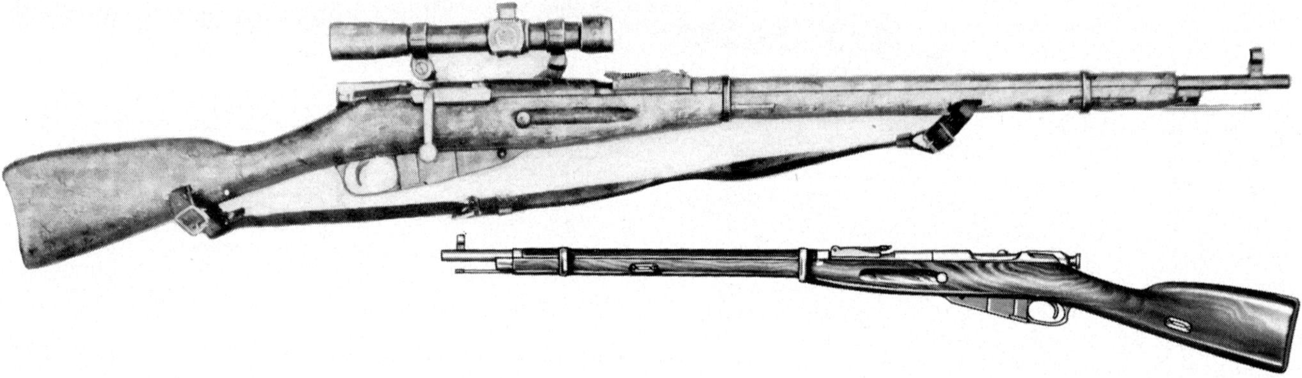

German designation 7.62 mm Gew 254(r) und 256(r)
Original designation *Vintovka obr. 1891/30 g*
Calibre/cartridge 7.62 mm × 54R
Magazine capacity 5 rounds
Length 1232 mm
Length of barrel 729 mm
Weight unloaded 4.25 kg
Muzzle velocity 860 m/sec
Original manufacturer Various Soviet State arsenals

Remarks: Modified and updated version of obr. 1891. Standard Red Army service rifle, produced in very large numbers until 1944. Captured guns mostly stockpiled but some retained by Russian and other local units fighting alongside German troops on the Eastern Front. Others issued to various second-line and local garrison units, and finally Volkssturm. Sniper variant: specially selected for accuracy, with elongated, bent-down bolt handle and 3.5 × PU telescope. Popular with German front-line troops.

7.65 mm Gewehr 261(b)

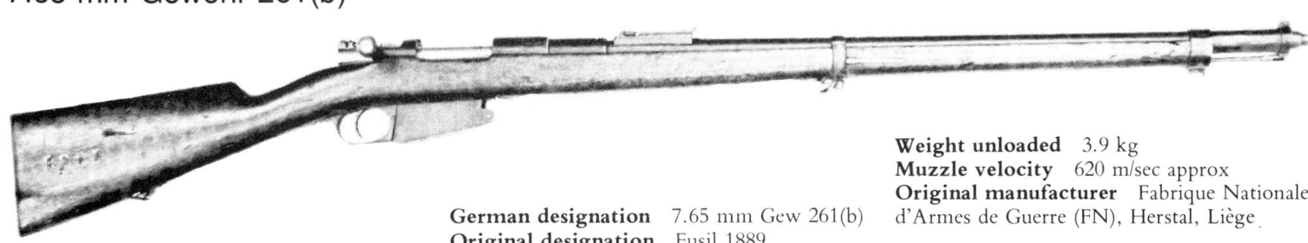

German designation 7.65 mm Gew 261(b)
Original designation Fusil 1889
Calibre/cartridge 7.65 mm Mauser
Magazine capacity 5 rounds
Length 1277 mm
Length of barrel 779 mm
Weight unloaded 3.9 kg
Muzzle velocity 620 m/sec approx
Original manufacturer Fabrique Nationale d'Armes de Guerre (FN), Herstal, Liège

Remarks: Standard Belgian rifle in 1940. Large numbers captured by German forces and subsequently widely used because of familiar Mauser bolt action.

7.65 mm Gewehr 263(b)

German designation 7.65 mm Gew 263(b)
Original designation Fusil 36
Calibre/cartridge 7.65 mm Mauser
Magazine capacity 5 rounds
Length 1096 mm
Length of barrel 600 mm
Weight unloaded 3.93 kg
Muzzle velocity 725 m/sec
Original manufacturer Fabrique Nationale d'Armes de Guerre (FN), Herstal, Liège

Remarks: Modernised version of Belgian Mauser Modell 1889. Substantial numbers used by German forces to equip various second-line formations; also training depots and (1944–45) Volkssturm.

7.7 mm Gewehr 281(e)

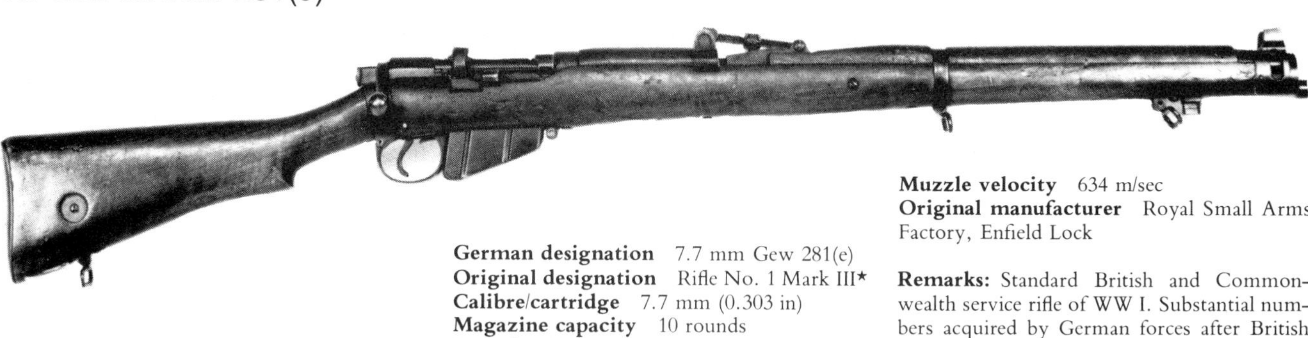

German designation 7.7 mm Gew 281(e)
Original designation Rifle No. 1 Mark III*
Calibre/cartridge 7.7 mm (0.303 in)
Magazine capacity 10 rounds
Length 1133 mm
Length of barrel 640 mm
Weight unloaded 3.93 kg
Muzzle velocity 634 m/sec
Original manufacturer Royal Small Arms Factory, Enfield Lock

Remarks: Standard British and Commonwealth service rifle of WW I. Substantial numbers acquired by German forces after British evacuation from France in 1940 and issued to various local garrison units. From 1944 also distributed to Volkssturm formations.

---

8 mm Gewehr 301(f), (g) and (j)

German designation 8 mm Gew 301(f), (g) oder (j)
Original designations (f): Fusil d'Infanterie mle 1886 transformé 1893. Fusil 86/45; (g): Lebel 86/93; (j): Puska 8 mm M 86
Calibre/cartridge 8 mm Lebel 1886 rimmed
Magazine capacity 8 rounds
Length 1303 mm
Length of barrel 798 mm
Weight unloaded 4.24 kg
Muzzle velocity 725 m/sec
Original manufacturers State arsenals, St. Etienne, Chatellerault, Tulle

Remarks: Standard French service rifle of WW I years, produced in very large numbers; usually known simply as 'Lebel'. Post WW I thousands sold abroad, mainly to Balkan countries. Beginning 1941 stocks of captured Lebels issued to many German second-line units, Home Defence Flak detachments, training depots, and finally Volkssturm.

8 mm Gewehr 302(f), (g) and (j)

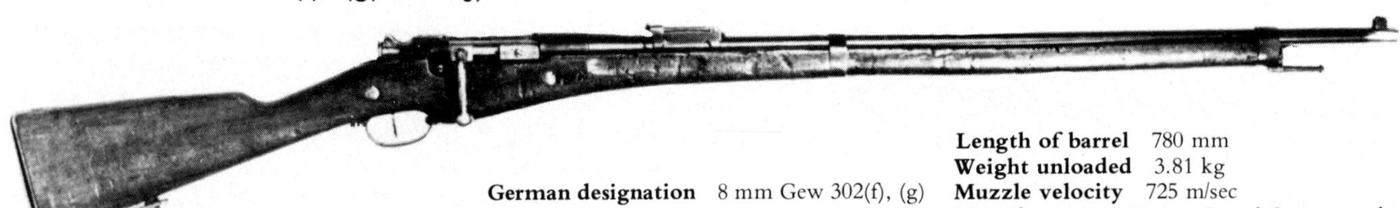

German designation 8 mm Gew 302(f), (g) oder (j)
Original designations (f): Fusil d'Infanterie mle 1907 transformé 1915. Fusil 07/15; (g): Lebel 07/15; (j): Puska 8 mm M 7/15F
Calibre/cartridge 8 mm Lebel rimmed
Magazine capacity 3 rounds
Length 1306 mm
Length of barrel 780 mm
Weight unloaded 3.81 kg
Muzzle velocity 725 m/sec
Manufacturers Various French State arsenals

Remarks: Known as the 'Berthier', this rifle supplemented Lebel rifles in French forces during WW I. After 1941–42 substantial numbers taken into German service and issued to various second-line units and finally Volkssturm.

---

8 mm Gewehr 303(f)

German designation 8 mm Gew 303(f)
Original designation Fusil mle 1886 Racroché 1935. Fusil 86-R-35
Calibre/cartridge 8 mm Lebel rimmed
Magazine capacity 3 rounds
Length 959 mm
Length of barrel 450 mm
Weight unloaded 3.556 kg
Muzzle velocity 634 m/sec
Original manufacturer Various French State arsenals

Remarks: Belated French attempt beginning 1935 to produce some form of 'modern' service rifle by updating old Lebels. After 1941–42 some numbers issued to various second-line German units, and eventually Volkssturm.

---

8 mm Gewehr 304(f) and (j)

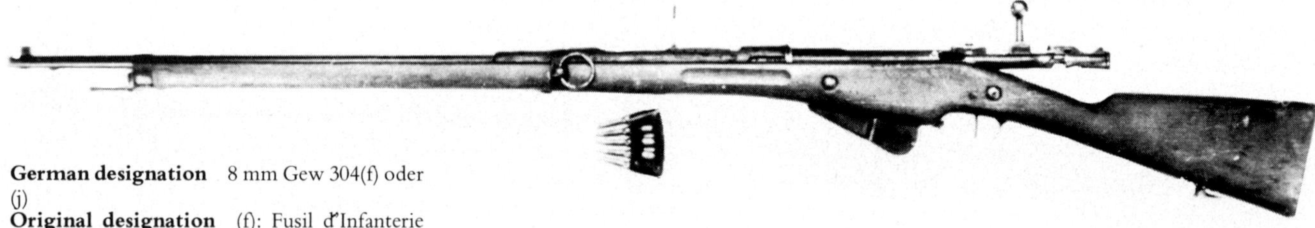

German designation 8 mm Gew 304(f) oder (j)
Original designation (f): Fusil d'Infanterie mle 1916. Fusil 1916; (j): Puska 8 mm M 16 F
Calibre/cartridge 8 mm Lebel rimmed
Magazine capacity 5 rounds
Length 1306 mm
Length of barrel 780 mm
Weight unloaded 4.195 kg
Muzzle velocity 725 m/sec
Manufacturers Various French State arsenals

Remarks: Version of mle 07/15 with increased magazine capacity. Produced in large numbers during WW I and afterwards. Captured rifles issued after 1942 to various German second-line units, Home Defence Flak detachments, and finally Volkssturm.

---

8 mm Gewehr 305(f) and (g)

German designation 8 mm Gew 305(f) oder (g)
Original designations (f): Fusil mle 1907 dit Colonial. Fusil mle 1907; (g): Lebel 07
Calibre/cartridge 8 mm Lebel rimmed
Magazine capacity 3 rounds
Length 1306 mm
Length of barrel 780 mm
Weight unloaded 3.8 kg
Muzzle velocity 725 m/sec
Original manufacturer St. Etienne, Chatellerault, Tulle; State arsenals

Remarks: Berthier design, originally intended for colonial service. Produced in large numbers during early part of WW I. German use of captured guns limited to training depots and Volkssturm.

8 mm Gewehr 307(j)

German designation 8 mm Gew 307(j)
Original designation Puska 8 mm M 93 MR; Mannlicher Modell 1893
Calibre/cartridge 8 mm Mannlicher
Magazine capacity 5 rounds
Length 1230 mm
Length of barrel 730 mm
Weight unloaded 3.8 kg
Muzzle velocity 730 m/sec
Original manufacturer Österreichische Waffenfabrik-Gesellschaft, Steyr

Remarks: Old Romanian service rifles bought by Yugoslavia during 1920s. German use of captured weapons limited to local garrison units and Volkssturm.

---

8 mm Gewehr 311(d)

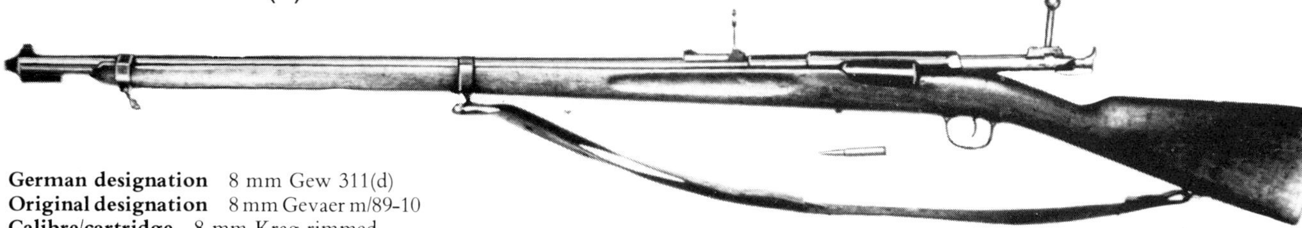

German designation 8 mm Gew 311(d)
Original designation 8 mm Gevaer m/89-10
Calibre/cartridge 8 mm Krag rimmed
Magazine capacity 5 rounds
Length 1330 mm
Length of barrel 840 mm
Weight unloaded 4.2 kg
Muzzle velocity 750 m/sec
Original manufacturer Haerens Tojhus, Copenhagen

Remarks: Norwegian-designed Krag-Jorgensen rifle adopted by Danish Army in 1889. Standard Danish service weapon in 1940. From 1942 issued to local German garrison units and various training depots.

6.5 mm Karabiner 408(i)

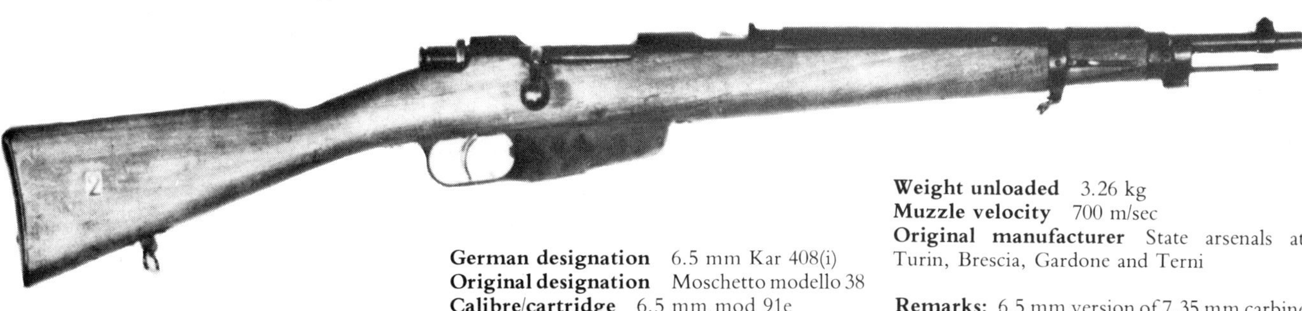

German designation 6.5 mm Kar 408(i)
Original designation Moschetto modello 38
Calibre/cartridge 6.5 mm mod 91e
Magazine capacity 6 rounds
Length 919 mm
Length of barrel 451 mm
Weight unloaded 3.26 kg
Muzzle velocity 700 m/sec
Original manufacturer State arsenals at Turin, Brescia, Gardone and Terni

Remarks: 6.5 mm version of 7.35 mm carbine modello 1938 manufactured after 1940. Some rebored to 7.92 mm calibre by Germans in 1943–44.

6.5 mm Karabiner 409(i)

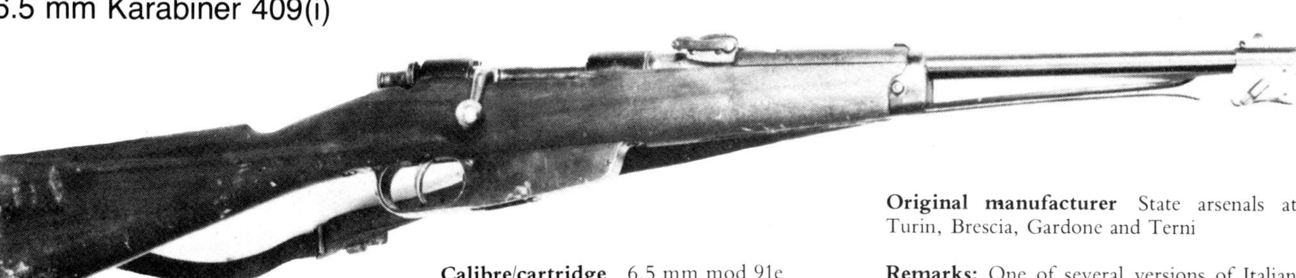

German designation 6.5 mm Kar 409(i)
Original designation Moschetto modello 91 per Cavalleria
Calibre/cartridge 6.5 mm mod 91e
Magazine capacity 6 rounds
Length 920 mm
Length of barrel 450 mm
Weight unloaded 3.16 kg
Muzzle velocity 700 m/sec
Original manufacturer State arsenals at Turin, Brescia, Gardone and Terni

Remarks: One of several versions of Italian Mannlicher-Carcano carbines with fixed folding bayonet. Large numbers issued to various Volkssturm units, also some Home Defence Flak detachments and other second-line formations.

6.5 mm Karabiner 410(i)

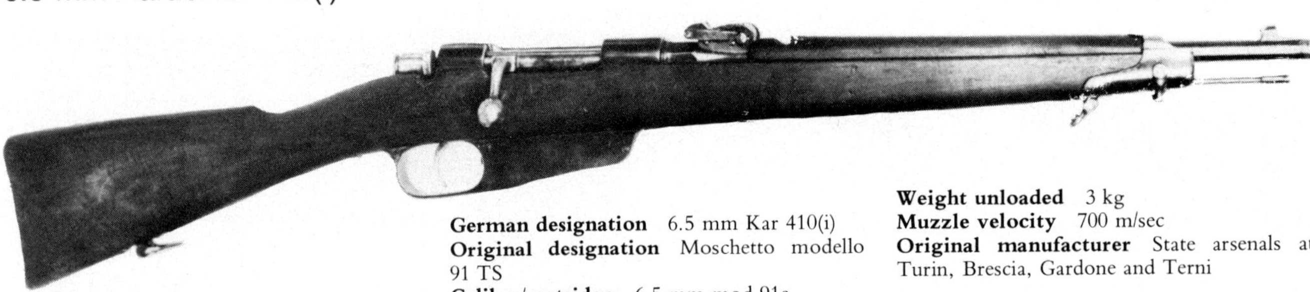

German designation 6.5 mm Kar 410(i)
Original designation Moschetto modello 91 TS
Calibre/cartridge 6.5 mm mod 91e
Magazine capacity 6 rounds
Length 920 mm
Length of barrel 450 mm
Weight unloaded 3 kg
Muzzle velocity 700 m/sec
Original manufacturer State arsenals at Turin, Brescia, Gardone and Terni

Remarks: 'TS' indicates 'Truppe Speciali'; used knife-type bayonet. Issued to local auxiliary units, and Volkssturm.

6.5 mm Karabiner 411(h)

German designation 6.5 mm Kar 411(h)
Original designation Karabijn aantal 1
Calibre/cartridge 6.5 mm Mannlicher rimmed
Magazine capacity 5 rounds
Length 940 mm
Length of barrel 448.6 mm
Weight unloaded 3.5 kg
Original manufacturer Hembrug Arsenal

Remarks: Kar 411–414(h) were all Dutch-produced carbines based on Mannlicher Modell 1895. They differed only in minor details, such as sling swivel location, bayonets, and sights. Used in limited numbers by local German garrison troops.

6.5 mm Karabiner 412(h)

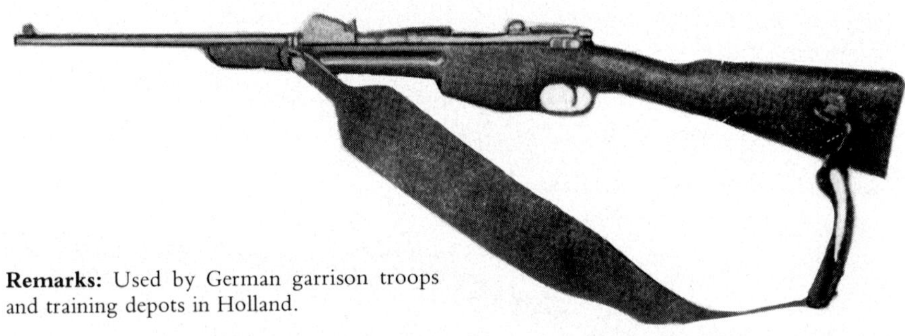

German designation 6.5 mm Kar 412(h)
Original designation Karabijn aantal 1 OM en NM
Calibre/cartridge 6.5 mm Mannlicher rimmed
Magazine capacity 5 rounds
Length 951 mm
Length of barrel 448.6 mm
Weight unloaded 3.26 kg
Original manufacturer Hembrug Arsenal

Remarks: Used by German garrison troops and training depots in Holland.

6.5 mm Karabiner 413(h)

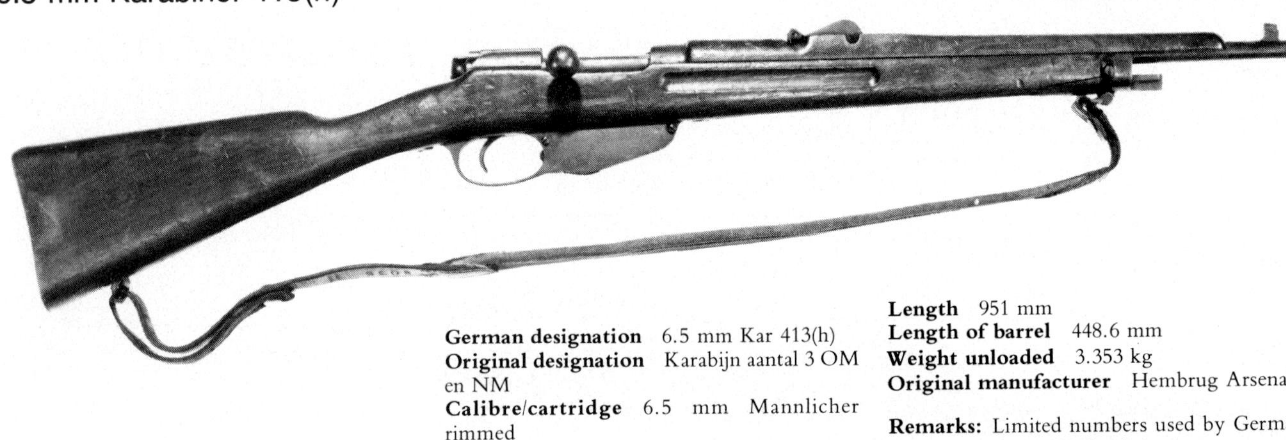

German designation 6.5 mm Kar 413(h)
Original designation Karabijn aantal 3 OM en NM
Calibre/cartridge 6.5 mm Mannlicher rimmed
Magazine capacity 5 rounds
Length 951 mm
Length of barrel 448.6 mm
Weight unloaded 3.353 kg
Original manufacturer Hembrug Arsenal

Remarks: Limited numbers used by German garrison troops in Holland.

6.5 Karabiner 414(h)

German designation 6.5 mm Kar 414(h)
Original designation Karabijn aantal 4 OM en NM
Calibre/cartridge 6.5 mm Mannlicher rimmed
Magazine capacity 5 rounds
Length 951 mm
Length of barrel 448.6 mm
Weight unloaded 3.367 kg
Original manufacturer Hembrug Arsenal

Remarks: Limited numbers used by local German garrison troops.

6.5 mm Karabiner 411(n)

German designation 6.5 mm Kar 411(n)
Original designation Kavalerikarabin m/1894
Calibre/cartridge 6.5 mm × 55 Mauser rimless
Magazine capacity 5 rounds
Length 1015 mm
Length of barrel 520 mm
Weight unloaded 3.4 kg
Muzzle velocity 785 m/sec
Original manufacturer Konsberg Vapenfabrik

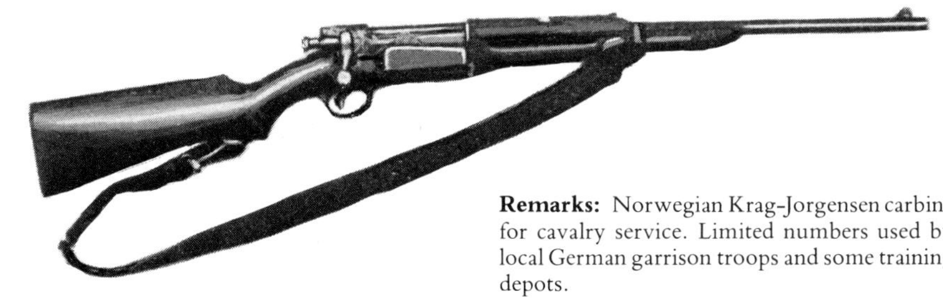

Remarks: Norwegian Krag-Jorgensen carbine for cavalry service. Limited numbers used by local German garrison troops and some training depots.

6.5 mm Karabiner 412(n)

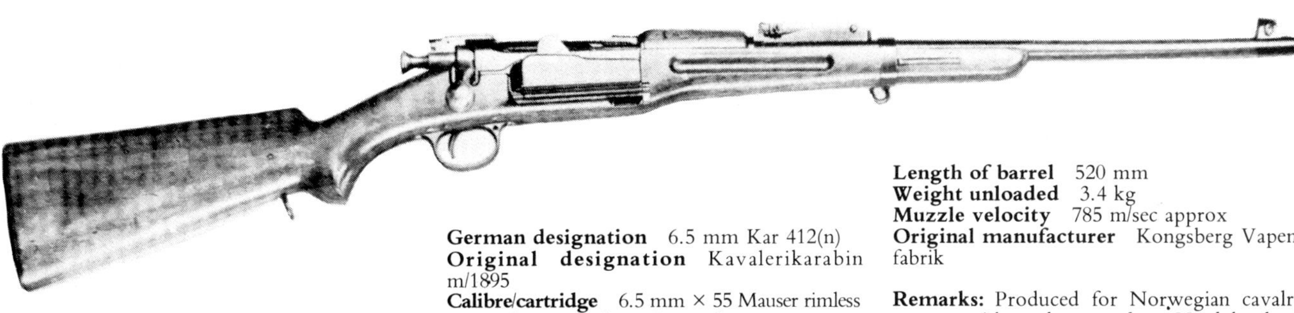

German designation 6.5 mm Kar 412(n)
Original designation Kavalerikarabin m/1895
Calibre/cartridge 6.5 mm × 55 Mauser rimless
Magazine capacity 5 rounds
Length 1015 mm
Length of barrel 520 mm
Weight unloaded 3.4 kg
Muzzle velocity 785 m/sec approx
Original manufacturer Kongsberg Vapenfabrik

Remarks: Produced for Norwegian cavalry troops; without bayonet lug. Used by local German garrisons and training depots.

6.5 mm Karabiner 413(n)

German designation 6.5 mm Kar 513(n)
Original designation Ingeniorkarabin m/1904
Calibre/cartridge 6.5 mm × 55 Mauser rimless
Magazine capacity 5 rounds
Length 1015 mm
Length of barrel 520 mm
Weight unloaded 3.8 kg
Muzzle velocity 785 m/sec
Original manufacturer Kongsberg Vapenfabrik

Remarks: Produced for Norwegian engineer troops; without bayonet lug. Used by German garrisons and training depots in Norway.

6.5 mm Karabiner 414(n)

German designation 6.5 mm Kar 414(n)
Original designation Artillerikarabin m/1907
Calibre/cartridge 6.5 mm × 55 Mauser rimless
Magazine capacity 5 rounds
Length 1015 mm
Length of barrel 520 mm
Weight unloaded 3.8 kg
Muzzle velocity 785 m/sec
Original manufacturer Kongsberg Vapenfabrik

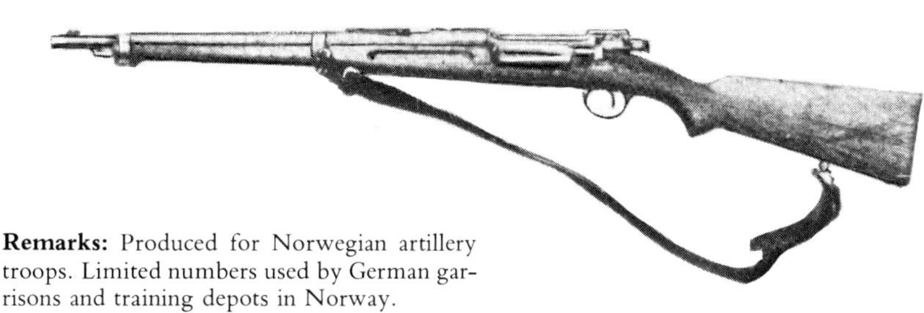

Remarks: Produced for Norwegian artillery troops. Limited numbers used by German garrisons and training depots in Norway.

6.5 mm Karabiner 415(n)

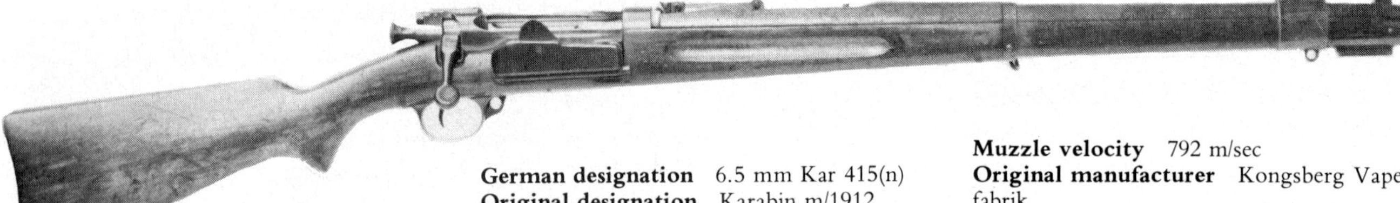

German designation 6.5 mm Kar 415(n)
Original designation Karabin m/1912
Calibre/cartridge 6.5 mm × 55 Mauser rimless
Magazine capacity 5 rounds
Length 1107 mm
Length of barrel 610.7 mm
Weight unloaded 4 kg
Muzzle velocity 792 m/sec
Original manufacturer Kongsberg Vapenfabrik

Remarks: Slightly longer than other Krag-Jorgensen carbines. Used in greater numbers by German garrison troops in Norway than more specialised variants of this basic design.

6.5 mm Karabiner 416(i) and (j)

German designation 6.5 mm Kar 416(i) oder (j)
Original designations (i): Moschetto modello 91/24; (j): Karabini 6,5 mm M 91 i
Calibre/cartridge 6.5 mod 91e
Magazine capacity 6 rounds
Length 920 mm
Length of barrel 450 mm
Weight unloaded 3 kg
Muzzle velocity 700 m/sec
Original manufacturer State Arsenals at Turin, Brescia, Gardone and Terni

Remarks: Carbine version of Gew 214(i), the Fucile mod 91, originally produced in 1924. Featured knife-type bayonet. Large numbers of captured and requisitioned guns issued to Volkssturm units; some also carried by transport drivers.

7.35 mm Karabiner 430(i)

German designation 7.35 mm Kar 430(i)
Original designation Moschetto modello 38
Calibre/cartridge 7.35 mm PA mod 38
Magazine capacity 6 rounds
Length 919 mm
Length of barrel 451 mm
Weight unloaded 3.23 kg
Muzzle velocity 731 m/sec
Original manufacturer State afsenals at Turin, Brescia, Gardone and Terni

Remarks: 7.35 mm version of 6.5 mm mod 1938 carbine or Kar 408(i) with attached folding bayonet. After 1943 some rebored by Germans to 7.92 mm. Mostly local German service use.

7.65 mm Karabiner 451(b)

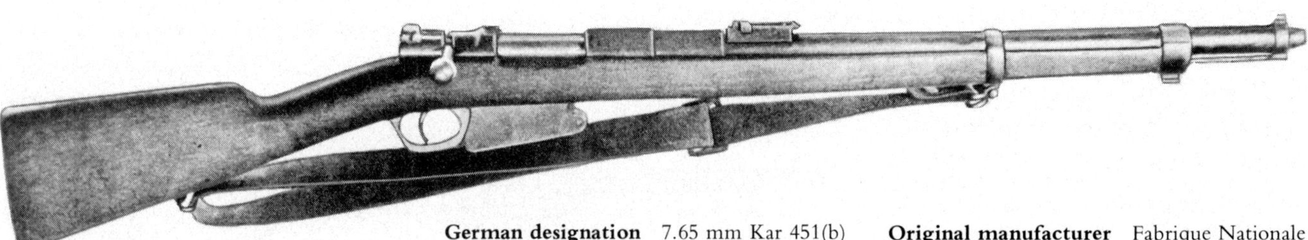

German designation 7.65 mm Kar 451(b)
Original designation Carabine 1889
Calibre/cartridge 7.65 mm Mauser
Magazine capacity 5 rounds
Length 1045 mm
Length of barrel 550 mm
Weight unloaded 3.6 kg
Muzzle velocity 579 m/sec approx
Original manufacturer Fabrique Nationale d'Armes de Guerre, Herstal, Liège

Remarks: Carbine version of Mauser Modell 1889, the Gew 261(b). Was built in four variants, with minor differences. Limited German service use.

7.65 mm Karabiner 453(b)

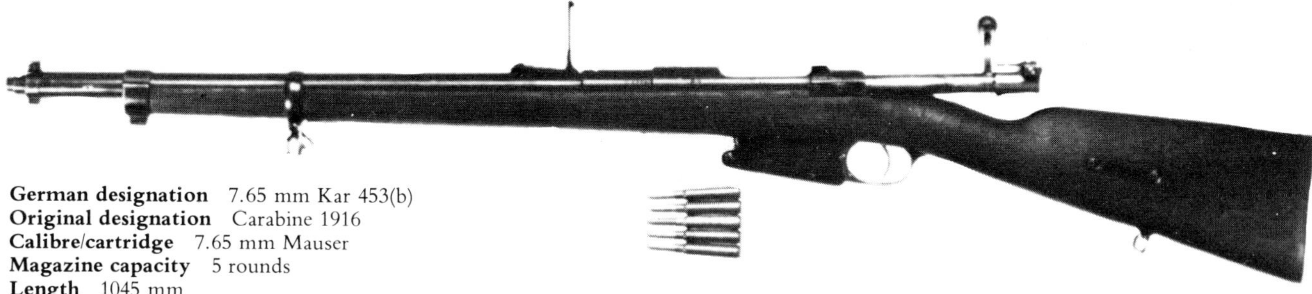

German designation 7.65 mm Kar 453(b)
Original designation Carabine 1916
Calibre/cartridge 7.65 mm Mauser
Magazine capacity 5 rounds
Length 1045 mm
Length of barrel 550 mm
Weight unloaded 3.6 kg
Muzzle velocity 579 m/sec approx
Original manufacturer Fabrique d'Armes de L'Etat (later FN), Herstal, Liège

Remarks: Conversion of Modell 1889 carbine. German use limited to some local garrisons, training depots and Volkssturm.

7.62 mm Karabiner 454(r)

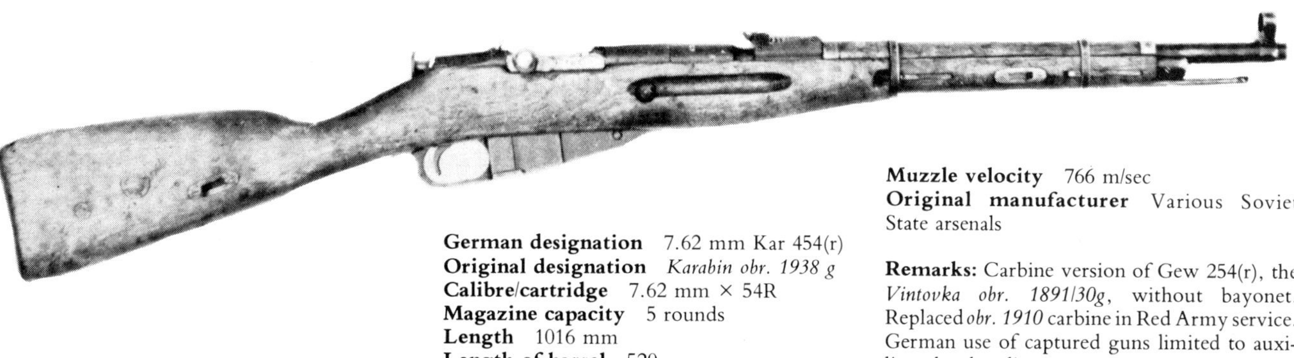

German designation 7.62 mm Kar 454(r)
Original designation *Karabin obr. 1938 g*
Calibre/cartridge 7.62 mm × 54R
Magazine capacity 5 rounds
Length 1016 mm
Length of barrel 520 mm
Weight unloaded 3.6 kg
Muzzle velocity 766 m/sec
Original manufacturer Various Soviet State arsenals

Remarks: Carbine version of Gew 254(r), the *Vintovka obr. 1891/30g*, without bayonet. Replaced *obr. 1910* carbine in Red Army service. German use of captured guns limited to auxiliary local police, garrison units, and Volkssturm.

7.62 mm Karabiner 457(r)

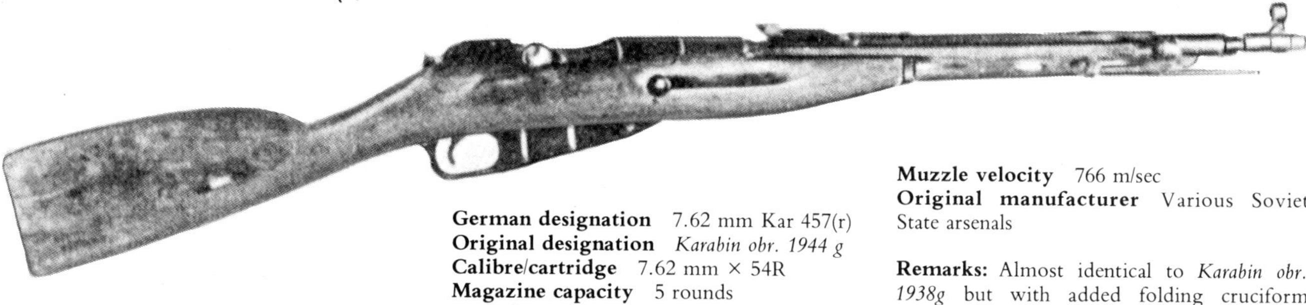

German designation 7.62 mm Kar 457(r)
Original designation *Karabin obr. 1944 g*
Calibre/cartridge 7.62 mm × 54R
Magazine capacity 5 rounds
Length 1020 mm
Length of barrel 515 mm
Weight unloaded 3.9 kg
Muzzle velocity 766 m/sec
Original manufacturer Various Soviet State arsenals

Remarks: Almost identical to *Karabin obr. 1938g* but with added folding cruciform bayonet and various minor modifications to ease large-scale production. Found only limited German use.

7.9 mm Karabiner 497(p)

German designation 7.9 mm Kar 497(p)
Original designation Karabinek 91/98/25
Calibre/cartridge 7.92 mm × 57
Magazine capcity 5 rounds
Length 1100 mm
Length of barrel 600 mm
Weight unloaded 3.7 kg
Muzzle velocity 753 m/sec approx
Original manufacturer Polish State Arsenals Warsaw and Radom

Remarks: Polish modification of WW I rifles – Russian Mosin-Nagant action with Mauser fittings. Captured guns issued to various German second-line troops, drivers, garrison units, training depots and Volkssturm.

8 mm Karabiner 506(d)

German designation 8 mm Kar 506/1, /2, /3, /4(d)
Original designations 506/1: 8 mm Fodfolkskarabin m/89-24; 506/2: 8 mm Artilleriekarabin m/89-24; 506/3: 8 mm Ingeniorkarabin m/89-24; 506/4: 8 mm Rytterkarabin m/89-24
Calibre/cartridge 8 mm Krag rimmed
Magazine capacity 5 rounds
Length 1100 mm
Length of barrel 600 mm
Weight unloaded 4 kg
Muzzle velocity 700 m/sec approx
Original manufacturer Haerens Tojhus, Copenhagen

Remarks: All these Krag-Jorgensen carbines differed only in minor details, such as swing swivel location, shape of bolt handle, sights, etc. German service use limited to some local garrisons and training depots in Denmark.

8 mm Karabiner 551 and 552(f)

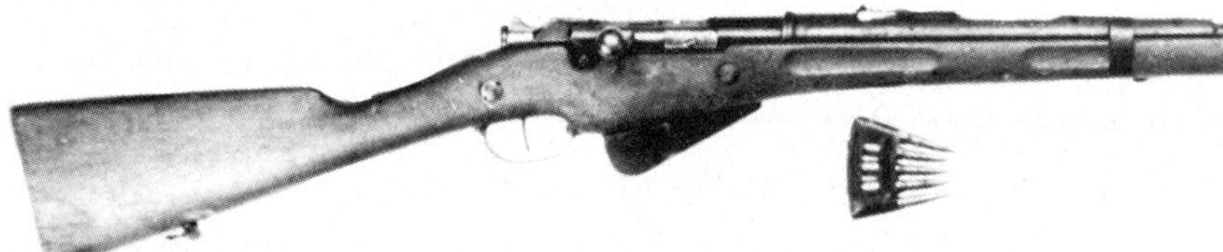

German designation 8 mm Kar 551 und 552(f)
Original designation Mousqueton mle 1890 et 1892
Calibre/cartridge 8 mm 1886 Lebel rimmed
Magazine capacity 3 rounds
Length 945 mm
Length of barrel 450 mm
Weight unloaded 3.1 kg
Muzzle velocity 634 m/sec
Original manufacturer St. Etienne, Chatellerault, Tulle and State arsenals

Remarks: These two carbines were almost identical Berthier designs. German service use confined to some garrisons, training depots and Volkssturm.

8 mm Karabiner 553(f)

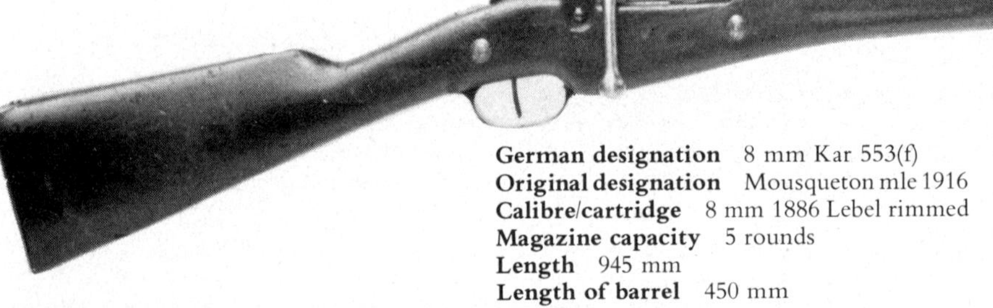

German designation 8 mm Kar 553(f)
Original designation Mousqueton mle 1916
Calibre/cartridge 8 mm 1886 Lebel rimmed
Magazine capacity 5 rounds
Length 945 mm
Length of barrel 450 mm
Weight unloaded 3.25 kg
Muzzle velocity 634 m/sec
Original manufacturer St. Etienne, Chatellerault, Tulle and State arsenals

Remarks: Similar to mle 1890 et 1892 but with increased magazine capacity. German use confined to some garrison units in France, training depots and Volkssturm.

8 mm Gewehr 98(ö), 306(g), (i), (j) and 294(j)

German designations 8 mm Gew 98(ö), 306(g), (i) oder (j) und 294(j)
Original designations (ö): 8 mm Repetier-Gewehr Modell 1895; (g): 8 mm M 95/24; (i): Fucile 'Mannlicher' 95; (j): Puska 8 mm M 95 M, Puska 7.9 mm M 95 M, Puska 7.9 mm 95/24
Calibre/cartridge 8 mm × 56; (294(j)): 7.92 mm × 57
Magazine capacity 5 rounds
Length 1270 mm
Length of barrel 765 mm
Weight unloaded 3.7 kg
Muzzle velocity 619 m/sec
Original manufacturer Österreichische Waffenfabrik-Gesellschaft, Steyr

Remarks: Standard service rifle of Austro-Hungarian Empire armies during WW I. After 1918 large numbers handed over or sold to Italy, Yugoslavia and Greece; modified version also sold to Bulgaria. Mannlicher bolt action.

8 mm Stützen 95(ö), 7.9 mm Karabiner 494(g), 8 mm Kar 505(g), (i) and (j)

German designations 8 mm Stützen 95(ö); 8 mm Kar 505(g), (i) oder (j); 7.9 mm Kar 494(g)
Original designations (ö): 8 mm Repetier-Stützen-Gewehr Modell 1895; (494(g): 'Mannlicher' 95/24; (505(g)): 'Mannlicher' 95; (i): Moschetto 'Mannlicher' 95; (j): Karabini 8 mm M 95 M
Calibre/cartridge 8 mm × 56 or 7.92 mm × 57
Magazine capacity 5 rounds
Length 1005 mm
Length of barrel 500 mm
Weight unloaded 3.2 kg
Muzzle velocity 580 m/sec
Original manufacturer Österreichische Waffenfabrik-Gesellschaft, Steyr

Remarks: Shortened version of Mannlicher Modell 1895 originally intended for various specialised arms such as signallers, drivers and engineers. After 1939 many issued to German second-line and garrison units as well as military police; also training depots.

Automatic rifles

The term 'automatic rifle' is generally given to what are really self-loading rifles where the weapon automatically loads itself after each pull of the trigger. This form of rifle became possible with the advances in design and metallurgy by about 1880 when the Austrian arms designer von Mannlicher built a few prototypes, but these weapons were too fragile for service use. Early in WW I Mauser produced small numbers of automatic rifles based on a design dating back to 1898, and these were used – together with some Mexican Mondragon self-loading rifles – during the early days of aerial warfare, but when flexible machine guns became standard fittings on aircraft these early automatic rifles were relegated to trench warfare where they soon proved to be too heavy, complex and fragile for the rigours of infantry combat and the automatic rifle fell from favour with German arms designers.

In many ways this fall from grace was an unfortunate mistake, for the tactical lessons of WW I all pointed towards the need for some form of automatic rifle to equip most of the infantry. But instead the German staffs issued specifications for conventional bolt-action rifles which led to the adoption of the *Kar 98b* and *Kar 98k*. On the other hand, the German military were not the only ones to make the same error, for by 1939 the USA was the only major nation in the process of arming her troops with a self-loading rifle.

Until 1939 little consideration was given towards producing an automatic rifle in Germany apart from a few 'paper' designs and the aquisition of a few foreign models for study. In fact it was not until 1941 that the invasion of the Soviet Union brought the German soldier into direct contact with appreciable numbers of Russian troops armed with automatic rifles. These weapons were Simonov and Tokarev designs which had been produced and delivered in substantial numbers from 1936 onwards. The more important were the *Tokarev obr. 1938* and *obr. 1940* and these made such an impression on the German troops that they immediately called for similar weapons. At that time there was nothing on the stocks in Germany and the best their supply services could do was to produce small numbers of captured French Fusil Mitrailleur RSC mle 1918.

As automatic rifles they had all the usual early drawbacks of weight and lack of strength and indeed they were so unreliable that in the mid-1930s the French Army began a programme of converting them into ordinary bolt-action rifles. But sufficient numbers remained in their original form and were impressed into temporary German service in 1941. By the end of 1941 a large quantity of Soviet automatic rifles were also in front-line German service, many of which remained in use until 1945.

Although the lack of automatic rifles came as a rather nasty surprise to the German soldier in 1941, some design work on a German equivalent had in fact begun in 1940, but at that time on a very low priority. Walther and Mauser both submitted prototypes, of which the Mauser *7.92 mm Gew 41(M)* was soon rejected and limited production of Walther products, the *7.92 mm Gew 41 (W)*, began in 1941. Both designs used a variation of the Danish Bang system which trapped part of the gases driving a bullet out of the muzzle and used them to operate pistons to move the mechanism and open the bolt. This system had its drawbacks, not the least of which was the number of precision machining operations needed in manufacture, and thus the unit cost was relatively high. Most of the *Gew 41s* produced were sent to the Eastern Front where they proved to be adequate weapons but no more.

Detailed examination of captured Soviet Tokarev automatic rifles showed that their gas-operated mechanism was superior to the Bang system used on the *Gew 41*. It was also soon realised that only a limited amount of detail design work was needed to convert the Walther design to take the Tokarev mechanism, and the result was the *7.92 mm Gew 43*. This was a much more viable self-loading rifle and was manufactured in large numbers. Lessons learned from the early production of the *Gew 41* included a reduced number of machine operations and the introduction of stamping techniques to rifle design for the first time. Other time and cost savers were the use of laminated wood for the stocks and furnishings and forgings wherever possible. In 1944 a slightly shorter version, the *Kar 43*, went into production and this change involved the introduction of even simpler production methods.

The *Gew 41* and *43* both fired the standard German service cartridge, the *Gewehr Patrone 98*. This cartridge was evolved for use with the old bolt-action *Gew 98* and was thus a relatively powerful bullet/charge combination. Like many other cartridges of its day it had been designed for long-range infantry warfare, and firing it from an automatic rifle called for the use of heavy and strong – and rather expensive – lock and bolt mechanisms. It was at this stage that the careful evaluation and analysis of combat reports strongly indicated the need for a change.

These combat reports once more emphasised the fact that most infantry engagements took place at ranges well under 400 metres and at that range the full potential of the standard infantry rifle/cartridge combination was being wasted. As a result a proposal was put forward for a new infantry rifle designed to fire a new, less powerful cartridge. Development work on new ammunition of this kind had been carried out since 1934, first by Genschow (trade name: GECO), with official contract awarded to Polte of Madgeburg in 1938. By late 1940 this new cartridge was ready. Initially known as the *7.9 mm Infanterie kurz Patrone*, it represented a standard 7.92 mm cartridge case shortened and necked down to take a lighter bullet with a lead or mild steel core. This shortened case carried a lighter propellant load and thus the range was much shorter than the original 7.92 mm round, but at short ranges the striking power was unimpaired.

Specifications for a new type of infantry weapon termed 'machine carbine' to fire this new cartridge were put out to the German armament industry already on 18 April 1938. The reduced charge of the new cartridge enabled it to be fired on full automatic which would considerably increase the firepower of the infantry at ranges in excess of the sub-machine gun which would have been its nearest equivalent, while the influence of the sub-machine gun was also seen in the production methods to be employed. The design of the new weapon was to embody as few machining operations as possible while stampings and forgings were employed wherever feasible. Two firms built prototypes, Walther and Haenel.

Both machine carbines appeared to be remarkably similar in design and both used almost identical gas-operated mechanisms. The Walther design, the *MKb 42(W)* was produced in relatively small numbers as troop trials in Russia with both weapons soon indicated the superiority of the Haenel design, the *MKb 42(H)*. This machine carbine was designed by Hugo Schmeisser and proved itself in combat when one of the first troop trial batches was parachuted into a surrounded infantry formation on the Eastern Front. After that initial success infantry units everywhere demanded this new weapon and although about 8,000 *MKb 42(H)* guns were produced, all in the remarkably short time of three months, they were issued only to elite formations. Slight alterations were made to the basic design as a result of battle experience and the result was the *MK 43*.

At this stage the heavy hand of Hitler descended and further production and development was prevented. However, so sure were the German weapon designers that there was a future for their progeny that production went ahead with a change of name to *Maschinen-Pistole 43 (MP 43)*. The main reasons for Hitler's intervention were probably the huge stocks and supply systems built up to provide conventional 7.92 mm ammunition, and the

increased burden placed on the German industry with the production of a new type of ammunition.

However, by the end of 1943 the demands from the field were becoming so urgent that even Hitler had to concede, and large-scale production got under way at last, this time as the *MP 44* which was virtually identical to the *MP 43*. In December 1944 this designation was once more changed to *Sturmgewehr 44* (*StG 44*), partly for political reasons and partly to indicate more clearly the actual designed role of this new weapon.

The *MP 43* family were in fact the first of what are now known as assault rifles. An infantry unit armed with these new weapons had a huge increase in firepower compared to a similar conventionally-armed unit, and this firepower increase, coupled with mobile warfare, also brought about certain changes in infantry tactics. The soldier in action could use the semi-automatic setting to fire at distant targets when in defence, while in the attack the change to full automatic greatly increased the offensive 'shock'. To further increase the versatility of the *MP 43* most were fitted with grenade launching cups, some had provision for telescopic sights, and 1945 saw the introduction of the first infra-red night sights, the *Zielgerät 1229 'Vampir'* which was used in action in small numbers before the war ended. Another very odd accessory was the *'Krummlauf'* curved barrel fitting. This attachment enabled the *MP 43/1* to fire at angles of up to 90 degrees from the line of sight, with complex mirror aiming devices to enable the gunner to see 'round the corner'. Exactly what tactical role this odd fitting was supposed to fulfil is now difficult to determine; streetfighting is one possibility. There were two basic versions evolved by Haenel and Rheinmetall: the *StG 44 V* which used a barrel with a 30 or 40 degree curve, and the *StG 44 P* with a 90 degree curve. However, such were the handling problems with the *StG 44 P* that its use was restricted to ball and socket mountings on armoured vehicles. Both versions had special prism sights, but the cost and complexity of this unusual weapon arrangement were high and few, if any, reached the front line.

The last of the *MP 43* series was a weapon that had several designations among which were *Gerät 06H*, *MP 45(M)* and *StG 45(M)*. Cheap and simple as the *MP 43* series was, it was thought that the weapon could be made simpler and quicker still, and as the war ended experimental work on the *StG 45(M)* was in an advanced state at the Mauser-Werke AG. By March 1945 experimental prototype weapons, using a two-part bolt with a roller delay device, were being test-fired, and when the war ended the design and development team nucleus managed to keep together and continue work, first in France and then in Spain. The end results of their endeavours are the *CETME 58* rifle and the *Heckler & Koch G3*, both of which are the present-day service rifles of Spain and Germany respectively.

The *MP 43* series was not the only automatic rifle designed to use the 7.92 mm '*kurz*', or shortened rifle cartridge. One other weapon that reached production, of a sort, was the *Volkssturmgewehr 1-5*. As its name implied, this was a weapon designed for use by the Volkssturm and as the production situation was growing more chaotic due to the Allied bombing and advances into Germany late in 1944 and early 1945, some form of easily-produced infantry weapon was badly needed. Such was the disruption of German industry by early 1945 that the actual production of the *VG 1-5* was made the responsibility, not of a centralised arms procurement department, but of the local Party Gauleiters. OKH and *Wa Prüf 2* oversaw the early design stages of the weapon, which is usually accredited to the Gustloff-Werke at Suhl. Thereafter mass production was scheduled to take place under Party supervision at several centres, among which were Mauser at Oberndorf, Rheinmetall at Sömmerda, Gustloff at Suhl, Walther at Zella-Mehlis, Steyr at the Österreichwerke, the Spreewerke and G. Appel in Berlin, and at numerous small sub-contractor concerns. But very few of the new rifles were actually completed, and there were many local variations of the basic design. The *VG 1-5* was a simple crude weapon, and a form of hybrid sub-machine gun and automatic rifle. Most of the components were simple cheap stampings and remotely-manufactured sub-assemblies. The mechanism was unusual in that when fired some of the propellant gases were directed into a piston round the barrel. These gases held the bolt mechanism forward long enough for the bullet to leave the muzzle (the same principle was used on the *Volkspistole*) until the internal pressures had dropped to a safe level for the re-cocking cycle to commence. Most *VG 1-5* examples captured were semi-automatic only, but some were intended for full automatic fire. Very few appear to have been issued, which was probably just as well: due to the low standard of manufacture their service lives would have been short indeed.

An off-shoot from the *MP 43* development was the *7.92 mm Fallschirmgewehr 42* (*FG 42*). This design had nothing directly to do with the *MP 43*, only that the early success of that weapon led to demands from the Luftwaffe parachute arm for a similar automatic rifle. But OKL were unwilling to adopt the shortened rifle round and issued a specification, during 1940, for a weapon that could fire the conventional 7.92 mm ammunition. Mauser, Krieghoff, Walther, Rheinmetall and Gustloff were all invited to submit designs but only Rheinmetall and Krieghoff had produced prototypes by 1942. Of these, the design by Louis Stange of Rheinmetall-Borsig AG of Düsseldorf was judged the better and selected for service, but by that time Rheinmetall had no spare production capacity and Krieghoff of Suhl were awarded the production contract.

In many ways the *FG 42* was one of the most remarkable small-arm designs to emerge from WW II. Like many other Rheinmetall products it bristled with novel features such as side-mounted magazine, a folding bayonet, and a light pressed-metal bipod. The mechanism was not an original design but a successful combination of several which enabled the *FG 42* gunner to direct automatic fire accurately, using a conventional round with a weapon weighing only about 4.5 kg. By the time the *FG 42* began to leave the production lines the German *Fallschirmjäger* formations were in gradual decline and being deployed more and more as ordinary infantry, with the result that the *FG 42* was often used as a light machine gun, a role for which it was not intended. Relatively few were made – most sources quote about 7,000 – and delivered in three versions, a result of the Rheinmetall/Krieghoff parentage. The first had a steel butt and a sloping pistol grip, but the bulk of the production run featured wooden butts and conventional pistol grips, the two models differing only in detail. All three versions also proved expensive to manufacture. After 1945 many features of the *FG 42* were incorporated into various new small arms designs, one of which was later to become the modern American M60 general-purpose machine gun.

However, such was the demand for automatic rifles from German infantry units after 1941 that it always exceeded supply and delivery. The *MP 43* series was usually issued to elite front-line units, with priority to the Waffen-SS, and most went to the Eastern Front. To meet the shortfall in supply, captured weapons were pressed into service wherever possible, but as the self-loading rifle was as novel a weapon to most of the Allies as it was to the Germans the use of such weapons was limited to captured Soviet Tokarev and, to a lesser extent, American M1 Garand rifles. Small numbers of captured American M1 and M1A1 carbines were also used, especially during the 1944–45 Ardennes offensive.

At this point it would be as well to consider some of the more general small arms research still in progress in 1945. Despite the priorities given to mass production, almost every German arms concern was engaged in research of some sort or another, much of it unofficial. Before the war ended a great deal of official research was carried out in evolving small arms ammunition that did not use metal cartridge cases; a consumable case was developed experimentally in 8 mm and a solid case-less propellant in 7.92 mm. These experiments were directed towards

German paratroops in action with the FG 42

reducing the substantial calls made by small-arms manufacturers on basic raw materials, especially copper, which was in ever-increasing supply problem state. By 1944 much of the basic 7.92 mm conventional and '*kurz*' ammunition was supplied with steel cartridge cases covered with a thin lacquer coating to improve feed lubrication. Despite this coating the feed problems, especially on the *MP 43* family, were increased, and stoppages were frequent.

To ease the lead supply problems many 7.92 mm bullets had their lead cores replaced by soft steel cores, and by 1945 work was advanced on a solid steel bullet with an integral driving band. Explosive ammunition in 9 mm, 7.92 mm conventional and '*kurz*' was in an advanced state of development by 1945, despite the terms of the Hague Convention of 1900, and for possible future weapon developments 7.92 mm cases suitable for electric ignition were also evolved. But perhaps the oddest ammunition development was a new 9 mm round intended for rocket propulsion; it also incorporated spin stabilisation. The firm responsible for this venture was Deutsche Waffen- und Munitionsfabriken AG, Lübeck.

In the weapons field it was intended that the *MP 43* would eventually become the standard infantry weapon and the *Gew 43* the normal sniper equipment. Extensive trials had indicated that any future sniper rifle should use a new 7 mm round but the vast stocks of 7.92 mm ammunition still on hand prevented any change in the foreseeable future. Silencers for sniper rifles were in a state of continuous development. To produce a really efficient silencer rubber baffles were needed, but by 1945 the rubber raw material situation was so acute that metal baffles had to be used, and a great deal of work went on to discover the most efficient arrangement. Other work was in progress to discover the best form of muzzle brake and flash eliminator or compensator for the *MP 43* series.

One entirely new weapon in the design stage with many arms concerns was a simple bolt-action rifle to take the '*kurz*' round. As far as can be determined none of these designs were ever made, apart from the odd crude 'last ditch defence' gun.

Considering how late the Germans were in developing the automatic rifle in all its various forms, it must be said that to produce a weapon as revolutionary as the *MP 43* together with its novel ammunition, all in a space of a few years and under war-time conditions, was a remarkable achievement. Had more of these weapons been available to the German troops in the field the outcome of many infantry engagements might have differed drastically, even though the eventual result of the war would have remained the same.

7.92 mm Gewehr 41(W)

7.92 mm Gew 41(W). Produced and issued for troop trials in 1941, the Walther self-loading rifle was adopted as a standard weapon in 1942

German designation 7.92 mm Gew 41(W)
Calibre/cartridge 7.92 mm × 57
Magazine capacity 10 rounds
Length 1124 mm
Length of barrel 546 mm
Weight unloaded 4.7 kg
Muzzle velocity 745 m/sec
Rate of fire (semi-auto): 40 rpm
Manufacturer Carl Walther Waffenfabrik AG, Zella-Mehlis

Remarks: Designed in 1940–41 and accepted for service in preference to Mauser proposal. In action not very successful due to fouling and corrosion caused by its modified Bang system. Only limited numbers made and delivered in two versions, as Modell 41 and Modell 41(W), differing in minor details. Was also known as SG 41. Design served as basis for Gew 43.

7.92 mm Gewehr 41(M)

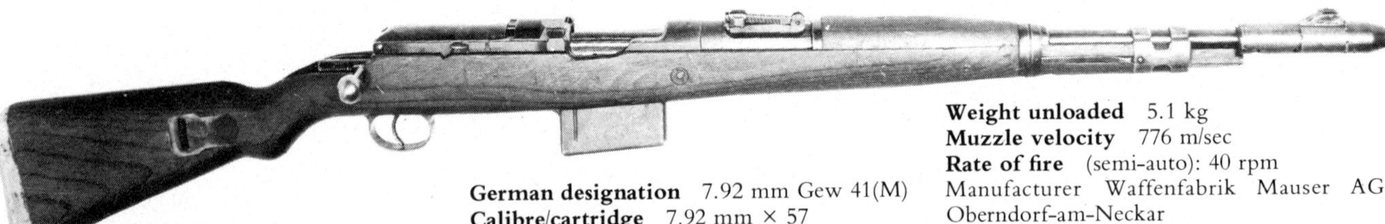

7.92 mm Gew 41(M), also tested in 1941 but not accepted for service

German designation 7.92 mm Gew 41(M)
Calibre/cartridge 7.92 mm × 57
Magazine capacity 10 rounds
Length 1175 mm
Length of barrel 552.5 mm
Weight unloaded 5.1 kg
Muzzle velocity 776 m/sec
Rate of fire (semi-auto): 40 rpm
Manufacturer Waffenfabrik Mauser AG, Oberndorf-am-Neckar

Remarks: Issued for troop trials late in 1941 but rejected in favour of Gew 41(W).

7.92 mm Gewehr 43 and 7.92 mm Karabiner 43

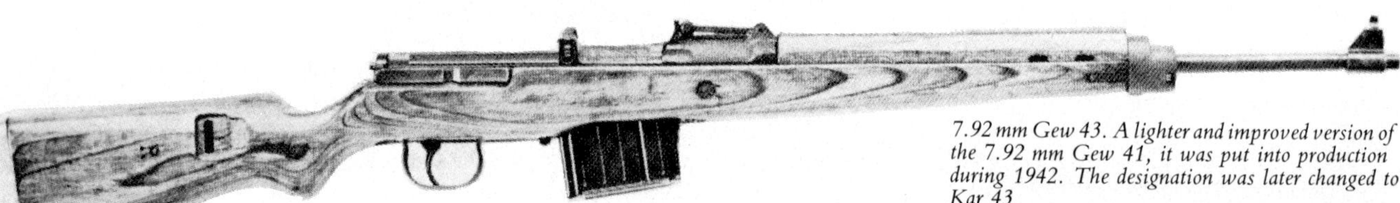

7.92 mm Gew 43. A lighter and improved version of the 7.92 mm Gew 41, it was put into production during 1942. The designation was later changed to Kar 43

German designation 7.92 mm Gew 43; 7.92 mm Kar 43
Calibre/cartridge 7.92 mm × 57
Magazine capacity 10 rounds

	Gew 43	Kar 43
Length	1117 mm	1067 mm
Length of barrel	549 mm	500 mm
Weight unloaded	4.4 kg	4.1 kg

Muzzle velocity 745 m/sec
Rate of fire (semi-auto): 40 rpm
Manufacturers Carl Walther Waffenfabrik AG, Zella-Mehlis; Walther-Werke, Buchenwald; Berlin-Lübecker Masch.-Fabrik, Berlin; Waffenwerke Brünn (formerly ČZ, Brno); Mauser-Werke AG, Oberndorf-am-Neckar; Gustloff-Werke, Suhl

Remarks: Modification of basic Gew 41(W) mechanism to a gas-operated semi-automatic system based on captured Soviet Tokarev rifles. Produced in large numbers; late series with plastic instead of wood furniture. Used many pressings, castings and forgings; minimum amount of machined parts. Many Gew 43s fitted with telescope mounting for use as sniper rifles. Kar 43 introduced in 1944 was simply a shorter version of basic weapon.

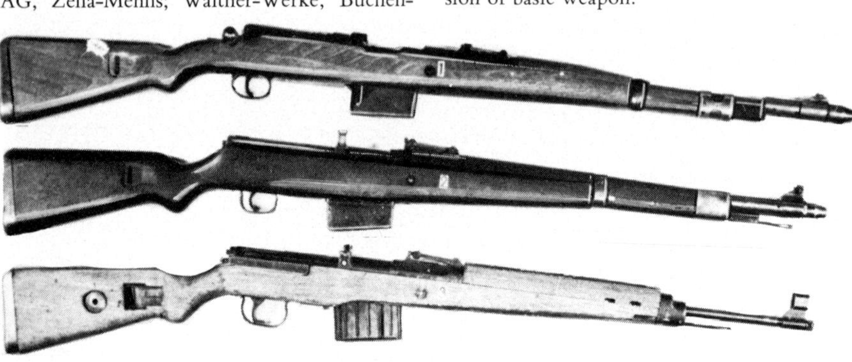

◄ *Comparison of the three 7.92 mm self-loading rifles.* Top, *Gew 41(M)*, centre, *Gew 41(W)*, and below, *Gew 43*

*Kar 43 with the ZFK 43 (*Zielfernrohr für Karabiner 43*) scope*

7.92 mm Maschinenkarabiner 42(H)

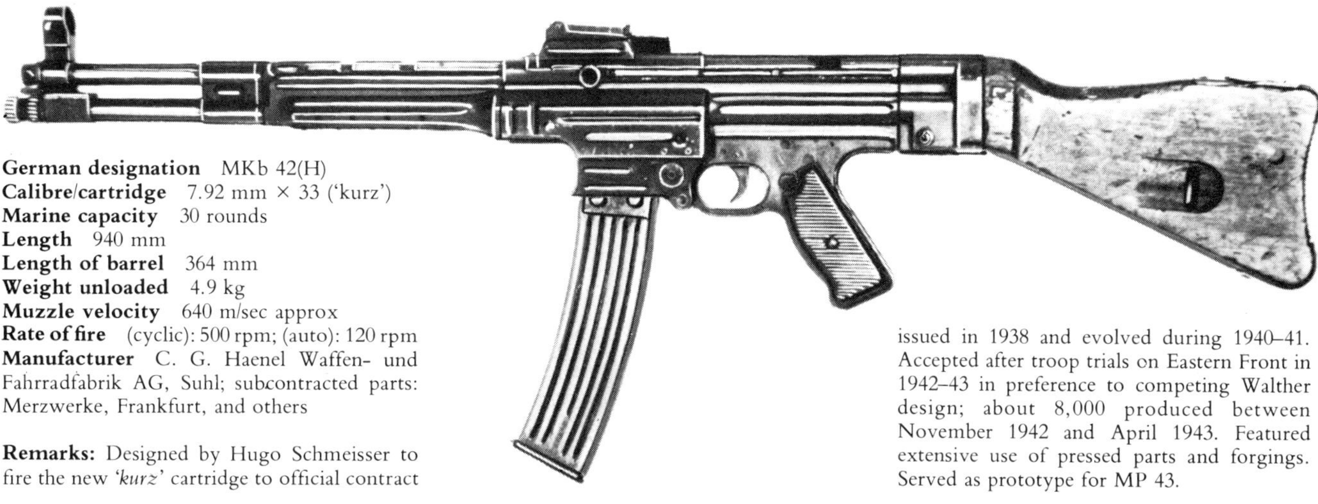

German designation MKb 42(H)
Calibre/cartridge 7.92 mm × 33 ('kurz')
Marine capacity 30 rounds
Length 940 mm
Length of barrel 364 mm
Weight unloaded 4.9 kg
Muzzle velocity 640 m/sec approx
Rate of fire (cyclic): 500 rpm; (auto): 120 rpm
Manufacturer C. G. Haenel Waffen- und Fahrradfabrik AG, Suhl; subcontracted parts: Merzwerke, Frankfurt, and others

Remarks: Designed by Hugo Schmeisser to fire the new *'kurz'* cartridge to official contract issued in 1938 and evolved during 1940–41. Accepted after troop trials on Eastern Front in 1942–43 in preference to competing Walther design; about 8,000 produced between November 1942 and April 1943. Featured extensive use of pressed parts and forgings. Served as prototype for MP 43.

7.92 mm Maschinenkarabiner 42(W)

MKb 42(W) with grenade launcher

German designation 7.92 mm MKb 42(W)
Calibre/cartridge 7.92 mm × 33 ('kurz')
Magazine capacity 30 rounds
Length 933.5 mm
Length of barrel 409 mm
Weight unloaded 4.43 kg
Muzzle velocity 650 m/sec
Rate of fire (cyclic): 600 rpm; (auto): 120 rpm
Manufacturer Carl Walther Waffenfabrik AG, Zella-Mehlis

Remarks: Second selective-fire weapon designed around the new *'kurz'* cartridge. Development commenced privately in 1940 to compete with Haenel. Official contract awarded in January 1941, prototypes completed by July 1942. Combat trials on Eastern Front 1942–43 revealed unsuitability of chosen annular piston system and development abandoned. Total about 4,500 guns produced.

7.92 mm Maschinenpistole 43, 43/1 and 44; 7.92 mm Sturmgewehr 44

German designation 7.92 mm MP 43, 43/1, 44; 7.92 mm StG 44
Calibre/cartridge 7.92 mm × 33 ('kurz')
Magazine capacity 30 rounds
Length 940 mm
Length of barrel 419 mm
Weight unloaded 5.22 kg
Muzzle velocity 685 m/sec
Rate of fire (cyclic): 500 rpm; (auto): 120 rpm
Manufacturers C. G. Haenel Waffen- und Fahrradfabrik AG, Suhl; Erma (Erfurter Maschinenfabrik B. Geipel GmbH, Erfurt;) Mauser-Werke AG, Oberndorf-am-Neckar; Carl Walther Waffenfabrik AG, Zella-Mehlis; Sauer-Gruppe, Suhl; Steyr-Werke, Steyr. Subcontractors included: Germania-Werke, Würtembergische Metallwarenfabriken, Progress-Werke, Lux, Adolf Rössler, L. W. Zeug- u. Metallwarenfabriken and I. G. Anschutz

Remarks: Evolved from MKb 42(H), with first series completed in July 1943. First combat use on Eastern Front late in 1943. Produced in several basically similar variants, of which MP 43/1 had provision for a screw-on grenade launcher muzzle cup. In spring 1944 designation altered to MP 44 and in December 1944 – to StG 44, without changes in basic construction. Limited number altered to accept 'Krummlauf' curved barrel attachments with special sights. In early 1945 some fitted and used with *Zielgerät 1229 'Vampir'* infra-red night sight.

Comparison of the MP 43 series. Top, *MP 43,* centre, *MP 43/1, and* below, *MP 44* (Sturmgewehr 44)

◄ *Commander of a heavy machine gun armed with the MP 43*

Sturmgewehr 44 *(StG 44)*

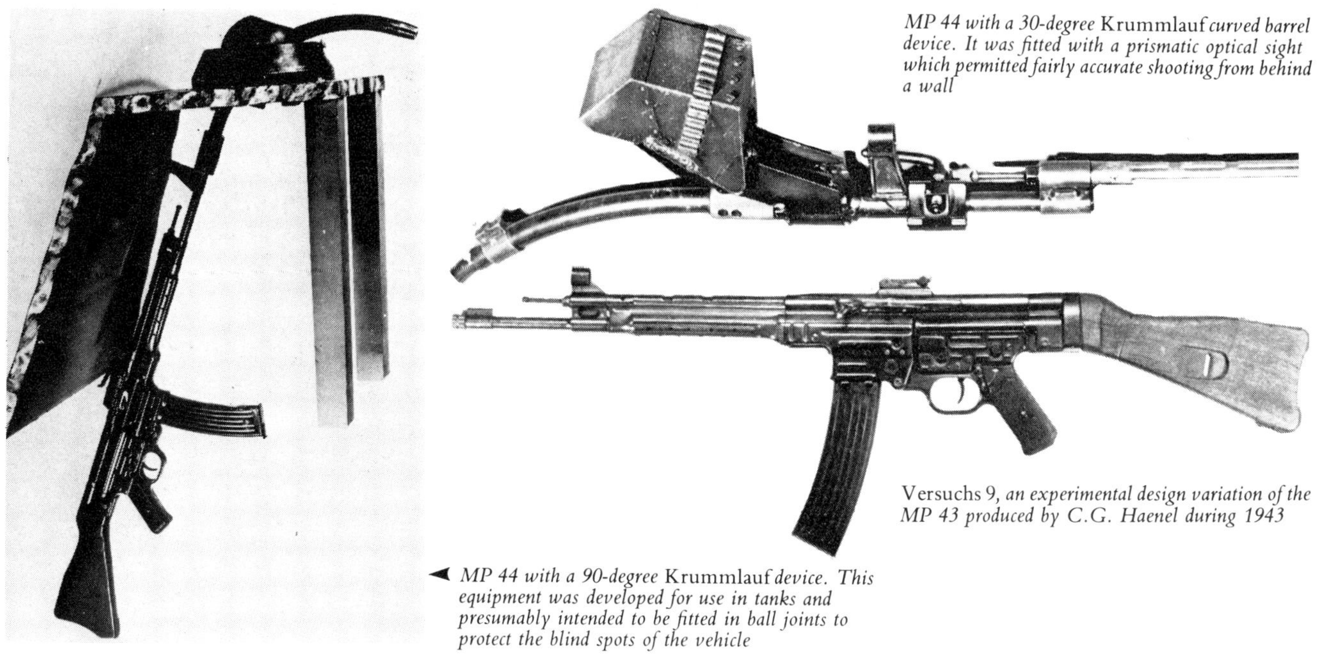

MP 44 with a 30-degree Krummlauf *curved barrel device. It was fitted with a prismatic optical sight which permitted fairly accurate shooting from behind a wall*

Versuchs 9, an experimental design variation of the MP 43 produced by C.G. Haenel during 1943

◄ *MP 44 with a 90-degree* Krummlauf *device. This equipment was developed for use in tanks and presumably intended to be fitted in ball joints to protect the blind spots of the vehicle*

7.92 mm Sturmgewehr 45(M) or Gerät 06(M)

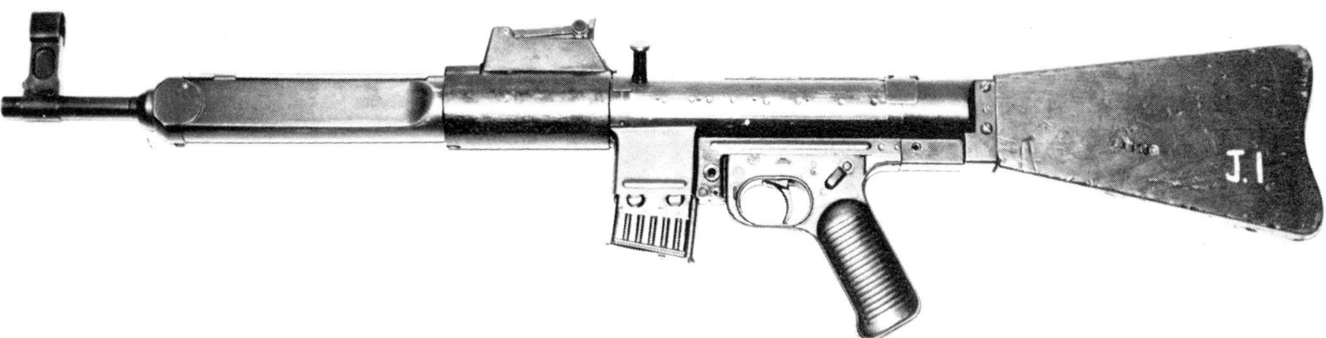

German designation 7.92 mm StG 45(M); Gerät 06(M)
Calibre/cartridge 7.92 mm × 33 ('kurz')
Magazine capacity 30 rounds
Length 893 mm
Length of barrel 400 mm
Weight unloaded 3.71 kg
Muzzle velocity 685 m/sec
Rate of fire (cyclic): 350–450 rpm
Manufacturer Mauser-Werke AG, Oberndorf am-Neckar

Remarks: An experimental lightweight selective-fire weapon featuring novel roller-locked retarded blowback system; was also known as MP 45(M). Designed as intended replacement for MP 43/StG 44 series but only prototypes completed before end of WW II. Development finalised postwar in Spain resulting in CETME 58 and in Germany as Heckler & Koch G3, both accepted as standard service weapons.

Another version of the 7.92 mm StG 45

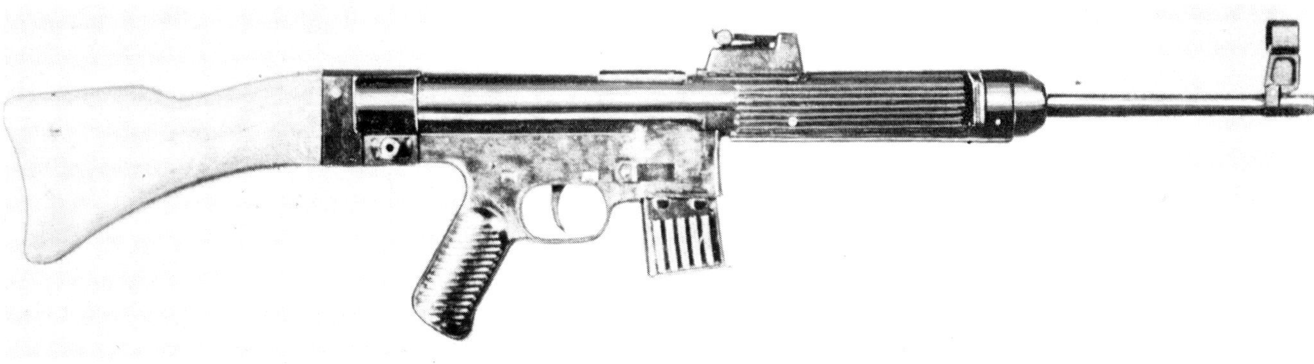

7.92 mm StG 45(M)

7.92 mm Volkssturmgewehr 1-5

German designation 7.92 mm VG 1-5
Calibre/cartridge 7.92 mm × 33 ('kurz')
Magazine capacity 30 rounds
Length 885 mm
Length of barrel 378 mm
Weight unloaded 4.62 kg
Muzzle velocity 770 m/sec
Rate of fire (semi-auto): 30 rpm
Intended manufacturers Mauser-Werke AG, Oberndorf-am-Neckar; Rheinmetall-Borsig, Sömmerda; Gustloff-Werke, Suhl; Carl Walther Waffenfabrik AG, Zella-Mehlis; Steyr-Werke, Steyr; Spreewerke, Berlin; G. Appel, Berlin

Remarks: Designed by Dr Barnitzke of Gustloff-Werke based on 1943 project work and intended as cheap mass-produced self-loading weapon. Breech locked by diverting part of exhaust gases forward to deter bolt until after each bullet had cleared muzzle. Extensive use of pressed parts and weldings; standard MP 43/44 magazine. First series completed late in 1944. Dispersed and hurried manufacture reflected in many minor variations revealed by captured and recovered guns. Some experimental models also completed as fully automatic weapons.

7.92 mm Fallschirmjägergewehr 42

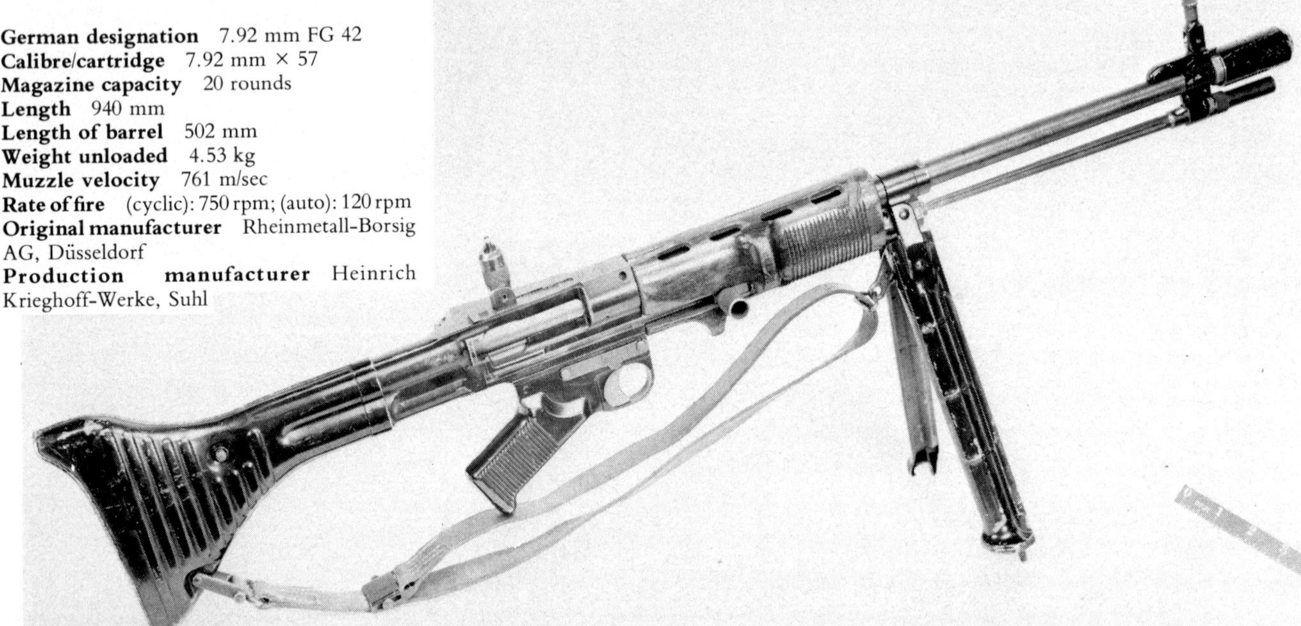

German designation 7.92 mm FG 42
Calibre/cartridge 7.92 mm × 57
Magazine capacity 20 rounds
Length 940 mm
Length of barrel 502 mm
Weight unloaded 4.53 kg
Muzzle velocity 761 m/sec
Rate of fire (cyclic): 750 rpm; (auto): 120 rpm
Original manufacturer Rheinmetall-Borsig AG, Düsseldorf
Production manufacturer Heinrich Krieghoff-Werke, Suhl

7.92 mm Fallschirmjägergewehr 42 *(FG 42). This was the first version with steel butt and sloping pistol grip*

Remarks: Evolved by Rheinmetall to 1940 Luftwaffe requirement. Accepted in preference to other proposals and ordered in 1942. Design made wide use of pressed and forged parts for light weight. Produced for parachute troops in three variants, two of which differed only in minor details. About 7,000 guns delivered before end of WW II but tactical impact reduced by frequent use in light machine gun role instead of as an assault rifle as intended.

FG 42 (*first version*) *with bayonet extended*

FG 42(*second version*) *with bipod legs folded back. This weapon has been fitted with the ZF 42 scope*

7.92 mm FG 42 (*second version*) *with wooden butt and redesigned pistol grip. The bipod legs have been moved forward*

7.62 mm Selbstladegewehr 257(r)

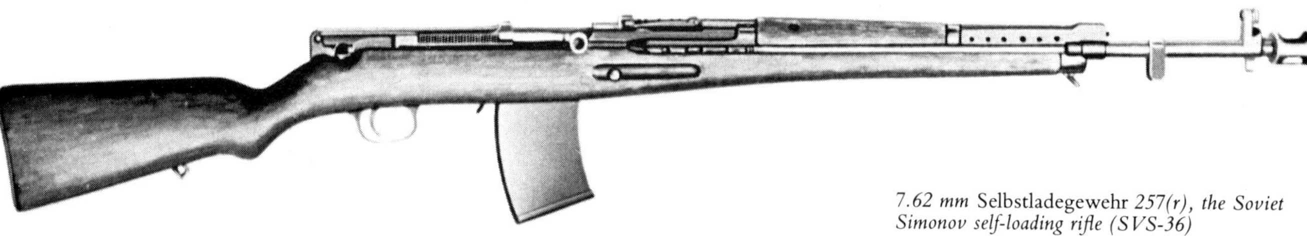

7.62 mm Selbstladegewehr *257(r), the Soviet Simonov self-loading rifle (SVS-36)*

German designation 7.62 mm SlGew 257(r)
Original designation *Avtomaticheskaya Vintovka Simonova obr. 1936 g* (AVS-36)
Calibre/cartridge 7.62 mm × 54R
Magazine capacity 15 rounds
Length 1233 mm
Length of barrel 614 mm
Weight unloaded 4.05 kg
Muzzle velocity 840 m/sec
Rate of fire (semi-auto): 40 rpm
Manufacturer Various Soviet State arsenals

Remarks: Selective-fire rifle designed by S. G. Simonov, introduced into Red Army service in 1936 but officially withdrawn already in 1938. Was tried out during Civil War in Spain. Only small numbers captured by German troops on Eastern Front and used locally.

7.62 mm Selbstladegewehr 258(r)

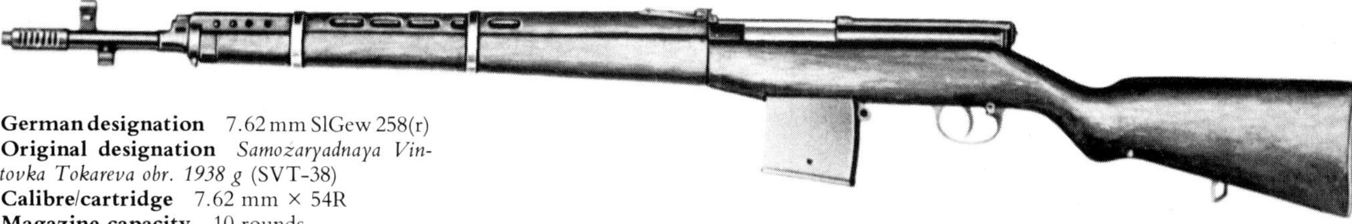

7.62 mm SlGew 258(r), the Soviet Tokarev self-loading rifle obr.1938 g

German designation 7.62 mm SlGew 258(r)
Original designation *Samozaryadnaya Vintovka Tokareva obr. 1938 g* (SVT-38)
Calibre/cartridge 7.62 mm × 54R
Magazine capacity 10 rounds
Length 1222 mm
Length of barrel 625 mm
Weight unloaded 3.95 kg
Muzzle velocity 830 m/sec
Rate of fire (semi-auto): 30 rpm
Manufacturer Various Soviet State arsenals

Remarks: Designed by F. V. Tokarev. Accepted in 1938 but proved too fragile for service use and progressively withdrawn. Limited numbers captured and used by German second-line and auxiliary local troops on Eastern Front.

7.62 mm Selbstladegewehr 259(r)

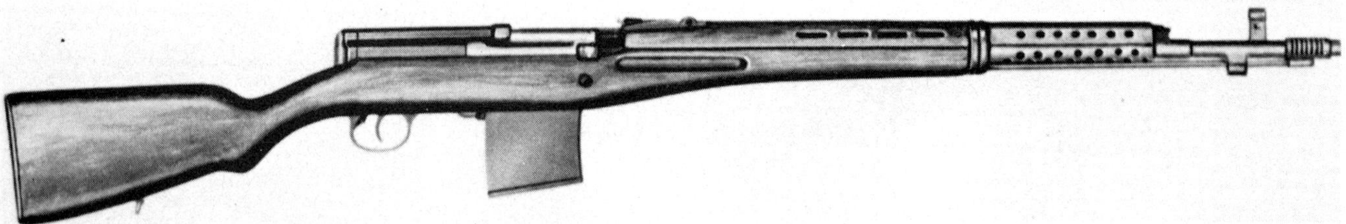

German designation 7.62 mm SlGew 259(r)
Original designation *Samozaryadnaya Vintovka Tokareva obr. 1940 g* (SVT-40)
Calibre/cartridge 7.62 mm × 54R
Magazine capacity 10 rounds
Length 1222 mm
Length of barrel 625 mm
Weight unloaded 3.89 kg
Muzzle velocity 830 m/sec
Rate of fire (semi-auto): 30 rpm
Manufacturer Various Soviet State arsenals

7.62 mm SlGew 259(r), the Soviet Tokarev obr. 1940 g *(SVT-40)*

Remarks: Evolved from Tokarev SVT-38 after combat experience during Soviet-Finnish war 1939–40. Although not an unqualified success in service, many captured examples used by German troops on Eastern Front in 1941–42 in preference to Gew 41(W); substantial numbers remained in German service, particularly with local volunteer troops, until end of WW II. Selected rifles fitted with telescopic sights also used on both sides. Gas operation system greatly influenced the design of Gew 43.

A German mountain troop sniper with the SVT-40 and PU scope

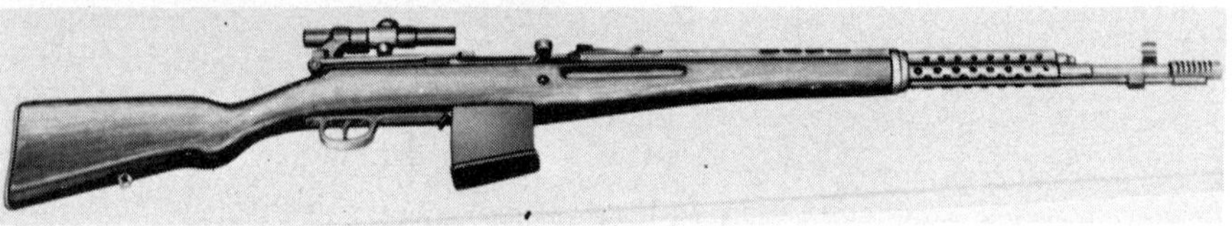

7.62 mm Selbstladegewehr 251(a)

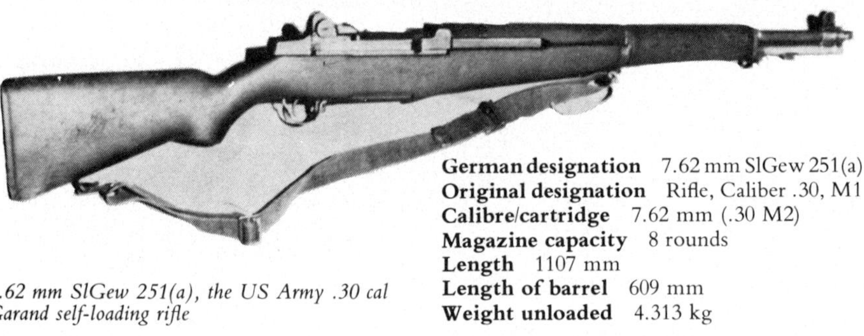

7.62 mm SlGew 251(a), the US Army .30 cal Garand self-loading rifle

German designation 7.62 mm SlGew 251(a)
Original designation Rifle, Caliber .30, M1
Calibre/cartridge 7.62 mm (.30 M2)
Magazine capacity 8 rounds
Length 1107 mm
Length of barrel 609 mm
Weight unloaded 4.313 kg
Muzzle velocity 855 m/sec
Rate of fire (semi-auto): 30 rpm
Original manufacturer Springfield Armory, Massachusetts, USA
Other manufacturers Harrington & Richardson Arms Co., Worcester; Winchester Repeating Arms Co., New Haven, Conn.; International Harvester Co.
Remarks: Designed by J. C. Garand and adopted by US military authorities in 1936. Standard American service rifle in WW II, known universally as 'Garand'. Use of captured examples by German troops mainly localised.

8 mm Selbstladegewehr 310(f)

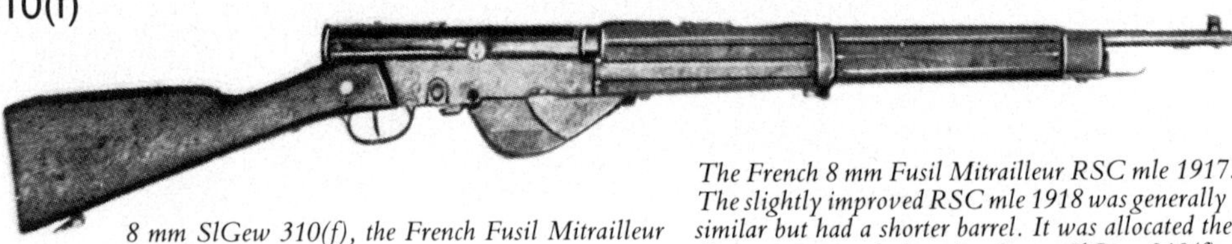

8 mm SlGew 310(f), the French Fusil Mitrailleur RSC mle 1918 or

The French 8 mm Fusil Mitrailleur RSC mle 1917. The slightly improved RSC mle 1918 was generally similar but had a shorter barrel. It was allocated the German service designation 8 mm SlGew 310(f)

German designation 8 mm SlGew 310(f)
Original designation Fusil Mitrailleur RSC mle 1918
Calibre/cartridge 8 mm Lebel
Magazine capacity 5 rounds
Length 1110 mm
Length of barrel 600 mm
Weight unloaded 4.7 kg
Muzzle velocity 665 m/sec
Manufacturer Manufacture d'Armes de St. Etienne

Remarks: Improved RSC mle 1917 with shorter barrel; Mannlicher-type magazine. Despite its age and basic unsuitability for modern warfare a limited number of these captured rifles were issued to German troops on Eastern Front in 1941–42.

7.62 mm Selbstladekarabiner 455(a)

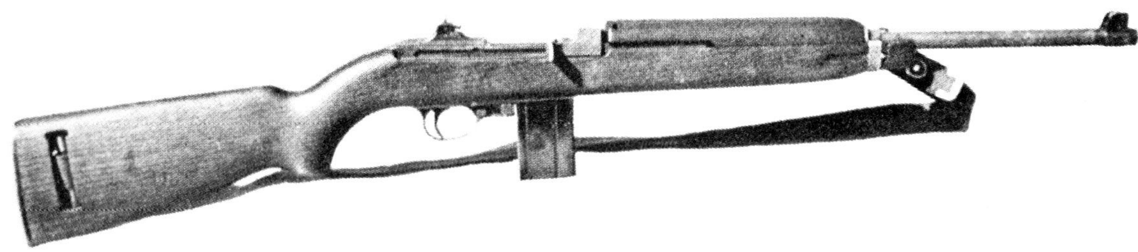

7.62 mm SlKb 455(a), the US Army .30 cal Carbine M1

German designation 7.62 mm SlKb 455(a)
Original designation Carbine, Caliber .30, M1
Calibre/cartridge .30 Short Rifle M1
Magazine capacity 15 or 30 rounds
Length 904 mm
Length of barrel 457 mm
Weight unloaded 2.36 kg
Muzzle velocity 600 m/sec
Rate of fire (semi-auto): 40 rpm
Manufacturers Winchester Repeating Arms Co. and various American armament and industrial concerns

Remarks: Light self-loading weapon evolved by Winchester in 1940, ordered into production in October 1941 and manufactured in very large numbers. Because of special ammunition German service confined to localised use; also carried by special German infiltration groups in NW Europe in 1944–45.

A Luftwaffe Field Division soldier armed with the American M1 carbine

Sub-machine guns

The sub-machine gun was very much a product of the trench warfare of 1914–1918 and was evolved from the trench raiding tactics of that conflict. The idea of the early sub-machine gun inventors was to produce what an American designer called a 'trench broom' which would sweep trenches clear of the enemy. This new weapon had to fire fully automatically, but to ensure that it could be small enough to do so and still be light and handy enough for one man to use the choice of ammunition had to be limited to pistol calibre. This meant that the sub-machine gun had only a short range but in the close confines of trench warfare this mattered little.

The first practical sub-machine gun was designed by Louis Schmeisser in 1917 and produced by Theodor Bergmann Waffenbau AG at Suhl. The place of manufacture gave the weapon its generic title of 'Bergmann' but officially it was the *Maschinenpistole 18* (*MP 18*). The *MP 18* was an important landmark in the course of small arms design, not only because it was the first sub-machine gun but also because it made an important impact, although it was not fully realised at the time, on infantry tactics. The weapon which is usually credited as being the first sub-machine gun, the Italian Villar-Perosa, was used tactically as a form of light machine gun, but the *MP 18* was intended from the outset as a one-man weapon. The first examples were built to take the 32-round 'snail' magazine used on versions of the *P 08* pistol but this was not very successful, and as soon as post-war conditions allowed it was replaced by what was to become the conventional 'in-line' box magazine. By 1918 some 30,000 *MP 18*s had been made and their impact on trench warfare, although limited by the numbers used, was profound.

After 1918 all further development of the sub-machine gun was forbidden in Germany under the terms of the Versailles Peace Treaty and the remaining *MP 18*s passed from general Army use to the police. It was they who modified the magazine and the designation then became the *MP 18/I*. Some commercial models were made in Belgium and Switzerland, mainly for export, but after 1939 substantial numbers were still available in Germany to be impressed and issued to second-line units of the German Army and the military police. These elderly 'Bergmann' sub-machine guns remained in use until 1945.

The *MP 18/I* set a trend that was soon followed by the *MP 28/II*. A product of the C. G. Haenel Waffen- und Fahrradfabrik AG at Suhl, it was basically a selective-fire version of the *MP 18* (which was fully automatic). The *MP 28/II* was made in a wide variety of calibres for export all over the world, and many were produced under licence by Pieper at Herstal (Liège) in Belgium. After about 1934 numbers were manufactured for the German Army and many were used in action in Spain between 1936 and 1939. After that the *MP 28/II* was relegated to second-line duties, with their inventory augmented by captured examples from Belgium and the Balkans.

Meanwhile, as Haenel was gradually building up the sub-machine gun market, other German firms also found ways to circumvent the Treaty restrictions by quietly involving their foreign subsidiaries in sub-machine gun design and manufacture. For example, the firm of Rheinmetall Metallwaren- und Maschinenfabrik of Düsseldorf had obtained an interest in the Swiss firm of Solothurn, who in their turn had an interest in the Austrian Österreichische Waffenfabrik-Gesellschaft at Steyr. Via this tortuous arrangement Rheinmetall were able to produce sub-machine guns based on the *MP 18/I* for a number of customers. One of these was the Austrian Army, later to come under the German sphere of influence so that the German Army took over yet another Bergmann-based sub-machine gun as the *MP 34(ö)*. This too went to many second-line and police units where its logistics back-up was complicated by there being three separate types of 9 mm ammunition needed to suit various models.

Rheinmetall was not the only weapon manufacturer contravening the terms of the Versailles Treaty; after 1919 Bergmann also took an interest in a foreign firm, this time in Denmark. Bergmann undertook an agreement with Schultz and Larsen Gewärfabrik of Otterup to produce a version of the *MP 18/I* known as the *BMK-32*. The *BMK-32*, a redesigned *MP 18/I* evolved by Theodor Bergmann, had the magazine to the right instead of the more usual left. The *BMK-32* was adopted for the Danish Army in small numbers but after 1933 production switched to the Bergmann plant in Berlin and there, as the *MP 34* and *MP 34/I*, this sub-machine gun was made in a variety of calibres and barrel lengths for export. The *MP 35* was similar but the main production version was the *MP 35/I* which was adopted by the Waffen SS and the entire output went to them.

Yet another pre-war German sub-machine gun was the *MP(Erma)*. This weapon was originally made in small numbers by the Vollmer factory at Biberach/Würtemberg during the early 1920s. In about 1930 production was taken over by the Erma-Werke at Erfurt who increased the output for the new German Army. The *MP(Erma)* saw action in Spain and during the early war years but afterwards it was relegated to second-line units. In several ways it was a remarkable design in that it heralded the production methods that later sub-machine guns were to embody. Thus, the receiver was a simple tube and the bolt and spring were covered by a telescopic housing which kept out dirt. A number of these guns were also sold to France and Yugoslavia (from where they later came under German control) and the type was also built under licence in Spain.

All the above-mentioned sub-machine guns employed the same method of operation, namely the blow-back principle. As they all used the relatively low-powered pistol ammunition the forces generated on firing were fairly weak and could be contained by the inertia of a heavy bolt pushed into the firing position by a spring. The forward inertia of this bolt would initially overcome the forces produced when the round was fired and only when the bullet was clear of the muzzle would the bolt be blown back onto the trigger sear. If the trigger was kept pulled the bolt would then go forward, picking up and chambering a round during its forward progress and firing it, usually with a fixed firing pin, just before the round was fully chambered.

These conventional methods took a rude shock when the OKW issued a specification for a new sub-machine gun more suited to the needs of modern mobile warfare. The specification was met by a remarkable design put forward by Erma-Werke of Erfurt which was adopted as the *MP 38*. The *MP 38* retained the blow-back mechanism and telescopic housing used on the *MP(Erma)* but the most important change related to the manufacturing method. The design used no wooden furniture, which was replaced by plastic, and a folding metal butt was fitted; a 32-round in-line box magazine was inserted below the receiver. Production began in 1938 and continued until 1940.

The *MP 38* was one of the most successful and influential sub-machine gun designs to emerge from the WW II years. Not only did it exactly suit the modern mobile tactics evolved by the German Army, but its production methods, although not yet fully developed, pointed the way ahead that all later sub-machine guns were to follow. Gone were the time-honoured procedures of gunmaking; in their place came modern mass-production methods involving sub-contractors. No longer was weapon manufacture a task for specialist gunmakers but any firm with a few machine tools could produce parts to be assembled at some central point. The *MP 38* did not involve all these processes but they were introduced with its successor, the *MP 40*.

The *MP 40* was developed from the *MP 38* primarily because the original model was too time-consuming to manufacture. To make the *MP 38* many lengthy machining operations were neces-

sary, and after 1939 sub-machine guns were needed in large numbers, and quickly. The *MP 40* embodied all the processes that were later to become the norm, namely stampings, forgings and cast components. Sub-contracting, as described above, was widely employed and the *MP 40* was made all over Germany and in several occupied countries. Contrary to general belief, the *MP 40* was more expensive than the *MP 38* – it cost RM 60, as against RM 57 for the *MP 38*.

Throughout the war the *MP 38* and *MP 40* were the two main German sub-machine gun types in service. The *MP 40* was produced in very large numbers (some sources quote well over 1 million) and its use spread to all arms and formations of the German armed forces. During its service life some minor changes were embodied but basically the first *MP 40* looked very much like the last. An early change to the *MP 38* came after combat experience in 1939 in Poland showed that when the bolt was forward on the base of a round a jolt or knock would cause the gun to fire. The answer was a simple angled notch cut into the receiver into which the bolt was placed. When this modification (which was retrofitted to all existing weapons and cut into production) was made, the designation was changed to *MP 38/40*. Another modification was the *MP 40/II* which represented an attempt to increase the magazine capacity to match that of the Soviet PPSh 41 sub-machine gun. On the *MP 40/II* a slide arrangement was fitted to the magazine housing holding an extra magazine which could be slid into place when needed. It was not a great success and appears to have been little used.

One of the best compliments paid to the *MP 38* and *MP 40* was that they were almost as well used by the enemies of Germany as they were by the Germans themselves. The very practical Russians employed large numbers of captured *MP 38*s and *MP 40*s as front-line weapons, and both types were also preferred by the various partisan and underground forces whenever they could be captured.

One odd throwback based on the *MP 40* was the *MP 41* produced by Haenel. This sub-machine gun used the receiver, barrel and magazine of the *MP 40* fitted to a conventional wooden stock similar to that used on the *MP 28/II*. Only small numbers of this weapon appear to have been made, and it is one of the minor mysteries of the German sub-machine gun scene as to where they went. They do not appear to have been used by any German units, so it would seem likely that the *MP 41* was destined for some police or militia formations under German control in one of the occupied countries.

By late 1944 the Allied bombing and gradual loss of industries until then working for the German war effort were having a serious effect on the arms supply situation. As with so many other weapons, sub-machine guns were needed in large numbers, but not even the production methods introduced on the *MP 40* could satisfy the demand. A stop-gap had to be sought, and was found in the shape of the British Sten Mk 2. This simple weapon, rapidly produced in 1940 to quickly re-equip the British armies after the Dunkirk evacuation, represented gun design reduced to the simplest form with tubes and welds replacing the more carefully machined parts, and in desperation the Germans took the Sten as their starting point. They decided against the side-mounted magazine housing and replaced it with their more usual vertical feed attachment to take a standard *MP 40* magazine, but otherwise the Sten was followed quite closely. The result, called *MP 3008*, was hastily manufactured in several German arms centres and extensive use was made of sub-contractors, but by the time the war ended only a limited number had passed into actual service. Most of these guns were crudely finished and their service life would have been very short.

But the British Sten also featured in another very odd facet of German sub-machine gun production. During 1944, just as the sub-machine gun supply situation was getting serious, the Mauser Werke at Oberndorf were ordered to produce exact copies of the Sten. So exact were these copies that even the British markings were reproduced. It would appear that these weapons, known as *Gerät Potsdam*, were intended for some form of clandestine or guerilla warfare, but all these plans, whatever they were, never materialised, and the *Gerät Potsdam* was never issued. Apart from what happened to the many thousands of these guns produced, the biggest mystery regarding this oddity is why it was felt necessary to go to such lengths as making copies of the Sten when large numbers of the original article were already in German hands.

A German NCO armed with a later type PPSh-41 (note the simple rear sight)

One very interesting prototype that appeared in 1943 was the *Erma EMP 44*. It would appear to have been an exercise in how simple a sub-machine gun could be made. All the parts were steel tubes and stampings, while the gun used the same dual magazine arrangement as was tried on the *MP 40/II*.

The *Erma EMP 44* was not put into production, but by late 1944 numbers of captured Soviet PPSh 41 sub-machine guns were being converted to take 9 mm barrels and *MP 40* magazines. Despite the obvious reason for this modification it would also seem to have caused some unecessary delays because large numbers of Soviet sub-machine guns of all types were in regular use by the German troops in their original 7.62 mm calibre, and ammunition supply was no problem: apart from the fact that it was almost the same as the 7.63 mm Mauser cartridge, large stocks of captured ammunition were readily available.

Whenever possible, when the Germans took over factories already manufacturing sub-machine guns they kept them in production for their occupation forces. This happened in France, where the St. Etienne plant was taken over at a time when supplies of the Mitraillette mle 38 were just getting into full swing for the French Army. It was kept in production under German supervision for the German forces, and some were even supplied to the Vichy French. In Czechoslovakia, the Československá Zbrojovka Koucy (ZK) were producing the vz.383 at Brno. This type was also kept in production for the Waffen-SS and some were exported to Bulgaria.

Although Italy was not nominally under German control until after 1943, the troops based in Italy often used Italian equipment. This included numbers of the superb Beretta sub-machine guns and also some FNA-B made at Brescia.

While the first-line formations of the German forces used the *MP 38* and *MP 40*, the rest of the German troops had to make do with whatever they could get hold of. WW I veterans were carried on the many irksome and dangerous guard duties in occupied countries, as were captured British Stens, American Thompsons, various Soviet sub-machine guns and even elderly Berettas. For all the success of the *MP 38* and *MP 40* there were never enough of them to satisfy the seemingly endless demands by the overstretched German forces, a situation repeated with every class of the German armament.

9 mm Maschinenpistole 18/I

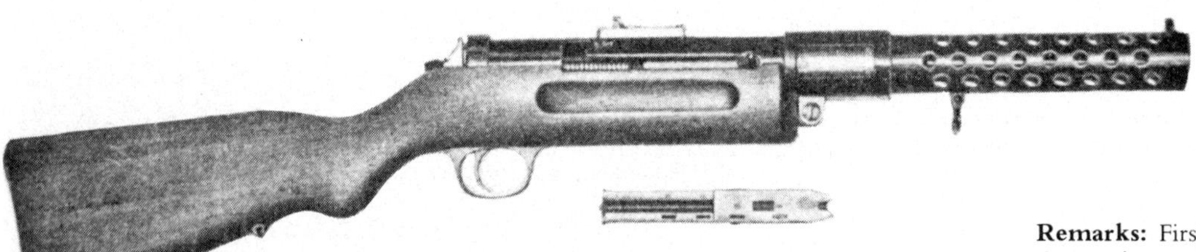

German designation 9 mm MP 18/I
Calibre/cartridge 9 mm Parabellum
Magazine capacity 32 round box
Length 815 mm
Length of barrel 200 mm
Weight 4.177 kg
Muzzle velocity 381 m/sec
Rate of fire (cyclic): 400 rpm
Original manufacturer Theodor Bergmann Waffenbau AG, Suhl

Remarks: First practical sub-machine gun designed to 1917 specifications by Louis Schmeisser. Fully automatic fire only. 50,000 ordered, about 10,000 delivered by end of WW I. Initially with drum, later box magazine. Subsequently German police service and commercial variants. During WW II mostly second-line and police use.

9 mm Maschinenpistole 28/II and 740(b)

German designation 9 mm MP 28/II oder 740(b)
Original designation (b): Mitraillette 34
Calibre/cartridge 9 mm Parabellum
Magazine capacity 20, 32 or 50 round box
Length 813 mm
Length of barrel 200 mm
Weight 4 kg
Muzzle velocity 381 m/sec
Rate of fire (cyclic): 500–600 rpm
Original manufacturer C. G. Haenel Waffen- und Fahrradfabrik AG, Suhl
Manufacturer (b): Anciens Etablissements Pieper, Herstal

Remarks: Originally commercial version in several calibres of MP 18/I. Manufacture transferred to Belgium in 1920s and evolved as mle 34. German MP 28/II identical to late-production Belgian weapon. Used during Spanish Civil War 1936–39; subsequently by German garrison troops, police and security forces.

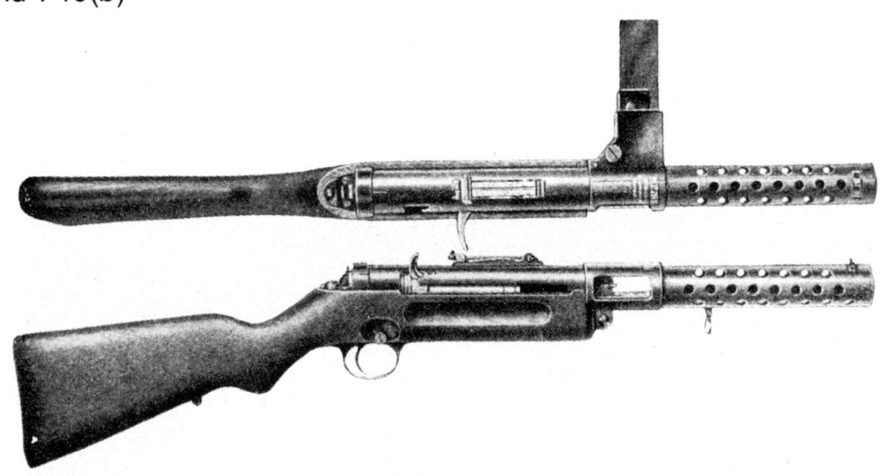

9 mm MP 28/II. Basically a modified MP 18/I, it was a selective fire weapon instead of automatic only capacity of the earlier model

9 mm Maschinenpistole 34/I, 35/I and 741(d)

German designation 9 mm MP 34/I, 35/I oder 741(d)
Original designation (d): BMK 32
Calibre/cartridge 9 mm Parabellum BMK-32; 9 mm Bayard
Magazine capacity 32 round box
Length 840 mm
Length of barrel 200 mm
Weight 4.05 kg
Muzzle velocity 381 m/sec
Rate of fire (cyclic): 650 rpm
Original manufacturer Carl Walther Waffenfabrik AG, Zella-Mehlis (MP 34/I); Junker & Ruh AG, Karlsruhe (MP 35/I)
Manufacturer (d): Schultz-Larsen Gewaerfabrik, Otterup

Remarks: BMK-32 represented Bergmann design licence-built in Denmark. Radial rear sight, bayonet mounting. Most taken over by German forces in 1940 and used locally. MP 34/I differed only slightly from BMK-32. First offered for commercial sale early in 1934, later in military form. Short- and long-barrelled variants. MP 35/I was slightly modified MP 34. Bulk of production delivered to SS units before and during WW II.

MP 35/I carried by an Army assault troop leader ►

9 mm MP 35/I. This weapon was used almost exclusively by SS troops ▼

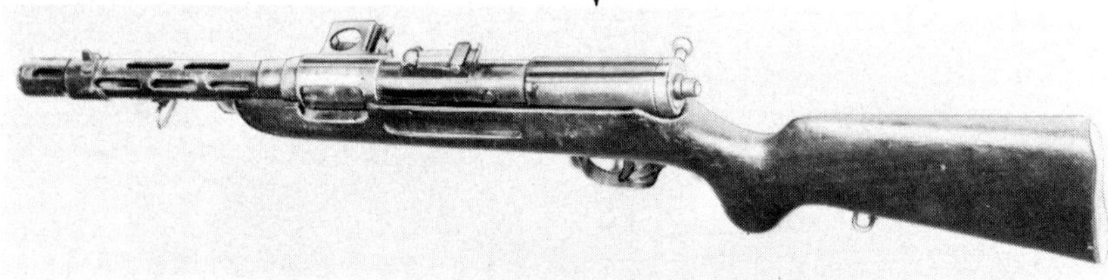

9 mm Maschinenpistole Erma, MPE and MP 740(f)

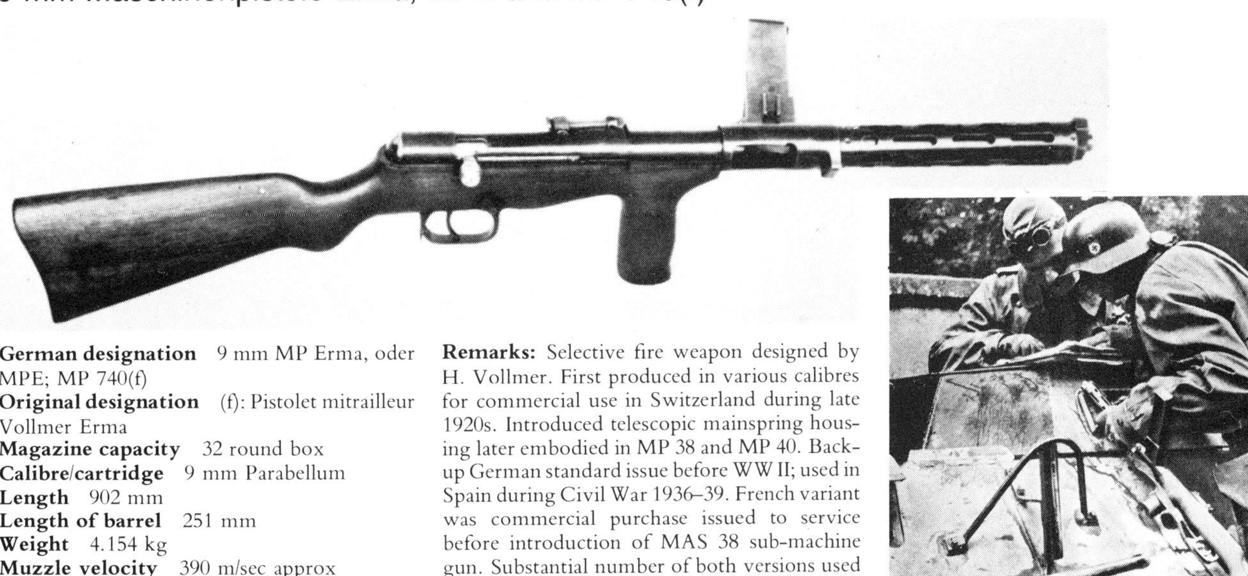

German designation 9 mm MP Erma, oder MPE; MP 740(f)
Original designation (f): Pistolet mitrailleur Vollmer Erma
Magazine capacity 32 round box
Calibre/cartridge 9 mm Parabellum
Length 902 mm
Length of barrel 251 mm
Weight 4.154 kg
Muzzle velocity 390 m/sec approx
Rate of fire (cyclic): 500 rpm
Original manufacturer Erfurter Maschine- und Werkzeugfabrik B. Geipel GmbH, Erfurt (Erma)

Remarks: Selective fire weapon designed by H. Vollmer. First produced in various calibres for commercial use in Switzerland during late 1920s. Introduced telescopic mainspring housing later embodied in MP 38 and MP 40. Back-up German standard issue before WW II; used in Spain during Civil War 1936–39. French variant was commercial purchase issued to service before introduction of MAS 38 sub-machine gun. Substantial number of both versions used by German forces during WW II.

German despatch rider armed with the MP Erma ➤

---

9 mm Maschinenpistole 38 and 38/40

German designation 9 mm MP 38 und 38/40
Calibre/cartridge 9 mm Parabellum
Magazine capacity 32 round box
Length of stock (extended): 833 mm
(folded): 630 mm
Length of barrel 251.5 mm
Weight 4.086 kg
Muzzle velocity 390 m/sec
Rate of fire (cyclic): 500 rpm
Original manufacturer Erma-Werke, Erfurt

Remarks: Designed to advanced specifications formulated in 1936 37. Revolutionary sub machine gun design using only steel and plastics, and fitted with folding butt. Automatic fire only. Adopted for service in August 1938, in production until 1940. MP 38/40 had additional safety slot at forward bolt position.

Assault troops armed with the MP 38/40

9 mm Maschinenpistole 40

German designation 9 mm MP 40
Calibre/cartridge 9 mm Parabellum
Magazine capacity 32 round box
Length of stock (extended): 833 mm
(folded): 630 mm
Length of barrel 251.5 mm
Weight 4.027 kg
Muzzle velocity 390 m/sec
Rate of fire (cyclic): 500 rpm
Original manufacturer Erma-Werke, Erfurt

Production manufacturers Erfurter Maschinenfabrik B. Geipel GmbH, Erfurt; Steyr-Werke, Steyr-Linz; C. G. Haenel Waffen- und Fahrradfabrik AG, Suhl; Steyr Daimler Puch AG, Steyr; Merz-Werke, Frankfurt; National Krupp Registrierkassen GmbH, Berlin.

Remarks: Improved MP 38; automatic fire only. Adopted for service in April 1940, production commenced in July 1940. Early series with smooth, later with ribbed magazine holder, but basically unchanged throughout production life. Output totalled over 1 million (1940 – 113,700, 1941 – 239,000, 1942 – 231,500, 1943 – 234,300, 1944 – 228,600).

9 mm Maschinenpistole 40/II

MP 40/II, showing the twin magazines in the sliding housing

9 mm MP 40/II, the basic MP 40 with a dual holder for two standard MP 40 box magazines in a sliding housing. As one magazine was emptied the other was moved over to continue firing

German designation 9 mm MP 40/II
Calibre/cartridge 9 mm Parabellum
Magazine capacity 2 × 32 round box
Length of stock (extended): 833 mm
(folded): 630 mm
Length of barrel 251.5 mm
Weight 4.54 kg
Muzzle velocity 390 m/sec
Rate of fire (cyclic): 500 rpm

Original manufacturer Erma-Werke, Erfurt; Steyr-Werke, Steyr (1943)

Remarks: An attempt to increase MP 40 ammunition capacity based on experience on Eastern Front. Initially known as *Gerät 3004*. First produced in July 1943 but proved unsuccessful in combat.

9 mm Maschinenpistole 41

German designation 9 mm MP 41
Calibre/cartridge 9 mm Parabellum
Magazine capacity 32 round box
Length 863.5 mm
Length of barrel 251.5 mm
Weight 3.7 kg
Muzzle velocity 390 m/sec
Rate of fire (cyclic): 500 rpm
Magazine C. G. Haenel Waffen- und Fahrradfabrik AG, Suhl

Remarks: Combined features of MP 28/II (stock, firing system, selector) and MP 40 (receiver, barrel design, bolt system). Limited production; exact service use uncertain.

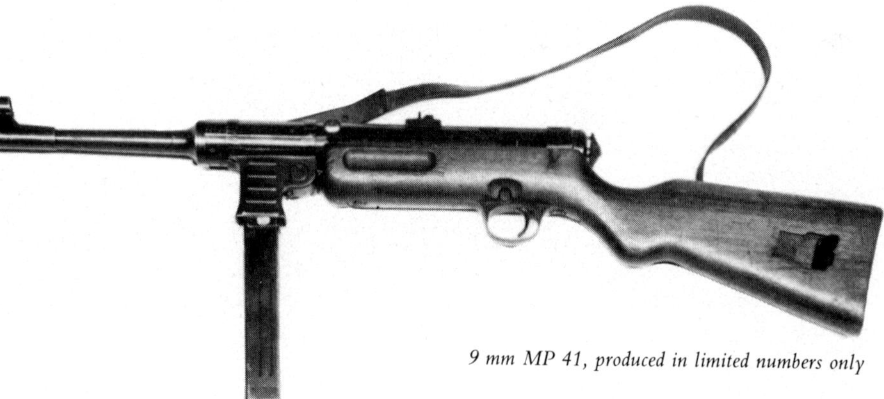

9 mm MP 41, produced in limited numbers only

9 mm Maschinenpistole 3008 and Volksmaschinenpistole

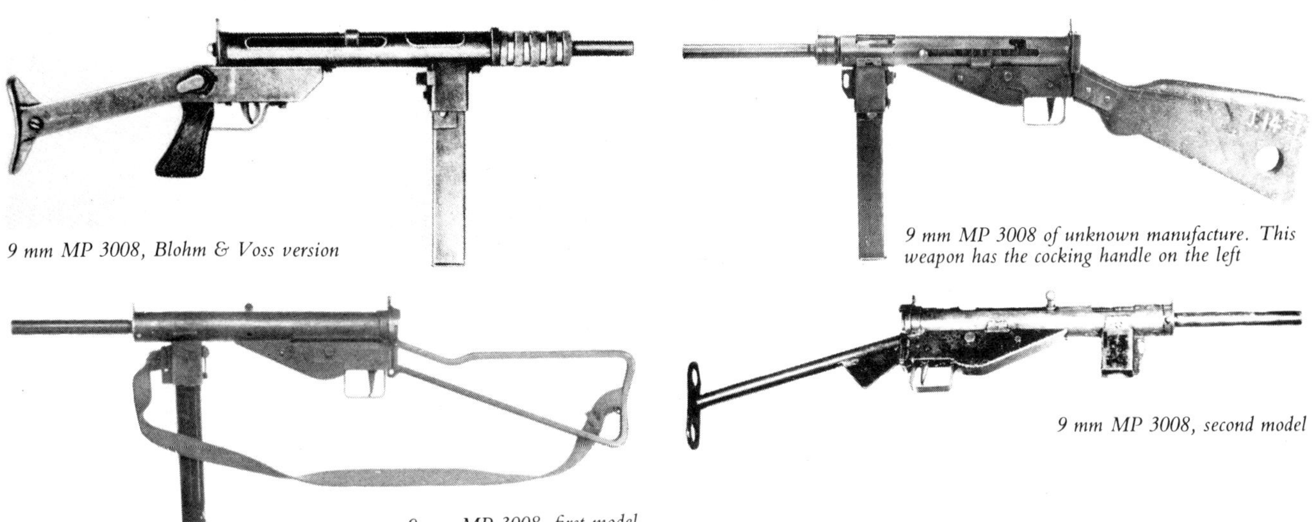

9 mm MP 3008, Blohm & Voss version

9 mm MP 3008 of unknown manufacture. This weapon has the cocking handle on the left

9 mm MP 3008, second model

9 mm MP 3008, first model

German designation 9 mm MP 3008, VolksMP
Calibre/cartridge 9 mm Parabellum
Magazine capacity 32 round box
Length 795 mm
Length of barrel 198 mm
Weight 2.95 kg
Muzzle velocity 381 m/sec
Rate of fire (cyclic): 500 rpm
Manufacturers Mauser-Werke AG, Oberndorf-am-Neckar; C. G. Haenel Waffen- und Fahrradfabrik AG, Suhl; Erfurter Maschinenfabrik B. Geipel GmbH, Erfurt (Erma), and others

Remarks: German emergency copy of British Sten Mk 2, initially known as '*Gerät Neumünster*'. Selective fire weapon. Design prepared in Aug-Nov 1944, production commenced Jan 1945, deliveries in March 1945. Large-scale production planned, and reportedly some 10,000 made before end of WW II. Numerous variations on basic design due to large number of sub-contractors. Limited service and possibly SD (security police) use from late March 1945.

---

9 mm Maschinenpistole Erma 44

German designation 9 mm MPE 44 oder EMP 44
Calibre/cartridge 9 mm Parabellum
Magazine capacity 2 × 32 round box
Length 721 mm
Length of barrel 250 mm
Weight 3.623 kg
Muzzle velocity 390 m/sec
Rate of fire (cyclic): 500–600 rpm approx
Manufacturer Erma-Werke, Erfurt

Remarks: This simplified but well finished 1943 design incorporated several specialised features, such as vertical hollow pistol grip, possibly for fixed installations. Fully automatic fire only. Produced in prototype form only; no known German service use.

Gerät Potsdam and 9 mm Maschinenpistole 749(e)

German designation Gerät Potsdam oder 9 mm MP 749(e)
Original designation (e): Machine Carbine, 9 mm Sten Mark 2
Calibre/cartridge 9 mm Parabellum
Magazine capacity 32 round box
Length 762 mm
Length of barrel 197 mm
Weight (unloaded): 3.3 kg
(loaded): 3.66 kg
Muzzle velocity 366 m/sec
Rate of fire (cyclic): 540 rpm
Manufacturer (Gerät Potsdam): Mauser-Werke AG, Oberndorf-am-Neckar; (e): Numerous British small-arms and industrial concerns

Remarks: Slightly shorter and lighter version of Sten Mk 1 evolved as 1940 British measure to provide large number of sub-machine guns quickly and cheaply. *Gerät Potsdam* was a direct German copy of Sten Mk 2 produced in 1944 and intended for some form of German guerilla warfare. Production totals very from 10,000 to 30,000 according to references used.

9 mm Maschinenpistole 34(ö)

9 mm MP 34(ö), the Austrian Steyr-Solothurn taken into service by the German Army

German designation 9 mm MP 34(ö)
Original designation Steyr-Solothurn S1-100; Austrian Army MP 34
Calibre/cartridge 9 mm Steyr, 9 mm Parabellum and 9 mm Mauser
Magazine capacity 32 round box
Length 851 mm
Length of barrel 198 mm
Weight 4.04 kg
Muzzle velocity (9 mm Mauser): 414 m/sec
Rate of fire (cyclic): 500–650 rpm
Original manufacturer Österreichische Waffenfabrik-Gesellschaft, Steyr

Remarks: Selective fire weapon designed by Louis Stange of Rheinmetall and originally built by Solothurn in Switzerland. Adopted by Austrian Army (9 mm Mauser) and police (9 mm Steyr, later 9 mm Parabellum) in 1934. Commonly known as 'Steyr-Solothurn'. Complete stocks taken into German service after annexation of Austria in 1938. Used mainly by German police forces during WW II.

9 mm Machinenpistole 383(t)*

German designation 9 mm MP 383(t)*
Original designation Kulometna Pistole ZK vz. 383
Calibre/cartridge 9 mm Parabellum
Magazine capacity 30 round box
Length 900 mm
Length of barrel 325 mm
Weight 4.27 kg
Muzzle velocity 426 m/sec approx
Rate of fire (cyclic): 500 or 700 rpm
Original manufacturer Československa Zbrojovka Koucky, Brno, later part of Waffenwerke Brünn

Remarks: Designed by Josef and Frantisek Koucky and produced in three slightly different versions. Selective fire weapon, noted for its good design features and accuracy. All output under German control delivered to Waffen-SS.

* designation unconfirmed

7.62 mm Maschinenpistole 715(r)

German designation 7.62 mm MP 715(r)
Original designation *Pistolet-Pulemyot Degtyarova obr. 1940 g* (PPD-40)
Calibre/cartridge 7.62 mm Type P
Magazine capacity 71 round drum
Length 780 mm
Length of barrel 260 mm
Weight 3.6 kg
Muzzle velocity 500 m/sec
Rate of fire (cyclic): 600–800 rpm
Manufacturers Various Soviet State arsenals

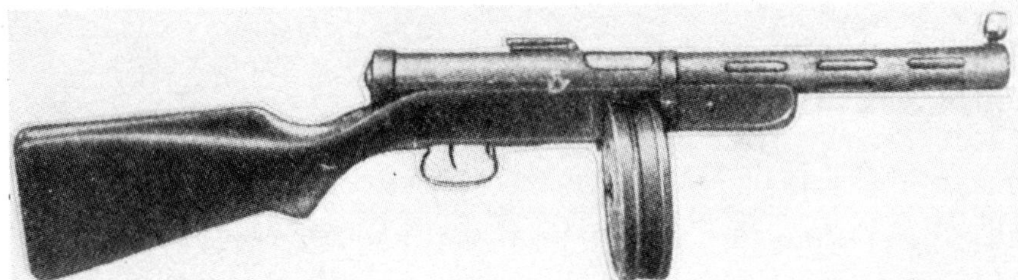

Remarks: Evolved from PPD-34/38. Well-made weapon produced 1940–41. After German invasion found not suitable for large-scale mass production and replaced by PPSh-41. Only relatively small numbers used by German forces.

7.62 mm Maschinenpistole 716(r)

German designation 7.62 mm MP 716(r)
Original designation *Pistolet-Pulemyot Degtyarova obr. 1934$38 g*
Calibre/cartridge 7.62 mm Type P
Magazine capacity 25 round box or 71 round drum
Length 777 mm
Length of barrel 273 mm
Weight 3.74 kg
Muzzle velocity 500 m/sec
Rate of fire (cyclic): 800 rpm
Manufacturer Various Soviet State arsenals

Remarks: First Soviet sub-machine gun designed by Vassili A. Degtyarov. Incorporated certain features of other early foreign weapons. Original 25 round box magazine found insufficient during Soviet-Finnish war in 1939–40 and hurriedly replaced by sturdy 71 round drum. Production terminated before WW II. Only limited numbers in German service.

7.62 mm Maschinenpistole 717(r)

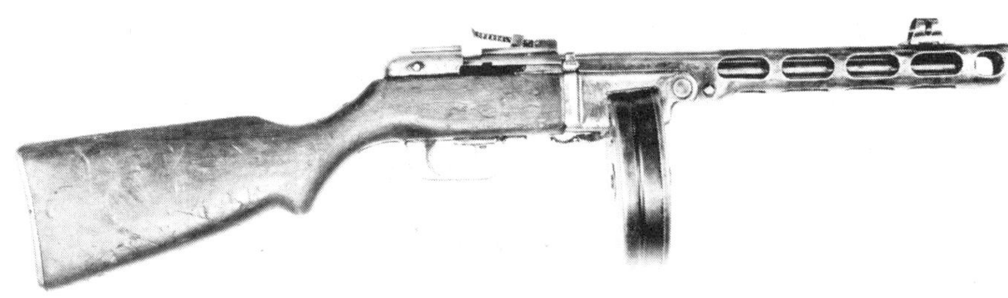

7.62 mm MP 717(r), the famous Shpagin PPSh-41. Captured weapons of this type were very popular with German front-line troops in the East

German designation 7.62 mm MP 717(r)
Original designation *Pistolet-Pulemyot Shpagina obr. 1941 g* (PPSh-41)
Calibre/cartridge 7.62 mm Type P
Magazine capacity 71 round drum or 35 round curved box
Length 840 mm
Length of barrel 269 mm
Weight 3.5 kg
Muzzle velocity 500 m/sec approx
Rate of fire (cyclic): 900–1000 rpm
Manufacturers Various Soviet State arsenals.

Remarks: Simple and most reliable weapon designed by Georgi S. Shpagin. Only few moving parts, twin triggers for individual/automatic fire, simple hinged L-shaped sight. Produced in very large numbers throughout WW II. Widespread use by German troops. In 1944–45 some converted by Germans to 9 mm Parabellum cartridge.

7.62 mm Maschinenpistole 719(r)

German designation 7.62 mm MP 719(r)
Original designation *Pistolet-Pulemyot Sudayeva obr. 1943 g* (PPS-43)
Calibre/cartridge 7.62 mm Type P
Magazine capacity 35 round curved box
Length of stock (extended): 820 mm
(folded): 623 mm
Length of barrel 254 mm
Weight 3.04 kg
Muzzle velocity 500 m/sec
Rate of fire (cyclic): 600–700 rpm
Manufacturers Various Soviet State arsenals

Remarks: Improved version of PPS-42 designed by Alexei I. Sudayev and produced in Leningrad during German siege. Automatic fire only. Substantial numbers used by German front-line troops on Eastern Front.

MP 719(r) with its metal stock folded back

7.65 mm Maschinenpistole 722(f)

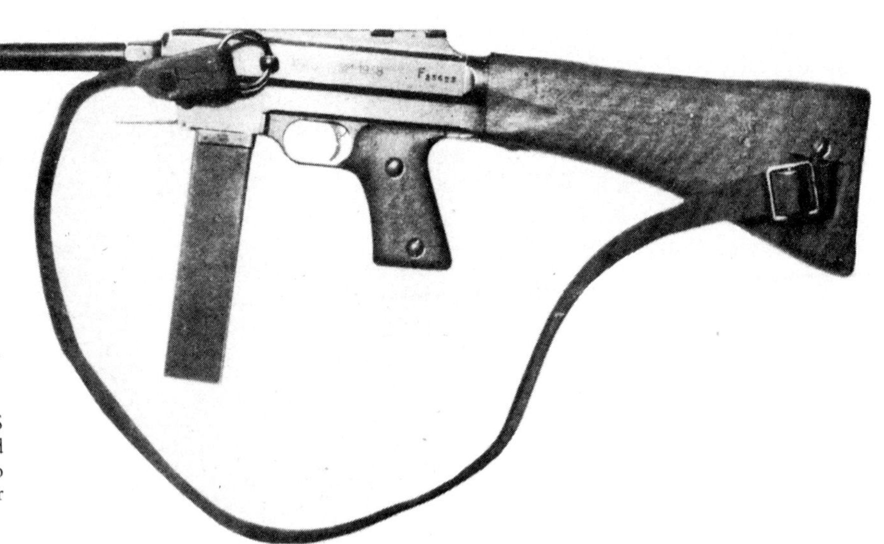

German designation 7.65 mm MP 722(f)
Original designation Mitraillette MAS mle 38
Calibre/cartridge 7.65 mm long (MAS 35 pistol)
Magazine capacity 32 round box
Length 623 mm
Length of barrel 224 mm
Weight 2.87 kg
Muzzle velocity 350 m/sec
Rate of fire (cyclic): 600–700 rpm
Manufacturer Manufacture d'Armes de St. Etienne, St. Etienne

Remarks: Evolved from experimental MAS 35. An accurate short-range weapon machined from solid metal, and unusual in having no spring-buffered sear. Kept in production for German forces until 1944.

9 mm Maschinenpistole 738(i)

German designation 9 mm MP 738(i)
Original designation Moschetto automatico modello 38/42
Calibre/cartridge 9 mm Parabellum
Magazine capacity 20 or 40 round box
Length 800 mm
Length of barrel 200 mm
Weight 3.27 kg
Muzzle velocity 381 m/sec
Rate of fire (cyclic): 550–600 rpm
Original manufacturer Pietro Beretta SpA, Gardone

Remarks: Basically simplified mod 38A for ease of wartime production. Progressive developments modello 38/43 had smooth barrel covering and modello 38/44 was made almost completely from stamped and forged parts. German service use mainly in Italy.

Beretta modello 38/42 sub-machine gun used by the German troops during the late period of the war in Italy

9 mm Maschinenpistole 739(i)

German designation 9 mm MP 739(i)
Original designation Moschetto automatico modello 38A
Calibre/cartridge 9 mm mod. 38A and 9 mm Parabellum
Magazine capacity 10, 20 or 40 round box
Length 947 mm
Length of barrel 320 mm
Weight 3.945 kg
Muzzle velocity (9 mm modello 38A): 450 m/sec; (9 mm Parabellum): 420 m/sec
Rate of fire (cyclic): 550 rpm
Manufacturer Pietro Beretta SpA, Gardone

Beretta modello 38A, third version, with a 20 round magazine

Remarks: Selective fire weapon with twin triggers. Issued to German units in North Africa and in Italy.

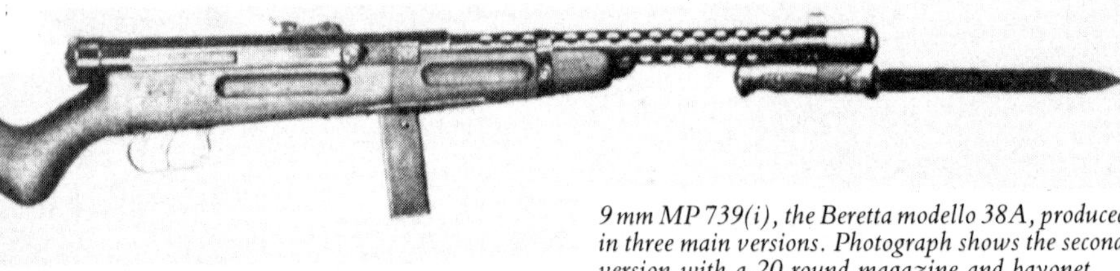

9 mm MP 739(i), the Beretta modello 38A, produced in three main versions. Photograph shows the second version with a 20 round magazine and bayonet

9 mm Maschinenpistole 746(d)

German designation 9 mm MP 746(d)
Original designation (d): Suomi M.42
Calibre/cartridge 9 mm Parabellum
Magazine capacity 20 or 50 round box or 70 round drum
Length 870 mm
Length of barrel 315 mm
Weight 4.676 kg
Muzzle velocity 399 m/sec
Rate of fire (cyclic): 800–900 rpm
Original manufacturer Tikkakosi O/Y; Sakara, Finland

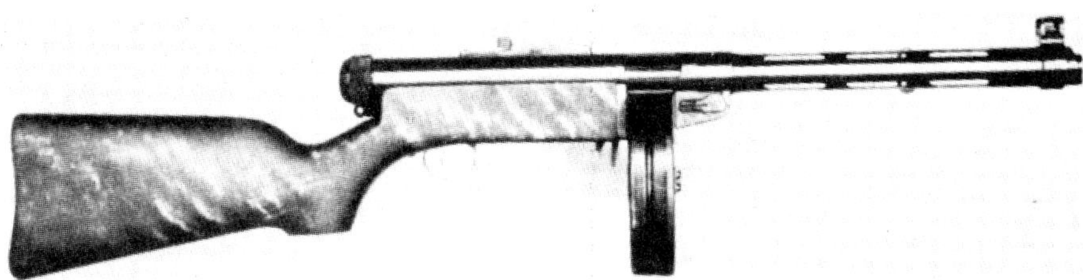

9 mm MP 746(d), the Finnish 'Suomi' m/31 built under licence as the M.42 by Madsen for the Danish Army and impressed into German service

Remarks: Improved M.31 used most successfully by Finnish troops in 1939–40. Automatic fire only. One of the most accurate sub-machine guns made. Small numbers used by German occupation troops in Denmark.

9 mm Maschinenpistole 751(e)

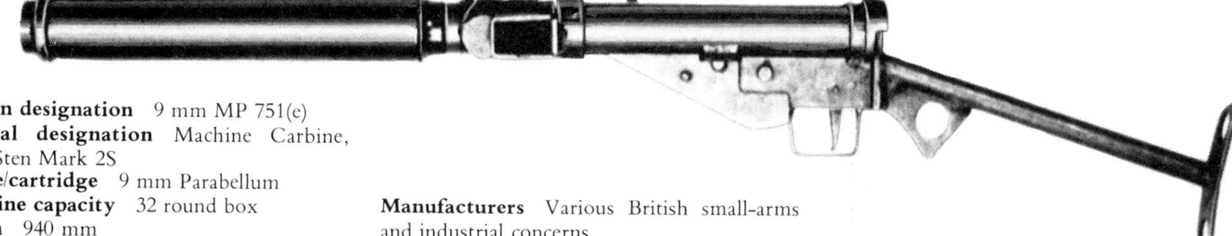

German designation 9 mm MP 751(e)
Original designation Machine Carbine, 9 mm Sten Mark 2S
Calibre/cartridge 9 mm Parabellum
Magazine capacity 32 round box
Length 940 mm
Length of barrel 91.7 mm
Weight empty 3.72 kg
Weight loaded 4.1 kg
Muzzle velocity 305 m/sec
Rate of fire (cyclic): Single shot but capable of 450 rpm

Manufacturers Various British small-arms and industrial concerns

Remarks: British Sten Mk 2 with shorter barrel and silencer, specially evolved for SOE and underground operations in Europe. Captured guns used by Germans in various counter-resistance measures.

9 mm MP 751(e), the British Sten Mk 2 with silencer attachment

9 mm Maschinenpistole(i)*

German designation 9 mm MP(i)*
Original designation Pistola mitriaglice FNA-B modello 1943
Calibre 9 mm
Magazine capacity 20 or 40 round box
Length of stock (extended): 790 mm
(folded): 526 mm
Length of barrel 198 mm
Weight 3.2 kg
Muzzle velocity 373 m/sec
Rate of fire (cyclic): 400 rpm
Manufacturer Fabbrica Nazionale d'Armi, Brescia

Remarks: Kept in production for German forces after 1943

* designation unconfirmed

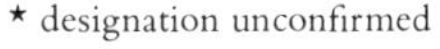

FNA-B with its stock and magazine folded

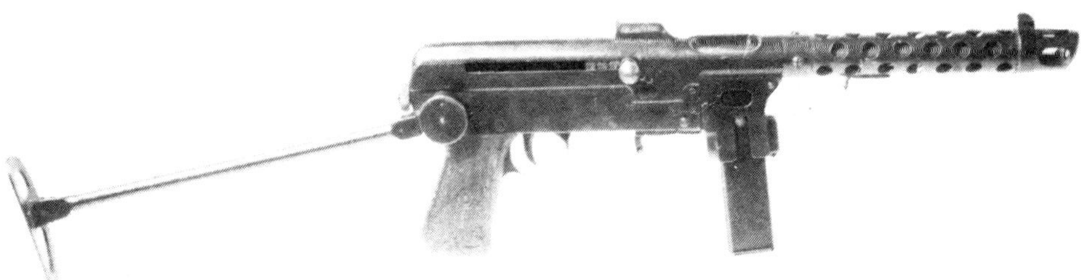

9 mm MP(i), the Italian FNA-B Modello 1943

11.43 mm Maschinenpistole 760(e), (j) and (a)

German designation 11.43 mm MP 760(e), (j) oder (a)
Original designations Commercial: Thompson Model 1928; Military: Gun, Submachine, Cal .45, Thompson, M1928A1 and Gun, Submachine, Cal .45, Thompson, M1
Calibre/cartridge 11.43 mm (.45 ACP)
Magazine capacity 20/30 round box or 50 round drum
Length 857 mm
Length of barrel 267 mm
Weight 4.88 kg
Muzzle velocity 280 m/sec
Rate of fire (cyclic): 600–725 rpm
Original manufacturer Colt Patent Firearms Manufacturing Corp., Hartford; Savage Arms Corp. Utica; both for Auto-Ordnance Corp., Bridgeport

11.43 mm MP 760(e), the US Thompson Model 1928 with a 50 round drum magazine

Remarks: Improved Mod. 1921, originally intended for commercial sale only. Substantial number purchased by Britain in 1939–40. Universally known as 'Tommy Gun' but too heavy and expensive for effective military service and production. German use localised by available ammunition.

11.43 mm Machinenpistole 761(f)

German designation 11.43 mm MP 761(f)
Original designation Thompson Model 1921
Calibre/cartridge 11.43 mm (.45 ACP)
Magazine capacity 20 round box or 50/100 round drum
Length 835 mm
Length of barrel 266 mm
Weight 4.5 kg
Muzzle velocity 280 m/sec
Rate of fire (cyclic): 800 rpm
Original manufacturer Colt Pat. Firearms Mfg. Co., Hartford, Connecticut, for Auto-Ordnance Corporation, New York

Remarks: Original model of sub-machine gun designed by Gen. John T. Thompson and engineers Eickhoff and Payne. A quantity purchased by France in 1939. Only limited German service use.

11.43 mm MP 761(f), the US Thompson Commercial Model 1921 with a 20 round drum magazine

Light machine guns

Like the sub-machine gun, the light machine gun was a result of the unique tactical problems set by the trench warfare of WW I. When it began the strength of the infantry lay in its massed rifle fire, but within a few short but bloody months the machine gun dominated the battlefield. The machine gun of 1914 was an efficient man-killer but it was usually held at battalion level and fired from carefully selected fixed points. The use of fixed points was dictated by the weight of the machine gun and its heavy tripod: any ideas of rapidly moving machine guns around the rutted battlefields of the Western Front were soon put down by the sheer physical problems involved. Even the introduction of sledge mountings and the like made no tactical difference. Another problem came with the control of the machine guns. As they were held at battalion level, individual sections or companies often saw the need for a machine gun, but by the time this need was noted at battalion level the situation had changed. This need for a machine gun capable of quick and easy carriage and suitable for use at company or lower level was soon realised by the German tactical planners and by the end of 1915 a specification for what would eventually be known as a light machine gun was issued.

The new specification was answered by a conversion of the existing *MG 08*. This Maxim machine gun was the backbone of the German infantry but, as stated above, it was heavy and not very mobile. From it evolved the *MG 08/15* which was the same basic weapon with a butt, simple bipod, smaller water jacket, pistol grip, different sights and a lightened receiver. For a light machine gun the *MG 08/15* was heavy, but as it was based on an existing weapon, training and logistics were simplified. It was first issued to front-line troops in 1916. After 1918 large stocks were retained by the reduced German Army and in 1939 there were still substantial quantities on hand for issue to second-line and garrison formations. Many were also fitted on anti-aircraft mountings. In 1919 large numbers of *MG 08/15* were delivered or transferred to Belgium and Poland, but most of these were recovered by the German forces in 1939 and 1940. Many ex-German *MG 08/15*s were also recaptured in the Balkan campaign in spring 1941.

The next German effort in this field was the *MG 08/18*, a much lighter air-cooled weapon evolved after experience gained with the *luMG 08/15* air-cooled aircraft machine gun. The *MG 08/18* was produced and issued in insufficient numbers to make any impact on infantry thinking at the time, but its lighter weight and air-cooled features were noted by the post-war tactical analysts. They decided that any future light machine gun would have to be air-cooled too, but any development of automatic weapons was precluded by the terms of the Versailles Treaty. It was also obvious that development of an air-cooled machine gun would involve time and money, and a stop-gap solution was found in modifying a number of existing water-cooled Dreyse machine guns to air-cooling. The result was the *MG 13*, and this rather long and bulky weapon served as the standard light machine gun of the German Army during the late 1920s and early 1930s. Soon after better weapons became available the *MG 13* was withdrawn from service and relegated to training; by 1939 most had been sold on the second-hand arms market, mainly to Portugal, while the remainder were issued to German occupation units in the Channel Islands after 1940.

Before continuing with the mainstream of German machine gun development, one odd episode needs to be recorded. This was the attempt of the German firm of Knorr-Bremse, normally an engineering enterprise dealing with brake linings, to interest the German Army in a Swedish light machine gun known then as the 6.5 mm LH33. This Swedish design was not remarkable but it was promoted with considerable vigour and rechambered to take the German 7.92 mm cartridge. This happened in 1935 when the future standard German machine gun had already been decided, and the Knorr-Bremse offering was rejected. But it would seem that some political strings were pulled and the Waffen-SS bought a few. These guns were designated *MG 35* but only a limited number appear to have been put into service.

As the terms of the Versailles Treaty forbade machine gun development in post-1919 Germany the German armament firms resorted to secret agreements with foreign armament concerns to conduct research and development for them. One of the most productive foreign link-ups was that between Rheinmetall and the Swiss Solothurn organisation. Waffenfabrik Solothurn AG carried out a great deal of work for Rheinmetall during the 1920s, and among the more important results of their collaboration was the Solothurn *MG 29*. This light machine gun was based on the requirements then still being formulated by the German tactical planners, and embodied most of the features later to be included in all German machine guns. The design was all 'in-line', in that the gun followed a straight line from the muzzle to the butt. As the barrel of this air-cooled machine gun had to be changed frequently the design also incorporated a quick barrel-change device. Rate of fire was quite high, and a 25-round box magazine was fitted. The *MG 29* was thought to have room for improvement so a later model, the *MG 30*, was offered to the German Army, who turned it down but requested further development. Numbers of the *MG 30* were sold to the Austrian and Hungarian armies instead.

However, the Luftwaffe took an interest in the *MG 30* and asked for an aircraft version to be developed. The result was the *MG 15*. By this time, 1932, Rheinmetall were undertaking most of the work themselves in Germany at their Düsseldorf factories. The *MG 15* was adopted as one of the 'standard' Luftwaffe machine guns, but after 1940 it became obvious that the days of single-barrel rifle-calibre flexible aircraft machine guns were over and large numbers were released for ground use. Many were converted to anti-aircraft weapons on various mountings, but as many others were made available to the German ground forces and fitted with rather unsatisfactory bipods and shoulder stocks to become light machine guns. The same fate befell large numbers of the *MG 17*, a modified fixed Luftwaffe version of the *MG 15*. By 1944 many of these guns had been replaced by heavier calibre weapons and were adopted for ground use, usually in the anti-aircraft role. Mauser also contributed to this category of redundant aircraft weapons with their *MG 81*. Introduced in 1939 when the *MG 15* was still in production, it was the first Luftwaffe machine gun to be mounted in flexible pairs on aircraft. By late 1944 many of these coupled guns became available after disbandment of bomber units and were issued for ground use as anti-aircraft weapons. In this role the *MG 81* was frequently coupled in quadruple mountings although most were left in their twin-barrel configuration.

But to return to the mainstream of German machine gun development. When the OKH turned down the *MG 30* it was not because the design was unsuitable but because they had a more futuristic concept in mind. Once again tactical analysis was used to determine future weapon specifications and this time it revealed the need for what has become known as the general purpose machine gun. The split of machine gun types into light and heavy was seen as an anomaly, and the intention was that one type of machine gun could fulfil both roles. The air-cooled machine gun equipped with a rapid barrel-change device and fitted to a heavy tripod was foreseen as the future heavy machine gun, while the same weapon could also be fitted with a bipod and an alternative box feed device for use as a light weapon. The standardisation of one type for both roles was an obvious attraction.

By the early 1930s Rheinmetall was not the only German firm

taking an interest in air-cooled machine guns, for Mauser also had several prototypes under development. When the specification for the new machine gun was issued it was realised that both firms offered weapons with attractive features but that no single firm's product had an overall advantage over the other. The obvious result was to amalgamate the best of both and this was done under the design leadership of Louis Stange of Rheinmetall. However, Mauser retained overall control of the project and later became the main production outlet for the new machine gun, known as the *MG 34*. In both the heavy and the light role the *MG 34* proved to be an excellent weapon. As this section deals with light machine guns alone only that version will be mentioned. As a light machine gun the *MG 34* was fitted with a bipod and a 75-round drum magazine taken from the *MG 15*. It soon became the standard section weapon and German infantry tactics until about 1944 were centred around the section machine gun. A wide range of accessories was gradually introduced and the *MG 34* could be easily adapted to a number of roles. It was issued to all branches of the German armed service and used with various accessories, one of which was out of the ordinary. This was an odd periscopic device that enabled the *MG 34* to be fired from below cover. How many were made is not known, but it must rate as one of the more useless of the many bizarre German weapon developments.

Production of the *MG 34* was carried out by a number of firms and subcontractors under the leadership of Mauser-Werke AG at Berlin, but it was production that finally spelt the demise of the *MG 34*. In their eagerness to obtain the best weapon available, the *Heereswaffenamt* (Army Weapons Board) had selected a superb machine gun but one that was difficult and expensive to produce. Each basic *MG 34* cost the German forces RM 310. By late 1940 it was appreciated that the *MG 34* would have to be replaced by something better suited to large-scale production. An additional impetus was provided by combat experiences on the Eastern Front which indicated the need for an increased quantity of faster-firing machine guns. Mauser adapted the basic *MG 34* design as the *MG 34S* and *MG 34/41* but both were deemed unsatisfactory, despite their increased rates of fire. The ideal solution had to combine advanced design features with simplified production methods introduced with the *MP 40*

le MG 42 in action with the German mountain troops

sub-machine gun.

The result was one of the finest machine guns ever made, the *MG 42*. The need for an eventual *MG 34* replacement had been foreseen as far back as 1937. The *MG 42* originated with a design from Grossfuss of Döbeln which combined the basic *MG 34* layout with a number of features taken from Polish and Czech design studies. A series of trials with various experimental guns led to the *MG 39/41*, later to be standardised as the *MG 42*, by which time Mauser had once again taken over the project leadership and organised the production of this new weapon in a number of centres. As with the *MP 40*, the time-honoured gunmakers' methods were abandoned in favour of easily turned-out spot welds, stampings and drop forgings. By 1942, when the first *MG 42s* were delivered to the German Africa Corps, production was the main headache for the German war planners. The demands of war on several fronts created a massive problem for German industry, and the *MG 42* exactly suited the moment. Not only was it easier and quicker to produce but it was also, at RM 250, cheaper than the *MG 34*. But these savings did not mean much to the soldier in the field: he realised that the *MG 42* was a superb weapon, and he called for ever-increasing quantities. On every front, the German troops soon appreciated the reliability and handling of the *MG 42*. It had a simple and quick barrel change, a very high rate of fire that often sounded like tearing of linoleum, and was easy to maintain. Surprisingly, the opportunity was not taken to adapt existing *MG 34* accessories for the *MG 42*, although some parts were interchangeable. A whole new range of *MG 42* accessories was developed, although for the light machine gun version the only item needed was the light bipod manufactured from steel stampings. Another part sometimes used on the *MG 42* was the so-called winter trigger. The basic design has stood the test of time: in slight modified form, the *MG 42* is still being produced for the West German armed forces (as the *MG 3*), and can thus claim the distinction of being one of the few WW II weapons not rendered obsolete by modern developments.

Good as the *MG 42* was, the German designers did not rest on their laurels and when the war ended they were hard at work designing a replacement, the *MG 45*. This machine gun did not get further than the prototype stage although many of its features were incorporated into several postwar designs.

As with other weapons, even these high production rates of both standard German machine guns still could not meet the constantly increasing demand, and many thousands of captured guns had to be impressed in service. Numerically, one of the most important of these non-German designs was the Czech ZB vz. 26, produced for the Czech Army and exported from late 1926 onwards. Large numbers of these excellent light machine guns were acquired when the Germans took over Czechoslovakia in 1938–39 and allocated the designation *MG 26(t)*. Many of these weapons were used during the 1939 and 1940 campaigns, together with the later vz. 30 under its German designation *MG 30(t)*. By 1942 both types were gradually withdrawn from front-line service and relegated to rear echelons and anti-partisan forces, as were most of these Czech light machine guns captured from Yugoslav troops in 1941.

Units based in France after 1940 were often issued with French Chatellerault light machine guns. There were three principal versions, the mle 1924 (small numbers only), the mle 1924/29 and the mle 1931. The latter was an adaptation of the mle 1924/29 for use in tanks and the fortresses of the Maginot Line. It had a side-mounted drum magazine of 150-round capacity and a strangely-shaped butt stock. The mle 1931 was frequently adapted for anti-aircraft defence. Other French machine guns used by the German troops in the light role included the Hotchkiss mle 1922 and 1926, numbers of which also came to the German armoury by way of Greece, and small numbers of the elderly mle 1909, some of which were left behind by the retreating British forces at Dunkirk. The Germans also made use of the infamous Fusil-Mitrailleur mle 1915, the unloved Chauchat. That this gun was impressed at all, even if the issue was limited to some garrison troops, serves as a measure of the desperate German shortage of machine guns. The Chauchat was an early (1915) and unsuccessful French attempt to produce a light machine gun. Designed and developed in a great hurry, the Chauchat was involved in a number of shady contract deals which resulted in a flood of these weapons even when it was realised that the basic design was unsound, and in action the guns proved unreliable. But so many were manufactured that they were still in service in 1940, and quite a number were issued to the Volkssturm in 1944–45. Another French machine gun impressed by the Germans in the light role was the Darne mle 1922. Designed as an aircraft weapon, in some ways it heralded the later simple production methods embodied in the *MG 42*. A simple and reliable weapon, the Darne was adapted by the Germans for installation in fortifications and used in some numbers.

Elsewhere in Europe the Germans kept the Danish Madsen light machine gun in production until 1942, and it was adopted for service as the *7.92 mm MG (Madsen)*. In addition, the Germans also impressed a wide variety of other Madsen light machine guns of various models and calibres from captured Danish, Norwegian, French, Dutch and Yugoslav stocks. Others came from the regions of the former Baltic States after the invasion of the Soviet Union. Units based in Italy were often issued with the Breda modello 30, while other second-line and garrison units were armed with captured British Bren guns, a weapon developed from the Czech vz. 26. Quite a few Lewis guns were also acquired by the Germans from the post-Dunkirk British booty, from Holland, France, and later the Baltic States of the Soviet Union. Many Browning automatic rifles fell into the German hands after the Polish campaign, and their stocks were augmented a year later by the licence-built French 7.65 mm Fusil-Mitrailleur 1930. Those captured weapons were used by German garrison troops all over Europe, with most of them ending up as part of the Volkssturm armoury. Captured Soviet light machine guns were also impressed in service, but their use was largely confined to the Eastern Front. Most important of these weapons was the *Pulemyot Degtyarova Pekhotny*, or DP. Special versions of this machine gun evolved for tanks and aircraft were also put into service by the Germans.

It is true that the vast bulk of the German front-line units were armed with the indigenous *MG 34* and *MG 42*, but one must not forget the great array of other light machine guns used by the second-line and garrison troops. Their range, as with so many impressed weapon types, was huge, and a mute testimony to Germany's basic state of unpreparedness for prolonged large-scale warfare.

7.92 mm Maschinengewehr 08/15, 7.65 mm leMG 125(b) and 7.9 mm leMG 145(j)

German designations 7.92 mm leMG 08/15 und 145(j); 7.65 mm leMG 125(b)
Original designations (j): Leki-Mitralez 7,9 mm M 8/15 M; (b): Mitrailleuse 'Maxim' légère
Calibre/cartridge 7.92 mm × 57; (b): 7.65 mm × 54
Type of feed 50, 100 or 250 round fabric belts
Length 1400 mm
Length of barrel 720 mm
Weight complete 18 kg
Muzzle velocity 900 m/sec
Rate of fire (cyclic): 450 rpm
Original manufacturers Königliche Gewehr- und Munitionsfabrik Spandau; Deutsche Waffen- und Munitionsfabriken AG, Berlin

Remarks: World War I adaptation of MG 08 to provide some form of light machine gun to suit conditions of infantry warfare on Western Front. After 1939 many still remained in service with second-line and garrison units; substantial numbers were also mounted for close-range AA defence. Late 1944 remainder issued to Volkssturm.

7.92 mm Maschinengewehr 13

◄ *MG 13 used for training purposes*

7.92 mm MG 13 with a 25 round box magazine and a 75 round saddle drum magazine ▲

German designation 7.92 mm leMG 13
Calibre/cartridge 7.92 mm × 57
Type of feed 25 round box or 75 saddle drum magazine
Length 1341 mm
Length of barrel 720 mm
Weight 11.43 kg
Muzzle velocity 823 m/sec
Rate of fire (cyclic): 550 rpm
Original manufacturer Rheinische Metallwaren- und Maschinenfabrik, Sömmerda

Remarks: Evolved by Louis Stange from an earlier Dreyse design to provide a modern air-cooled selective-fire light machine gun for German Army. Officially adopted 1932, and principal German weapon of this class until 1936 when superseded by MG 34. Afterwards most MG 13s sold to Portugal and Spain but some remained in German service (mainly with garrison units) until 1945.

7.92 mm Maschinengewehr Knorr-Bremse

German designation Knorr-Bremse MG 35
Calibre/cartridge 7.92 mm × 57
Type of feed 20 round box magazine
Length 1308 mm
Length of barrel 691 mm
Weight 10 kg
Muzzle velocity 792 m/sec
Rate of fire (cyclic): 490 rpm
Manufacturer Knorr-Bremse AG, Berlin-Lichtenberg

Remarks: Evolved by Knorr-Bremse in 1935 from Swedish 6.5 mm LH 33. Officially rejected by German military but limited number sold to Waffen-SS. From 1943 onwards issued to some 'foreign' Waffen-SS units.

7.92 mm Maschinengewehr 30

MG 30 in action

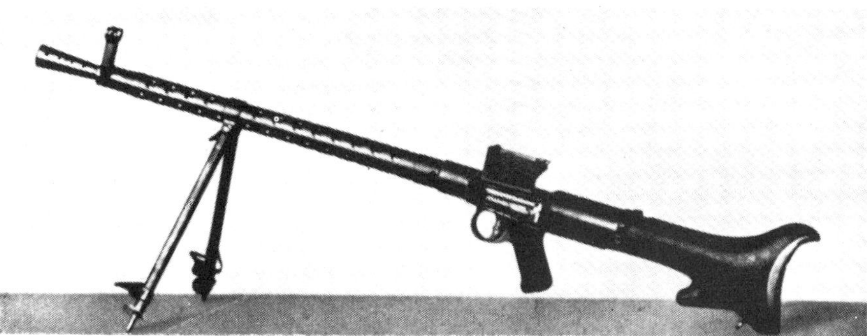

7.92 mm MG 30

German designation 7.92 mm MG 30
Original designation Solothurn MG 30
Calibre/cartridge 7.92 mm × 57
Type of feed 25 round box magazine
Length 1174 mm
Length of barrel 596 mm
Weight 7.7 kg
Muzzle velocity 760 m/sec
Rate of fire (cyclic): 800 rpm
Original manufacturer Waffenfabrik Solothurn AG, Solothurn

Remarks: A German-Swiss design featuring novel quick-change barrel system. Adopted by Austrian Army in 1930; also Hungarian Army (as 31M). Austrian equipment taken over by German forces 1938–39. Available MG 30s used mainly for guard duties and training.

7.92 mm Maschinengewehr 15

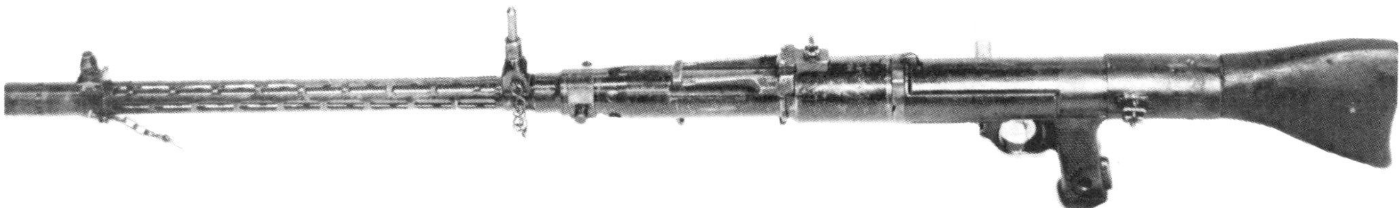

German designation 7.92 mm MG 15 oder leMG 15
Calibre/cartridge 7.92 mm × 57
Type of feed 75 round saddle drum
Length 1334 mm
Length of barrel 595 mm
Weight 12.7 kg
Muzzle velocity 755 m/sec
Rate of fire (cyclic): 850 rpm
Manufacturer Rheinmetall AG, Düsseldorf

Remarks: Aircraft machine gun evolved from Solothurn MG 30. Introduced in service 1932 and standard Luftwaffe flexible machine gun for most of WW II. Large numbers made available and adapted for ground use from 1943 onwards. Widely used by Luftwaffe field divisions, various Army garrison and guard units and training establishments. From late 1944 issued to Volkssturm.

Fliegerabwehr Pivot für MG 15 *(Fla-Pivot 15). This device consisted of a clamp with three jaws that could be attached to a wooden post. Similar equipment was produced for the MG 17 (Fla-Pivot 17) and the MG 81 (Fla-Pivot 81)*

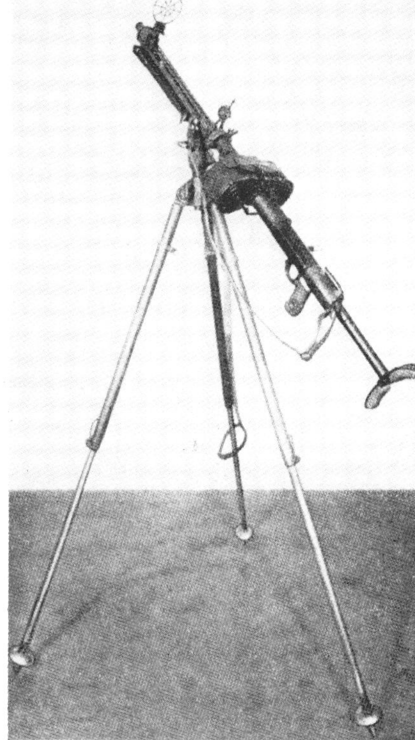

MG 15 on a tripod mount for AA defence

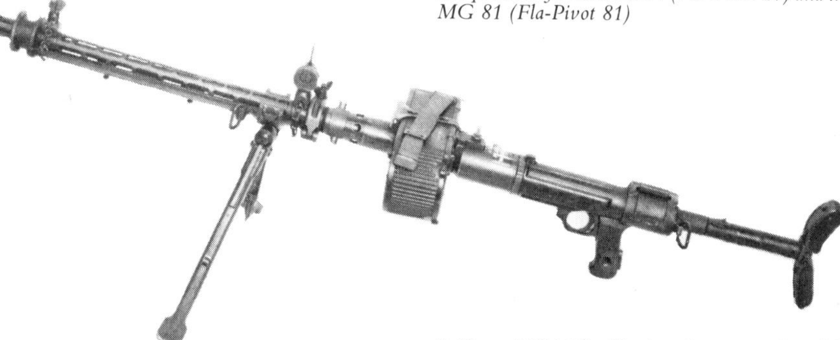

7.92 mm MG 15 flexible aircraft weapon adapted for infantry service

7.92 mm Maschinengewehr 34

MG 34 with belt feed and the bipod in rear position

MG 34 in action

German designation 7.92 mm leMG 34
Calibre/cartridge 7.92 mm × 57
Type of feed 75 round saddle drum or 50 and 250 round disintegrating link metal belts
Length 1219 mm
Length of barrel 627 mm
Weight with bipod 11.5 kg
Muzzle velocity 755 m/sec
Rate of fire (cyclic): 900 rpm
Manufacturer Mauser-Werke AG, Berlin and Oberndorf-am-Neckar; Gustloff-Werke, Suhl; Maget, Berlin-Tegel; Steyr-Daimler-Puch AG, Steyr; Waffenwerke Brünn (Brno)

7.92 mm MG 34 with a 50 round belt drum and the bipod in forward position

Remarks: Design evolved by Louis Stange, based on earlier experimental models. Accepted for service in 1934, with production deliveries commencing 1936. Standard issue to German Army, Waffen-SS and other first-line formations. Remained in production until 1945; used a variety of mounts and accessories. Notable as the first true general-purpose selective fire machine gun.

7.92 mm Maschinengewehr 34/41 and 34S

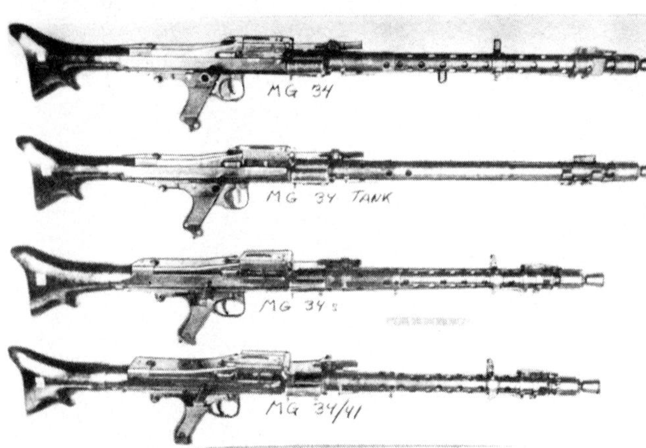

Comparison of the four MG 34 variations

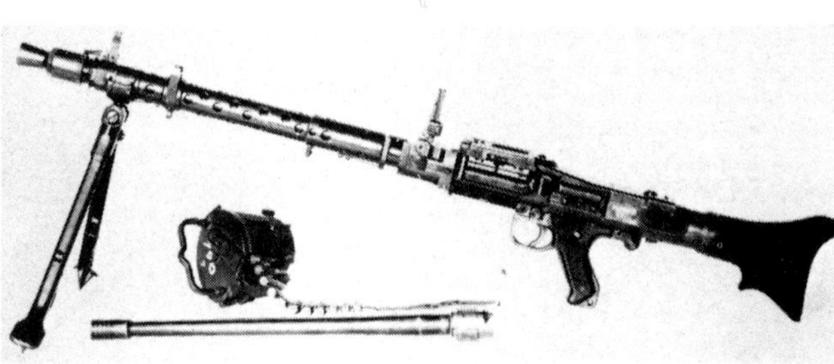

7.92 mm MG 34S

7.92 mm MG 34/41

German designation 7.92 mm MG 34/41; 7.92 mm MG 34S
Calibre/cartridge 7.92 mm × 57
Type of feed 50 or 250 round disintegrating link metal belts
Length 1135 mm
Length of barrel 500 mm
Weapon (weapon only): 11.42 kg
Muzzle velocity 730 m/sec approx
Rate of fire (cyclic): MG 34S: 1200 rpm; MG 34/41: 1500 rpm
Manufacturer Mauser-Werke AG, Berlin

Remarks: 300 examples of fully-automatic MG 34/41 used for troop trials on Eastern Front during 1942 but not adopted for service. MG 34S represented an experimental design only. Both machine guns featured shorter barrels than MG 34.

7.92 mm Maschinengewehr 42

leMG 42 with belt feed

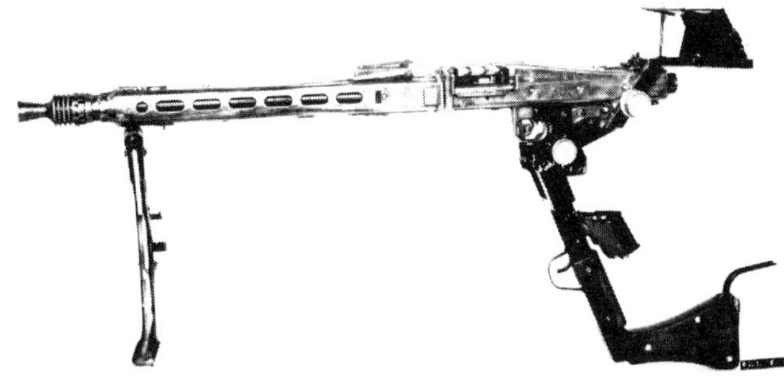

The 'Dezetgerät' device attached to the leMG 42

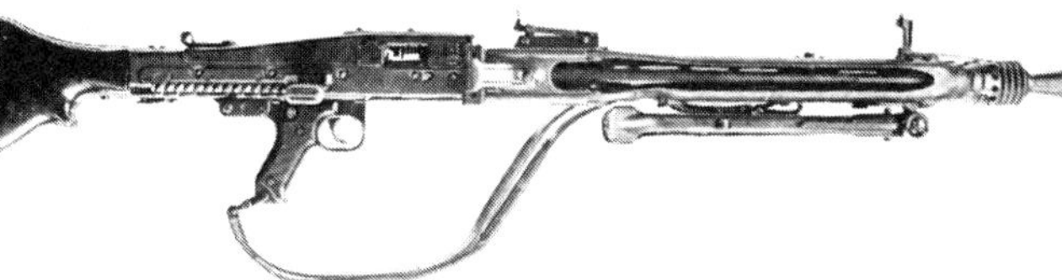

leMG 42 with its front bipod folded back. Note the cut-out in the barrel jacket for quick barrel change ➤

German designation 7.92 mm leMG 42
Calibre/cartridge 7.92 mm × 57
Type of feed 50 and 250 round disintegrating link metal belts
Length 1230 mm
Length of barrel 530 mm
Weight with bipod 11.6 kg
Muzzle velocity 820 m/sec
Rate of fire (cyclic): 1500 rpm
Manufacturers Mauser-Werke AG, Berlin; Grossfuss, Döbeln/Sachsen; Maget, Berlin-Tegel; Steyr-Daimler-Puch AG, Steyr; Gustloff-Werke, Suhl

Remarks: Originally known as MG 39/41; automatic fire only. Designed by Dr Grunow of Grossfuss-Werke, who also evolved novel simple and most economical manufacturing methods. Entered service in 1942; over 750,000 produced by 1945. Like MG 34, used various mountings and accessories. Probably the most remarkable machine gun design ever evolved, which introduced recoil-operated roller locking system and fast barrel change.

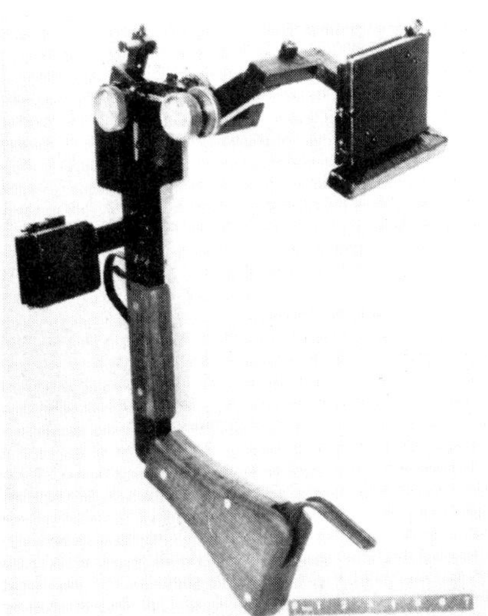

Deckungszielgerät für leMG 34 u.42 *(Dezetgerät). This device enabled the gunner to use the weapon from a trench or dugout without exposing himself to enemy fire*

leMG 42 in action on the Eastern Front

7.92 mm Maschinengewehr 17

German designation 7.92 mm MG 17
Calibre/cartridge 7.92 mm × 57
Type of feed 250 round disintegrating link metal belts
Length 1213 mm
Length of barrel 600 mm
Weight 12.55 kg
Muzzle velocity 760 m/sec
Rate of fire 1100 rpm
Manufacturer Rheinmetall-Borsig AG, Sömmerda

Remarks: Standard Luftwaffe fixed machine gun incorporating MG 15-type locking ring system; electric (solenoid) firing. In 1944–45 large numbers made available for ground use.

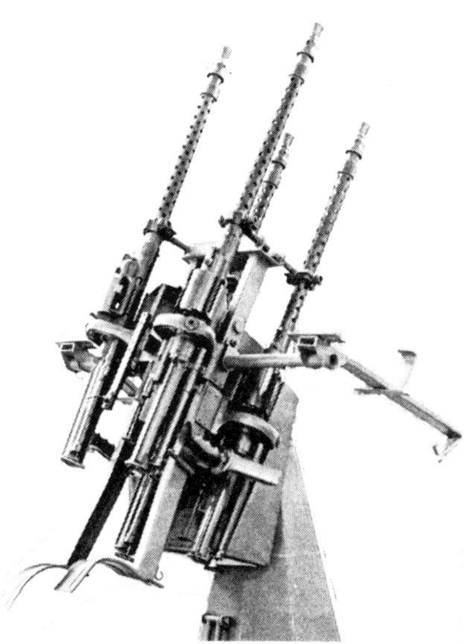

Surplus 7.92 mm MG 17 aircraft machine guns on a quadruple AA mount

7.92 mm Maschinengewehr 81

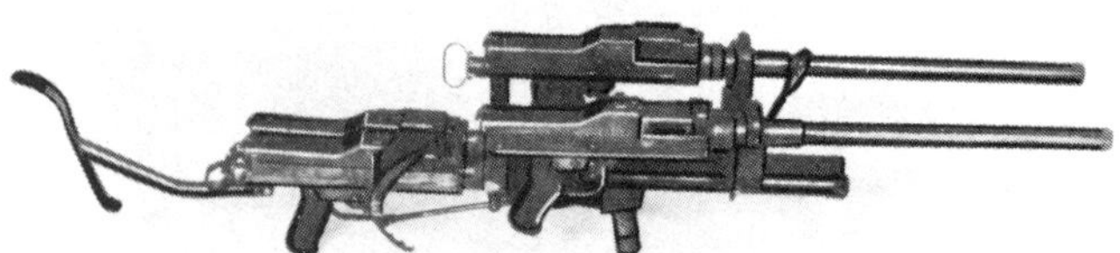

MG 81 on a quadruple mount for AA defence

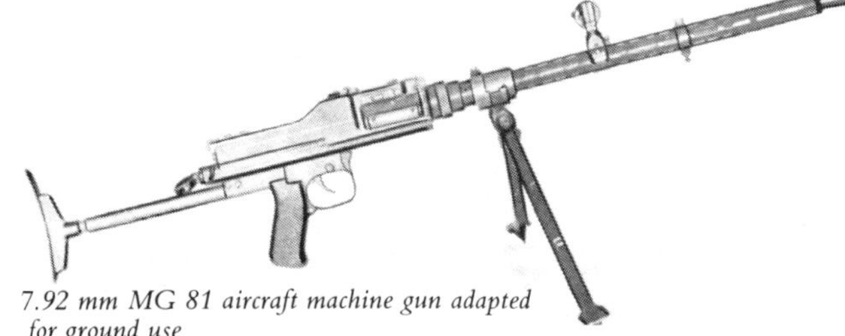

7.92 mm MG 81 aircraft machine gun adapted for ground use

German designation 7.92 mm MG 81
Calibre/cartridge 7.92 mm × 57
Type of feed 250 round disintegrating link metal belts
Length 889 mm
Length of barrel 476 mm
Weight 6.3 kg
Muzzle velocity 710 m/sec
Rates of fire 1600 rpm
Manufacturers Mauser-Werke AG, Berlin and Oberndorf-am-Neckar; Krieghoff/Suhl; Waffenwerke Brünn (Brno)

Remarks: 1939 Mauser development based on MG 34, intended as replacement for MG 15 on multi-seat Luftwaffe aircraft. After reduction of bomber formations in 1944 large numbers made available for ground use. Some fitted with bipods for infantry service (Volkssturm) but most used on coupled or quadrupled mountings for close AA defence. Also fitted on light German navel vessels and U-boats.

Ground version of the MG 81 being demonstrated by a Luftwaffe Field Division armourer

7.92 mm Maschinengewehr 26(t) and 146/1(j)

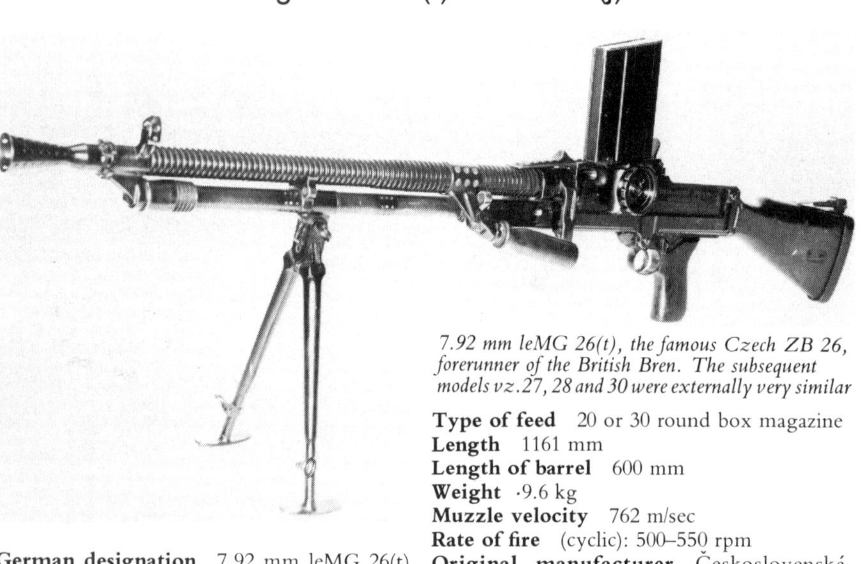

7.92 mm leMG 26(t), the famous Czech ZB 26, forerunner of the British Bren. The subsequent models vz.27, 28 and 30 were externally very similar

German designation 7.92 mm leMG 26(t) oder 146/1(j)
Original designations (t): 7.92 mm Kulomet vz. 26; (j): Puska-Mitralez 7.9 mm M 26
Calibre/cartridge 7.92 mm × 57
Type of feed 20 or 30 round box magazine
Length 1161 mm
Length of barrel 600 mm
Weight ·9.6 kg
Muzzle velocity 762 m/sec
Rate of fire (cyclic): 500–550 rpm
Original manufacturer Československá Zbrojovka, Brno

Remarks: Evolved by Václav and Emmanuel Holek in 1924 as progenitor of a long series of most successful light machine guns, including British Bren. Adopted as standard by Czech Army. Large numbers used by German troops and military police.

An leMG 26(t) clamped to a wooden post for AA defence purposes

7.92 mm Maschinengewehr 30(t) and 148(j)

German designation 7.92 mm leMG 30(t) oder 148(j)
Original designations (t): Kulomet vz. 30; (j): Puska-Mitralez 7.9 mm M 37
Calibre/cartridge 7.92 mm × 57
Type of feed 30 round box magazine
Length 1200 mm
Length of barrel 600 mm
Weight 9.5 kg
Muzzle velocity 762 m(sec
Rate of fire 500–550 rpm
Original manufacturer Československá Zbrojovka, Brno

Remarks: Almost identical to MG 26(t), Czech ZB vz. 26. Most successful light machine gun design of pre-WW II years, adopted in 24 countries. Also developed into British 'Bren'. Yugoslav gun represented a special export model designated lehky kulomet ZB 30J. Many used by German forces, particularly military police and anti-partisan formations.

leMG 30(t) in action, France, 1944

7.92 mm Maschinengewehr 30(t)

German designation 7.92 mm MG 30(t)
Original designation Kulomet vz. 30
Calibre/cartridge 7.92 mm × 57
Type of feed 250 or 500 round belts
Length 1023 mm
Length of barrel 721 mm
Weight 10.8 kg
Muzzle velocity 770 m/sec approx
Rate of fire 900–1000 rpm
Manufacturer Československá Zbrojovka, Brno

Remarks: Aircraft gun converted for ground use. In German service used in limited numbers for training and second-line/garrison duties by Luftwaffe ground personnel.

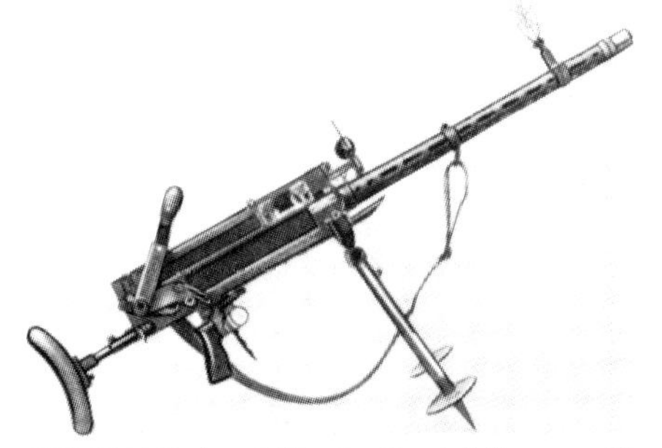

MG 30(t) Erd und Fla, *the Czech 7.9 mm vz.30 aircraft machine gun adapted to ground defence or AA role*

MG 30(t) on an improvised AA socket mount

6.5 mm Maschinengewehr 099(i)

German designation 6.5 mm leMG 099(i)
Original designation Fucile Mitriagliatori Breda modello 30
Calibre/cartridge 6.5 mm M91–95
Type of feed 20 round charger from fixed magazine
Length 1230 mm
Length of barrel 450 mm
Weight 10.6 kg
Muzzle velocity 630 m/sec
Rate of fire (cyclic): 450–500 rpm
Manufacturer Societa Anonima Ernesto Breda, Brescia

Remarks: One of first air-cooled machine guns with quick-change barrel, evolved from Breda mod 1924 via mod 1928. Standard light machine gun of Italian Army in WW II. Numbers issued to German Afrika Korps; also used by German units in Italy 1943–45.

6.5 mm Maschinengewehr 100(h)

6.5 mm leMG 100(h), British version of the Lewis gun produced in Holland as the M.20

Captured M.20 in action, May 1940.

German designation 6.5 mm leMG 100(h)
Original designation Mitrailleur M 20
Calibre/cartridge 6.5 mm M93
Type of feed 50 and 97 round drum magazines
Length 1260 mm
Length of barrel 654 mm
Weight less magazine 13 kg
Muzzle velocity 730 m/sec approx
Rate of fire (cyclic): 450 rpm
Manufacturer Birmingham Small Arms Company, Birmingham; Hembrug Arsenal, Holland

Remarks: Basically US Lewis in 6.5 mm calibre. Adopted by Dutch Army in 1920 and still standard light machine gun in 1940. Captured guns issued to some German occupation units, including those on Jersey.

6.5 mm Maschinengewehr 102(n) and 103(n)

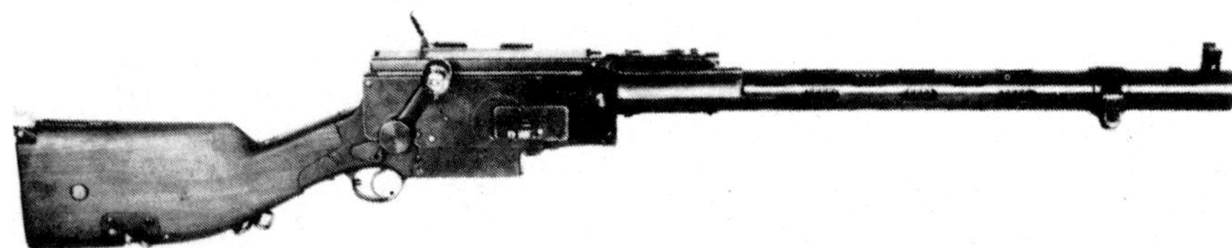

6.5 mm leMG 102(n)

German designations 6.5 mm leMG 102(n) und 103(n)
Original desgnations 102Zn): Maskingevaer m/1r (system Madsen) mg m/14; 103(n): Maskingevaer m/22 (system Madsen) mg m/22
Calibre/cartridge 6.5 mm M94
Type of feed 25 round box magazine
Length 1170 mm
Length of barrel 588 mm
Weight 10.2 kg
Muzzle velocity 700 m/sec approx
Rate of fire (cyclic): 500 rpm
Manufacturer Dansk Rekyt-Riffel Syndikat A/S 'Madsen', Copenhagen

Remarks: Small numbers used by German occupation troops in Norway.

6.5 mm leMG 103(n) used by a German coastal artillery soldier in Norway

6.5 mm Maschinengewehr 104(g) and 7.9 mm leMG 152(g)

German designations 6.5 mm leMG 104(g); 7.9 mm leMG 152(g)
Original designation Hotchkiss Model 1926
Calibres/cartridge 6.5 mm × 55 and 7.92 mm × 57
Type of feed 25 round metal trays
Length 1220 mm
Length of barrel 550 mm
Weight 9 kg
Muzzle velocity 745 m/sec approx
Rate of fire (cyclic): 450–500 rpm
Manufacturer Société de la Fabrication des Armes à Feu Portatives Hotchkiss et Cie, St. Denis

Remarks: Commercial Hotchkiss design which retained generally unsatisfactory strip-feed mechanism. Produced with bipods and tripods; some also sold to Czechoslovakia. Only limited German service use.

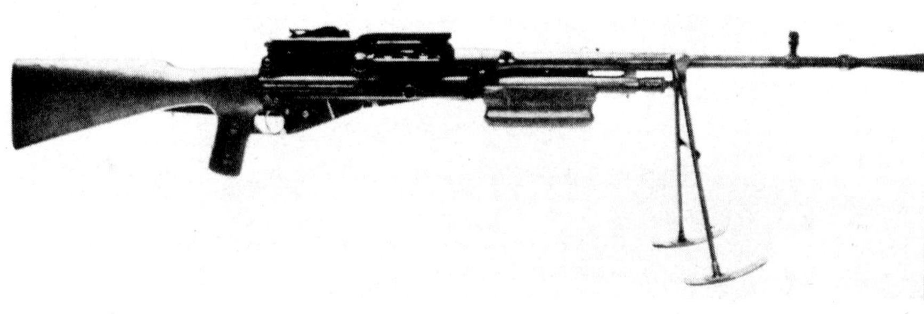

8 mm Maschinengewehr 105(f)

German designation 8 mm leMG 105(f)
Original designation Fusil-mitrailleur Hotchkiss mle 1922
Calibre/cartridge 8 mm mle 1886D (some made in 6.5 mm)
Type of feed 15 or 30 round steel strips
Length 1350 mm
Length of barrel 700 mm
Weight 9.22 kg
Muzzle velocity 760 m/sec approx
Rate of fire 300–600 rpm
Manufacturer Société de la Fabrication des Armes à Feu Portatives Hotchkiss et Cie, St. Denis

Remarks: German service use limited to garrison units in France.

8 mm leMG 105(f)

leMG Hotchkiss(t), Czech version of the French Hotchkiss mle 1922, with its bipod folded back

6.5 oder 8 mm Maschinengewehr 106(f)

German designation 6.5 oder 8 mm leMG 106(f)
Original designation Fusil-mitrailleur Darne mle 1922
Calibre/cartridge 6.5 mm × 55 and 8 mm mle 1886D
Type of feed 100 or 250 round steel belts
Length 1120 mm
Length of barrel 600 mm
Weight 9.7 kg
Muzzle velocity 740 m/sec approx
Rate of fire (cyclic): 650 rpm
Manufacturer Unceta y Compania, Guernica for R. et P. Darne et Cie, St. Etienne

Remarks: Originally designed by Regis and Pierre Darne for aircraft armament. Converted guns often used by German forces for coastal and light AA defence. Notable as probably the simplest and cheapest machine gun built, but efficient in use.

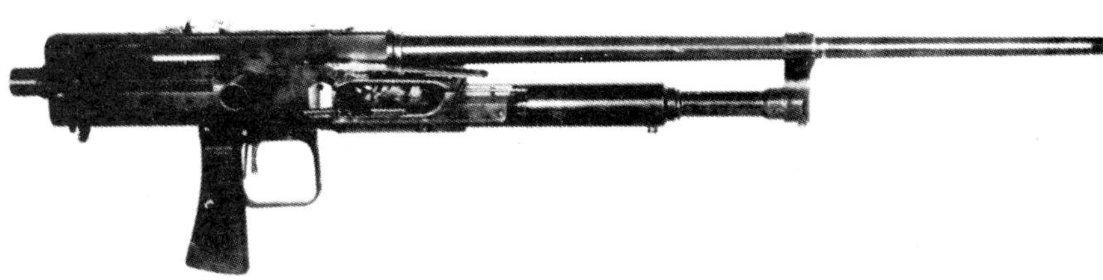

6.5 oder 8 mm leMG 106(f), the French Darne aircraft machine gun. Adapted for ground use it was fitted with a variety of crude wooden butts and bipods by the troops in the field

8 mm Maschinengewehr 107(f)

German designation 8 mm leMG 107(f)
Original designation Fusil-mitrailleur 'Lewis' mle 1924
Calibre/cartridge 8 mm mle 1886D; some also in 6.5 mm × 55
Type of feed Box magazine
Length 1140 mm
Length of barrel 600 mm
Weight 8.5 kg
Muzzle velocity 730 m/sec
Rate of fire (cyclic): 450 rpm
Manufacturer Uncertain. May have been licence-built in France by Société d'Armes Lewis/St. Denis

Remarks: Aircraft guns adapted for ground use. Fitted with a bipod and monopod under shoulder butt. Only limited German service use, mainly by garrison troops in France.

7.5 mm Maschinengewehr 116(f)

German designation 7.5 mm leMG 116(f)
Original designation Fusil-mitrailleur mle 1924/29
Calibre/cartridge 7.5 mm mle 1929
Type of feed 25 round box magazine
Length 1007 mm
Length of barrel 500 mm
Weight 8.93 kg
Muzzle velocity 820 m/sec
Rate of fire (cyclic): 450–600 rpm
Manufacturer Manufacture d'Armes de Chatellerault; Manufacture d'Armes de St. Etienne

Remarks: Selective fire weapon developed from Chatellerault mle 1924. Introduced new 7.5 mm cartridge in French service. After initial problems very popular with French troops and produced in large numbers. Substantial numbers in German service after 1940, mostly with various garrison and military police units.

7.62 mm Maschinengewehr 120(r)

German designation 7.62 mm leMG 120(r)
Original designation *Pulemyot Degtyarova Pekhotnii obr. 1928 g* (DP)
Calibre/cartridge 7.62 mm × 5rR
Type of feed 47 round drum
Length 1266 mm
Length of barrel 605 mm
Weight loaded 9.2 kg
Weight empty 8.5 kg
Muzzle velocity 840 m/sec
Rate of fire (cyclic): 500–600 rpm
Manufacturer Various Soviet State arsenals

Russian volunteer unit in German service traning with captured Soviet DP light machine guns, the leMG 120(r) ►

Remarks: First Soviet machine gun, adopted in 1927. Subsequently produced in very large numbers in several versions, including WW II modification DPM. Standard Red Army light machine gun until 1945. Substantial numbers in German service. Occasionally used as immediate front-line weapons but mostly issued to Russian and other Eastern formations fighting on German side. Hardy and most reliable weapon.

7.65 mm Maschinengewehr 127(b)

German designation 7.65 mm leMG 127(b)
Original designation Fusil-Mitrailleur 1930
Calibre/cartridge 7.65 mm × 54
Type of feed 20 round box magazine
Length 1150 mm
Length of barrel 560 mm
Weight 9.3 kg
Muzzle velocity 620 m/sec approx
Rate of fire (cyclic): 250–500 rpm
Manufacturer Fabrique Nationale d'Armes de Guerre (FN), Herstal, Liège

Remarks: Slightly modified US 1918A1 (BAR) automatic rifle manufactured in Belgium under Browning licence. Some produced with quick-change barrels and special tripods. Only limited German service.

7.7 mm Maschinengewehr 136(e) and (g)

German designation 7.7 mm leMG 136(e) oder (g)
Original designation Gun, Machine, Hotchkiss Mks 1 and 1*
Calibre/cartridge 7.7 mm (.303 Br)
Type of feed 24 or 30 round metal strips
Length 1190 mm
Length of barrel 600 mm
Weight 11.7 kg
Muzzle velocity 762 m/sec approx
Rate of fire (cyclic): 400 rpm
Manufacturer Royal Small Arms Factory, Enfield Lock

7.7 mm leMG 136(e) oder (g), the British-built version of the French Hotchkiss mle 1908

Remarks: Hotchkiss mle 09 built under licence in UK during WW I and later adapted as tank gun. Was carried on light British tanks in France 1939–40 and most left behind during retreat to Dunkirk. German use limited to coastal emplacements and few garrison units.

7.7 mm Maschinengewehr 137(e)

German designation 7.7 mm leMG 137(e)
Original designation Gun, Machine, Lewis, 0.303-in Mk 1
Calibre/cartridge 7.7 mm (.303 Br)
Type of feed 47 or 97 drum magazine
Length 1283 mm
Length of barrel 667 mm
Weight 11.8 kg
Muzzle velocity 745 m/sec
Rate of fire (cyclic): 450–550 rpm
Manufacturer Birmingham Small Arms Company, Birmingham

7.7 mm leMG 137(e). The same designation covered captured British and Canadian-built weapons

Remarks: Designed by Col I. N. Lewis in USA but first manufactured in quantity in Belgium 1915. Became principal light machine gun of British Army in WW I. Lewis guns left behind by British forces in France 1940 stockpiled until issued to Volkssturm 1944–45.

7.7 mm Maschinengewehr 138(e)

7.7 mm leMG 138(e), the Bren, a British development of the Czech ZB.26/34 series of light machine guns ▲

Captured Bren mounted on a Dreifuss 34 *AA tripod for the MG 34*

German designation 7.7 mm leMG 138(e)
Original designation Gun, Machine, Bren, 0.303-in Mks 1 and 2
Calibre/cartridge 7.7 mm (.303 Br)
Type of feed 29 round box magazine
Length 1156 mm
Length of barrel 635 mm
Weight (Mk 1): 10.05 kg; (Mk 2): 10.6 kg
Muzzle velocity 744 m/sec
Rate of fire (cyclic): 500–540 rpm
Manufacturer Royal Small Arms Factory, Enfield Lock. Also in Canada by John Inglis Ltd, Toronto

Remarks: Originated as British development of Czech ZB 26. Production in UK commenced in 1937. Usually known simply as 'Bren' which stood for Brno-Enfield. Captured guns often used as 'battlefield' issue, but also armed various occupation troops. Notable as one of the best light machine guns of WW II.

7.9 mm Maschinengewehr 154/1(p), 154/2(p) and (28(p))*

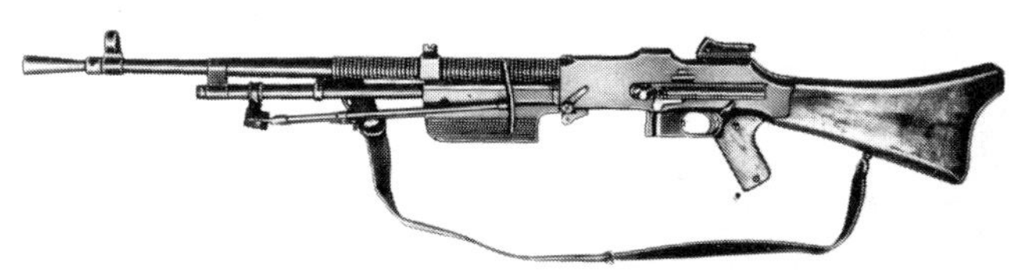

7.9 mm leMG 154/1(p), the Polish-built version of the American BAR

7.9 mm leMG 154/2(p)

German designation 7.9 mm leMG 154/1(p), 154/2(p) oder 28(p)
Original designation 7.92 mm reczny karabin maszynowy wz. 28
Calibre/cartridge 7.92 mm × 57
Type of feed 20 round box magazine
Length 1215 mm
Length of barrel 610 mm
Weight 9.5 kg
Muzzle velocity 815–853 m/sec
Rate of fire (cyclic): 600 rpm

* Colloquial designation – not confirmed

Remarks: Used same system as US Browning M1918A2 (BAR): leMG 154/1 and /2 differed only in sights and shape of butt. German service use limited by available stocks of captured weapons.

8 mm Maschinengewehr 156(f), (g) and (j), 7.65 mm leMG 126(b) and 7.9 mm leMG 147(j)

German designations 8 mm leMG 156(f), (g) oder (j); 7.65 mm leMG 126(b); 7.9 mm leMG 147(j)
Original designations (f): Fusil-mitrailleur mle 1915, Chauchat; (g): 7.8 mm 'Gladiator'; (j): Puska-Mitralez 8 mm M 15; (b): Fusil-Mitrailleur 15–27; 147(j): Puska-Mitralez 8 mm M15
Calibres/cartridges 7.65 × 54, 7.92 × 57 and 8 mm mle 1886D
Type of feed 20 or 25 round curved box magazines
Length 1143 mm
Length of barrel 470 mm
Weight 9.2 kg
Muzzle velocity 700 m/sec approx
Rate of fire (cyclic): 250–300 rpm
Manufacturers Various French State arsenals

Remarks: One of the first automatic weapons manufactured almost entirely of stampings, tubular and lathe-turned components. Was called 'automatic rifle' by French. Despite unsatisfactory long recoil` mechanism and poor manufacturing standards produced in very large numbers between 1915 and 1918 and subsequently widely sold abroad. Also produced in slightly modified form as mle 1929. Initials C.S.R.G. on gun stood for designers Col. Chauchat/Suterre/Riberolle and factory mark 'Gladiator'. German use of captured weapons limited to occupation troops and various foreign contingents.

8 mm Maschinengewehr 157(f)

German designation 8 mm leMG 157(f)
Original designation Fusil-mitrailleur Madsen mle 1922
Calibre/cartridge 8 mm mle 1886D
Type of feed 20 round box magazine
Length 1160 mm
Length of barrel 450 mm
Weight 8.8 kg
Muzzle velocity 700 m/sec approx
Rate of fire 500 rpm
Manufacturer Dansk Rekyt-Riffel Syndikat A/S Madsen, Copenhagen

Remarks: Slightly modified basic Madsen lmg to French contract. Only limited use by French and later German troops.

8 mm Maschinengewehr 158(d) and 159(d)

German designation 8 mm leMG 158(d) und 159(d)
Original designations 158(d): Rekytgevaer M 1903/24; 159(d): Rekytgevaer M 1924
Calibre/cartridge 8 mm rimless
Type of feed 20 round box magazine
Length 1145 mm
Length of barrel 596 mm
Weight 10.25 kg
Muzzle velocity 700 m/sec approx
Rate of fire 450 rpm
Manufacturer Dansk Rekyt-Riffel Syndikat A/S 'Madsen', Copenhagen

Remarks: Kept in production for German forces until 1942. Used by German occupation troops in Denmark and elsewhere.

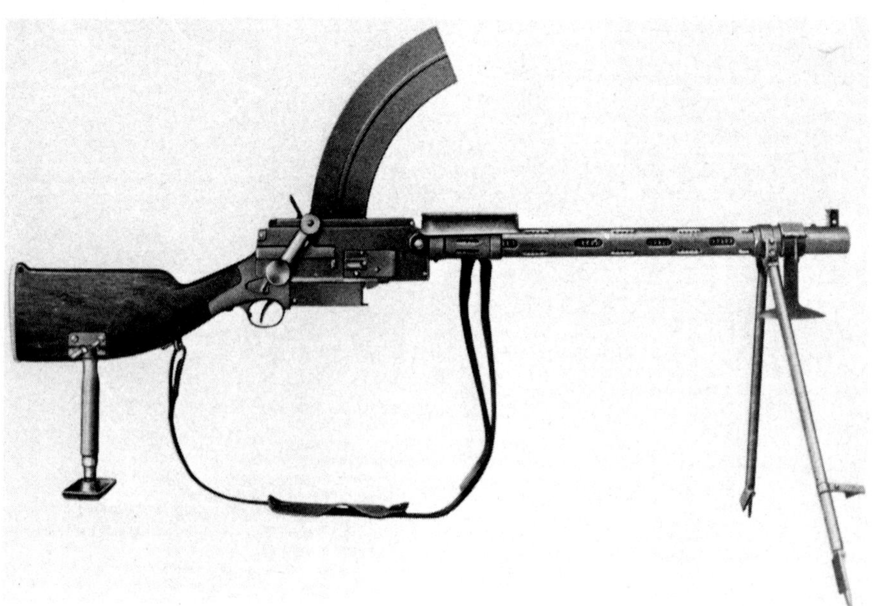

8 mm leMG 158(d); the leMG 159(d) was very similar

7.92 mm Maschinengewehr Madsen

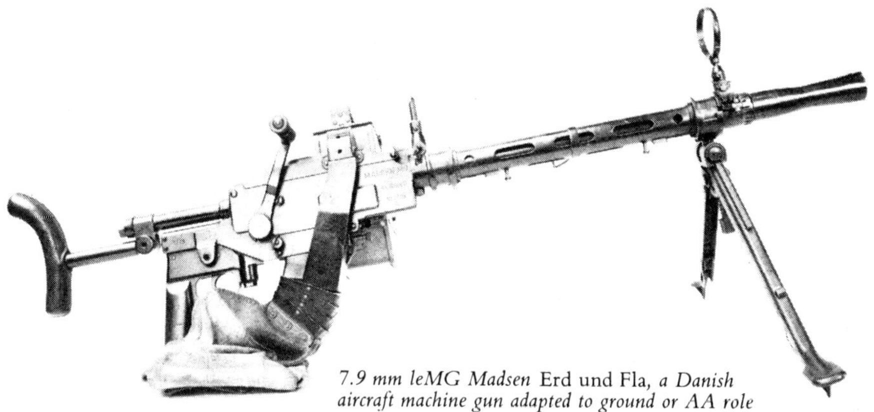

7.9 mm leMG Madsen Erd und Fla, *a Danish aircraft machine gun adapted to ground or AA role*

leMG Madsen Erd und Fla *with claw attachment for mounting on a wooden post for AA role*

German designation 7.92 mm MG Madsen
Calibre/cartridge 7.92 mm × 57
Type of feed 100 round belt
Length with stock 1150 mm
Length of barrel 477.5 mm
Weight 10 kg
Muzzle velocity 700 m/sec
Rate of fire 1100 rpm

Manufacturer Dansk Rekyt-Riffel-Sindikat A/S 'Madsen', Copenhagen

Remarks: Redesign of basic Madsen machine gun to belt feed was carried out under German control during 1941–42 and production commenced in 1942.

7.5 mm Kampfwagen-Maschinengewehr 311(f)

Kpfw MG 311(f) on an improvised AA mount ◄ *manned by a Cossack volunteer*

German designation 7.5 mm KpfwMG 311(f)
Original designation Mitrailleuse de 7.5 mm mle 1931
Calibre/cartridge 7.5 mm mle 1929
Type of feed 150 round side-mounted drum
Length 1030 mm
Length of barrel 600 mm
Weight empty 11.8 kg
Muzzle velocity 850 m/sec
Rate of fire (cyclic): 750 rpm
Manufacturers Manufacture d'Armes de Chatellerault

Remarks: Dismounted French tank machine gun often used in anti-aircraft role by German troops in France.

French volunteers in German service training with the Kpfw MG 311(f). These light machine guns have been fitted with bipods and shoulder stocks

7.62 mm Kampfwagen-Maschinengewehr 320(r)

German designation 7.62 mm KpfwMG 320(r)
Original designation *Pulemyot Degtyareva Tankovii* (DT)
Calibre/cartridge 7.62 mm × 54R
Type of feed 60 round drum
Length 1181 mm
Length of barrel 597 mm
Weight on bipod 12.7 kg
Muzzle velocity 840 m/sec
Rate of fire 550–600 rpm
Manufacturer Various Soviet State arsenals

Remarks: Version of standard Soviet DP light machine gun with pistol grip and telescoping shoulder stock for installation in tanks and armoured cars. Fitted with bipod for infantry service, was used as such on a limited scale by German troops.

Heavy machine guns

Of all the various machines of war that have been introduced during the 20th century, none has been more destructive of human life than the heavy machine gun. The slaughter inflicted by the machine gun during the years 1914–18 has still, even after 60 years, not removed the burden of economic and social changes imposed by the dominance of the machine gun over the dreadful battlefields of WW I. So many men were killed and so many economies hopelessly disrupted by the machine gun that it can safely be said that in some way or another the lives of all of us have been affected. The years 1939–45 once again emphasised the important role of the machine gun but the more fluid, mobile tactics in use tended to reduce its dominance. The role of the machine gun seemed to pass to the portable light machine gun carried by the infantry sections, but whenever conditions dictated static warfare the heavy machine gun once more emerged as one of the most important weapons on the battlefield.

The public image of the heavy machine gun is usually that of a heavy, water-cooled weapon, capable of prodigious fire rates and possessing an amazing reliability in action. This definition would almost exactly suit the heavy machine guns of the WW I years, and particularly the German *MG 08*. Almost as soon as Sir Hiram Maxim started to demonstrate his machine gun in Europe the German military authorities obtained examples and began to experiment. The German Army and Navy both took the type into service and the ultimate Army version was the *MG 08*. This model used the basic Maxim action virtually unchanged and the usual mounting was the large and heavy *Schlitten 08* sledge. In action, the *MG 08* was usually carefully emplaced during the many WW I campaigns as its weight prevented rapid moves. Time after time the *MG 08* was responsible for breaking up Allied attacks and only the advent of the tank, a weapon specifically designed to counter the machine gun, marked its decline. After 1919 large numbers of *MG 08s* were delivered to other countries as war reparations, but a substantial quantity still remained in German hands, both officially with the 100,000-man Reichswehr and police, and unofficially, hidden away from the Treaty commissions. As a result, after 1933, when the German armed forces once again began to expand, there were sufficient number of *MG p8*s on hand to equip garrisons, fortress and training units. The basic sledge mounting was altered to incorporate a pole support for anti-aircraft use and many of these guns served to provide some form of air defence for Luftwaffe airfields, bridge crossings and fixed fortifications. By 1941 the German inventory had been augmented by *MG 08*s taken over from Poland, Belgium, Yugoslavia and the former Baltic States; most of these guns were former WW I reparations.

A contemporary ot the *MG 08* was the *Schwarzlose* machine gun. This Austro-Hungarian design served alongside the *MG 08* during many WW I battles and after 1919 became widely distributed among the new states formed from the wreckage of the old Empire. The Schwarzlose used a different blow-back system to that of the Maxim but it was also water-cooled and at first sight similar to the *MG 08*. After the annexation of Austria in 1938 many Schwarzlose machine guns gradually found their way into German service, and their numbers were increased over the next few years by others captured in Poland, Holland, Yugoslavia and Greece. Later Italy added to this total, while Bulgaria, Hungary and Romania used the Schwarzlose machine gun while they were German allies.

The origins of the general-purpose machine gun have already been outlined in the section on light machine guns, so the early history of the *MG 34* and *MG 42* will not be mentioned here. Both these types were used as heavy machine guns. For this role their bipods were replaced by rather complex tripods which could be quickly altered to incorporate an upright pole for anti-aircraft purposes. For long-range heavy machine gun work a dial sight was an optional extra. Other accessories included a *Zwillingslafette 34* or *42* twin mounting for mobile AA defence and a *Zwillingsockel* for fixed AA installations on ships and fortifications. Other special mountings for fortifications included complex (and expensive) ball supports, eventually replaced by a simplified fortress mounting designed for firing through loopholes. For the use of AFVs special ball mountings were designed, and there was even a remote-control mounting for use atop the fighting compartments of assault guns. Most of these special mountings could take either the *MG 34* or *MG 42* but many accessories fitted only one type.

The *MG 34* and *MG 42* bore the brunt of the heavy machine gun work required by the German forces. Strangely enough, there was no German equivalent of the American Browning 0.5-in heavy machine gun used by the ground forces, although similar weapons were produced for the Luftwaffe. Some of these eventually found their way into use by the various ground troops, but only as extemporised weapons, as the heavier calibre Luftwaffe machine guns were needed for aircraft installations. One of these ex-Luftwaffe weapons was the Rheinmetall *13 mm MG 131*; substantial numbers of flexibly-mounted guns of this type became available after disbandment of many bomber units late in 1944. These *MG 131s* were easily adapted for ground use, although the variant intended for fixed installations on aircraft was not so easily converted. Most *MG 131s* were used on improvised AA mountings.

Another ex-Luftwaffe automatic weapon that was used by the ground forces in some numbers from late 1944 onwards was the *15 mm MG 151/15*. Surplus guns of this type were issued to the front-line troops as support weapons and employed either in the static role or in triple mountings on the *SdKfz 251/21* half-tracks. For AA use the *MG 151/15* was fitted on single and twin mountings, some of them hastily assembled from steel tubing. Most of these were issued to the Volkssturm. Another extemporised ground mounting for the *MG 151* was the carriage of the *sPzB 41* anti-tank rifle, numbers of which had become available after the original barrels had worn out.

There were no other German heavy machine guns. Several experimental models, mostly intended for the Luftwaffe, were produced but not developed. These included the *12.7 mm T 14 210, 13 mm MG 110, 13 mm MG 215, 15 mm MG 210, 15 mm HF 15* and the *16 mm ML 16*.

As with other weapons, the demands of the German armed forces increasingly outstripped the ability of the German industry to supply heavy machine guns and the only solution was to make the best possible use of captured foreign equipment.

Among the earliest of these 'foreign additions' were Czech machine guns, the more important of which was the ZB vz. 53. When this modern design was first issued to the Czech Army in 1937 it was one of the best machine guns produced in Central Europe and a welcome addition to the German armoury when they took over Czechoslovakia in 1938–39. This Czech machine gun had an air-cooled barrel with a quick-change fitting and was belt-fed using the standard German 7.92 mm ammunition. As a result it was adopted by the German Army as the *MG 37(t)* and became first-line issue, being widely used wherever German troops were deployed. Many of these machine guns had been produced by the Czechs in special fortress mountings for the defence of Sudetenland, and after 1941 most of these special installations were removed and emplaced along the Channel Coast. In 1941 additional Czech-made vz. 53 machine guns were captured in Yugoslavia.

Another Czech heavy machine gun much favoured by the Germans was the 15 mm ZB vz. 60. It was an enlarged version of the ZB vz. 53 and thus easily assimilated into the German arsenal. This gun, designated *MG M 38(t)*, was kept in production for the

An sMG 34 in a coastal defence emplacement with the tripod legs folded upwards and the butt removed

German Navy and Luftwaffe, and a special version was evolved mounted on a steel-wheeled carriage. However, most vz. 60 heavy machine guns in German service were used as AA weapons mounted on heavy tripods.

German forces in Italy were frequently issued with Italian machine guns. This had the advantage of easy spares and ammunition access but these positive points were often negated by the fact that Italian machine guns were not very reliable and were prone to stoppages. Types used included the Mitriaglice Fiat 14 and 14/35 and the Mitriaglice Breda modello 37, allocated the German designations *MG 200(i), MG 255(i)* and *MG 259(i)* respectively. The Breda mod. 37 was the most modern and the most favoured, but all three types depended on lubricated ammunition and were thus prone to clogging by dirt and dust.

Numerically among the more important of captured weapons were the French machine guns, especially the various Hotchkiss models, a design dating back to before WW I. For its age the air-cooled Hotchkiss was a sturdy reliable weapon, and it was exported widely. It used an unusual feed system in which the cartridges were fed into the gun in metal strips each holding 24 or 30 rounds. Large numbers of these guns fell into the German hands in 1940 and more were added to their armoury from the campaigns in Poland, Belgium, Norway and Yugoslavia. Most were issued to garrison units in the countries concerned, but substantial numbers were also used to bolster the defence along the Channel coast.

Another French machine gun, captured in smaller quantities, was the Mitrailleuse St. Etienne mle 1907. It had proved less successful than the Hotchkiss design and as a result had usually been issued to colonial forces. However, a number of these guns had been incorporated in the Maginot Line defences and after 1940 many of these fortress installations were moved by the Germans to their 'Atlantic Wall'. Additional guns of this type were acquired in 1941 during the German campaigns in Yugoslavia and Greece.

Among the more prized French acquisitions was the 13.2 mm Hotchkiss mle 1930 machine gun. This heavy weapon was a modern design which could be adapted to a variety of roles, including single and twin AA mountings. As the mle 1930 had been exported widely additional guns of this type were captured in Poland, Yugoslavia and Greece. Allocated the German designation *MG 271(f)*, numbers of these heavy machine guns were used by the German forces for AA defence and also deployed along their Atlantic Wall.

More heavy machine guns were added to the German armoury from other sources, each one bringing along its own peculiarities and adding to the training and logistics problems. The Polish campaign produced an American design, the Colt commercial version of Browning M1917 heavy machine guns, impressed by the Germans as the *MG 249(p)* or *MG 30(p)*. A number of similar guns were acquired in Norway and allocated the designation *MG 245(n)*. Heavy machine guns of a different system and calibre came into German hands after the evacuation of British forces from France in 1940 – large numbers of the Vickers Mk 1. The invasion of the Soviet Union added vast stocks of Soviet equipment to the German armoury, including some more Vickers machine guns from the former Baltic States – which, together with the British Army issue, were impressed as the *MG 230* – and very large numbers of the old Soviet PM1910 *Maksim* machine gun on a wheeled carriage. Substantial numbers of these were taken into German service as *MG 216(r)* and issued to various second-line, garrison and later Volkssturm formations. Quite a few were also deployed along the 'Atlantic Wall'. Later in the war heavy machine guns of yet another system and calibre were captured from the American troops and used by the German ground forces, often against their former owners.

One odd item that must be mentioned here as it will fit into no other category is the *Zielfeuer-Gerät 38*. To add realism to training exercises the German military went to the extent of developing and producing a device to fire special lubricated cartridges fed from a 70-round magazine. The *ZG 38* was fired using a wire leading to a trigger, and the device was usually fixed to some form of hard point. Although it was meant for training the *ZG 38* also had the potential of a tripwire warning device or, as it could fire automatically, as a form of battlefield decoy.

7.92 mm Maschinengewehr 08, 7.65 mm sMG 221(b) and 7.9 mm sMG 248(j), (p) and (r)

German designations 7.92 mm sMG 08; 7.65 mm sMG 221(b); 7.9 mm sMG 248(j), (p) und (r)
Original designations (b): 7.65 mm Mitrailleuse 'Maxim'; (j): Mitralez 7,9 mm M 8 M; (p): Maxsim 08; (r): Ex-Baltic and Polish weapons
Calibre/cartridge 7.92 mm × 57; (b): 7.65 mm M30
Type of feed 250 round fabric belts
Length 1175 mm
Length of barrel 720 mm
Weight of gun 20 kg
Weight of *Schlitten 08* 33 kg
Weight of *Dreifuss 16* 28 kg
Muzzle velocity 900 m/sec
Rate of fire 300–450 rpm
Original manufacturer Königliche Gewehr- und Munitionsfabrik, Spandau; Deutsche Waffen- und Munitionsfabriken AG, Berlin

Remarks: Standard German heavy machine gun of WW I years, designed around basic Maxim mechanism. Post-1918 large numbers delivered abroad as war reparations; some also sold commercially. Repossessed guns, together with German sMG 08s, used by many second-line and garrison units throughout WW II. Also issued to Volkssturm.

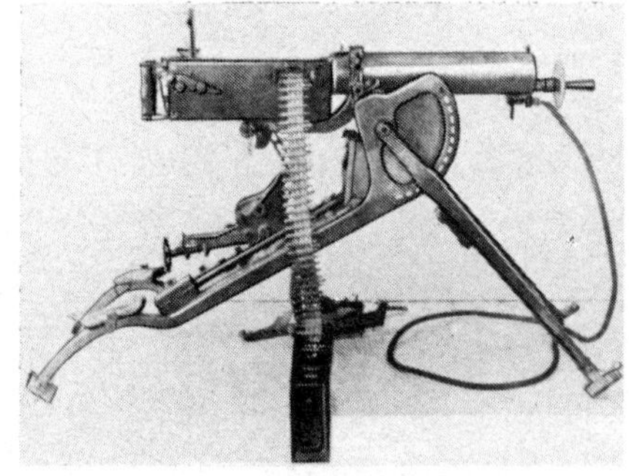

7.92 mm sMG 08 on Schlitten 08 *sledge mount*

sMG 08 with AA adapter

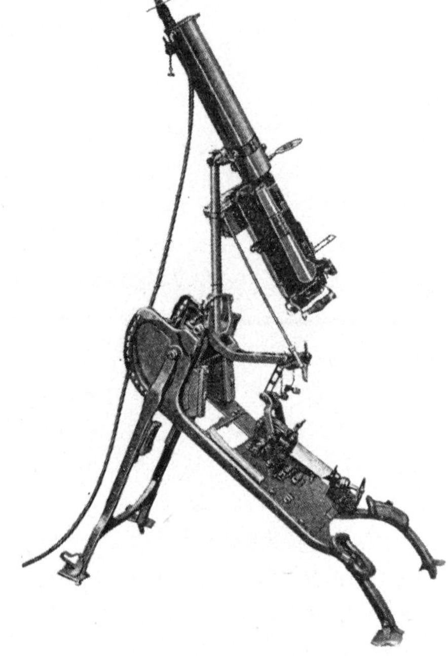

sMG 08 in position as a harbour AA defence weapon

7.92 mm Maschinengewehr 34

German designation 7.92 mm sMG 34
Calibre/cartridge 7.92 mm × 57
Type of feed 250 round disintegrating link metal belts
Length 1219 mm
Length of barrel 627 mm
Weight of gun 11 kg
Weight of *Dreibein 34* 6.75 kg
Weight of *Lafette 34* 23.6 kg
Muzzle velocity 755 m/sec
Rate of fire (cyclic): 900 rpm
Manufacturers Mauser-Werke AG, Berlin; Gustloff-Werke, Suhl; Maget, Berlin-Tegel; Steyr-Daimler-Puch AG, Steyr; Waffenwerke, Brünn

Remarks: Heavy variant of MG 34 general-purpose machine gun. In production and service from 1936 to 1945, using a great variety of mountings and accessories.

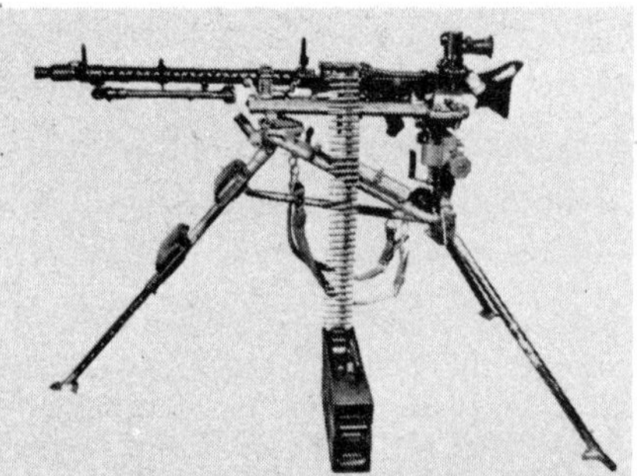

sMG 34. Note the two pads on the front leg which rested against the bearers' back when the mount was folded and carried by means of a sling, and the container for two spare barrels

An sMG 34 in a coastal defence emplacement with the tripod legs folded upwards and the butt removed

sMG 34 ready for action

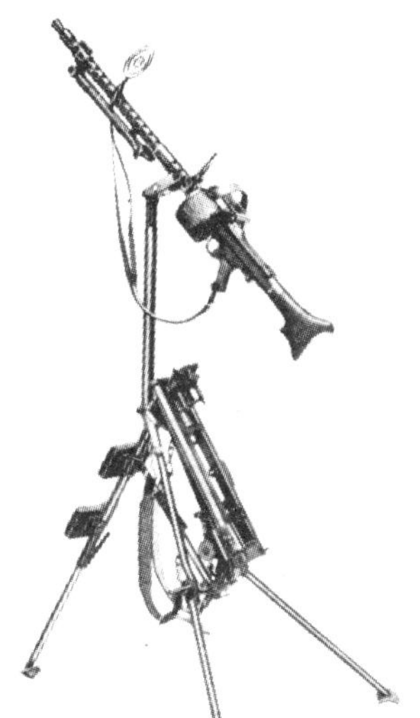

Lafette 34 *mount with AA adapter*

MG 34 on the lightweight Dreifuss 34 *AA tripod. The suspended ammunition drums were to give the mount greater stability. The* Dreifuss 34 *could also be used for firing at ground targets*

MG 34 on the Dreifuss 34 *positioned for the AA defence of a crossroad on the Eastern Front. Note the 50 round drum*

Close-up of an MG 34 on the AA adapter, Western Front, 1939

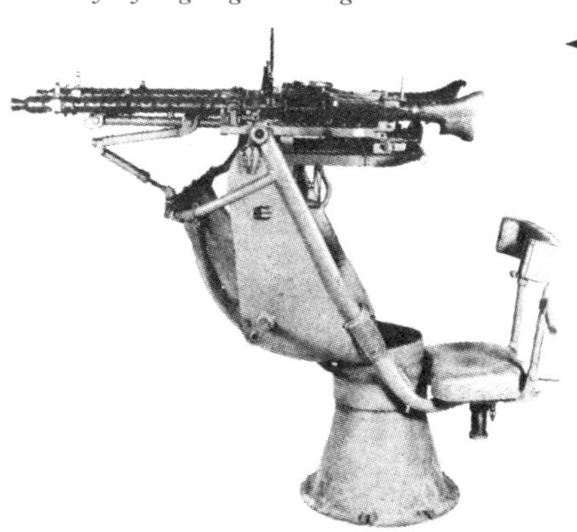

◄ Zwillingslafette 36, *a pedestal AA mount for two MG 34s. This mount could also be used for ground defence or carried in a single-axle limber* (MG-Doppelwagen 36) *for AA protection of infantry on the march*

MG 34 on a fortress loophole mount

MG-Doppelwagen. *This was towed by a motor vehicle or, with limber, horse-drawn. The normal ground tripods were carried strapped at the rear of the limber*

Zwillingslafette 36 *in static defence role. Note the special aircraft speed right sight*

7.92 mm Maschinengewehr 42

German designation 7.92 mm sMG 42
Calibre/cartridge 7.92 mm × 57
Type of feed 50 and 250 round disintegrating link metal belts
Length 1230 mm
Length of barrel 530 mm
Weight of gun 10.6 kg
Weight of *Lafette 42* 20.5 kg
Muzzle velocity 820 m/sec
Rate of fire (cyclic): 1500 rpm
Manufacturers Mauser-Werke AG, Berlin and Oberndorf-am-Neckar; Grossfuss, Döbeln/Sachsen; Maget, Berlin-Tegel; Steyr-Daimler-Puch AG, Steyr; Gustloff-Werke Suhl

Remarks: Heavy version of MG 42 general-purpose machine gun, in service from 1942. Generally acknowledged as probably the best machine gun design ever developed.

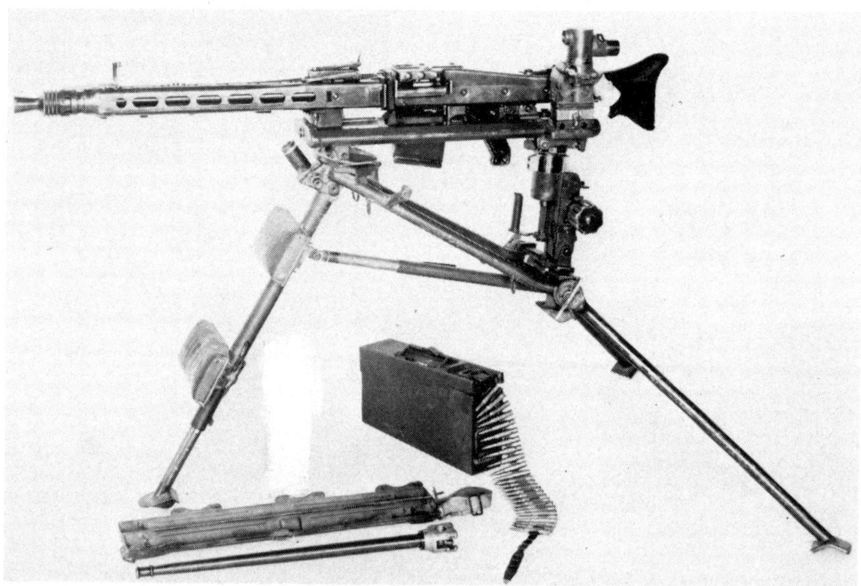

7.92 mm sMG 42 on the Lafette 42 *for heavy machine gun role. Note the optic sight and container for two spare barrels* ►

sMG 42 in a reinforced dugout in a swampy sector of the Eastern Front. The crew are wearing mosquito netting on their helmets

An sMG 42 emplaced for coastal defence

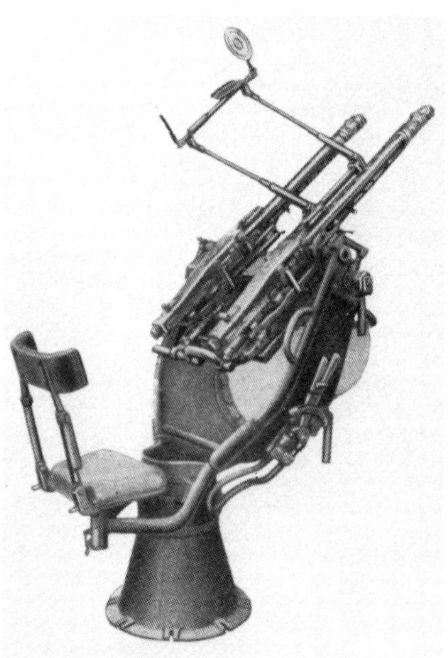

sMG 42 on the Zwillingslafette 36 *AA mount*

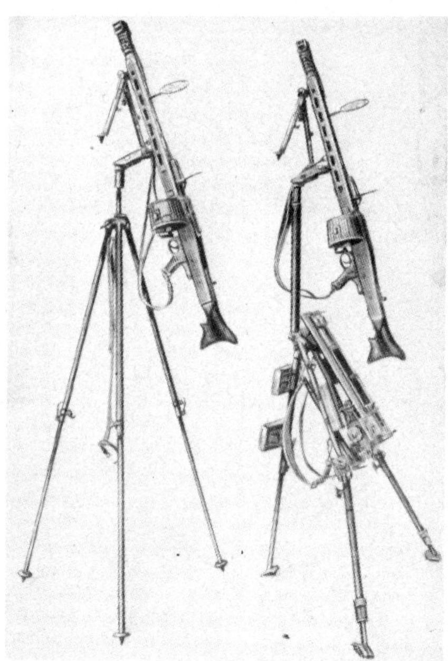

MG 42 AA adapters. Left, the lightweight Dreifuss 34, *right, the* Lafette 42

7.92 mm Maschinengewehr 45 or 42V

German designation 7.92 mm MG 45 oder 42V
Calibre/cartridge 7.92 mm × 57
Type of feed 50 and 250 round disintegrating link metal belts
Weight (approx): 8 – 9 kg
Rate of fire (cyclic): 1800 rpm
Design centre Rheinmetall-Borsig AG, Unterlüss (not confirmed)
No further data available

Remarks: Gradual development of MG 42, hence MG 42V designation, but using retarded-blowback mechanism with bolthead rollers (as in StuG 45 and several post-war weapon designs). First prototype made June 1944 and by May 1945 about ten prototypes had been built and tested – each with minor variations. Testing was successful but no production resulted due to end of hostilities.

13 mm Maschinengewehr 131

German designation 13 mm MG 131
Calibre/cartridge 13 mm
Type of feed 100 and 250 round disintegrating link metal belts
Length 1168 mm
Length of barrel 550 mm
Weight of gun 16.6 kg
Muzzle velocity 710–750 m/sec
Rate of fire (cyclic): 900–950 rpm
Manufacturer Rheinmetall-Borsig AG, Sömmerda

Remarks: Electrically operated Luftwaffe aircraft weapon for fixed and flexible installations. Surplus guns made available for ground use in 1944–45 and employed as heavy support weapons, and in light AA role, either singly or mounted in pairs or threes.

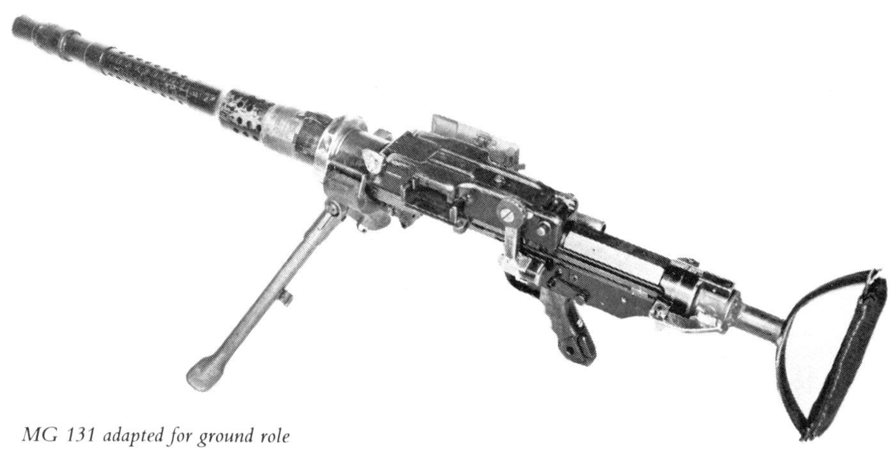

MG 131 adapted for ground role

15 mm Maschinengewehr 151/15

German designation 15 mm MG 151 oder 151/15
Calibre/cartridge 15 mm
Type of feed 50 round disintegrating link metal belts
Length 1916 mm
Length of barrel 1254 mm
Weight 42 kg
Muzzle velocity 790 m/sec
Rate of fire (cyclic): 700 rpm
Manufacturer Mauser-Werke AG, Berlin

Remarks: Electrically operated Luftwaffe aircraft weapon for fixed and flexible installations of similar design to MG 131. Surplus guns made available for ground use in 1944–45 and mostly employed in light AA role. Special wheeled carriage with balloon tyres evolved for heavy ground support use but not used in service.

Fliegerabwehr-Sockellafette für 1 MG 151 *(Fla SL 151), an emergency solution for mounting surplus MG 151 aircraft machine guns for static AA defence*

Fla SL 151D pedestal mounting with 3 × MG 151 aircraft weapons for defence against low-level air attacks. The inner armoured shield carries the reminder: 'Observe – Camouflage'

◄ *MG 151 aircraft machine gun adapted for ground role as a heavy infantry support weapon*

8 mm Maschinengewehr 07/12(ö), 7.9 mm sMG 247(j) and 8 mm sMG 261(i)

German designations 8 mm sMG 07/12(ö) oder (oe); 7.9 mm sMG 247(j); 8 mm sMG 261(i)
Original designations (ö): Maschinengewehr Model 07/12; (j): Mitralez 7,9 mm M 07/12 S; (i): 07/12
Calibres/cartridges 7.92 × 57 and 8 mm M 1893
Type of feed 250 round fabric belt
Length 1066 mm
Length of barrel 526 mm
Weight of gun 19.9 kg
Weight of tripod 19.8 kg
Muzzle velocity 620 m/sec
Rate of fire (cyclic): 400–500 rpm
Manufacturer Österreichische Waffenfabrik-Gesellschaft, Steyr

Remarks: Designed by Andreas Wilhelm Schwarzlose in 1902. Adopted by Austro-Hungarian Army in 1907. Standard A-H heavy machine gun in WW I and widely sold or delivered abroad as war reparations afterwards. Many used by Italians in WW II. German service use mainly by garrison and other second-line troops.

8 mm sMG 07/12(ö) with flame damper

sMG 248(j), the Yugoslav Mitralez 7.9 mm M 8M

7.92 mm Maschinengewehr 37(t) and 7.9 mm sMG 246(j)

German designation 7.92 mm sMG 37(t) oder 246(j)
Original designations (t): Kulomet vz. 37; (j): Mitralez M 40
Calibre/cartridge 7.92 mm × 57
Type of feed 100 or 200 round metal link belts
Length 1105 mm
Length of barrel 678 mm
Weight 18.86 kg
Muzzle velocity 792 m/sec
Rate of fire 500 or 700 rpm
Manufacturer Československá Zbrojovka, Brno; later Waffenwerke Brünn

Remarks: Adopted as standard Czech heavy machine gun one year before German occupation. Kept in production for German forces during WW II. Basic vz. 37 also served as pattern for the British Besa tank machine gun. Widespread German service use.

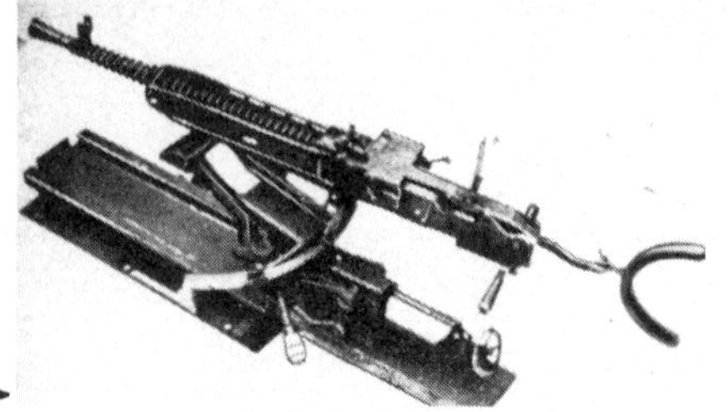

sMG 37(t) on fortress loophole mount ➤

Maschinengewehr M38(t) Kal. 15 mm, 15 mm Fliegerabwehr-Maschinengewehr 39 and 490(j)

German designations MG M38(t) Kal. 15 mm; 15 mm FlaMG 39; 15 mm FlaMG 490(j)
Original designations (t): ZB vz. 60; (j): Mitralez 15 mm M 38
Calibre/cartridge 15 mm
Type of feed 40 round metal link belts
Length 2050 mm
Length of barrel 1400 mm
Weight of gun 55 kg
Weight of wheeled tripd 203 kg
Muzzle velocity 860–970 m/sec
Rate of fire 420 rpm
Manufacturer Československá Zbrojovka, Brno, later Waffenwerke Brünn

Remarks: Numbers of M38 guns taken over after German occupation of Czechoslovakia in 1938–39, others captured in Yugoslavia 1941. Almost identical AA version kept in production under German supervision during WW II years. Same 15 mm vz. 60 also manufactured in UK as Besa Mk 1 tank weapon. German service use of all models only in light AA role.

sMG M38(t) on a modified carriage with pneumatic tyres

sMG M38(t) on its original carriage

6.5 mm Maschinengewehr 200(i) and (j)

German designation 6.5 mm sMG 200(i) oder (j)
Original designations (i): Mitriaglice Fiat 14; (j): Mitralez 6,5 mm (i)
Calibre/cartridge 6.5 mm M 91–95
Type of feed 50 round hopper magazine
Length 1720 mm
Length of barrel 650 mm
Weight of gun 17 kg
Weight of tripod 23 kg
Muzzle velocity 760 m/sec
Rate of fire (cyclic): 450–500 rpm
Manufacturer Fiat SpA, Torino

Remarks: Designated by Bethal Abiel Revelli in 1908, manufactured by Fiat, and generally known as Fiat-Revelli. Standard Italian heavy machine gun in WW I, remained in service and also exported afterwards. Many modified in 1935 as Fiat (Revelli) 14/35. Only limited number of this older version used by German troops in North Africa and Italy.

6.5 mm Maschinengewehr 201(n) and 7.9 mm sMG 240(n)

German designations 6.5 mm sMG 201(n); 7.9 mm sMG 240(n)
Original designations 201(n): Hotchkiss's 6.5 mm mitraljose m/98; 240(n): Hotchkiss's 7.9 mm mitraljose m/98t
Calibres/cartridges 6.5 M 94 and 7.92 mm
Type of feed 30 round metal strips
Length 1450 mm
Length of barrel 750 mm
Weight (approx): 28 kg
Weight of tripod (approx): 25 kg
Muzzle velocity (6.5 mm): 790 m/sec; (7.92 mm): 640 m/sec
Rate of fire 500 rpm
Original manufacturer Société de la Fabrication des Armes à Feu Portatives Hotchkiss et Cie., St. Denis

Remarks: Two basically similar Norwegian variants of Hotchkiss mle 1898 evolved from original 1898 design sold by Austro-Hungarian Cav. Capt Baron Adolf von Odkolek. Only limited numbers in German service, used by occupation troops in Norway.

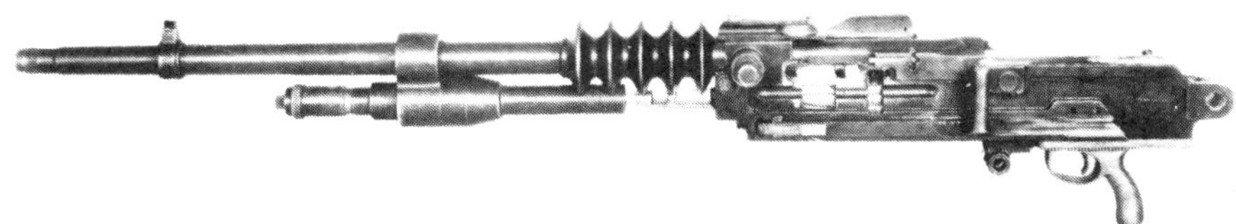

7.9 mm sMG 240(n) without its tripod and shoulder pad attachment

7.62 mm Maschinengewehr 216(r)

German designation 7.92 mm sMG 216(r)
Original designation *Stankovy Pulemyot Maksima obr. 1910 g,* or SPM
Calibre/cartridge 7.62 mm × 54R
Type of feed 250 round fabric belts
Length 1107 mm
Length of barrel 720 mm
Weight of gun 23.8 kg
Weight of carriage mounting 45.2 kg
Weight of quad AA mounting (less guns): 234 kg; (with guns): 460 kg
Muzzle velocity 800 m/sec
Rate of fire (cyclic): 520–600 rpm
Manufacturer Various Tsarist and Soviet State arsenals

Remarks: 1910 Russian version of Maxim machine gun on wheeled Sokolov carriage with frontal armour plate. In production in slightly modified form until 1943. Very large numbers captured by German forces and, despite its age and weight, issued to various garrison and coastal defence units. Also served with Russian and other local formations fighting on German side on Eastern Front.

Captured Soviet mobile quadruple Maksim *AA mount operated by a German crew*

◄ *7.62 mm sMG 216(r) without the usual armour plate*

German troops with a Soviet Maksim obr.1910 *on a sled shortly after its capture in March, 1943* ►

7.62 mm Maschinengewehr 218(r)

German designation 7.62 mm sMG 218(r)
Original designation *Pulemyot obr. 1939 g*
Calibre/cartridge 7.62 mm × 54R
Type of feed 50 and 250 round cloth or metal link belts
Length 1171 mm
Length of barrel 721 mm
Weight of gun 13.55 kg
Weight of tripod and shield 19.2 kg
Muzzle velocity 863 m/sec
Rate of fire 520–580 or 1020–1180 rpm
Manufacturer Various Soviet State arsenals

Remarks: Complex design by Vassili A. Degtyarov with two rates of fire. Withdrawn from production by late 1941 or early 1942. Only limited German service use, mostly as a local battlefield weapon.

7.62 mm sMG 218(r) on the original tripod without an armour plate

sMG 218(r) on the improved lightweight tripod and with frontal armour plate

7.62 mm Maschinengewehr 43(r)*

German designation 7.62 mm sMG 43(r)*
Original designation *Stankovy Pulemyot Goryunova obr. 1943 g* (GS–43)
Calibre/cartridge 7.62 mm × 54
Type of feed 50 round metal link belts
Length 1120 mm
Length of barrel 719 mm
Weight complete 40.4 kg
Weight of gun 13.8 kg
Muzzle velocity 800 m/sec
Rate of fire (cyclic): 600–700 rpm
Manufacturer Various Soviet State arsenals

Soviet Goryunov heavy machine gun 1944 series. The SG-43 was similar

* Designation unconfirmed

Remarks: Designed by Pyotr M. Goryunov late in 1942 as intended replacement for elderly PM1910. In general service by autumn 1943. Although impressed by German troops whenever some were captured, it did not receive an official German 'foreign weapon' designation and combat use appears to have been confined to Eastern Front.

7.65 mm Maschinengewehr 220(b)

German designation 7.65 mm sMG 220(b)
Original designation Mitrailleuse 'Hotchkiss'
Calibre/cartridge 7.65 mm M 30
Type of feed 30 round metal strips
Length 1400 mm
Length of barrel 760 mm
Weight 25 kg
Weight of tripod 25 kg
Muzzle velocity 700 m/sec approx
Rate of fire 450 rpm
Manufacturer Société de la Fabrication des Armes à Feu Portatives Hotchkiss et Cie., St. Denis

Remarks: Version of Hotchkiss machine gun modified to chamber Belgian cartridge. German service use mainly by local occupation troops; some also issued to coastal defence units.

7.7 mm Maschinengewehr 230(e), (r) and 231(h)

German designations 7.7 mm sMG 230(e) oder (r); 7.7 mm sMG 231(h)
Original designations (e): Gun, Machine, Vickers, 0.303 in, Mk 1; (r): Requisitioned from occupied Baltic States by USSR Prov. Soviet designation: 7.7 *mm Vikkers*; (h): Mitrailleur M 18
Calibre/cartridge 7.7 mm (.303 Br)
Type of feed 250 round fabric belts
Length 1156 mm
Length of barrel 721 mm
Weight of gun with water 18 kg
Weight of tripod 22 kg
Muzzle velocity 744 m/sec
Rate of fire (cyclic): 450–500 rpm
Manufacturer Various Royal Ordnance Factories in UK. Vickers', Sons and Maxim, Crayford

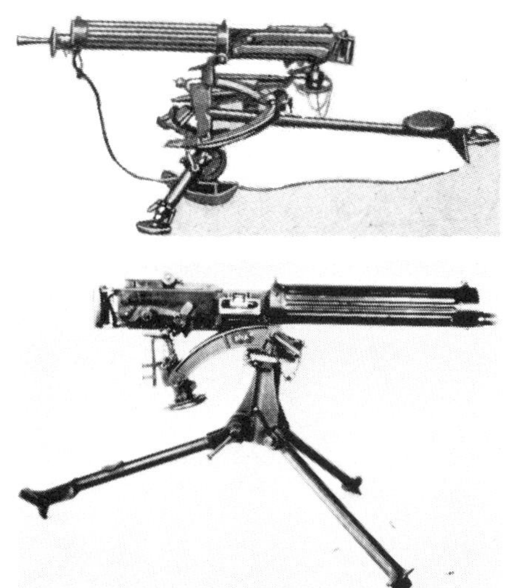

7.7 mm sMG 231(h)

7.7 mm sMG 230(e), the standard Vickers Mk 1

Remarks: Originally known as Vickers-Maxim. Adopted in slightly modified form by British Army in 1912 and served almost unchanged as standard British heavy machine gun in WW I and WW II. Notable for its robustness and reliability. Also widely sold abroad. Large numbers acquired by German forces during Western campaign in 1940, and Greece and Baltic area in 1941. Most captured Vickers guns issued to garrison and coastal defence units, but some also used actively by various local auxiliary formations.

7.9 mm Maschinengewehr 241(h), 242(h) and 243(h)

German designations 7.9 mm sMG 241(h), 242(h) und 243(h)
Original designations 241(h): Mitrailleur M 08; 242(h): Mitrailleur M 08/13; 243(h): Mitrailleur M 08/15
Calibre/cartridge 7.92 mm × 57
Type of feed 215 and 250 round belts
Length 1200 mm
Length of barrel 700 mm
Weight of gun 24 kg
Muzzle velocity 700 m/sec approx
Rate of fire 500 rpm
Original manufacturer Österreichische Waffenfabrik-Gesellschaft, Steyr

Remarks: Standard Schwarzlose machine guns adopted by Dutch armed forces. Heavy version had tripod with half-circle segment for movement. In German service used by occupation troops, some also transferred to other second-line duties, and a number issued to Volkssturm.

7.9 mm Kavallerie-Maschinengewehr 244(h)

German designation 7.9 mm KavMG 244(h)
Original designation Cavaleriemitrailleur M 08/15
Calibre/cartridge 7.92 mm × 57
Type of feed 215 and 250 round belts
Length 1030 mm
Length of barrel 530 mm
Weight 21.5 kg
Muzzle velocity 600 m/sec approx
Rate of fire (cyclic): 500 rpm
Original manufacturer Österreichische Waffenfabrik-Gesellschaft, Steyr

Remarks: Shorter and lighter version of Schwarzlose MG 07/12 adapted for cavalry use. Transported either in a small cart or suspended on a special horse harness. Original tripods were of British manufacture. In German service distributed to second-line and garrison troops; some also used by anti-partisan detachments in Balkan area.

7.9 mm Maschinengewehr 245/1(n) and 245/2(n)

German designation 7.9 mm sMG 245/1 und /2(n)
Original designations 245/1(n): Colt mitraljose m/29I; 245/2(n): Colt mitraljose m/29T
Calibre/cartridge 7.92 mm × 57
Type of feed 250 round belts
Length 1110 mm
Length of barrel 607 mm
Weight of gun with water 18.5 kg
Weight of tripod 28.4 kg
Muzzle velocity 854 m/sec
Rate of fire (cyclic): 500 rpm
Manufacturer Colt's Patent Firearms Manufacturing Co., Hartford, Connecticut

Remarks: Both weapons represented Colt commercial models of Browning M1917 machine gun. Although both variants were adopted by Norwegian forces and eventually passed into German hands, only m/29I (MG 245/1(n)) was chambered for standard German cartridge. German use limited to occupation troops in Norway.

7.9 mm Maschinengewehr 249(p) and 30(p)

German designation 7.9 mm sMG 30(p) oder 249(p)
Original designation CKM wz. 30
Calibre/cartridge 7.92 mm × 57
Type of feed 300 round belts
Length 1110 mm
Length of barrel 715 mm
Weight of gun 17 kg
Weight of tripod 26.5 or 22.5 kg
Muzzle velocity 760 m/sec
Rate of fire (cyclic): 600–700 rpm
Manufacturer Colt's Patent Firearms Manufacturing Co., Hartford, Connecticut

Remarks: Colt commercial model of Browning M1917 machine gun. German use limited to various anti-partisan troops and second-line formations.

sMG 30(p) with AA adapter in action

8 mm Maschinengewehr 255(i)

German designation 8 mm sMG 255(i)
Original designation Mitraglice Fiat modello 1914/35
Calibre/cartridge 8 mm mod. 35
Type of feed 300 round belts
Length 1265 mm
Length of barrel 650 mm
Weight of gun 17.2 kg
Weight of tripod 18.7 kg or 17.2 kg
Muzzle velocity 750 m(sec
Rate of fire (cyclic): 500–600 rpm
Manufacturer Fiat SpA, Torino

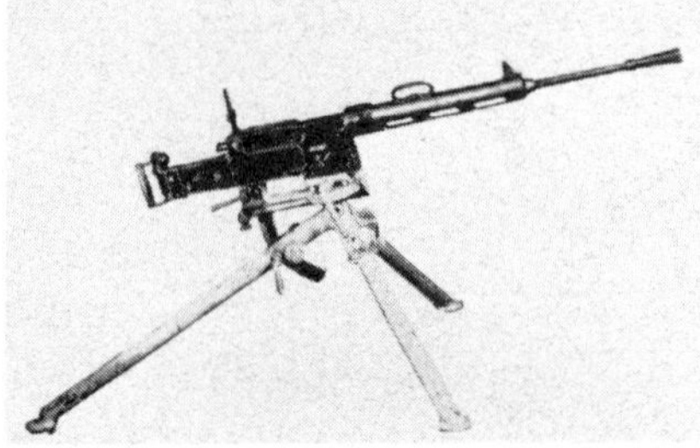

Remarks: Air-cooled version of Mitraglice Fiat 1914, also known as Revelli. Not very suitable for action under adverse conditions due to need to lubricate ammunition feed. Some used by German troops in North Africa and Italy.

8 mm Maschinengewehr 256(f), (g) and (j)

German designation 8 mm sMG 256(f), (g) und (j)
Original designations (f): Mitrailleuse 'St. Etienne' mle 1907; (j): Mitralez 8 mm M 7/15; (g): Model 1907
Calibre/cartridge 8 mm mle 1886D
Type of feed 24 or 30 round metal strips
Length 1180 mm
Length of barrel 710 mm
Weight 23.8 kg
Weight of tripod 26.5 kg
Muzzle velocity 700 m/sec
Rate of fire 400–600 rpm
Manufacturer Manufacture d'Armes de St. Etienne, St. Etienne

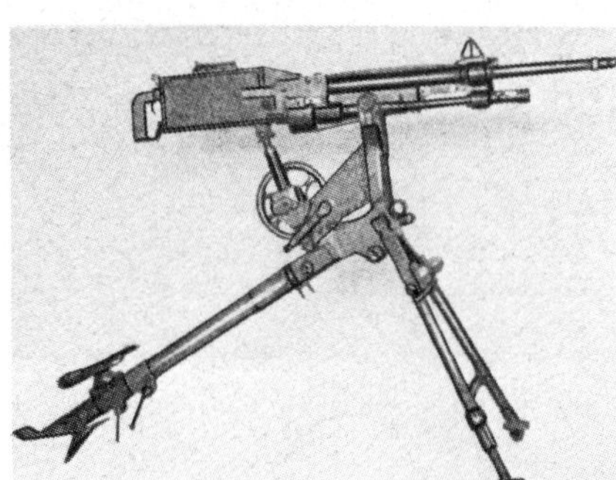

Remarks: Evolved from mle 1905 designed at State Arsenal Puteaux. Basically only minor modification of Hotchkiss machine gun, and not very successful. Gradually replaced in French service already during WW I. Limited numbers impressed by Germans for coastal defence purposes in France.

8 mm Maschinengewehr 257(f), (j), (p) and 7.9 mm sMG 238(p)

German designations 8 mm sMG 257(f), (j) oder (p); 7.9 mm sMG 238(p)
Original designations (f): Mitrailleuse 'Hotchkiss' mle 1914; (j): Mitralez 8 mm M 14 H; (p): Ciezki karabin maszynowy wz. 14 (Hotchkiss)
Calibres/cartridges 7.92 × 57 or 8 mm mle 1886D
Type of feed 24 or 30 round metal strips or 249 round belt of strips joined together
Length with flash hider 1300 mm
Length of gun 1270 mm
Length of barrel 775 mm
Weight 23.6 kg
Weight of tripod 25 kg
Muzzle velocity 725 m/sec approx
Rate of fire 400 rpm
Manufacturer Société de la Fabrication des Armes à Feu Portatives Hotchkiss et Cie., St. Denis

Remarks: Standard French heavy machine gun of WW I, still used in that role in 1939–40. Heavy and bulky, but a reliable weapon in action. German use of captured guns mostly for coastal defence in France.

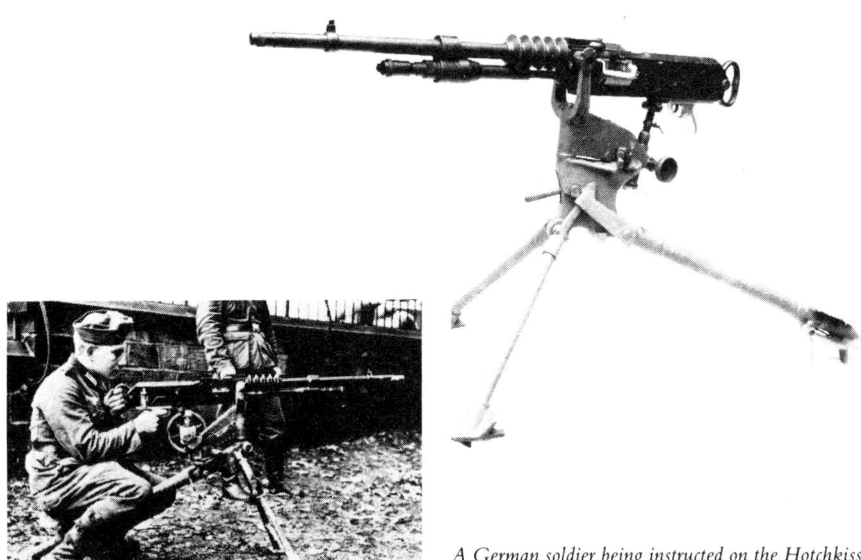

A German soldier being instructed on the Hotchkiss mle 1914

---

8 mm Maschinengewehr 259(i)

German designation 8 mm sMG 259(i)
Original designation Mitraglice Breda modello 37
Calibre/cartridge 8 mm mod. 35/38
Type of feed 20 or 25 round metal strips
Length 1270 mm
Length of barrel 780 mm
Weight 19.4 kg
Weight of tripod 18.8 kg
Muzzle velocity 780 m/sec
Rate of fire (cyclic): 440–460 rpm
Manufacturer Societa Anonima Ernesto Breda, Brescia

Remarks: Standard Italian heavy machine gun of WW II. A very reliable weapon, despite unusual ammunition feed. Used mainly by German units in Italy.

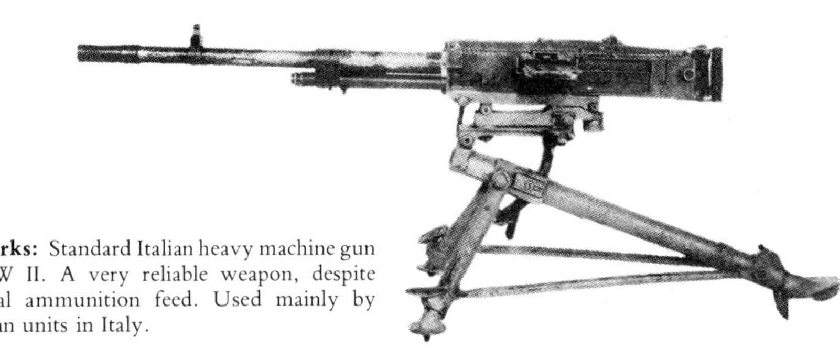

12.7 mm Maschinengewehr 268(r)

German designation 12.7 mm sMG 268(r)
Original designation *12.7 mm DShK obr. 1938 g*
Calibre/cartridge 12.7 mm × 107
Type of feed 50 round steel belts
Length 1602 mm
Length of barrel 1002 mm
Weight of gun 34 kg
Weight of carriage and shield 134 kg
Muzzle velocity 850 m/sec
Rate of fire 550–600 rpm
Manufacturer Various Soviet State arsenals

Remarks: Designed by Vasili A. Degtyarov and Georgi S. Shpagin as heavy support weapon and tank machine gun, but also adapted for anti-tank and AA use. As support weapon fitted on a wheeled carriage with frontal armour plate. Notable as first Soviet heavy machine gun built in quantity. German use limited only by ammunition supply.

Captured 12.7 mm DShK in German service ►

A version of the DShK heavy machine gun with shoulder pads ▼

13.2 mm Maschinengewehr 271(f)

German designation 13.2 mm sMG 271(f)
Original designation Mitrailleuse Hotchkiss de 13.2 mm mle 1930
Calibre/cartridge 13.2 mm Hotchkiss
Type of feed 30 round box magazine or 15 or 20 round metal strips
Length 1670 mm
Length of barrel 1000 mm
Weight of gun 37.5 kg
Weight of wheeled carriage 155 kg
Weight of tripod 43 kg
Weight of single AA mounting 160 kg
Weight of twin AA mount (approx): 300 kg
Muzzle velocity 800 m/sec
Rate of fire 450–480 rpm
Manufacturer Société de la Fabrication des Armes à Feu Portatives Hotchkiss et Cie., St. Denis

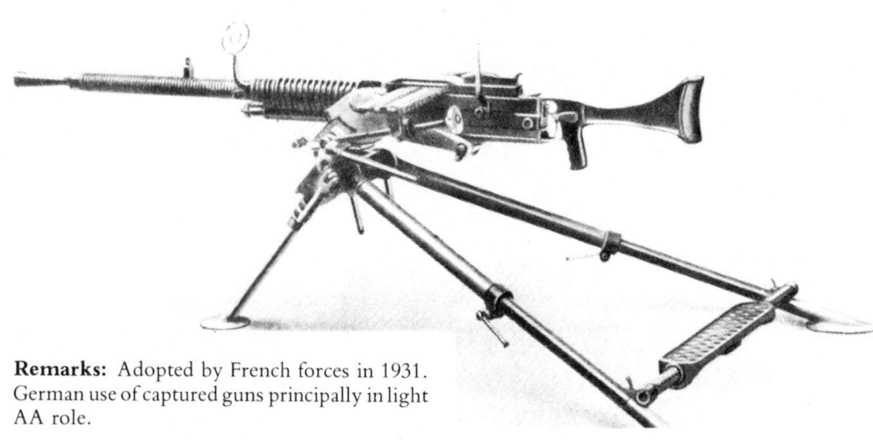

Remarks: Adopted by French forces in 1931. German use of captured guns principally in light AA role.

7.9 mm Kampfwagen-Maschinengewehr 341(e)

German designation 7.9 mm KpfwMG 341(e)
Original designation Gun, Machine, Besa, 7.92 mm Mks 1 to 3
Calibre/cartridge 7.92 mm × 57
Type of feed 225 round belts
Length 1110 mm
Length of barrel 736 mm
Weight 21.15 kg
Muzzle velocity 823 m/sec
Rate of fire (cyclic): 450 or 700–750 rpm
Manufacturer Birmingham Small Arms Co. Ltd., Redditch

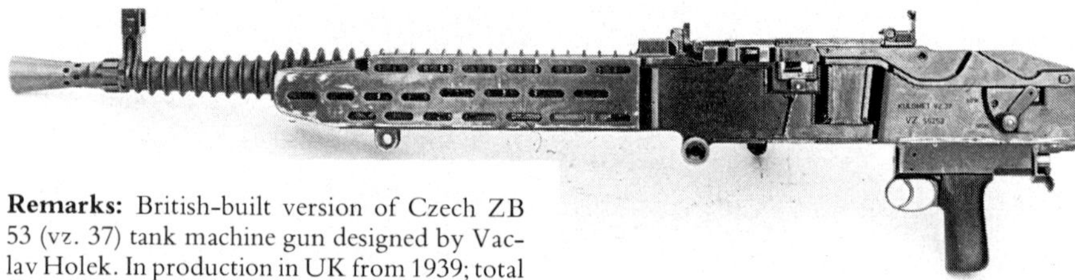

Remarks: British-built version of Czech ZB 53 (vz. 37) tank machine gun designed by Vaclav Holek. In production in UK from 1939; total of 59,322 made in three versions during WW II. Unusual for British service by being chambered for German cartridge. Only localised German use, mainly in fixed installations.

7.92 mm KpfwMG 341(e). Illustration shows the original Czech version

8 mm Kampfwagen-Maschinengewehr 350(i)

German designation 8 mm KpfwMG 350(i)
Original designation Mitraglice Breda modello 38 per carro armati
Calibre/cartridge 8 mm mod. 35
Type of feed 24 round box magazine
Length 897.5 mm
Length of barrel 600 mm
Weight 16.3 kg
Weight of tripod 23 kg
Muzzle velocity 770 m/sec
Rate of fire (cyclic): 600 rpm
Manufacturer Societa Anonima Ernesto Breda, Brescia

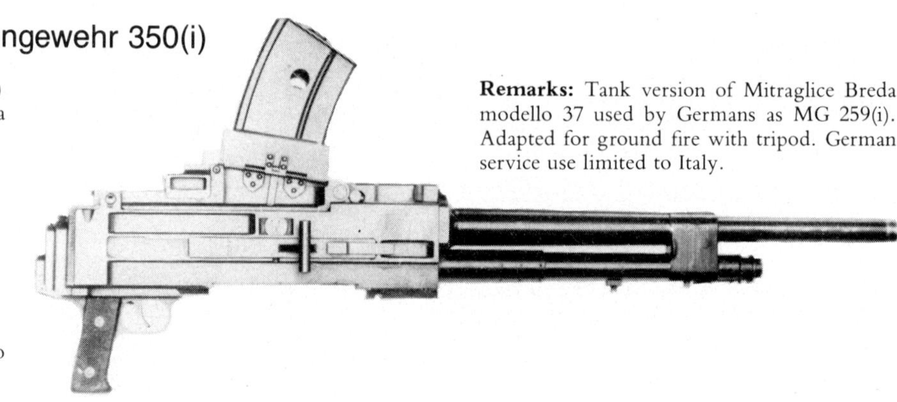

Remarks: Tank version of Mitraglice Breda modello 37 used by Germans as MG 259(i). Adapted for ground fire with tripod. German service use limited to Italy.

15 mm Kampfwagen-Maschinengewehr 376(e)

German designation 15 mm KpfwMG 376(e)
Original designation 15 mm Besa Tank Machine Gun Mk 1
Calibre/cartridge 15 mm
Type of feed 25 round steel belts
Length 2050 mm
Length of barrel 1463 mm
Weight 56.5 kg
Muzzle velocity 819 m/sec
Rate of fire (cyclic): 400–450 rpm
Manufacturer Birmingham Small Arms Co. Ltd., Redditch

Remarks: British licence-built version of Czech ZB vz. 60. Production commenced in 1939, first deliveries May 1940. Total of 3.218 made by BSA during WW II. Only limited German use.

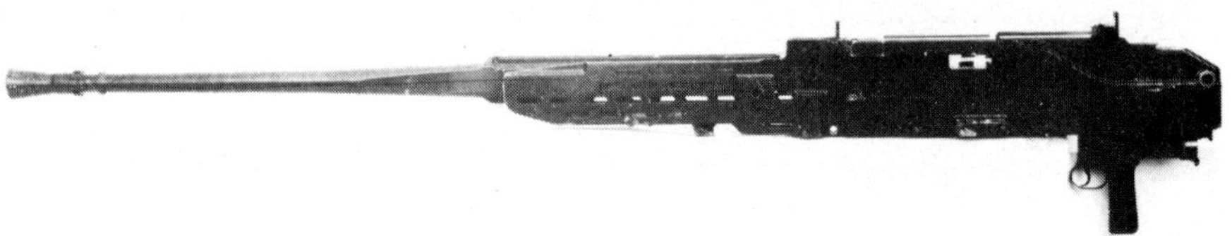

ML 16

German designation (design): ML 16
Calibre 16 mm
Type of feed 20 round box magazines
Number of barrels 8
Weight 180 kg
Muzzle velocity 900 m/sec
Rate of fire 300 rpm
Manufacturer Mauser-Werke AG

Remarks: A multi-barrel light AA weapon intended for German ground forces. Still in experimental stage when hostilities ended in Europe.

Zielfeuer Gerät 38

Remarks: The ZfG 38 was originally intended as an infantry training weapon which could be used to fire accurately along fixed lines above the heads of infantry gaining battle experience. In a more warlike role it was also used as a 'booby-trap' device which could be fired via a lanyard attachment to either the front or the rear of the trigger mechanism fixed above the receiver. The operation appears to have been purely blow-back, an assumption which is supported by the fact that the seventy 7.92 mm × 57 rounds in the magazine had to be specially lubricated. No other data available.

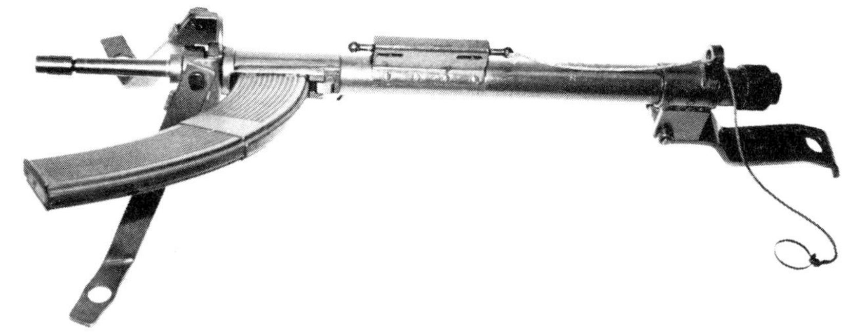

Anti-tank rifles

Germany was the first country to develop and use an anti-tank rifle. The first tanks on the Somme battlefields of late 1916 came as a very unpleasant surprise to the German Army and to provide some form of defence against the new weapon, Mauser produced a large 13 mm single-shot rifle firing steel shot. As a form of anti-tank defence the '*Tank-Gewehr*' was quite successful but its large powerful cartridge (which was used as the basis for the later Browning 0.5 in cartridge) produced a very heavy recoil and firing it needed considerable nerve. Many broken collar-bones were inflicted on the gunners, but as the result was often a perforated tank the side-effects were thought worthwhile.

After 1919 the anti-tank rifle fell into a temporary decline until the mid-1930s. Once again the question of some form of infantry defence against armour arose and one of the answers produced was the anti-tank rifle. Rheinmetall-Borsig appear to have been one of the first in this field, and seem very likely to have been the originators of the *PzB 38*, which was a rifle firing a 13 mm cartridge necked down to 7.92 mm. The 7.92 mm bullet used a steel core and some projectiles even included a small lachrymatory capsule intended to incommode the occupants of a pierced tank. This chemical feature was soon proved useless but it provides once again an example of the German tactical designers seizing upon unlikely but novel weapons. The *PzB 38* itself was a satisfactory weapon but it was complex. For example, the breech mechanism resembled that of a miniature artillery piece, and was opened after firing by the recoil forces. Such a complex device was soon proved to be not worth the effort involved in producing it and carrying it around.

The next anti-tank rifle was the *PzB 39*. Although it cannot be stated for certain, Gustloff of Suhl appear to have been the originators of this design, but some were produced by Rheinmetall and Steyr-Daimler-Puch. The *PzB 39* was a much simpler weapon than its predecessor and fired single shots only. The breech was opened by lowering the pistol grip and extra rounds were held in a box on the right of the receiver. At first, the same steel-cored 7.92 mm bullets were fired as those used on the *PzB 38*, but the Polish campaign of 1939 produced examples of the projectiles fired by the Polish Marosczek anti-tank rifles. These projectiles had a tungsten core and despite the general tungsten shortage in Germany the new core was adopted as its denser metallic structure made armour penetration more certain.

But almost as soon as large numbers of the *PzB 39* entered service, tank armour increased in thickness too. Despite the new tungsten cores the relatively light projectiles could not penetrate the armour belts used on most contemporary tanks and the anti-tank rifle fell into gradual disuse as a combat weapon. Many were passed to second-line units but even there they were no more than an awkward load and were returned to their depots and scrapped. From late 1941 onwards numbers of the *PzB 39* were converted into grenade launchers by fitting a '*Schiessbecher*' grenade cup to the muzzle. In this form they became the *GrB 39* and could fire the normal run of German rifle grenades.

But the story of German anti-tank rifles did not end with the *PzB 39*. German money invested in the Swiss Solothurn concern had influenced the Swiss designers also to develop an anti-tank rifle and in 1941 the Solothurn *M SS 41* was produced. This weapon was a very advanced design, and its German influence was notable in the use of an *MG 34* bipod. A semi-automatic action was used along with a six-round magazine. By the time the *M SS 41* was ready for service the anti-tank rifle was on its way out as a viable weapon and very few were produced. Some were used in North Africa, but thereafter they were relegated to second-line units; small numbers are believed to have been issued to the Volkssturm in late 1944.

Another Swiss product purchased by the Germans was the Solothurn *S 18-1100*, a large 20 mm weapon. Although relatively few were bought by Germany they were used as front-line weapons until 1942, as their heavier projectiles made them much

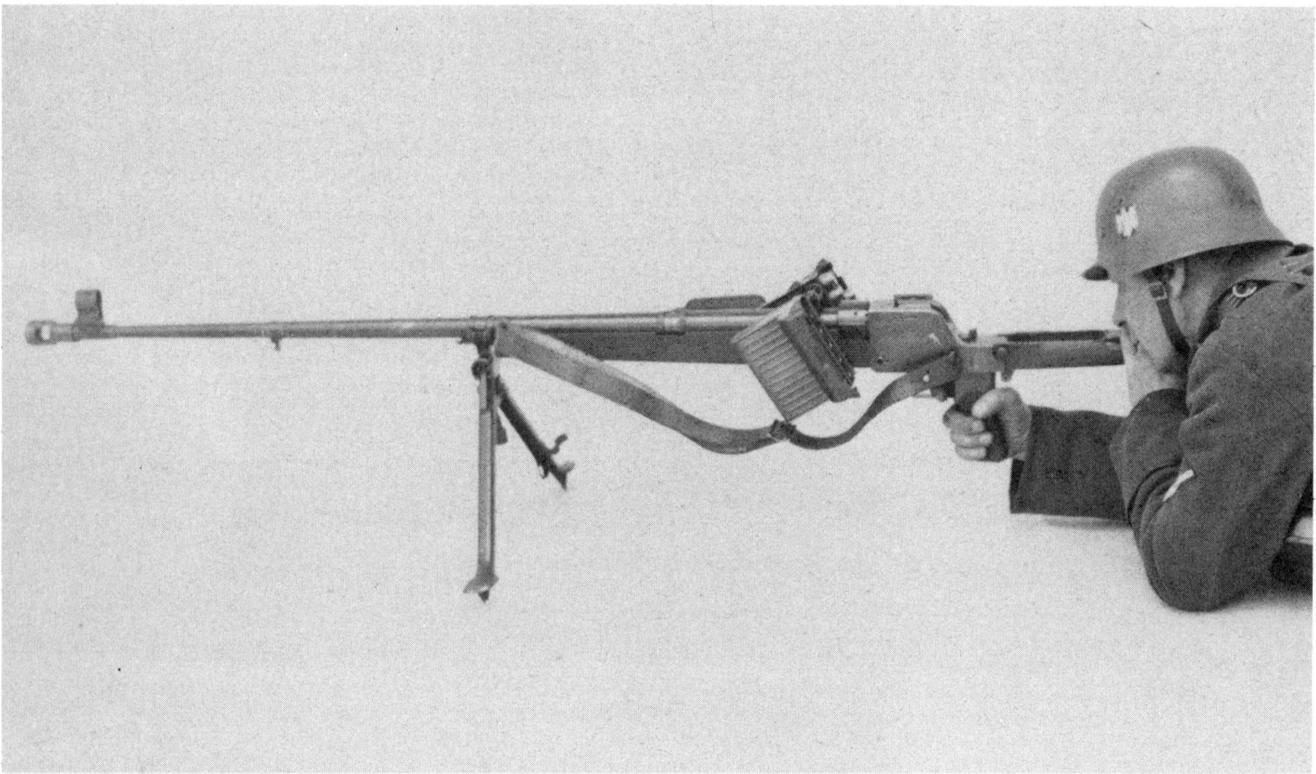

7.92 mm PzB 39, a simplified version of the PzB 38 with manually-operated ejection action

more effective against tanks. Their weight and bulk dictated the use of a low two-wheeled carriage. The bulk of these German weapons appear to have been passed over to the Italian Army, and the same seems to have happened to anti-tank rifles captured from Holland in 1940. Some were also delivered to Hungary and Romania while they were German allies on the Eastern Front.

German influence could also be seen in the Polish Marosczek anti-tank rifle, which was a much modified and lightened version of the old Mauser '*Tank-Gewehr*'. Many of these weapons, chambered to take the same 13/7.92 mm cartridge as used by the Germans, were captured during the 1939 Polish campaign and, as has already been stated, their tungsten-cored ammunition greatly influenced future German ammunition design. A substantial number of these anti-tank rifles were passed to the Italian Army after 1941.

Other captured weapons included British Boys anti-tank rifles acquired after the Dunkirk withdrawal, and later large numbers of Soviet anti-tank rifles. These Soviet rifles were of two types, the PTRD 1941 and PTRS 1941. Both fired the same 14.5 mm projectile and were thus much more viable weapons than the smaller 7.92 mm rifles. The Soviets continued to keep the anti-tank rifle in service until 1945 and the Germans also used as many captured examples on the Eastern Front as they could obtain. Both rifles were long, heavy devices but proved very effective against light armour and soft-skin vehicles.

Despite the gradual decline of the anti-tank rifle the Germans still continued to develop and produce a large number of experimental weapons. One of these was the Mauser *EW 141,* or *Gerät 318*, which fired the conventional 13/7.92 mm round, although by the time it was produced tungsten cores had been introduced. About the same time a design study for the *PzB 40* was initiated. Walther (Zella-Mehlis), Gustloff (Suhl) and Krieghoff (Suhl) all submitted prototypes but there was no production contract. From this proposal was evolved the *PzB 243,* again using the 13/7.92 mm round, and Gustloff and Krieghoff both produced prototypes. The Waffenwerke Brünn (Brno) also submitted a prototype, but by the time all the submissions were ready the need for a rifle firing the normal calibre bullet had passed. This did not stop development, and the 13 mm *PzB 244* was submitted. However, in the meantime it was decided to produce a 15 mm weapon to be known as the *PzB 42*. Again, Gustloff, Krieghoff and Waffenwerke Brünn all entered prototypes, but the whole series was abandoned some time in 1942.

As always with German research, the lack of a tactical need did not preclude experimentation. Tapered barrels were tried on some *PzB 39* conversions; these barrels tapered from 14 to 10 mm. A F. Janecek at the Waffenfabrik Prag (Prague) carried out research using barrels tapered from 11 to 7.92 mm and 15 to 11 mm. Tapered barrels were also tried on the *EW 141*. Other experiments were carried out into using more efficient muzzle brakes on anti-tank rifles.

Perhaps the most unlikely experimental work was directed to producing an anti-tank machine gun. This would have been the *MG 141*, using the normal 13/7.92 mm cartridges. Although this machine gun would have been of little use against tanks it would have made a powerful heavy machine gun and given the Germans a useful counter to the Browning 0.5 in and Soviet DShK machine guns. Work on the *MG 141* project started in March 1940 and both Mauser and Gustloff submitted design studies but the whole idea appears to have been dropped.

7.92 mm Panzerbüchse 38

German designation 7.92 mm PzB 38
Calibre 7.92 mm
Calibre of cartridge case 13 mm
Length of stock (extended): 1615 mm
(folded): 1293 mm
Length of barrel 1085 mm
Weight 16.2 kg
Muzzle velocity 1140 m/sec
Armour penetration 25 mm at 300 m (30°)
Manufacturer Rheinmetall-Borsig AG, Düsseldorf

Remarks: Single-shot weapon with automatic ejector. Trial series of 1600 produced and used in service by German troops but not adopted by Army authorities. Fired hardened steel core ammunition; later also tungsten-cored rounds.

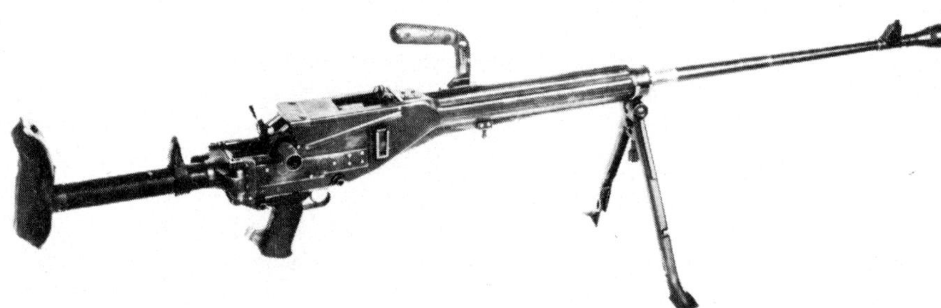

7.92 mm Panzerbüchse 38 *single-shot anti-tank rifle with automatic extraction mechanism*

7.92 mm Panzerbüchse 39

Rear view of the PzB 39 showing the breech and the two clipped-on ammunition boxes holding 10 rounds each

PzB 39 in the Western Desert

German designation 7.92 mm PzB 39
Calibre 7.92 mm
Calibre of cartridge case 13 mm
Length of stock (extended): 1620 mm
(folded): 1280 mm
Length of barrel 1085 mm
Weight 12.6 kg
Muzzle velocity 1140 m/sec
Armour penetration 25 mm at 300 m (30°)
Manufacturers Gustloff-Werke, Suhl; Rheinmetall-Borsig AG, Düsseldorf; Steyr-Daimler-Puch AG, Wien

Remarks: Single-shot weapon with sliding breech. Adopted for service without troop trials and ordered into large-scale production. Used tungsten-cored ammunition after 1939.

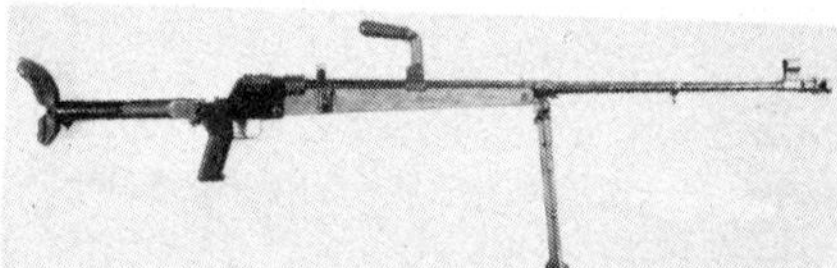

7.92 mm PzB 39, a simplified version of the PzB 38 with manually-operated ejection action

7.92 mm Granatbüchse 39

GrB 39 loaded with the grosse Gewehr-Panzergranate

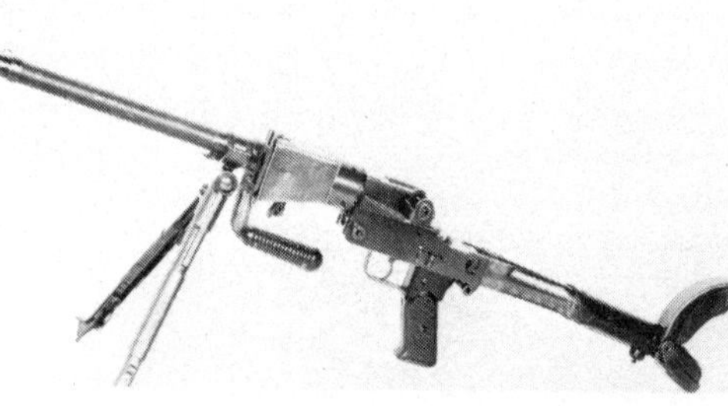

German designation 7.92 mm GrB 39
Calibre of launcher cup 30 mm
Length of stock (extended): 1232 mm
(folded): 908 mm
Length of barrel 613 mm
Length of barrel with cup 749 mm
Weight (approx): 10.44 kg
Effective range, moving targets 75 m
Effective range, static targets 125 m
Weight of anti-tank grenade 0.25 kg or 0.383 kg
Weight of HE grenade 0.255 kg

Remarks: Converted PzB 39 with 'Schiess-becher'.

7.92 mm Granatbüchse 39, *a modified PzB 39 fitted with the* Schiessbecher *grenade launcher. Three types of grenades could be fired from this equipment: the anti-personnel,* Gewehr-Sprenggranate *small anti-tank* Gewehr-Panzergranate *and the large anti-tank* grosse Gewehr-Panzergranate

7.92 mm M SS 41

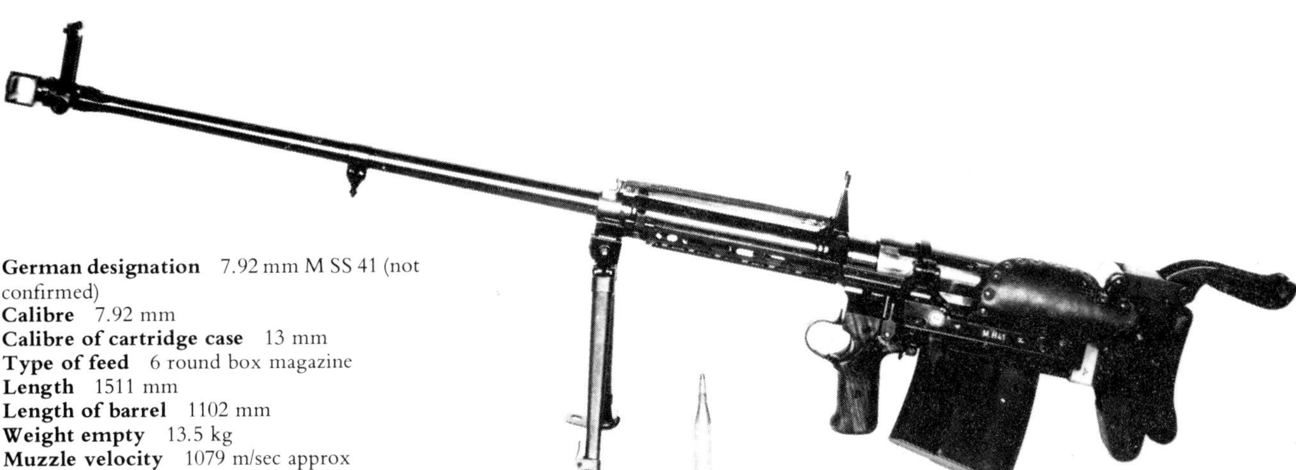

7.92 mm M SS 41 with a sample round

German designation 7.92 mm M SS 41 (not confirmed)
Calibre 7.92 mm
Calibre of cartridge case 13 mm
Type of feed 6 round box magazine
Length 1511 mm
Length of barrel 1102 mm
Weight empty 13.5 kg
Muzzle velocity 1079 m/sec approx
Armour penetration 20 mm at 300 m (0°) approx
Original manufacturer Waffenfabrik Solothurn AG, Solothurn, Switzerland

Remarks: Produced in Switzerland in 1941 to German specifications. Only small number delivered to Germany and used by German troops on Eastern Front and in North Africa. Designed to use MG 34 bipod as support.

7.92 mm Panzerbüchse 40(G)

German designation 7.92 mm PzB 40(G)
Calibre 7.92 mm
Type of feed 8 round magazine
Length in action 1660 mm
Length when carried 1460 mm
Length of barrel 1085 mm
Length of rifling 1000 mm
Weight 13.5 kg
Muzzle velocity 1150 m/sec
Manufacturer Gustloff-Werke, Suhl

Remarks: Experimental only.

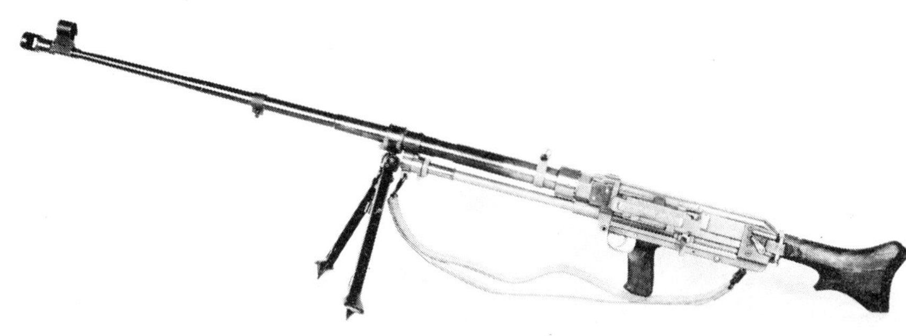

7.92 mm Panzerbüchse 40(K)

German designation 7.92 mm PzB 40(K)
Calibre 7.92 mm
Type of feed 8 round magazine
Length in action 1570 mm
Length when carried 1400 mm
Length of barrel 1085 mm
Length of rifling 1000 mm
Weight (approx): 14 kg
Muzzle velocity 1150 m/sec
Manufacturer Heinrich Krieghoff, Suhl

Remarks: Experimental only.

7.92 mm Panzerbüchse 40(W)

German designation 7.92 mm PzB 40(W)
Calibre 7.92 mm
Type of feed 8 round magazine
Length in action 1770 mm
Length when carried 1570 mm
Length of barrel 1085 mm
Length of rifling 1000 mm
Weight 14.5 kg
Muzzle velocity 1150 m/sec
Manufacturer Carl Walther Waffenfabrik AG, Zella-Mehlis

Remarks: Experimental only.

7.92 mm Panzerbüchse 41

German designation 7.92 mm PzB 41
Calibre 7.92 mm
Type of feed 8 round magazine
Length 1670 mm
Length of barrel 1085 mm
Length of rifling 1000 mm
Weight 12.5 kg
Weight of barrel (approx): 5 kg
Muzzle velocity 1150 m/sec
Manufacturer Mauser-Werke AG, Oberndorf-am-Neckar

Remarks: Experimental gas-operated self-loading weappn. Total of 14 made and tested in 1941; development stopped in October of same year. A 5.6 mm barrel was also produced and tested with this rifle, but its use appears to have been for training only.

EW 141 (Gerät 318)

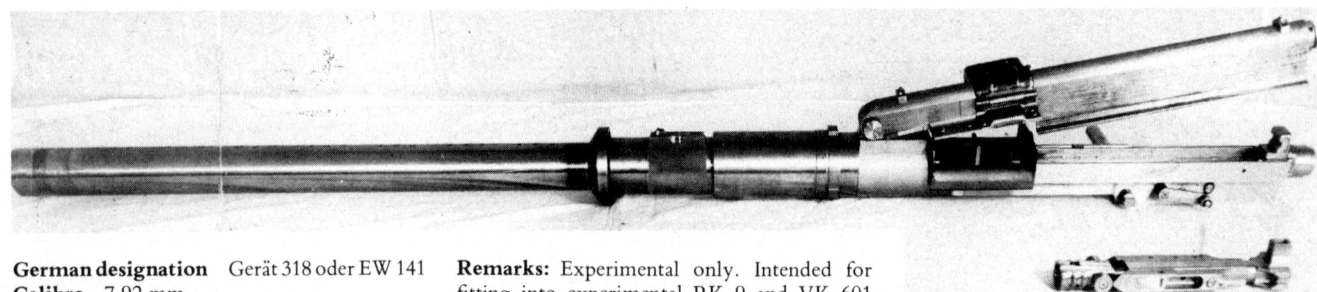

German designation Gerät 318 oder EW 141
Calibre 7.92 mm
Calibre of cartridge case 13 mm
Type of feed Metal belt
Length 1670 mm
Length of barrel 1085 mm
Weight 30.4 kg
Muzzle velocity 1170 m/sec
Armour penetration 25 mm at 300 m (30° angle)
Manufacturer Mauser-Werke AG, Berlin

Remarks: Experimental only. Intended for fitting into experimental RK 9 and VK 601 reconnaissance cars. Development started in July 1938 and weapon was approved for production, but none were manufactured. Although EW 141 was designed for armoured vehicles it could be dismounted onto a bipod for ground use.

Maschinengewehr 141(M)

German designation MG 141(M)
Calibre 7.92 mm or 13 mm
Type of feed Metal belt
Length 1815 mm
Length of barrel 1000 mm
Weight 7.92 mm: 25.5 kg; 13 mm: 27.3 kg
Muzzle velocity 7.92 mm; 1170 m/sec; 13 mm: 1340 m/sec
Rate of fire (cyclic): 850 rpm
Manufacturer Mauser-Werke AG, Berlin

Remarks: Experimental only. Development commenced during September 1937. Parallel MG 141(G) evolved by Gustloff-Werke of Suhl was essentially similar but equipped with a device to limit bursts to only six shots. First examples for comparison tests completed in March 1940; no production.

7.92 mm Panzerbüchse 35(p), 770(p) and (i)

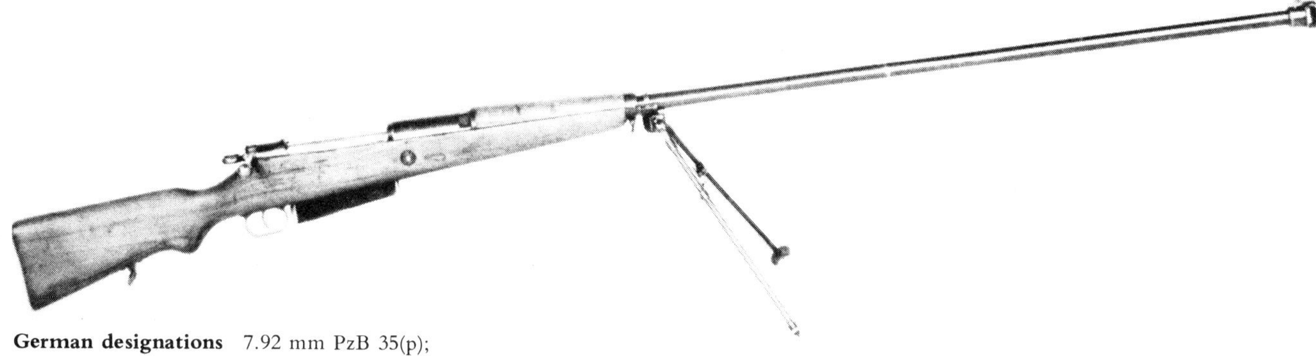

German designations 7.92 mm PzB 35(p); 7.92 mm PzB 770(p) oder (i)
Original designations (p): Karabin przeciwpancerny wz.35–Marosczek; (i): Fucile anticarro
Calibre 7.92 mm
Calibre of cartridge case 13 mm
Type of feed 4 round magazine
Length 1760 mm
Length of barrel 1200 mm
Weight 9.1 kg
Muzzle velocity 1280 m/sec
Armour penetration 25–33 mm at 300 m (0°)
Original manufacturer Fabryce Karabinow w Warsawie

Remarks: Developed in 1935; design based on German Mauser T-Gewehr of 1918. Fired tungsten-cored ammunition which was adopted by German forces after 1939. Barrel life limited to 200 rounds. In 1941–42 most ex-Polish weapons of this type handed over to Italian 8th Army on Eastern Front.

13.9 mm Panzerabwehrbüchse 782(e)

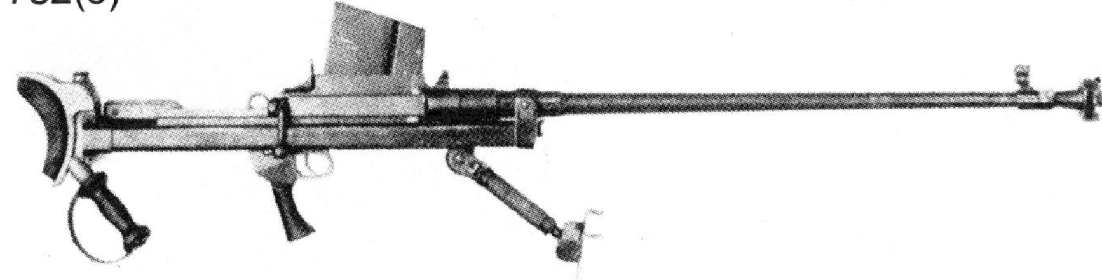

German designation 13.9 mm PzB 782(e)
Original designation Rifle, Anti-tank, 0.55-in Boys Mk 1
Calibre 13.97 mm
Type of feed 5 round magazine
Length 1614 mm
Length of barrel 915 mm
Weight 16.56 kg
Muzzle velocity 990 m/sec
Armour penetration 21 mm at 300 m (0°)
Manufacturer Royal Small Arms Factory, Enfield Lock

Remarks: Developed in 1934; originally known as Stanchion Gun. Used steel core ammunition. Effective only against lighter armoured vehicles; obsolete by 1941. Only limited German service use.

14.5 mm Panzerabwehrbüchse 783(r)

German designation 14.5 mm PzB 783(r)
Original designation *Protivotankovoye Ruzhye Degtyarova obr. 1941 g* (PTRD-41)
Calibre 14.5 mm
Type of feed Single shot
Length 2020 mm
Length of barrel 1350 mm
Weight 17.3 kg
Muzzle velocity 1012 m/sec
Armour penetration 30 mm at 100 m (0°); 27.5 mm at 300 m (0°); 25 mm at 500 m (0°)
Manufacturer Various Soviet State arsenals

Remarks: Manual bolt action. Fired Type B 32 M1932 and BS 41 M1941 armour-piercing ammunition. Effective only against lighter German tanks and armoured personnel carriers. Most PTRD-41s gradually withdrawn during 1942 but certain number remained in service with selected detachments. Only limited German service use on Eastern Front.

14.5 mm Panzerabwehrbüchse 784(r)

German designation 14.5 mm PzB 784(r)
Original designation *Protivotankovoye Ruzhye Simonova obr. 1941 g* (PTRS-41)
Calibre 14.5 mm
Type of feed 5 round magazine
Length 2108 mm
Length of barrel 1216 mm
Weight 20.9 kg
Muzzle velocity 1012 m/sec
Armour penetration 30 mm at 100 m (0°); 27.5 mm at 300 m (0°); 25 mm at 500 m (0°)
Manufacturer Various Soviet State arsenals

Remarks: Semi-automatic action. Fired API Type BS 41 M1941 armour-piercing ammunition. Effective only against lighter German tanks and armoured personnel carriers. After 1941 used mainly as support weapon in street fighting and by partisans against road/rail transport.

---

2 cm Panzerabwehrbüchse 785(s), (i) and (h)

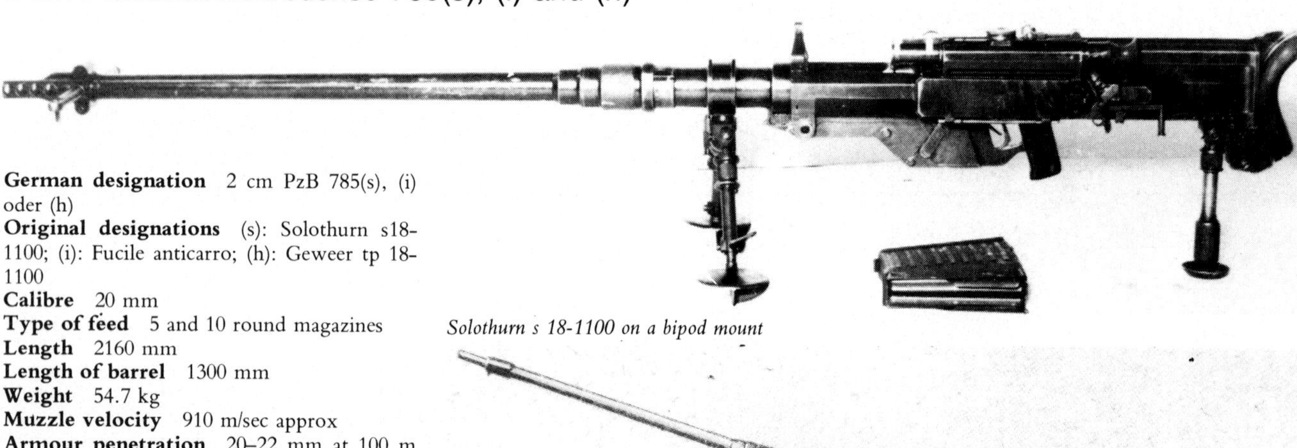

Solothurn s 18-1100 on a bipod mount

German designation 2 cm PzB 785(s), (i) oder (h)
Original designations (s): Solothurn s18-1100; (i): Fucile anticarro; (h): Geweer tp 18-1100
Calibre 20 mm
Type of feed 5 and 10 round magazines
Length 2160 mm
Length of barrel 1300 mm
Weight 54.7 kg
Muzzle velocity 910 m/sec approx
Armour penetration 20–22 mm at 100 m (30°); 15–18 mm at 300 m (30°)
Manufacturer Waffenfabrik Solothurn AG, Solothurn, Switzerland

Remarks: Well-designed semi-automatic weapon produced for export in 1936; also known as M.36. Could be fired from two-wheel carriage or a small bipod. Only limited numbers in German service.

s 18-1100 on a wheeled carriage

Anti-tank guns

It has always been one of the fundamental truths of warfare that whenever a new weapon appears, another weapon is developed to counter it. A typical example of this was seen when the first tanks lumbered across the Somme battlefields in 1916. In a very short time the tank was met by the first anti-tank rifles, massed field gun fire and the simple expedient of reversing rifle bullets. As the first tanks had relatively thin armour these expedient measures were quite successful, but combat experience soon led to an increase in tank armour, and better quality armour, so that the long term answer could only be a specialised type of gun delivering a solid hard shot at high velocities. This new type of gun became the anti-tank gun, and from the appearance of the first anti-tank gun there began a technological race of anti-tank gun versus armour.

Forbidden by the Versailles Treaty to develop her own armoured vehicles, Germany initially studied instead the problem of anti-tank defence and came up with what was a very good answer for that time. Starting in 1924, Rheinmetall-Borsig developed a 37 mm anti-tank gun and the first examples were issued for service in 1928. Known as the *3.7 cm Pak L/45* it was designed for horse traction although for short distances it was intended that men, using special harnesses, would drag the gun into and out of position. For its day, the *3.7 cm Pak* was an excellent weapon. It was low, light, easy to conceal and could penetrate the armour of almost any tank then in service. The basic design greatly influenced anti-tank gun design elsewhere (the American 37 mm Antitank Gun M3 was an example of this), and many were exported.

By the early 1930s it had become obvious that horse traction was on its way out as the petrol engine showed itself to be more suitable for military tasks and the *3.7 cm Pak* was adapted to take new magnesium-alloy wheels fitted with pneumatic tyres. By 1934 the new model began to enter service and the old spoke-wheel guns were withdrawn and updated. The new version was designated the *3.7 cm Pak 35/36* and, like the earlier model, it was widely sold abroad to such nations as Italy, Turkey, Holland and even the Soviet Union, where the design was copied and eventually enlarged to 45 mm calibre. In Germany it was built in very large numbers, and by 1941 over 15,000 had been delivered from the production lines of Rheinmetall-Borsig and numerous other contractors. The *Pak 35/36* was 'blooded' in Spain (and was adopted by the Spanish Nationalist Army) where it acquitted itself well, and it also performed well enough in the early campaigns of 1939 and 1940. But it was during the campaign in France in May 1940 that it became apparent that the *Pak 35/36* was unable to penetrate the armour of many of the British and French tanks, and by 1941, when the *Pak 35/36* encountered the Soviet T-34, it was painfully obvious that the 3.7 cm gun had become obsolete. It simply did not have the power or shot weight to penetrate the ever-increasing armour carried by the new tanks, and so it was progressively withdrawn and issued for a while to various second-line and training units. Many had their barrels replaced by a variety of 75 mm barrels to become infantry guns while temporarily the useful life of the *Pak 35/36* was extended by the use of the *3.7 cm Stielgranate 41*. This was a stick bomb fitted over the muzzle of the gun and propelled by a blank cartridge. The bomb was fin-stabilised and had a hollow-charge warhead for use against armour and concrete obstacles. The range of this bomb was limited but it helped to make the *Pak 35/36* useful until the end of the war.

The approaching obsolescence of the *3.7 cm Pak 35/36* had been foreseen by the German military experts as early as 1936, and in 1937 work had begun on a new 50 mm anti-tank gun. Rheinmetall-Borsig were the designers and the first examples reached the troops in April 1940. The new gun, the *5 cm Pak 38*, was destined to become one of the most widely-used and versatile guns in the German armoury. The basic 5 cm gun was used not only as an anti-tank gun but also as a tank gun, coastal defence gun, aircraft gun and even an anti-aircraft gun. Like the *3.7 cm Pak 35/36*, the *5 cm Pak 38* was low and light and it used a light, split-trail carriage. For ease in manhandling an extra dolly wheel was added under the trail spades. A wide range of ammunition was developed for this gun, of which the most powerful anti-tank projectile was the tungsten-cored *AP40*. During the early days of the invasion of Soviet territories in 1941 this was the only projectile able to penetrate the armour of the T-34, but by that date it was also realised that the *Pak 38* would be unable to counter the next generation of tank armour and a new gun was already on its way into service.

This was the *7.5 cm Pak 40*. Once again, the German war planners had rightly foreseen a need for a gun heavier than the *5 cm Pak 38* and work on a replacement had begun in 1939. The first examples reached the field in late 1941 and from then on the *7.5 cm Pak 40* was one of the most important anti-tank guns in German service. As before, Rheinmetall-Borsig was the main designer and, to speed up development, the *Pak 40* was virtually a scaled-up *Pak 38*. There were many differences, one of the more important being that the use of various light alloys had to be abandoned in place of steel as a result of the raw material supply situation which was by then, 1940, beginning to cast its shadow. As a result the *Pak 40* was proportionally much heavier than the earlier guns and to further simplify production such items as the curved shield of the *Pak 38* were replaced by flat angled plates. A wide range of ammunition was developed for the *Pak 40* and it too was diversified into a variety of roles. By 1945 the basic *Pak 40* had become a tank gun, aircraft gun, coastal defence gun and even a light field gun. Many were placed on self-propelled '*Panzerjäger*' carriages, but it was as an anti-tank gun that the *Pak 40* made its mark. The *Pak 40* was hard-hitting, easy to conceal and, at RM12,000, relatively cheap.

The *Pak 40* was desperately needed during 1941 when the T-34 appeared on the Eastern Front battlefields and proved itself almost invulnerable to all the German anti-tank guns other than the *5 cm Pak 38* firing *AP 40* shot. As always in an emergency there were never enough of these guns when they were needed and the call went out for more powerful weapons. But the *Pak 40* took time to get into production and despite great efforts no appreciable numbers reached the field until late 1941. In the meantime, the Soviet threat was met by a typical German stop-gap solution. During the campaigns of 1940 large quantities of the venerable French mle 1897 field gun had fallen into German hands. Many of these were taken into service as they were, but large numbers were carefully stockpiled for possible future need. That need came in 1941 when it was decided to convert some seven hundred mle 1897 barrels into anti-tank guns. To add strength, four extra hoops were sweated on to the outside of the barrel and a Solothurn 'pepper-pot' muzzle brake was added. Most of these modified barrels were placed on *Pak 38* carriages but some were also combined with *Pak 40* carriages awaiting their own barrels held up on the production lines. The new combination was designated *7.5 cm Pak 97/38* or *97/40* and rushed to the front where they proved just adequate to counter the T-34. But the hasty improvisation was really too powerful for the carriage and it proved unstable when fired. As soon as possible it was replaced in front-line use and passed to various occupation and garrison formations. Most of the rounds fired from the *Pak 97/38* were captured Polish and French stocks but a few special hollow-charge rounds were also developed.

Another captured weapon that was converted into a German anti-tank gun was the Soviet 76.2 mm Field Gun obr. 1936. Large

numbers of these guns had been captured during the 1941 and 1942 battles as they were the standard Soviet field price and the Germans decided to adapt them for anti-tank use (not all were so converted; many were retained as light field guns). New ammunition was developed and the gun chambers were modied to suit. A *Pak 40*-type muzzle brake was added and the fire controls were altered for use by one gun-layer. On some examples the shield was removed. The revised gun was designated *7.62 cm Pak 36(r)* and soon became regarded as one of the best anti-tank guns ever devised. It was hard-hitting, mobile, easy to conceal and easy to service. An extra bonus for the Germans was that it cost them very little. Another Soviet gun that was similarly converted was the 76.2 mm Field Gun obr. 1939 which became the *7.62 cm Pak 39(r)*, but the numbers involved were much smaller.

At this point it would be as well to mention an off-shoot from the general run of anti-tank guns: the taper-bore guns. These guns all used the Gerlich principle to increase shot velocity, combining a tapered bore with flanged or 'skirted' projectiles. The idea was to use the initial explosion of the propellant charge to drive a full-size projectile down the barrel. As the projectile went down the tapered barrel the flanges on the sides were pressed into the body of the projectile. The initial pressure remained the same so as the projectile got nearer to the muzzle its velocity increased. When this increased velocity was combined with the greater density of a tungsten core the result was an increased penetration of the target armour. The Gerlich principle was seized upon by German armament designers as the answer to a multitude of problems and was applied to a number of different weapon types. One post-war listing shows no less than eighteen calibre/calibre combinations used in experiments ranging from 13/7.92 mm up to 240/210 mm, but it must be stated that many of these were purely experimental and not all of them were directed towards the anti-tank role. Considerable work was also carried out on the close cousin of the tapered bore, the coned bore, where only a portion of the bore is tapered (usually towards the muzzle).

The smallest of the tapered bore weapons to see service was the *2.8 cm sPzB 41*. Although this was really a large anti-tank rifle it is included here because of its calibre which tapered from an initial 28 mm to an emergent 20 mm. There were two types of carriage. The standard type used two large wheels and the second type, the *le Feldlafette 41*, was a much lighter tubular steel model with two small wheels. It was intended for use by airborne troops but by the time this version was ready the Luftwaffe '*Fallschirmjäger*' were hard put to recover from the heavy losses suffered during the capture of Crete and thereafter were unable to use the weapon in its intended role. As a result it took its place in the field alongside the conventional carriage. Both types continued in use until 1945 but their main reason for withdrawal from service was that the barrel life was restricted to only about 400–500 rounds. When the barrels were worn the carriages were usually scrapped but some were used to provide a form of carriage for *MG 151* machine guns issued to the Volkssturm late in 1944. Like the larger calibre taper bore guns they suffered from the shortage of tungsten needed for their projectiles caused by the lack of supplies of wolfram, the raw material for it. After 1940 the only source of this rare mineral was Spain and various South American states. The Allied blockade of Germany, largely by the Royal Navy, effectively prevented supplies of wolfram from reaching Germany in any appreciable amounts. The only supplies were brought in by blockade runners, often at very high cost, and by 1941 the tungsten shortage was so acute that a priority decision at the highest level had to be made. The decision was between tungsten for weapons or tungsten for machine tools. Not surprisingly the result was in favour of machine tools, without which there could be no weapons, and the service life of the taper bore guns was thus limited.

The other two guns using the taper bore were the *4.2 cm Pak 41* and the *7.5 cm Pak 41*. The former used the carriage of the *3.7 cm Pak 35/36*, and had a barrel tapering from 40.3 mm at the breech to 29 mm at the muzzle. The larger gun was a Krupp weapon that had several novel features. One was the barrel which was not strictly tapered but had about half of its length conventionally bored. Another novel feature involved the split trail legs which were attached directly to the shield to save weight. When it first arrived on the scene the superior performance of the *7.5 cm Pak 41* (the 75 mm at the chamber tapered eventually to 55 mm at the muzzle) seemed to indicate that it would supplant the *Pak 40* as the standard service weapon but other factors than the tungsten shortage dictated otherwise. One of the main reasons for the preference for the *Pak 40* was that although the *Pak 41* had a high muzzle velocity this velocity fell drastically at long ranges and accuracy and penetration suffered as a result. Only 150 *Pak 41*s were produced and they were issued to special anti-tank units only. When their barrels were worn (usually after about 400 rounds) and/or their ammunition was exhausted, the guns were scrapped but it is believed that some were adapted to take *Pak 40* barrels.

But to return to the conventional anti-tank gun development stream. By 1943 it was apparent that the next generation of heavy tanks would have armour much heavier than existing models. Already the Germans had the Tiger, which was almost invulnerable to the *7.5 cm Pak 40*, and the even more powerful Tiger II was already in an advanced development stage. It needed little planning to realise that the Allies were not far behind and a specification for a really heavy anti-tank gun was issued. That was in 1942. By that date Krupp of Essen had been involved in a prolonged development programme with 8.8 cm anti-aircraft guns while Rheinmetall had been given a contract for an advanced Flak gun which resulted in the *8.8 cm Flak 41*. Although this gun was an excellent performer it had many problems and Krupp were asked to provide a 'back-up' programme code-named *Gerät 42*. In typical Krupp fashion the designers not only provided for the anti-aircraft version but also made provision for a tank and anti-tank version. This was just as well for in late 1942 the Luftwaffe planning staff, G. L. Flak, issued new specifications which could not be met without extensive redevelopment of the *Gerät 42*. Consequently the *Gerät 42* was dropped and the Krupp design team concentrated on the anti-tank and tank version. During 1943 this was designated the *8.8 cm Pak 43* and the result was one of the finest anti-tank guns ever to see service. In action the *Pak 43* was able to knock out any tank pitted against it at ranges well in excess of any other anti-tank weapon. Situated on its low, easy-to-conceal cruciform carriage, the *Pak 43* was difficult to spot and its ease of handling made it a difficult opponent for Allied tank crews. The *Pak 43* reached the Eastern Front just in time to counter the heavy Soviet IS-1 and IS-2 tanks which could be knocked out only with great difficulty or lucky shots by other anti-tank guns. Despite its apparent simplicity the *Pak 43* was not easy to produce and it was demanding of high quality materials, and demand soon outstripped supply. Production was mainly carried out by Henschel at Kassel and Eisenwerke Weserhütte at Bad Oeynhausen; others were produced by Krupp at Essen until Allied bombing severely disrupted the assembly lines.

The slow production rate of the *Pak 43* led to yet another stop-gap production solution to cater for the enormous demands made from the field. It seemed that every field commander now urgently needed the *Pak 43*, so to provide some form of 8.8 cm anti-tank gun quickly the German production engineers evolved an exigency weapon of available components. The result was a mixture from various assembly lines. The split trail carriage of the *10.5 cm leFH 18* was adapted to take the wheels of the *15 cm sFH 18* (some sources state that the carriage came from the *10 cm K 41*). A new saddle was hastily contrived and a shield added. The barrel was a simplified *Pak 43* component produced by Rheinmetall-Borsig with a modified breech mechanism and simpler sighting arrangements. This hybrid was designated the

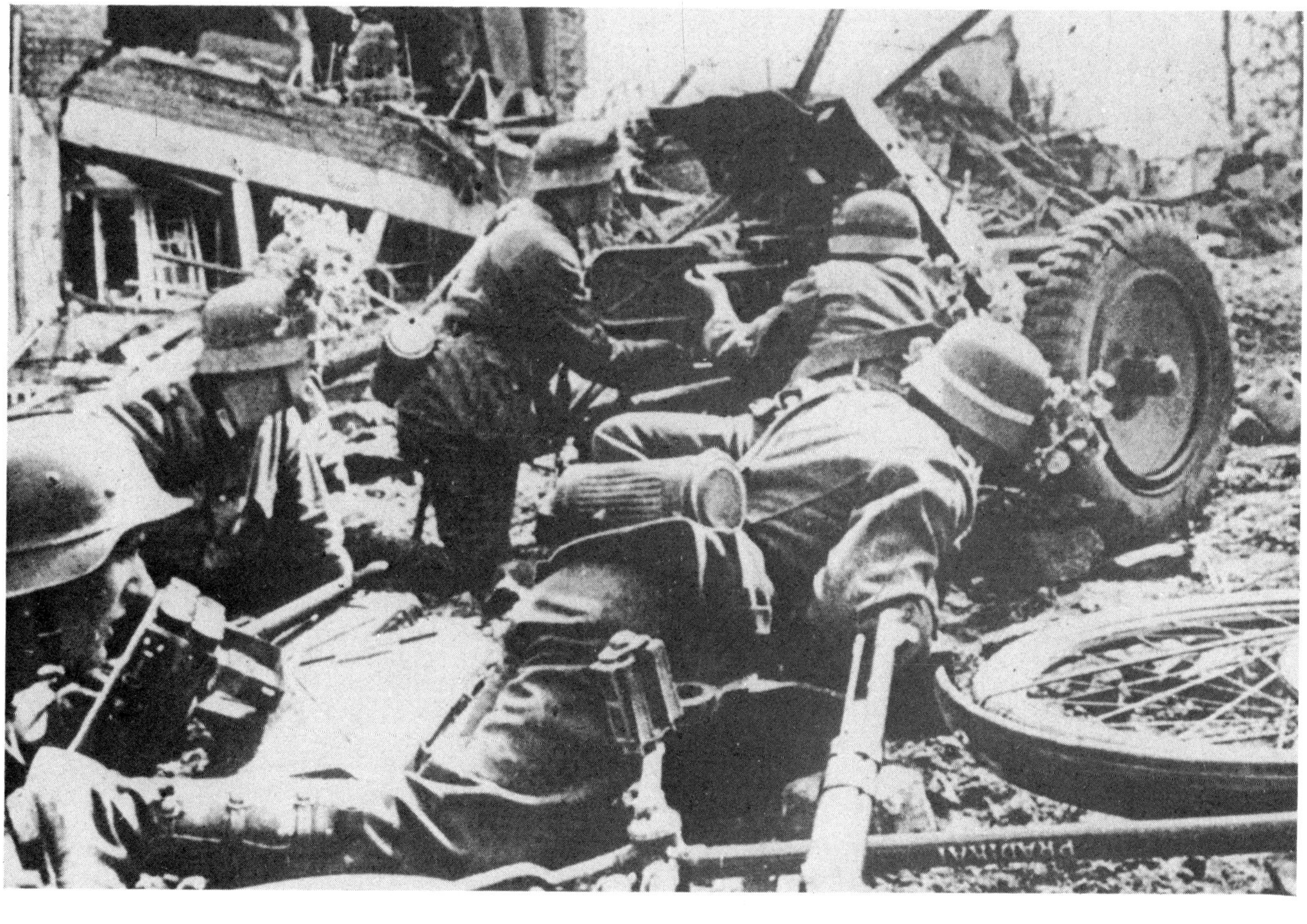

8.8 cm Pak 43/41 and hastily issued to all fronts. In service the *Pak 43/41* soon proved to be a high and awkward gun to use but it was every bit as effective as the *Pak 43*. In action, the *Pak 43/41* was as powerful as any gun anywhere but to move it was a major operation.

But when it came to sheer bulk the 8.8 cm guns were put in the shade by the ultimate in anti-tank guns, the 12.8 cm weapons. The reason for developing these large-calibre guns was to provide some form of weapon which could combine the attributes of a field gun with the power of an anti-tank gun. The specification for such a gun emerged from combat experience on the Eastern Front where most Soviet guns were used in both field and anti-tank roles. Krupp and Rheinmetall-Borsig entered prototypes for the contest and Krupp were awarded the production contract. The new gun was known under several designations, the anti-tank version being listed either as the *12.8 cm PjK 44*, *Pak 44* or *Pak 80*. As a field gun the same piece was known as the *12.8 cm K 44*. The carriages of both prototypes were massive complex affairs which involved a great deal of production capacity and therefore time. As a the battle-fronts were becoming increasingly hard-pressed the situation soon demanded another German stop-gap solution. The basis for this improvisation was the barrel of the *12.8 cm K 81*, a tank gun modified for use on the *Pak 44* or *K 44* carriage. These barrels were placed on captured French 155 mm GPF-T or Soviet 152 mm Gun-Howitzer obr. 1937 carriages. In this form, the few produced doubled as anti-tank guns or field pieces but the numbers involved were not large. The 'proper' *12.8 cm Pak 80* guns were built by Friedr. Krupp at the Bertha-Werke, Breslau.

By mid-1944 the need for anti-tank guns was so acute that nearly a third of German gun production was devoted to the manufacture of these weapons and several experimental guns were under development, one of which was the *7.5 cm Pak 44*. Although this gun featured a conical bore it was different from the earlier taper bore guns. On the *Pak 44* the first part of the barrel after the chamber tapered and the rest of the barrel used conventional straight rifling. As it was realised that the rifled part would wear quickly it was designed to be changed easily. The carriage was of an extremely light and advanced design. However, only prototypes were made before the war ended.

A more conventional gun was the *7.5 cm Pak 50* which combined a shortened *Pak 40* barrel with a revised muzzle brake on a *Pak 38* carriage. As it used already available parts the *Pak 50* was an obvious quick, easy and cheap way of providing an anti-tank gun for infantry units but few appear to have been produced and issued.

One of the most advanced design concepts to emerge from World War II was the '*Niederdruck*' or high-low pressure system. To produce a light anti-tank weapon which would be sparing in propellant and yet accurate up to useful combat ranges the German designers had to evolve some radical solutions and both Rheinmetall-Borsig and Krupp devoted considerable experimental facilities to the problem. Recoilless guns had proved to have many disadvantages and rockets were still too inaccurate and expensive to provide a realistic solution, but Rheinmetall produced an experimental weapon, the *8 cm PAW 600*. In this unorthodox weapon the projectile was a fin-stabilised shell not unlike a mortar bomb. The propellant charge was situated around the tail base and detonated in the normal way. But instead of the resultant propellant gases immediately driving the shell forward the gases were restricted from entering the space behind the shell by a plate perforated by a number of nozzles. The size and number of these nozzles prevented a rapid build-up of pressure and thus the main stresses were restricted to the relatively heavy breech section. By the time the gases had reached the barrel area the pressures were considerably lower but still high enough to produce a useful velocity. As a result the barrel and its carriage could be made fairly light and of considerably less weight than a conventional assembly. The relatively low velocity of the projectile made little difference to its anti-tank performance as the projectiles used hollow-charge warheads. By 1945 numbers of this weapon had been issued for troop trials as the *8 cm PWK 8 H 63*. An even larger weapon, the *10 cm PAW*, was also produced experimentally but of this very little is known. It is possible that this weapon was a Krupp counterpart to the Rheinmetall 8 cm gun.

Other design work was carried out at the Skoda-Werke at Pilsen in occupied Czechoslovakia. Most of this seems to have been purely experimental and devoted to producing self-loading devices for anti-tank guns. Among these were the *5, cm Automatische Pak 206/835*, the *5 cm Pak 208* (sometimes known as the *Pak 43*), the *6.6 cm Pak 5/800* and the similar *7.5 cm Pak 8/600*. At one time Krupp produced a taper-bore gun with an initial calibre of 45 mm, but only a single prototype appears to have been made.

The anti-tank gun field was not one where captured equipment was used to any great extent after 1941. Most of the guns captured until 1941 or 1942 were usually in the same performance class as the *3.7 cm Pak 35/36* and only a limited amount of French and Soviet equipment was used by the German troops. Many light guns were added to the 'Atlantic Wall' but they were used mainly for beach defences.

One of the more important captured guns was the 47 mm Böhler used by many nations as an anti-tank/pack/infantry gun. Numbers of these guns came to Germany from Austria, Italy and Holland, but by 1942 most seem to have been handed over to Romania or various foreign auxiliary formations. The 1940 campaign in Belgium produced numbers of the 47 mm FRC anti-tank gun, one of the more powerful European anti-tank weapons at that time, and many were later incorporated into the 'Atlantic Wall'. From Czechoslovakia came the 37 mm vz. 37 and 47 mm vz. 36, both of which were used by the German Army during the 1940 campaigns. Afterwards many were deployed along the Atlantic coast but some second-line units retained these guns and used them in 1944, often with the *Stielgranate* muzzle-loaded hollow-charge bomb. A number of anti-tank guns were also acquired in France. From there came the 25mm mle 1934 and 1937, both of which were widely used by German units based in France. Many were issued to coastal defence units but their combat value was very limited. Far more useful were the 47 mm mle 1937 and 1939. These two were issued mainly to units based in France and many of these guns were used against the Allies during the 1944 invasion. Some British 2 pr guns were sited in the 'Atlantic Wall' bunkers but most, like the Danish Madsen 37 mm guns, saw only limited used before being scrapped. The same fate seems to have overtaken the Rheinmetall 47 mm guns used by Holland. These were enlarged commercial versions of the *37mm Pak 35/36*.

The invasion of the Soviet Union in 1941 resulted in a vast amount of war booty. Among it were a number of 37 mm obr. 1930 anti-tank guns, the Soviet-built version of Rheinmetall Pak 35/36. Some of these were retained for a while with various units but most were eventually returned for conversion to infantry guns or simply scrapped. More valuable were the 45 mm obr. 1932 guns. These were based on the 37 mm Rheinmetall design but were far more useful to the front-line infantry units than the earlier version. Later in the fighting numbers of the long-barrelled 45 mm obr. 1942 wartime modification of the basic design also added to the fire power of German troops, as did the 57 mm obr. 1941 and 1943 dual purpose anti-tank/field guns.

The anti-tank gun field was one where the Germans always seemed to be one step ahead of the Allies. From 1940 onwards the German anti-tank guns, and their close relatives the tank guns, consistently placed Allied tank crews at a tactical disadvantage, but as far as numbers were concerned the usual situation arose of demand from the German troops far outstripping supply. As mentioned above this led to some desperate improvisations, but perhaps it was just as well for the Allies that such weapons as the superlative *8.8 cm Pak 43* were not met in greater numbers.

Appendix 1

Armour penetration tables for the 'standard' German anti-tank guns

2.8 cm sPzB 41

2.8 cm PzGr Patr 42. Weight 1.305 kg. **Muzzle velocity** 1402 m/sec

Range in metres	Penetration at 0° (mm)	Penetration at 30° (mm)
100	94	69
200	86	65
300	79	60
400	72	56
500	66	52
600	60	48
700	54	44
800	49	41

3.7 cm Pak 35/36

3.7 cm Pak PzGr. Weight 0.68 kg. **Muzzle velocity** 762 m/sec

Range in metres	Penetration at 0° (mm)	Penetration at 30° (mm)
200	56	42
400	51	38
500	48	36
600	46	34
3.7 cm Pak PzGr 40. Weight 0.354 kg. **Muzzle velocity** 1030 m/sec		
100	79	68
200	72	61
300	65	55
400	58	49
3.7 cm StielGr 41. Weight 8.5 kg. **Muzzle velocity** 110 m/sec		
200	180	—

4.2 cm Pak 41

4.2 cm PzGr 41. Weight 0.336 kg. **Muzzle velocity** 1256 m/sec

Range in metres	Penetration at 0° (mm)	Penetration at 30° (mm)
0	124	95
250	105	83
500	87	72
750	70	62
1000	60	53

5 cm Pak 38

5 cm Pak 38 PzGr. Weight 2.25 kg. **Muzzle velocity** 823 m/sec

Range in metres	Penetration at 0° (mm)	Penetration at 30° (mm)
0	99	73
250	88	67
500	78	61
750	69	56
1000	61	50
1200	53	45
1500	47	40
5 cm Pak 38 PzGr 40. Weight 0.975 kg. **Muzzle velocity** 1198 m/sec		
0	165	143
250	141	109
500	120	86
750	101	69
1000	84	55
1250	70	44

7.5 cm Pak 40

7.5 cm PzGr Patr 39. Weight 6.8 kg. **Muzzle velocity** 792 m/sec

Range in metres	Penetration at 0° (mm)	Penetration at 30° (mm)
0	149	121
500	135	106
1000	121	94
1500	109	83
2000	98	73
7.5 cm PzGr Patr 40. Weight 3.2 kg. **Muzzle velocity** 933 m/sec		
0	176	137
500	154	115
1000	133	96
1500	115	80
2000	98	66
2500	83	53

7.5 cm Pak 41

7.5 cm Pak 41 PzGr 41. Weight 2.48 kg. **Muzzle velocity** 1210 m/sec

Range in metres	Penetration at 0° (mm)	Penetration at 30° (mm)
0	245	200
250	226	185
500	209	171
750	192	157
1000	177	145
1250	162	133
1500	149	122
1750	136	111
2000	124	102

7.62 cm Pak 36(r)

7.62 cm PzGr 39. Weight 7.54 kg. **Muzzle velocity** 740 m/sec

Range in metres	Penetration at 0° (mm)	Penetration at 30° (mm)
0	133	108
500	120	98
1000	108	88
1500	97	79
2000	87	71
2500	78	64

8.8 cm Pak 43 and 8.8 cm Pak 43/41

8.8 cm PzGr 39/43. Weight 10.16 kg. **Muzzle velocity** 1000 m/sec

Range in metres	Penetration at 0° (mm)	Penetration at 30° (mm)
0	225	198
500	207	182
1000	190	167
1500	174	153
2000	159	139
2500	145	127
8.8 cm PzGr 40/43. Weight 7.3 kg. **Muzzle velocity** 1130 m/sec		
0	311	265
500	274	226
1000	241	192
1500	211	162
2000	184	136
2500	159	114

2.8 cm schwere Panzerbüchse 41

German designation 2.8 cm sPzB 41
Starting calibre 28 mm
Emergent calibre 20 mm
Length of piece 1714 mm
Length of barrel with m/b 1700 mm
Length of rifling 1370 mm
Weight with wheels 223 kg
Weight without wheels 162 kg
Traverse (horizontal): 90°; +45°: 30°
Elevation −5° to +45°
Muzzle velocity 1400 m/sec
Shot weight 0.124 kg
Shell weight 0.091 kg
Barrel life 500 rounds
Manufacturers (gun): Rheinmetall-Borsig AG, Düsseldorf; Mauser-Werke AG Oberndorf-am-Neckar. (carriage): AMBI/BUDD, Berlin

Remarks: First operational German AT weapon with Gerlich tapered-bore barrel. Approved for service November 1940. Very useful in 1941 but increased tank armour and shortage of tungsten for special ammunition gradually made it redundant.

2.8 cm schwere Panzerbüchse 41 *(sPzB 41), issued in 1941 to infantry and pioneer battalions*

If the time allowed, the sPzB 41 was removed from its wheeled carriage and fired from the mount

The sPzB 41 could also be used mounted on its ammunition trailer

Sonderanhänger für sPzB 41 *trailer*

sPzB 41 with its JF 8 infantry limber

2.8 cm schwere Panzerbüchse 41 auf leichter Feldlafette 41

German designation 2.8 cm sPzB le.F1 41
Starting calibre 28 mm
Emergent calibre 20 mm
Length of piece 1714 mm
Length of barrel with m/b 1700 mm
Length of rifling 1370 mm
Weight with wheels 139 kg
Weight off wheels 118 kg
Traverse 360°
Elevation −15° to +25°
Muzzle velocity 1400 m/sec
Shot weight 0.124 kg
Shell weight 0.091 kg
Barrel life 500 rounds
Manufacturers (gun): Rheinmetall-Borsig AG, Düsseldorf; Mauser-Werke, Oberndorf-am-Neckar. (carriage): Rheinmetall-Borsig AG, Düsseldorf; AMBI-BUDD, Berlin; G. Appel, Berlin-Spandau; Heidmann, Einbeck

Remarks: Similar to sPzB 41 but on special lightweight carriage reducing total weight by about 50%. Issued to German airborne formations 1941; later also used by infantry.

2.8 cm sPzB 41 auf leichter Feldlafette 41. *This equipment was dropped by parachute in three loads*

3.7 cm Panzerabwehrkanone 35/36, 153(n), 158(r) and 162(i)

German designations 3.7 cm Pak 35/36; 3.7 cm Pak 153(h), 158(r) und 162(i)
Original designation (h): 37 mm Rheinmetall; (r): *37 mm Protivotankovaya Pushka obr. 1930 g*; (i): Cannone controcarro da 37/45
Calibre 37 mm
Length of piece (L/45): 1665 mm
Length of barrel 1568 mm
Length of rifling 1308 mm
Weight in action – wooden wheels 330 kg
Weight in action – disc wheels 328 kg
Weight complete – disc wheels 440 kg
Traverse 59°
Elevation −8° to +25°
Muzzle velocity (AP): 760 m/sec; (AP40): 1030 m/sec; (StielGr 41): 110 m/sec
Maximum range 7000 m
Shot weight (AP): 0.68 kg
Shot weight (AP 40):0.354 kg
Shell weight (HE): 0.625 kg
Bomb weight (StielGr 41): 8.5 kg
Maximum range (StielGr 41): 365 m
Barrel life 4000–5000 rounds
Original manufacturer Rheinmetall Borsig AG, Düsseldorf

Remarks: One of the best AT guns until 1941. Development began during 1925, production of horse-drawn version commenced 1928, disc wheels 1934. Tested in action in Spain 1936 onwards. Widely exported and also licence-built abroad. Over 15,000 produced in Germany by 1941. Afterwards service life extended by use of 'Stielgranate' hollow-charge stick shells. Served in various capacities (also on SP mounts) until 1945.

Loading the Stielgranate 41

The Stielgranate 41. *This anti-tank projectile consisted of an egg-shaped head containing a hollow-charge warhead, a nose fuse and a base fuse. The cylindrical tail comprised a rod which fitted into the bore of the gun and a concentric perforated outer sleeve with six stabilising fins*

3.7 cm Pak 35/36, the standard German infantry anti-tank gun at the outbreak of World War II

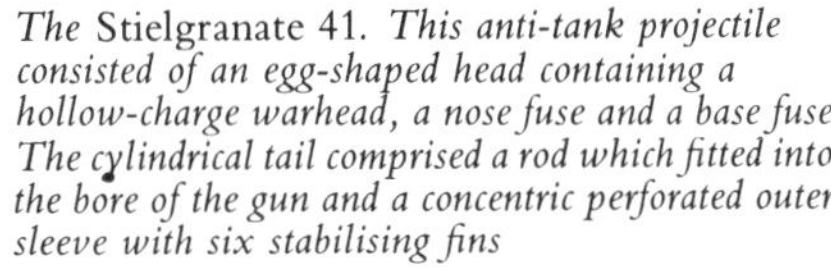

3.7 cm Pak leFLaf with parachute and other air landing equipment positioned under the fuselage of a Ju 52/3m transport

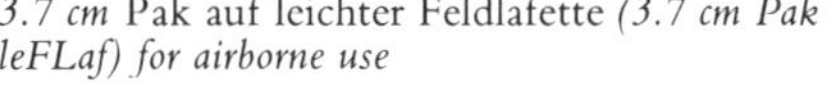

3.7 cm Pak auf leichter Feldlafette *(3.7 cm Pak leFLaf) for airborne use*

4.2 cm leichte Panzerabwehrkanone 41

German designation 4.2 cm lePak 41
Starting calibre 40.3 cm
Emergent calibre 29.4 mm
Length of piece (L/55.8): 2250 mm
Length of bore 2114 mm
Length of rifling 1700 mm
Weight in action 560 kg
Traverse 60°
Elevation −8° to +25°
Muzzle velocity 1265 m/sec
Maximum range 7000 m
Maximum a/t range 1000 m
Shot weight 0.336 kg
Shell weight 0.28 kg
Barrel life 1000 rounds

Remarks: Issued to German AT units in 1941 as second Gerlich tapered-bore gun to become operational. Effective at that time but due to increased tank armour and shortage of tungsten for special ammunition production ceased in 1942. Existing guns remained in service until available ammunition used up.

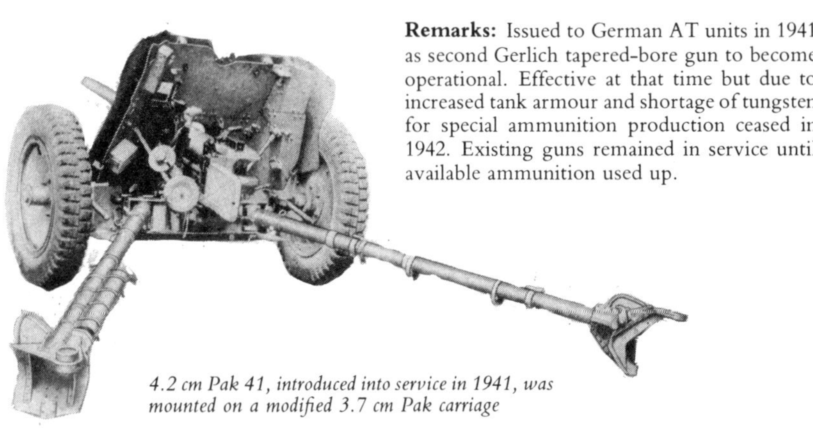

4.2 cm Pak 41, introduced into service in 1941, was mounted on a modified 3.7 cm Pak carriage

5 cm Panzerabwehrkanone 38

German designation 5 cm Pak 38
Calibre 50 mm
Length of piece (L/60): 3187 mm
Length of bore 2824 mm
Length of rifling 2381 mm
Weight in action 1000 kg
Weight transported 1062 kg
Traverse 65°
Elevation −8° to +27°
Muzzle velocity (PzGr 39): 835 m/sec; (AP 40): 1180 m/sec; (HE): 550 m/sec
Maximum range (HE): 2650 m
Shot weight 2.06 kg
Shot weight (AP 40): 0.925 kg
Shell weight 1.82 kg
Weight (StielGr 42): 8.2 kg
Maximum effective range (StielGr): 150 m
Barrel life 4000–5000 rounds
Original manufacturers Rheinmetall-Borsig AG, Düsseldorf

Remarks: Development work started during 1937, completed 1939. First issued to service April 1940; production ceased mid-1944. Well-designed and very efficient AT gun, served on all fronts throughout WW II. Using tungsten-core AP40 shells could penetrate nearly all types of heavy Allied tanks. Also mounted on a variety of SP carriages for mobile AT role.

5 cm Pak 38, also introduced into service in 1941 to replace the 3.7 cm Pak, became standard equipment until the end of the war. Modified with automatic loading the same basic gun was also adopted by the Luftwaffe as a fixed aircraft weapon

5 cm Pak 38 during a training shoot

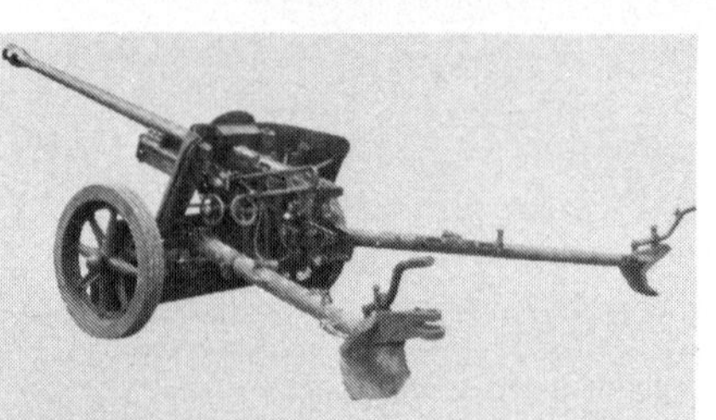

Rear view of the 5 cm Pak 38

7.5 cm Panzerabwehrkanone 40 and 7.5 cm Feldkanone 40

German designations 7.5 cm Pak 40; 7.5 cm FK 40
Calibre 75 mm
Length of piece (L/46): 3700 mm
Length of bore 3450 mm
Length of rifling 2461 mm
Weight complete 1500 kg
Weight in action 1425 kg
Elevation −5° to +22°
Traverse 65°
Muzzle velocity (PzGr 39 und 39AL): 750 m/sec; (PzGr 40): 930 m/sec; (SprGr): 550 m/sec
Maximum range (HE): 7680 m
Maximum range (AT): 1800 m
Shot weight (PzGr 39): 6.8 kg; (PzGr40): 4.1 kg
Shell weight (SpGr): 5.74 kg
Barrel life 6000 rounds
Manufacturers Rheinmetall-Borsig AG, Düsseldorf; Ardeltwerke, Eberswalde; Gustloffwerke, Weimar; Ostlandwerke, Königsberg

Remarks: Designed to OKH requirement issued in 1939. Entered service late 1941. Basically scaled-up Pak 38 but with various detail differences. Remained in production until 1945. Best of German standard AT guns, used on all fronts. Also mounted on a variety of SP carriages.

7.5 cm Pak 40. It was generally similar in appearance to the 5 cm Pak 38 but had a heavier muzzle brake and an angular rather than a curved gun shield

7.5 cm Pak 40 in alarm firing position direct from the march. Note the two gunners holding down the trail spades to minimise backward movement

Rear view of the 7.5 cm Pak 40

7.5 cm Panzerabwehrkanone 41

German designation 7.5 cm Pak 41
Starting calibre 75 mm
Emergent calibre 55 mm
Length of piece (L/57.6): 4320 mm
Length of bore 2950 mm
Length of rifling 2410 mm
Weight complete 1880 kg
Weight in action 1390 kg
Elevation −10° to +18°
Traverse 60°
Muzzle velocity (PzGr 41 HK): 1220 m/sec; (PzGr 41 W): 1230 m/sec; (SprGr): 900 m/sec
Maximum range (AT): 1800–2000 m
Shot weight (PzGr 41 HK): 2.6 kg
Shot weight (PzGr 41 W): 2.5 kg
Shell weight 2.65 kg
Barrel life 1000 rounds
Manufacturer Friedr. Krupp AG, Essen

Remarks: Designed to compete with Rheinmetall-Borsig Pak 40. Notable as largest of Gerlich tapered-bore AT guns built and used in action. Due to shortage of tungsten needed for special ammunition only 150 complete guns delivered. Some barrels replaced by 7.5 cm Pak 40 components when Pak 41 barrels wore out. Numbers involved were small.

Rear view of the 7.5 cm Pak 41

7.5 cm Panzerabwehrkanone 97/38 and 97/40

German designation 7.5 cm Pak 97/38; 7.5 cm Pak 97/40
Calibre 75 mm
Length of piece (L/36): 2720 mm
Length of bore 2489 mm
Weight complete 1270 kg
Weight in action 1190 kg
Elevation −8° to +25°
Traverse 60°
Muzzle velocity (AP): 570 m/sec; (HE): 450 m/sec
Maximum range (AT): 1900 m; (HE): 11,000 m
Shot weight 6.8 kg
Shell weight 4.8 kg

7.5 cm Pak 97/38, an emergency solution to face the more heavily armoured Soviet tanks

Remarks: Stop-gap measure to counter Soviet T-34 and KV-1 tanks. Total of 700 converted French Canon mle 1897 barrels combined with Pak 38 or (in small numbers) Pak 40 carriages. Entered service in late 1941 and used captured French and Polish ammunition, but also hollow-charge shells. Unstable when fired but served its purpose. Withdrawn after service debut of Pak 40 and relegated to second-line troops. Some converted barrels fitted on SP carriages.

Manhandling the 7.5 cm Pak 97/38. Note the auxiliary wheel under the trail spades

7.5 cm Pak 97/38 ready for action

7.5 cm Panzerabwehrkanone 50

German designation 7.5 cm Pak 50
Calibre 75 mm
Length of piece (L/30): 2245 mm
Length of rifling 1435 mm
Weight in action (approx): 1100 kg
Elevation −8° to +27°
Traverse 65°

Remarks: Shortened 7.5 cm Pak 40 barrel on 5 cm Pak 38 carriages. Only limited number so converted and issued to German infantry formations from late 1944 onwards.

◄ *7.5 cm Pak 50 was produced with two types of muzzle brakes, with five and three baffles (shown here)*

7.5 cm Pak 50, an emergency design for a light and more manoeuvrable anti-tank gun

7.62 cm Panzerabwehrkanone 36(r)

Another view of the 7.62 cm Pak 36(r)

7.62 cm Pak 36(r), a successful adaptation of a Soviet field gun

German designation 7.62 cm Pak 36(r)
Original designation *76.2 mm Pushka obr. 1936 g* (76–36)
Calibre 76.2 mm
Length of piece (L/51): 3895 mm
Length of rifling 2930 mm
Weight in action 1710 kg
Traverse 60°
Elevation −6° to +18°
Muzzle velocity (PzGr 40): 960 m/sec; (SprGr 39): 550 m/sec
Maximum range (HE): 10,400 m
Shot weight (PzGr 39): 7.54 kg; (PZGr 40): 4.15 kg
Shell weight 6.2 kg
Barrel life 6000 rounds

Remarks: German conversion of captured Soviet field gun. Originated 1942 as an emergency measure to provide additional AT guns capable of facing T-34 and KV-1 tanks. Introduced in service same year and soon acknowledged as one of the best AT guns in action. Large number in service; also on SP carriers (Marder II etc.).

8 cm Panzerabwehrwerfer 600, 8 cm Panzerwurfkanone 8H63

8 cm PAW 600 RV 3 on the Lafette Nr V2 *(Krupp prototype)*

8.1 cm PAW L/105. This heavier experimental version of the high-low pressure projector used components of the standard service equipment. Only a pilot model was built

8.1 cm PAW L/105 with limber

Another view of the 8 cm PAW 600 RV 3 on the Laf Nr V2 *(Krupp)*

The revolutionary high-low pressure 8 cm Panzerabwehrwerfer 600 *(8 cm PAW 600), Rheinmetall prototype*

A modified version of the 8 cm PAW 600 with strengthened tubular trails and a muzzle brake

German designations 8 cm PAW 600 (design designation); PWK 8H63 (troop trial designation)
Calibre 81.4 mm
Length of piece (w/o muzzle brake): 2941 mm
Weight complete 640 kg
Weight in action 630 kg
Traverse 55°
Elevation −6° to +32°
Muzzle velocity (hollow charge): 520 m/sec; (HE)-maximum: 420 m/sec
Maximum effective range (AT): 600 m
Maximum range (HE): 6200 m
Shell weight (hollow charge): 2.7 kg; (HE): 4.46 kg
Armour penetration 140 mm at 750 m (0°)
Manufacturers Rheinmetall-Borsig AG, Düsseldorf; Bückau-R. Wolf, Magdeburg

Remarks: Revolutionary high/low pressure gun. Ten issued for troop trials in late 1944; total of 260 completed and delivered by end of WW II. Some mounted on 5 cm Pak 38 carriages. Krupp involved in research stage and built a similar experimental weapon designated PAW 600 V5.

8 cm PWK 8H63 production version with modified 7.5 cm Pak 40 muzzle brake

8.8 cm Panzerabwehrkanone 43, 8.8 cm Kanone 43

8.8 cm Pak 43 on cruciform platform, ready for action

8.8 cm Pak 43 on pneumatic-tyred wheels

German designation 8.8 cm Pak 43 oder 8.8 cm K 43
Calibre 88 mm
Length of piece (L/71): 6610 mm
Length of piece (w/o muzzle brake) 6280 mm
Length of bore 6010 mm
Length of rifle 5125 mm
Weight complete 4750 kg
Weight in action 3650 kg
Traverse 360°
Elevation −8° to +40°
Muzzle velocity (PzGr 39/43): 1000 m/sec; (PzGr 40/43): 1130 m/sec; (SprGr 43): 950 m/sec
Shot weight (PzGr 39/43): 10.16 kg; (PzGr 40/43): 7.3 kg
Shell weight (SprGr 43): 9.4 kg
Maximum range (HE): 15,150 m
Barrel life 1200–2000 rounds
Manufacturers Friedr. Krupp AG, Essen; Henschel, Kassel; Weserhütte, Bad Oeynhausen

Remarks: Very advanced design evolved from projected 8.8 cm Pak 42. Featured semi-automatic breech and electrical firing circuit. Generally acknowledged as best AT gun produced during WW II, extremely hard-hitting at very long range. Also mounted on SP carriers (Nashorn etc.).

The Pak 43 could also be fired from its wheels if the direction of fire did not exceed 30 degrees either side of the longitudinal carriage leg

*8.8 cm Pak 43 was transported on two single-axle limbers (*Sonderanhänger 204*). This view shows the Sd Ah 204 with solid-tyred spoke wheels*

8.8 cm Panzerabwehrkanone 43/41

German designation 8.8 cm Pak 43/41
Calibre 88 mm
Length of piece (L/71) 6610 mm
Length of piece (w/o muzzle brake) 6280 mm
Length of bore 6010 mm
Length of rifling 5125 mm
Weight complete 4380 kg
Traverse 56°
Elevation −5° to 38°
Muzzle velocity (PzGr 39/43): 1000 m/sec; (PzGr 40/43): 1130 m/sec; (SprGr 43): 950 m/sec
Shot weight (PzGr 39/43): 10.16 kg; (PzGr 40/43): 7.3 kg

Shot weight (SprGr 43): 9.4 kg
Maximum range (HE) approx: 15,000 m
Barrel life 1200–2000 rounds
Manufacturer Rheinmetall-Borsig AG

Remarks: A very successful emergency solution to provide an efficient AT gun to meet increasing demands. Combination of 8.8 cm Pak 43 mounted on 10.5 cm leFH 18 carriage and 15 cm sFH 18 wheels; simplified Pak 43 barrel, dial sights. Intended as dual purpose AT/field gun but almost invariably used in AT role. Nicknamed 'Scheunentor' (barn door).

Pak 43/41 emplaced for coastal defence

Rear view of the Pak 43/41

12.8 cm Panzerabwehrkanone 44 and 80, 12.8 cm Kanone 44

German designations 12.8 cm Pak 44, K 44 oder Pak 80. Also referred to as 12.8 cm Pak 43 or PjK 44
Calibre 128 mm
Length of piece (L/55): 7023 mm
Length of bore 6623 mm
Length of rifling 5538 mm
Weight of gun 3353 kg
Weight of gun and carriage 10,160 kg
Traverse 360°
Elevation −7°51′ to +45°27′
Muzzle velocity (AP): 950 m/sec; (HE): 750 m/sec
Shot weight 28.3 kg
Shell weight 28 kg
Armour penetration 219 mm at 500 m (0°); 202 mm at 1000 m; 187 mm at 1500 m
Maximum range (HE): 24,410 m
Barrel life 1000–2000 rounds
Manufacturer Friedr. Krupp AG, Bertha-Werke, Breslau

Remarks: Designed during 1944 as a dual-purpose field and heavy anti-tank gun. Rheinmetall also built a prototype. Krupp Pak 80 had elevation limited from −7° to +15°.

12.8 cm Kanone 44 (Krupp) in travelling position

12.8 cm K 44 (Krupp) on its cruciform platform

Rheinmetall version of the 12.8 cm K 44 in firing position

12.8 cm K 44 (Rheinmetall)

12.8 cm Kanone 81/1

German designation 12.8 cm K 81/1
Calibre 128 mm
Length of piece (L/55): 7023 mm
Length of bore 6623 mm
Length of rifling 5538 mm
Weight in action 12,150 kg
Traverse 60°
Elevation −4° to −45°
Muzzle velocity (AP): 950 m/sec; (HE): 750 m/sec
Shot weight 28.3 kg
Shell weight 28 kg
Maximum range (HE) approx: 24,000 m

Remarks: K 81 was a version of K 44 intended for installation in tanks, but during late 1944 some 50 were converted for mounting on French Canon de 155 GPF-T carriages and used in AT/field gun role.

An 12.8 cm K 81/1 captured by the American troops

12.8 cm Kanone 81/1 was a dual-purpose anti-tank and field gun (see K 81/1 under Medium and Heavy Artillery)

12.8 cm Kanone 81/2

German designation 12.8 cm K 81/2
Calibre 128 mm
Length of piece (L/55): 7023 mm
Length of bore 6623 mm
Length of rifling 5538 mm
Weight in action (approx): 8200 kg
Traverse 40°
Elevation −4° to +45°
Muzzle velocity (AP): 950 m/sec; (HE): 750 m/sec
Shot weight 28.3 kg
Shell weight 28 kg
Maximum range (HE) approx: 24,000 m

Remarks: Another make-shift combination to provide a heavy AT gun, this time 12.8 cm K 81 barrel mounted on Soviet 152 mm Gun-Howitzer carriage obr. 1937 (ML-20). Only very limited number completed from late 1944 onwards.

12.8 cm Kanone 81/2 was also a dual-purpose anti-tank and field gun

12.8 cm K 81/2

5 cm automatische Panzerabwehrkanone 2.06/835

German designation 5 cm autPak 2.06/835 (Skoda Werke design designation)
Calibre 50 mm
Type of feed 5 round clips
Trunnion height 920 mm
Weight travelling 1350 kg
Weight in action 1300 kg
Traverse 65°
Elevation −12° to +20°
Muzzle velocity 835 m/sec
Shot weight 2.06 kg
Cyclic rate of fire 80–85 rpm, later 120 rpm
Armour penetration 70 mm – no range specified
Manufacturer Skoda Werke, Pilsen

Remarks: Development begun during 1940 but only a prototype was produced, mounted on 6.6 cm Pak 5/800 carriage.

Rear view of the 5 cm aut. Pak 2.06/835 showing the loading tray ➤

6.6 cm Panzerabwehrkanone 5/800

German designation 6.6 cm Pak 5/800 (Skoda Werke design designation)
Calibre 66 mm
Height of barrel 900 mm
Weight travelling 1560 kg
Weight in action 1500 kg
Traverse 65°
Elevation −12° to +20°
Muzzle velocity 800 m/sec
Shot weight 5 kg
Manufacturer Skoda Werke, Pilsen

Remarks: Experimental only. Apparently only one built. Used split trails and a firing pedestal which was lowered when trails opened.

7.5 cm Panzerabwehrkanone 6/860

German designation 7.5 cm Pak 6/860 (Skoda Werke design designation)
Calibre 75 mm
Height of barrel 950 mm
Weight travelling 2100 kg
Weight in action 2000 kg
Traverse 65°
Elevation −12° to +20°
Muzzle velocity 860 m/sec
Shot weight 6 kg
Manufacturer Skoda Werke, Pilsen

Remarks: Experimental only. Enlarged variant of 6.6 cm Pak 5/800. Only one built.

7.5 cm Panzerabwehrkanone 44

German designation 7.5 cm Pak 44
Starting calibre 75 mm
Emergent calibre 55 mm
Length of piece 1070 mm
Muzzle velocity (tungsten core shot): 1300 m/sec; (AP): 1500 m/sec
Shot weight 2.5 kg
Armour penetration (claimed): 120 mm at 2500 m (30°)
Barrel life, rifled portion 200 rounds
Manufacturer Rheinmetall-Borsig AG, Düsseldorf

Remarks: An experimental AT gun designed by Dr Grotsch. Under development 1942–45. Rifled part of barrel designed for easy change due to its known short life.

10 cm Panzerabwehrwerfer 600 and 10 cm Panzerwurfkanone 10H64

10.5 cm PAW on 5 cm Pak 38 carriage during trials early in September 1944

10.5 cm PAW on the Lafette V1, *a modified version of the basic weapon mounted on a special lightweight carriage*

Front view of the 10.5 cm PAW on 5 cm Pak 38 carriage

Rear view of the 10.5 cm PAW mit Lafette V1 *early in September 1944*

German designations 10 cm PAW 600 oder 10 cm PAW (PWK) 10H64
Calibre 105 mm
Weight in action 1035 kg
Weight of gun 275 kg
Traverse 60°
Elevation −5° to +30°
Muzzle velocity 550–600 m/sec
Shell weight 6.6 kg
Armour penetration 200 mm at 60°
Designer Friedr. Krupp AG, Essen

Remarks: Experimental Krupp design based on revolutionary high/low pressure principle. Project work started during 1944. Several prototypes completed by end of WW II, of which one mounted on Pak 38 carriage and another on a specially-designed Lafette V1 carriage. *See* also PAW 600.

2.5 cm Panzerabwehrkanone 112(f)

German designation 2.5 cm Pak 112(f)
Original designation Canon léger de 25 antichar SA-L mle 1934
Calibre 25 mm
Length of piece (L/72): 1800 mm
Length of bore 1600 mm
Weight in action 496 kg
Traverse 60°
Elevation −5° to +21°
Muzzle velocity 918 m/sec
Shot weight 0.32 kg
Maximum range 1800 m
Armour penetration 50 mm at 600 m
Manufacturer Hotchkiss et Cie, St. Denis

Remarks: Standard French Army light AT gun 1939–40; some also issued to BEF troops in France. Due to short range and light shot largely ineffective in combat. In German service issued only to some garrison and coastal defence units.

2.5 cm Panzerabwehrkanone 113(f)

German designation 2.5 cm Pak 113(f)
Original designation Canon léger de 25 antichar SA-L mle 1937
Calibre 25 mm
Length of piece (L/77): 1925 mm
Length of bore 1813 mm
Weight in action 310 kg
Traverse 37°
Elevation −10° to +26°
Muzzle velocity 900 m/sec
Shot weight 0.32 kg
Maximum range 1800 m
Armour penetration 50 mm at 600 m (not confirmed)
Manufacturer Atelier Puteaux

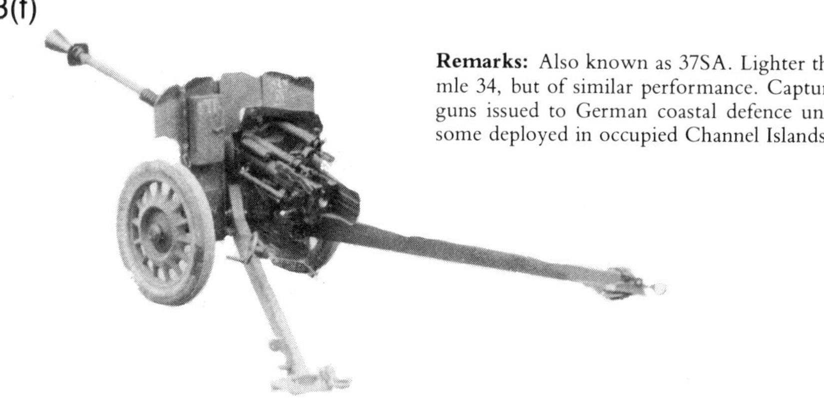

Remarks: Also known as 37SA. Lighter than mle 34, but of similar performance. Captured guns issued to German coastal defence units; some deployed in occupied Channel Islands.

3.7 cm Panzerabwehrkanone 37(t) and 156(j)

German designation 3.7 cm Pak 37(t) oder 156(j)
Original designation Škoda 37 mm kanon P.U.V. vz. 37
Calibre 37.2 mm
Length of piece (L/47.8): 1778 mm
Weight travelling 405 kg
Weight in action 370 kg
Traverse 50°
Elevation −8° to +26°
Muzzle velocity 750 m/sec
Shot weight 0.85 kg
Maximum range 5000 m
Maximum effective range 900 m
Manufacturer Škoda, Pilsen

Remarks: Efficient Czech AT gun evolved from earlier vz. 34. Widely exported. 'Standard issue' in German Army 1939–40. Subsequently deployed as 'back-up' weapon. Also used against partisans in Yugoslavia.

3.7 cm Pak 37(t) with the original wooden spoke wheels

3.7 cm Pak 37(t) with pneumatic wheels

3.7 cm Panzerabwehrkanone 36(p) and 157(d)

German designation 3.7 cm Pak 36(p) oder 157(d)
Original designations Swedish: 37 mm pakan m/34; (p): Armata przeciwpancerna wz. 36
Calibre 37 mm
Length of piece (L/45): 1665 mm
Weight travelling 930 kg
Weight in action 380 kg
Traverse 50°
Elevation −10° to +25°
Muzzle velocity 800 m/sec
Shot weight 0.7 kg
Maximum range 4500 m
Maximum effective range 1000 m
Armour penetration 37 mm at 400 m (30°); 40 mm at 600 m (0°)
Original manufacturer AB Bofors, Sweden

Remarks: During WW II used by Polish, Danish and Finnish armies, and also by British forces, as 'Ordnance Q.F. 37 mm Mk 1'. In German service issued only to second-line units.

3.7 cm Panzerabwehrkanone 164(d)

German designation 3.7 cm Pak 164(d)
Original designation Madsen Model 1935
Calibre 37 mm
Length of piece (L/60): 2220 mm
Weight travelling 540 kg
Weight in action 340 kg
Traverse 60°
Elevation −10° to +24°
Shot weight 0.8 kg
Muzzle velocity 900 m/sec
Maximum range 9000 m
Maximum effective range 1000 m
Manufacturer Dansk Industri Syndicat A/S 'Madsen'

Remarks: Very similar to 37 mm Swedish Bofors AT gun m/34. Only limited numbers in Danish service before 1940. No confirmed subsequent German use.

4.7 cm Panzerabwehrkanone 36(t) and 179(j)

German designation 4.7 cm Pak 36(t) oder 179(j)
Original designation Škoda 47 mm kanon P.U.V. vz. 36
Calibre 47 mm
Length of piece (L/43.4): 2040 mm
Weight complete 605 kg
Weight in action 590 kg
Traverse 50°
Elevation −8° to +26°
Muzzle velocity 775 m/sec
Shot weight 1.64 kg
Shell weight 1.5 kg
Maximum range 4000 m
Armour penetration 60 mm at 1200 m (homogeneous plate)
Manufacturer Škoda, Pilsen

Remarks: Requisitioned Czech AT guns used operationally by German troops 1939–40. A number also placed on PzKw I chassis to become first German 'tank hunters'. From 1941 gradually relegated to second-line service and coastal defences.

4.7 cm Panzerabwehrkanone 181(f) and 183(f)

German designation 4.7 cm Pak 181(f) und 183(f)
Original designation 181(f): Canon de 47 antichar SA mle 1937; 183(f): Canon de 47 antichar SA mle 1939
Calibre 47 mm
Length of piece (L/53): 2491 mm
Length of bore 2350 mm
Weight complete 1090 kg
Weight in action 1050 kg
Traverse 68°
Elevation −13° to +16°30′
Shot weight 1.725 kg
Muzzle velocity 855 m/sec
Maximum range 6500 m
Armour penetration 80 mm at 200 m
Manufacturer Atelier Puteaux

Remarks: Also known as SA 37 APX. Potentially effective AT gun; substantial number captured by German forces 1940. Some later issued to occupation troops in France and in service until 1944.

4.7 cm Pak 181(f) on the Atlantic coast

4.5 cm Panzerabwehrkanone 184(r) and 184/1(r)

German designations 4.5 cm Pak 184(r) und 184/1(r)
Original designation 184(r): *45 mm Protivotankovaya Pushka obr. 1930 g*; 184/1(r): *45 mm PTP obr. 1932 g*
Calibre 45 mm
Length of piece (L/46): 2072.5 mm
Length of bore 1975 mm
Length of rifling 184(r): 1688 mm; 184/1(r): 1650 mm
Weight travelling, tyred wheels 510 kg
Weight in action 425 kg
Traverse 60°
Elevation −8° to +25°
Muzzle velocity (AP): 760 m/sec; (HE): 335 m/sec
Shot weight 1.43 kg
Shell weight 2.15 kg
Maximum range 8870 m
Maximum range (HE): 4670 m
Armour penetration 38 mm at 900 m (30°)
Manufacturer Various Soviet State arsenals

4.5 cm Pak 184(r). Note the similarity with the German 3.7 cm Pak 35/36

Remarks: Basically scaled-up version of German 37 mm Pak 35/36. Many captured by German forces 1941–42 and almost invariably impressed in service. 45 mm obr. 1937 AT gun very similar. Slightly modified longer-barrelled variant remained in Soviet service until 1945.

4.5 cm Panzerabwehrkanone 184/6(r)

German designation 4.5 cm Pak 184/6(r)
Original designation *45 mm Tankovaya Pushka obr. 1938 g*
Calibre 45 mm
Length of piece (L/46): 2072 mm
Length of bore 1975 mm
Length of rifling 1650 mm
Weight in action 650 kg
Traverse 15°
Elevation −10° to +10°
Muzzle velocity 760 m/sec
Shot weight 1.43 kg
Maximum range 8870 m
Armour penetration 38 mm at 900 m (30°)
Manufacturer Various Soviet State arsenals

Remarks: Soviet 1942 emergency measure to provide an AT gun – 45 mm obr. 1938 tank gun on an extemporised field carriage suitable for horse traction. Only limited German use.

4.7 cm Panzerabwehrkanone 185(b)

German designation 4.7 cm Pak 185(b)
Original designation Canon de 47 antichars SA-FRC
Calibre 47 mm
Length of piece (L/33.6): 1579 mm
Length of bore 1438 mm
Weight in action 568 kg
Traverse 40°
Elevation −3° to +20°
Muzzle velocity 720 m/sec
Maximum effective range 1000 m
Shot weight 1.5 kg
Manufacturer Fonderie Royale de Canons, Liège

Remarks: An original Belgian design produced in limited numbers before WW II. Most captured guns issued to German occupation troops in Belgium; some emplaced in German-occupied Channel Islands.

4.7 cm Panzerabwehrkanone (Böhler), 177(i), 187(h) and 196(r)

German designations 4.7 cm Pak (Böhler), 177(i), 187(h) und 196(r)
Original designation Böhler Modell 1935; (i): Cannone anticarro e d'accompagnamento 47/32 modello 35; (h): Kanon van 4,7; (r): (provisionally) *47 mm PTP Boler*; Model 35B
Calibre 47 mm
Length of piece (L/35.8): 1680 mm
Length of bore 1525 mm
Length of rifling 1328 mm
Weight travelling 315 kg
Weight in action 277 kg
Traverse 62°
Elevation −15° to +56°
Muzzle velocity (AP): 630 m/sec; (HE): 250 m/sec
Maximum range (HE): 7000 m
Shot weight 1.44 kg
Shell weight 2.37 kg
Original manufacturer Gebr. Böhler, Kapfenberg, Austria
Italian licence manufacturer Terni-O.T.O

Remarks: An original 1935 Austrian design, produced in various versions and widely sold abroad. Despite lack of hitting power used operationally during early stages of WW II. Model 35B acquired during Soviet occupation of Baltic States in 1940–41 and some captured during German advance. Only localised German service use; most available Böhler guns later transferred to coastal defence and training. Some also issued as infantry guns.

4.7 cm Pak 177(i). This equipment could be fired from its wheeled carriage or from a ground position

4.7 cm Pak 177(i) in service with the Afrika Korps

4.7 cm Pak Böhler in ground position

4 cm Panzerabwehrkanone 192(e)

German designation 4 cm Pak 192(e)
Original designation Ordnance QF2 pr on Carriage 2 pr
Calibre 40 mm
Length of piece (L/52): 2081 mm
Length of bore 2000 mm
Length of rifling 1672 mm
Weight complete 840 kg
Weight in action 757 kg
Traverse 360°
Traverse from wheels 14° left to 9° right
Elevation −13° to +15°
Muzzle velocity 853 m/sec max
Shot weight 1 kg
Maximum range 7315 m
Maximum effective range 548 m
Armour penetration 60 mm at 183 m; 55 mm at 365 m; 51 mm at 548 m
Manufacturer Royal Ordnance Factory, Leeds

Remarks: Entered British Army service 1938 and standard equipment 1939–40. Rather heavy for its calibre and performance. Most British guns left in France 1940 later deployed as coastal/beach defence guns by German troops.

5.7 cm Panzerabwehrkanone 208(r)

5.7 cm Pak 208(r), 1943 production series

German designation 5.7 cm Pak 208(r)
Original designation *157 mm Protivotankovaya Pushka obr. 1941 g* (57-41 ZiS-2)
Calibre 57 mm
Length of piece (L/73): 4162 mm
Weight in action 1125 kg
Traverse 56°
Elevation −10° to +18°
Muzzle velocity 1020 m/sec
Shot weight 3.148 kg
Shell weight 4.2 kg
Maximum range (AP): 4000 m
Maximum range (HE): 5200 m
Armour penetration 140 mm at 500 m
Manufacturer Various Soviet State arsenals

Remarks: An efficient Soviet dual-purpose (AT/field gun) piece introduced in service in spring 1941. Later *obr. 1943* (ZiS-3) differed in having tubular trails on split trail carriage. Captured examples impressed in German service on Eastern Front and used until end of hostilities.

Light anti-aircraft guns

For the purpose of this book, light anti-aircraft guns will be taken as those with calibres between 20 mm and 40 mm with an operating ceiling between ground level and up to about 1500 metres.

The light anti-aircraft gun is yet another weapon that can trace its origins in the events of WW I. By 1918 war in the air had reached the stage where air-to-air combat was at a level of extreme sophistication compared with the first aggressive excursions of 1914 and 1915. Aircraft armament had already advanced to the use of automatic cannon replacing the earlier machine guns and handguns and some of these cannon were capable of firing light explosive shells. What was suitable for aircraft armament also made a viable light gun suitable for ground defence against aircraft and, as will be seen, the paths of aircraft guns and light anti-aircraft guns were often to cross one another.

Among the early aircraft cannon was a 20 mm weapon designed by Reinhold Becker, who submitted his first patent for such a gun as early as 1914. He produced these guns in his own factory at Willich-am-Rhein and by 1918 numbers were in use as heavy bomber defence guns. After 1918 Becker was reluctant to allow his design to be scrutinised by the various Treaty commissions, and formed an alliance with the Swiss Seebach Maschinenbau AG based near Zürich, usually known as Semag. All the Becker drawings and tooling were transferred to Switzerland and there the Becker gun was made and developed further under both the Becker and Semag trade names. Among the various versions sold were a 20 mm aircraft cannon and a 25 mm 'infantry' gun, and some sales were made to China and Japan. But the fortunes of the Semag/Becker consortium waned and the company was taken over by the Werkzeug Maschinenfabrik Oerlikon, based at the village of Oerlikon near Zürich. Oerlikon possessed both a new factory and a powerful sales team and the old Becker/Semag designs continued to be sold and progressively developed by the Oerlikon design team to the stage where the Oerlikon guns were sold widely and in large numbers. Most of the output was intended for aircraft use but an ever-growing number of light anti-aircraft guns, based closely on aircraft weapons, were also sold. Numbers of Oerlikon guns were sold to Germany where the ground version was designated the *2 cm Flak 28* and *29*. By 1939 most of these were used by the German Navy but later the numbers used were increased by the use of captured examples from Holland, France and elsewhere.

But Becker was not the only one to take his advanced weapon designs out of Germany in 1918. In that same year Heinrich Erhardt was the head of a Rheinmetall design team working on a light 20 mm cannon at Unterlüss. This cannon was virtually a scaled-up Dreyse machine gun using an almost identical action based on a Schmeisser design dating from 1907. As the war ended work on this promising gun came to an abrupt end but with an eye to the future all the working drawings, tooling, parts and models of the gun were secretly shipped to warehouses in neutral Holland where they remained until about 1925 when work was once more resumed, in secret and on a small scale. In 1929 there was an attempt by Rheinmetall to form a Dutch 'shadow' company but it did not materialise. Later the same year Rheinmetall were more successful when they took over the Swiss Solothurn AG which then switched from watchmaking to weapon development and production for the German concern. All the 1918 drawings and models were transferred to Switzerland and the Erhardt design was developed into the Solothurn MK-ST-5 and offered commercially (the MK-ST-11 was an aircraft version). During 1934 the gun was accepted by the German Navy for use on both single and twin mountings and in 1935 the same gun was accepted by the Luftwaffe as the *2 cm Flak 30*. Manufacture was switched from Switzerland to various factories in Germany and large scale mass production began.

By the late 1930s it had been realised that the *Flak 30* was not the perfect weapon that GL/Flak thought they needed. The rate of fire was too low and feed troubles were never entirely eliminated. By the time it was realised that further development was necessary Rheinmetall found themselves too involved in other work and the task was passed on to the Mauser team who carried out the required work under the unofficial designation of '*Flak 35*'. The resultant weapon featured the required increase rate of fire for which the carriage had to be modified (by Gustloff of Suhl). This new gun was offered as a *Flak 30* replacement at which point Rheinmetall once more entered the picture by offering the *2 cm MG C/35*, a modified aircraft gun, as an alternative. Mauser's submission was considered the more suitable and production began in 1939 as the *2 cm Flak 38*.

The *Flak 38* became the 'standard' German light anti-aircraft gun of all the German services from 1940 onwards. Combat experience in Poland and the Low Countries during 1940 revealed the shortcomings of the *Flak 30* and thereafter it was gradually supplanted, but never entirely replaced, by the *Flak 38*. In August 1944 there were 17,589 *Flak 30* and *38*s in service with the Luftwaffe alone. The German Navy used this gun on numerous mountings, including the *Doppellafette M 43* and *M 45*. Land mountings were produced in both mobile and static forms, and most of the single mountings were made at three main production centres – Benteler-Werke at Bielefeld, Brennabor-Werke at Brandenburg, and Gustloff-Werke at Suhl. The guns themselves were manufactured not only by the Mauser-Werke at Oberndorf, but also by five other concerns.

The most effective weapon using the basic *Flak 38* was undoubtedly the *2 cm Flakvierling 38*. Originally produced for the German Navy, this consisted of four *Flak 38* guns mounted on their sides on a single carriage. When fired, the amount of fire power produced was lethal to any low-flying aircraft likely to be caught in its sights and soon it established itself as the most feared of all the German light anti-aircraft guns. Allied aircrew became very wary of its capabilities, especially when the *Flakvierling* was employed in batteries. At sea the *Flakvierling* soon proved to be too heavy for light craft and U-boats so a new and lighter version was produced, the *Flakvierling 38/43*. This mounting became especially important on U-boats, for which purpose it was specially waterproofed, as the long-range Allied patrol aircraft gradually restricted the surface activities of submarines. One of these naval mountings was used for experiments with radar ranging during the closing months of the war. The gun was designated *2 cm Flakvierling 38/43F*, and the radar dish (the equipment was developed by Telefunken) was fixed in the centre of the four barrels.

To fully utilise the capabilities of the *Flakvierling 38* on land they were frequently emplaced statically on special Flak towers round important targets. It had been intended that only two guns were to be fired at any one time but this was soon disregarded and the full fire of all four barrels was directed at low-flying aircraft. Like all other light anti-aircraft guns, the *Flakvierling 38* could also be used against ground targets, and especially tanks. By 1944 it was a not uncommon practice to place *Flak 30* and *38* guns on extemporised ground mountings for use as anti-tank guns. Many of these mountings were simple crude affairs made from light steel tubing. For anti-aircraft use the *Flak 30* and *38* used different carriages but by 1943 Gustloff-Werke at Suhl had produced a carriage to take both guns.

Variants of the basic *Flak 38* included the *2 cm KwK 38* intended for use in light tanks and armoured cars. Another variant was the *2 cm GebirgsFlak 38* which had a rather unusual development history. This equipment was originally the *Gerät*

239 and used a new 2 cm gun and carriage developed by Gustloff of Suhl. A trials batch of twenty-five was produced but the results were not encouraging. The carriage proved unstable (it was intended for stripping down into several components as man-pack loads) and it was also felt that the provision of an extra weapon type, even for such a specialised role, was too extravagant. As a result the *Flak 38* was placed on the *Gerät 239* carriage and the *GebirgsFlak 38* entered service. Like the gun it replaced, it was really too heavy for the carriage and by February 1945 only 180 were in service. By that time these guns were also used by the Luftwaffe paratroop formations.

Other 20 mm guns that did not see service were the Navy *2 cm Flak 40* and the *Gerät 40*. The latter was a land version of the naval gun, and both were still under development as the war ended. By that time it was felt that 20 mm was too light a calibre for the increased speeds and armour of modern tactical aircraft, and to ensure the destruction of such a target 30 mm was to be considered the minimum. But by then the Allied air forces had almost complete air supremacy over the remaining German territories and any movement by the German land forces had to be covered by an 'umbrella' of anti-aircraft fire. The correspondingly greatly increased demands for light Flak guns could not be met by the output of German industry. This situation led to various makeshift expedients, chief among which were spare ex-Luftwaffe guns made available to the ground forces, including numbers of *MG 151/20* cannon, the 20 mm version of the Mauser *MG 151* mentioned in the section on heavy machine guns. Most of these were placed on makeshift carriages, usually in threes, to provide some form of airfield defence for the Luftwaffe.

Another aircraft gun that was considered for use as a possible Flak weapon was the *3 cm MK 108*. Despite several attempts, under the direction of Mauser, no suitable form of mounting for this large gun was evolved and anyway, the ammunition supply was restricted. Attention was focussed on the use of the *3 cm MK 103*, and once again the Gustloff-Werke were called upon to alter the carriage of the *Flak 38* as a possible expedient. The result was the *3 cm Flak 103/38* and mass production was commenced in November 1944. A four-barrel version became the *Flakvierling 103/38*. But by late 1944 German industry was in such a state that the large numbers expected could not be produced. Apart from that, the 3 cm gun was really too heavy for the carriages used, and in action the gun proved unstable and inaccurate. Skoda at Pilsen started to develop a twin-barrel version which was still on the drawing-board when the war ended.

Another 3 cm gun that did not get beyond the early development stage was the 3 cm *Flak 44*, a Navy weapon. The eventual contract went to Brünner-Waffenwerke who produced a prototype as the *MK 303(Br)*. In competition Mauser offered the *3 cm Flak M 44* (300 M), and Rheinmetall also produced a design. The war ended before any progress could be made.

Of the light anti-aircraft calibres used by the Germans, the second most important was 37 mm. This calibre was chosen because it offered a useful increase in range and shell weight over the 20 mm guns. Once again Rheinmetall were early in the field as a result of their Swiss Solothurn interests. Solothurn were used as the development agents for an early Rheinmetall design, the ST 10. In Switzerland this was developed into the Solothurn S10-100, and emerged as a scaled-up version of the *2 cm Flak 30*. This gun became the *3.7 cm Flak 18* and was issued for service with the German forces in 1935. But in 1936 production of this model ceased in favour of a new version, the *3.7 cm Flak 36*. The main reason for this short production run was the relative bulk of the *Flak 18* carriage which needed two two-wheeled axles to carry it. With the *Flak 36* this carriage was adapted for transport on a single axle, but it was not long before another change was introduced, a new mechanical sight, which involved some modification of the carriage. The changes involved were so extensive that a new designation was thought necessary and the *3.7 cm Flak 37* emerged.

The *3.7 cm Flak 18, 36* and *37* served in some numbers until 1945. In August 1944 there were 4211 in Luftwaffe service but by then many had been produced for static concrete emplacements. Another change came in 1940 when the chambers of the guns were altered to take ammunition with revised driving bands. The bulk of the production was carried out at three centres which also turned out 3.7 cm guns for the German Navy. One of these Navy guns was the *3.7 cm SKC/30*, many of which were used for coastal defence. A later product was the *3.7 cm Flak M 42*, based on the *Flak 36$37*.

Good as the *Flak 18, 36* and *37* were, by the late 1930s it was felt that the series lacked range when compared to the 2 cm equipments and their rate of fire was considered to be too low for the increasing speeds of modern aircraft. A new specification was put out to the three main armament concerns of Mauser, Krupp and Rheinmetall-Borsig. In view of the large stocks of 37 mm ammunition ready to hand it was stated that the existing round would have to be used so any increase in range was precluded from the start and all possible changes were concentrated on an increase in the rate of fire. Mauser produced their *Gerät 337 M* but this design did not enter the running which was soon centred on the Rheinmetall and Krupp proposals. The Krupp entry was the *Gerät 339* and it was this type that was chosen for production in 1942. At that time Krupp were overloaded with work and the production of this new gun was transferred to Dürkopp at Bielefeld. Setting up this production line involved a great deal of time and effort, but in the meantime Rheinmetall, who were far from happy at the outcome, produced a new version of their submission, the *Gerät 338*, which involved the experience gained in modern small-arms production. It had been intended from the start that the assembly of the new gun would involve as much simple sub-contracting work as possible, but the new Rheinmetall design went one better in using a great deal of simple stampings and sheet-metal components in place of heavy machined parts. Normally, the Krupp gun would have gone into production despite the new Rheinmetall design but in June 1942 the Krupp prototype revealed serious weaknesses in the carriage and GL/Flak changed their preference to the Rheinmetall submission. This did not suit Reichsminister Speer who insisted that the Krupp gun, on which a great deal of production effort had already been spent, should go ahead. OKL became involved and there developed a high-ranking and time-consuming wrangle between Speer and the Luftwaffe chiefs. In the meantime the Dürkopp line stood idle and only the successful trials of the Rheinmetall design showed that the Luftwaffe was correct in its stand. The Rheinmetall gun was standardised by the OKL as the *3.7 cm Flak 43*, and the Dürkopp lines had to be drastically altered to build the new gun. Again, precious time was lost and it was not until early 1944 that the first production series came off the line. By that time heads had rolled and the whole affair proved damaging to the already brittle relationships between the Luftwaffe and Reichsminister Speer's department.

In service the *Flak 43* soon proved to be a successful gun and Speer asked for ever-increasing numbers of them. A new assembly line was started at the Eisenwerke Weserhütte at Bad Oeynhausen. It was realised early on that an even higher rate of fire would be desirable and the simplest way to achieve this would be by doubling the number of barrels. The result was the *3.7 cm Flakzwilling 43* which had two superimposed barrels. In this form the gun was a rather bulky cumbersome load and the single axle carriage of the *Flak 43* had to be abandoned in favour of a new twin-axle carriage. The prototype was completed in 1943 and immediate large-scale production was ordered at six industrial centres. This large number of production centres was not only a reflection of the importance of the *Flakzwilling 43* but a measure of the ease with which firms not used to weapon production could turn out modern weapons that were designed from the start to include modern production techniques. But despite all the effort involved the number of guns actually delivered was disappointingly low. By February 1945 there were 1032 *Flak 43*s

Rheinmetall (Solothurn) 2 cm Flak Modell ST 5, *a naval weapon used for coastal AA defence*

in service of which only 380 were *Flakzwilling 43* guns.

Despite the dire need to concentrate on the production of existing weapons time and resources were also found to continue development. At one point Reichsminister Speer, who seems to have realised that his earlier stand was mistaken, ordered development of a *Flakvierling* version of the *Flak 43*. Not surprisingly, this was not proceeded with as by 1943 it had been realised that the decision not to proceed with new ammunition for the *Flak 43* was wrong and more powerful rounds were needed. A *Flakvierling 43* would have had the bulk and weight of a 8.8 cm gun with none of the attendant range advantages. Therefore work proceeded on new 3.7 cm guns with more modern ammunition, but none of the projects came to anything. The *Gerät 341* was a twin-barrelled weapon with the barrels side by side. As far as can be determined only four were actually built, but it was planned that the gun would use the *Flak 43* mechanism combined with new ammunition, and many sources state that it would have been mounted on a Panther tank chassis. Other sources quote the designation *Flak 45* for this project. Another project about which very little is known was *Fledermaus* which was under the control of Stübgen of Erfurt. Various naval mountings were introduced for the *Flak 43*, many of which were very advanced and included such things as protecting the gun-layer in a deck well which was raised and lowered to suit the angle of elevation. Both single and twin mountings were projected.

The only other 'German' light anti-aircraft gun that remains to be mentioned is the *4 cm Flak 28*, the well-known Swedish Bofors obtained in relatively small numbers before 1939. After the war started large quantities fell into German hands from Poland, Norway, the Low Countries, France, Yugoslavia and the aftermath of the Dunkirk evacuation. The German Flak troops grew to appreciate the qualities of this superb design and the weapon was kept in small-scale production at the Waffenfabrik Kongsberg in occupied Norway.

At this point it would be as well to consider a few aspects of the German light anti-aircraft scene that have not been covered. One of these concerns the gun sights. When the war started most light anti-aircraft sights were virtually miniature predictors which involved the use of clockwork or electrical mechanisms. Gradual change and development soon led to these sights growing so complex that they defeated their own object, and the costs involved were enormous. There began a change to simpler forms of gun sight. For a while reflector sights were used but they were gradually replaced by simple ring sights and the use of tracer for aiming. The gun layer simply fired in the general direction of a target and used the tracer to guide him to the required point. The use of tracer for aiming did away with all but the simplest of sighting arrangements and it also eliminated one of the gun detachment who normally used the hand-held rangefinder. By late 1944 gun crews were drastically reduced and light guns were even being seriously considered for remote control under the central guidance of fire control radar.

Another sphere that requires attention is the ammunition sector. During the war considerable strides were made in increasing the ammunition efficiency in even the smallest calibres. In 1939 the high explosive filling of a 3.7 cm shell was about 30 grams. By 1945 it had increased to 90 grams and this was combined with an efficient point contact fuse in addition to the normal 'safety' time fuse. This change was due to the introduction of thin-walled '*Minengranate*' which were capable of containing larger amounts of filling with little or no weight increases. Since it had been calculated that at least 75 grams of explosive were needed to ensure the destruction of an aircraft it was obvious that only calibres of 30 mm and above were capable of bringing down modern aircraft – smaller calibres had to rely on high fire rates, which explains why the *2 cm Flakvierling 38* was the only viable combination of 20 mm barrels that could still prove really effective in 1945. Most of this combat research was carried out by the Luftwaffe for air warfare but the results were also used in practice with ground weapons.

4 cm Flak(e). This was the British QF 2-pr Mk 8 naval gun adapted by German field armourers for static AA defence purposes

In the light anti-aircraft field the use of captured weapons was not so critical as it was with other types of weapon. Germany was generally well able to provide its own types of weapon and it was only in the last months of the war that the supply situation grew really desperate. Nevertheless, a large quantity of light anti-aircraft guns did fall into German hands as a result of the various conquests of 1939–42, and many of them were found to be useful weapons.

Like the 40 mm Bofors (*4 cm Flak 28*) mentioned earlier, this war booty was of international origin. In 1940, Denmark provided a selection of 20 mm Madsen guns. Of modern design, the Madsen had been produced in various versions and most were impressed into German service on Danish soil. The fall of France in the same year contributed large numbers of 25 mm Hotchkiss AA guns which generally remained in France under German control. In 1941–42 the Soviet gun manufactureing concerns involuntarily added their share with numbers of 37 mm obr. 1939 AA guns, many of which were modified to suit German standards and later used by various Reich defence Flak formations. In Italy, German troops took over numbers of 20 mm Breda and Scotti guns, and the 37 mm Breda AA guns were also impressed. All sorts of odd items were deployed along the Atlantic coast where the various garrison units spent some of their time converting otherwise unlikely weapons into useful light AA guns. A good example of this activity could be seen in land-mounted 40 mm Vickers naval AA guns, the famous 'pom-poms', taken off damaged or half-sunk British warships and subsequently put to German use. But this was hardly typical of the overall scene as the light anti-aircraft field was one in which the Germans needed little outside help until the war was nearly over.

Appendix 1

Light anti-aircraft guns in Luftwaffe service September 1942 to February 1945

1942	2 cm Flak 30.8	2 cm Flak-vierling 38	3.7 cm Flak 18.36	3.7 cm Flak 43	3.7 cm-Flak-zwilling 43
September	14,434	693	1760		
October	14,746	751	1792		
November	15,541	820	1789		
December	16,428	936	1926		
1943					
January	16,985	1062	2375		
February	17,327	1204	2107		
March	17,819	1225	2201		
April	18,294	1415	2275		
May	18,010	1439	2415		
June	18,029	1574	2474		
July	18,271	1697	2585		
August	18,637	1891	2743		
September	18,775	2017	2791		
October	...	...	...		
November	18,887	2149	3072		
December	...	...	...		
1944					
January	19,001	2602	3358		
February	19,465	2719	3548		
March	19,692	2825	3669		
April	19,626	3026	3723		
May	19,674	3107	3833		
June	18,808*	3414	3969		
July	17,702	3277	4024		
August	17,000	3424	3999	431	41
September	12,291	3136	2592	520	58
October	13,030	3577	3120	633	93
November	12,609	3464	2913	750	150
December	12,563	3701	2756	904	182
1945					
January	11,999	3806	2778	942	283
February	10,531	3768	2601	1032	380

* 904 guns handed over to Army

2 cm Flak 28 in action

Appendix 2

Foreign light anti-aircraft guns in Luftwaffe service August 1943 to February 1945

	2 cm Flak 28 Oerlikon	2 cm Flak Scotti	2 cm Flak Breda	2.5 cm Flak Hotchkiss	3.7 cm Flak M 39(r)	3.7 cm Flak Breda	4 cm Flak 28
1943							
August	836	—	—	488	531	-	304
September	777	—	—	514	569	-	289
October	881	—	—	529	612	-	298
November	834	34	5	532	666	48	258
December	861	34	10	556	642	62	234
1944						62	199
January	803	72	45	523	642	56	194
February	815	73	46	550	641	55	186
March	813	188	59	549	652	69	174
April	818	197	75	622	644	68	191
May	841	327	92	611	639	56	163
June	851	312	123	639	652	53	214
July	769	352	170	618	651	69	83
August	713	431	281	609	642	88	112
September	676	465	305	319	600	74	126
October	727	344	395	315	624	71	111
November	772	379	357	322	606	61	92
December	708	406	427	323	585		
1945							
January	677	496	469	314	532	57	32
February	714	497	423	296	489	55	34

2 cm Flugabwehrkanone 30

2 cm Flak 30 engaging ground targets

2 cm Flak 30 defending an oil storage plant

German designation 2 cm Flak 30
Prototype designation Solothurn S5-100
Calibre 20 mm
Length of piece 2300 mm
Length of bore (L/65): 1300 mm
Length of rifling 720 mm
Weight travelling 890 kg
Weight in action with sights 463 kg
Traverse 360°
Elevation −12° to +90°
Type of feed 20 round magazine
Muzzle velocity (AP): 830 m/sec; (HE): 900 m/sec
Shot weight 0.33 kg; (AP40): 0.14 kg
Shell weight 0.3 kg
Maximum effective ceiling 2200 m
Rate of fire (cyclic): 280 rpm; (practical): 120 rpm
Barrel life 10–12,000 rounds
Original manufacturer Rheinmetall-Borsig AG, Düsseldorf. (For later production *see* Flak 38)

2 cm Flak 30 on a ground mount

Remarks: Slightly modified Solothurn S5-100 gun. Adopted as 'standard' in Germany 1935 and during WW II widely used by all services. Also fitted on a variety of mobile mounts and employed in dual AA/light ground support role.

*2 cm Flak 30 on its special two-wheeled trailer (*Sonderanhänger 51*)* ►

---

2 cm Flugabwehrkanone 38

German designation 2 cm Flak 38
Calibre 20 mm
Length of piece 2252.5 mm
Length of bore (L/65): 1300 mm
Length of rifling 1158 mm
Weight travelling 860 kg
Weight in action 405 kg
Traverse 360°
Elevation −20° to +90°
Type of feed 20 round magazine
Muzzle velocity (AP/HE): 830 m/sec; (HE): 900 m/sec
Shot weight 0.33 kg
Shell weight 0.3 kg
Maximum effective ceiling 2200 m
Rate of fire (cyclic): 450 rpm; (practical): 220 rpm
Barrel life 10–12,000 rounds

2 cm Flak 38 mounted on a sled

Manufacturers (guns): Mauser-Werke AG, Oberndorf-am-Neckar; Havelwerke, Brandenburg; Gustloff-Werke, Meiningen; Röchling-Buderus, Wetzlar; Ostmark-Werke, Wien; Stübgen & Co., Erfurt; (carriages): Benteler-Werke, Bielefeld; Brennabor-Werke, Brandenburg; Gustloff-Werke, Suhl

2 cm Flak 38 on its trailer

Remarks: Improved Flak 30 with increased rate of fire. In service supplanted but never completely replaced earlier model. Widely used by all German services; also fitted on a variety of self-propelled mounts. Each basic Flak 38 cost RM3000; with standard carriage – RM6000.

---

2 cm Gebirgsflugabwehrkanone 38

German designation 2 cm GebFlak 38
Calibre 20 mm
Length of piece 2252.5 mm
Length of bore (L/65): 1300 mm
Length of rifling 1158 mm
Weight complete 315 kg
Weight in action 276 kg
Traverse 360°
Elevation −28° to +90°
Type of feed 20 round magazine
Muzzle velocity (AP/HE): 830 m/sec; (HE): 900 m/sec
Shot weight 0.33 kg
Shell weight 0.3 kg

2 cm Flak 38 Sondergeschütz für die Fallschirmtruppe. *Intended for use by airborne troops, this version was designed to be dropped by parachute*

Maximum effective ceiling 2200 m
Rate of fire (cyclic): 450 rpm; (practical): 220 rpm
Barrel life 10–12,000 rpm
Manufacturer (carriage): Gustloff-Werke, Suhl

Remarks: Basically a combination of Flak 38 barrel and new carriage evolved for unsuccessful Gerät 239 light dual-purpose gun. Entered service 1942, but produced only in small numbers; limited use by German mountain and airborne troops.

2 cm Geb Flak 38 with wheels removed, on a tripod mount

2 cm Flak 38 positioned under a Ju 52/3m transport aircraft with the parachute attached

A small auxiliary wheel was attached to the gun trails for easier manhandling

2 cm Flak 38 Sondergeschütz für die Gebirgstruppe. *This variant was designed for use by mountain troops against ground targets*

2 cm Flugabwehrkanonevierling 38

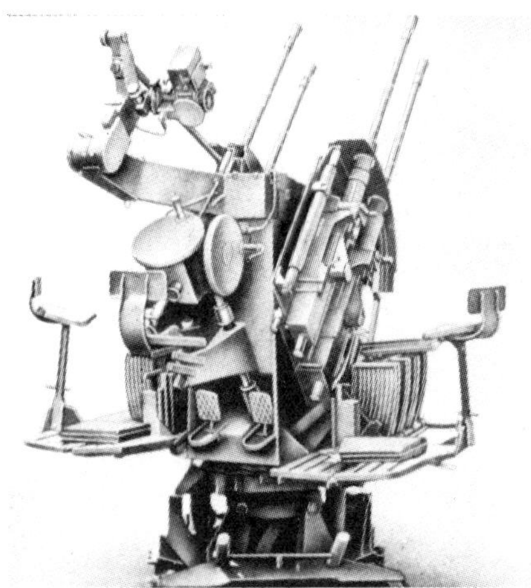

2 cm Flakvierling 38

2 cm Flakvierling on its special two-wheeled trailer (Sonderanhänger 52)

German designation 2 cm Flakvierling 38
Calibre 20 mm
Length of piece 2252.5 mm
Length of bore (L/65): 1300 mm
Length of rifling 1158 mm
Weight in action 1509 kg
Traverse 360°
Elevation −10° to +100°
Type of feed 4 × 20 round magazines
Muzzle velocity (AP/HE): 830 m/sec; (HE): 900 m/sec
Shot weight 0.33 kg
Shell weight 0.3 kg
Maximum effective ceiling 2200 m
Rate of fire (cyclic): 4 × 450 = 1800 rpm; (practical): 720–800 rpm
Barrel life 10–12,000 rounds
Manufacturers (carriage): Ostmark-Werke, Wien; Auto-Union AG, Chemnitz; Benteler-Werke, Bielefeld

Remarks: A most successful combination of four Flak 38 barrels on a modified carriage. Originally produced for Kriegsmarine in 1938, but ordered into production for Luftwaffe and ground forces 1940. Each complete Flakvierling 38 cost RM20,000. Became most effective light AA weapon, feared and respected by all low-flying Allied aircrews.

2 cm Flakvierling 38 on a 'High stand'.

2 cm Flakvierling 38 on a Flak tower. As a rule, the light guns were positioned on the lower shelves while the heavier AA guns were mounted on the top platforms. The height of these Flak towers ranged from 40 to 50 metres

2 cm Flakvierling 38 ready for action. Italy, 1943

2 cm Flugabwehrkanonevierling 38/43

German designation 2 cm Flakvierling 38/43
Calibre 20 mm
Length of piece 2252.5 mm
Length of bore (L/65): 1300 mm
Length of rifling 1158 kg
Weight complete with shield 2200 kg
Traverse 360°
Elevation −10° to +90°

Type of feed 4 × 20 round magazines
Muzzle velocity (HE): 900 m/sec
Shell weight 0.3 kg
Maximum effective ceiling 2200 m
Rate of fire (cyclic): 4 × 450 = 1800 rpm; (practical): 720–800 rpm
Barrel life 10–12,000 rounds
Manufacturer Rheinmetall-Borsig AG, Düsseldorf

Remarks: Specially waterproofed and sealed version of Flakvierling 38 intended for naval (U-boat) use. Not put into full-scale production due to change in U-boat tactics. Small number completed and deployed, mostly for shore AA defence. In April 1945 one gun, designated Flakvierling 38/43 'F', used for radar-controlled firing trials. Equipment included Telefunken radar scanner mounted centrally between four barrels.

Maschinengewehr 151/20

An unusual improvised MG 151/20 carriage for anti-tank use. This equipment was constructed with parts of the US Army. 50 cal M3 machine gun mount

German designation MG 151/20
Calibre 20 mm
Length of gun 1766 mm
Length of barrel 1104 mm
Weight of gun complete 42 kg
Traverse Dependent on installation
Elevation Dependent on installation
Type of feed Disintegrating metal link belt or drum magazine
Muzzle velocity (HE): 805–810 m/sec; (AP/HE): 705 m/sec

Shot weight 0.115 kg
Rate of fire (cyclic): 700 rpm
Manufacturer Mauser-Werke AG, Oberndorf-am-Neckar

Remarks: Ex-Luftwaffe aircraft cannon converted for ground use on a variety of mountings.

Fliegerabwehr-Sockellafette mit 3 MG 151 *(Fla SL 151) AA pedestal mount with 3 × MG 151/20 aircraft cannon*

3 cm Flugabwehrkanone 103/38

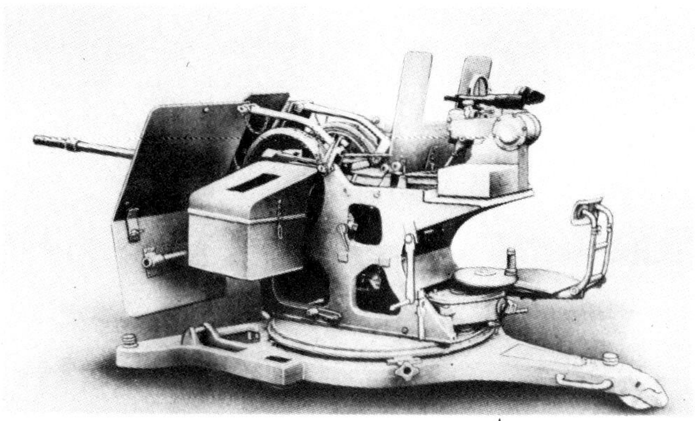

3 cm Flak 103/38. Note the fitting of both reflex AA and telescopic ground target sights on the same holder

German designation 3 cm Flak 103/38
Calibre 30 mm
Length of piece 2318 mm
Length of barrel (with muzzle brake): 1608 mm; (less muzzle brake): 1338 mm
Length of rifling 1159.7 mm
Weight travelling 879 kg
Weight in action 619 kg
Traverse 360°
Elevation −10° to +80°
Type of feed 30 or 40 round magazine (disintegrating metal links)
Muzzle velocity (HE): 800 m/sec; (Inc): 900 m/sec
Shell weight (HE): 0.815 kg
Maximum effective ceiling 1500–1600 m
Rate of fire (cyclic): 400 rpm; (practical): 250 rpm
Manufacturer Rheinmetall-Borsig AG, Düsseldorf; Gustloff-Werke, Suhl

Erd-Lafette für MK 103. *This was an emergency ground mount for the Mk 103 aircraft cannon consisting of a simple device that could be fixed on any strong wooden post* ➤

Remarks: Basically a combination of MK 103 aircraft cannon and 2 cm Flak 38 carriage. Developed in mid-1944, but proved an unsatisfactory, muzzle-heavy weapon. Also evolved in Flakvierling version. Mauser-Werke, Brunserwerke and Skoda also built prototypes, but none of these reached production stage.

3.7 cm Schiffskanone C/30 in Einheitslafette C/34

German designation 3.7 cm SKC/30
Calibre 37 mm
Length of piece (L/83): 3076 mm
Length of barrel 2962 mm
Travers 360°
Elevation −10° to +80°
Type of feed 6 round linked clips
Muzzle velocity 1000 m/sec
Shell weight 0.745 kg
Maximum effective ceiling 2000 m
Barrel life 3000 rounds
Manufacturer Rheinmetall-Borsig AG, Düsseldorf

Remarks: Designed as naval AA gun. Development commenced 1930, in service by 1935. During WW II used as dual-purpose AA and coastal defence weapon.

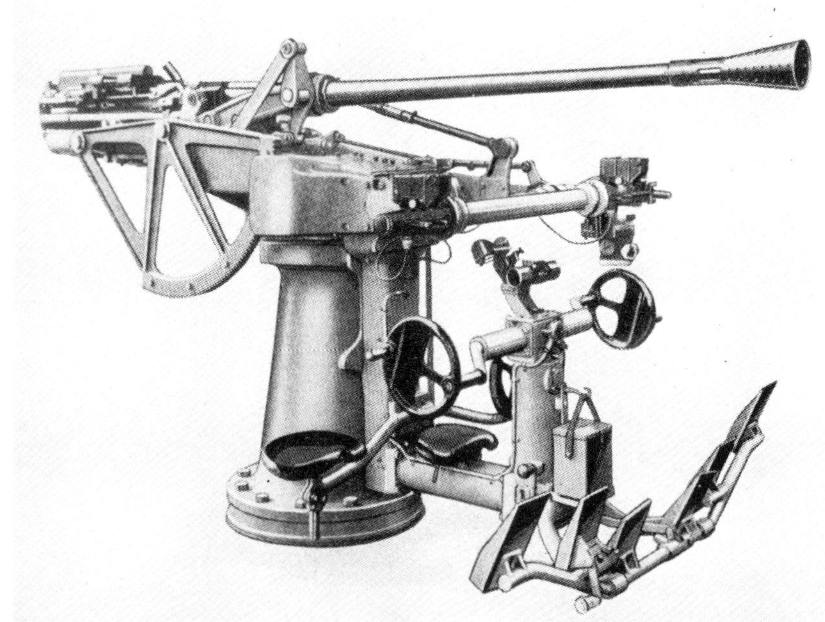

3.7 cm Schiffskanone *C/30 on a standard C/34 naval mount for static AA defence* ➤

3.7 cm Flugabwehrkanone 18

German designation 3.7 cm Flak 18
Prototype designation Solothurn S10-100
Calibre 37 mm
Length of piece 3626 mm
Length of bore (L/57): 2112 mm
Length of rifling 1826 mm
Weight travelling 3560 kg
Weight in action 1750 kg
Traverse 360°
Elevation −5° to +85°
Type of feed 6 round linked clips
Muzzle velocity (HE): 820 m/sec; (AP): 770 m/sec
Shell weight 0.625 kg
Shot weight 0.685 kg
Maximum effective ceiling 2000 m
Rate of fire (cyclic): 160 rpm; (practical): 80–100 rpm
Barrel life 8–10,000 rounds
Manufacturer Rheinmetall-Borsig AG, Düsseldorf

Remarks: Entered service in 1935, but due to various mechanical troubles produced only in

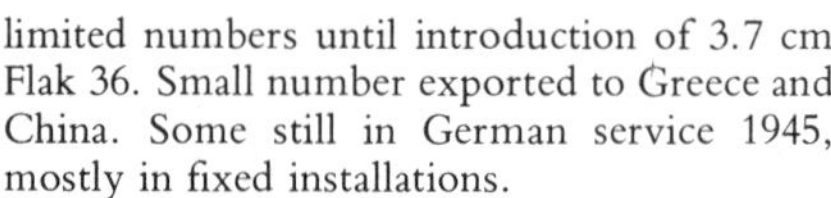
limited numbers until introduction of 3.7 cm Flak 36. Small number exported to Greece and China. Some still in German service 1945, mostly in fixed installations.

3.7 cm Flugabwehrkanone 36 and 37

Loading the 3.7 cm Stielgranate 41

The 3.7 cm Stielgranate 41 *hollow-charge projectile was used to increase the anti-tank capability of the 3.7 cm Flak 36. Note the improved armour protection for the crew*

A Waffen-SS 3.7 cm Flak 36 in action on the Eastern Front

3.7 cm Flak 36. Note the 21 'victory' rings on the barrel, a common German practice to record destroyed air or ground targets

German designation 3.7 cm Flak 36 oder 37
Calibre 37 mm
Length of piece 3626 mm
Length of bore (L/57): 2112 mm
Length of rifling 1826 mm
Weight travelling 2414 kg
Weight in action 1552 kg
Traverse 360°
Elevation −8° to +85°
Type of feed 6 round linked clips
Muzzle velocity (HE): 820 m/sec; (AP): 770 m/sec
Shell weight 0.625 kg
Shot weight 0.685 kg
Maximum effective ceiling 2000 m
Rate of fire (cyclic): 160 rpm; (practical): 80–100 rpm
Barrel life 8–10,000 rounds
Manufacturers Dürkopp-Werke, Bielefeld; DWM, Berlin-Borsigwalde; Skoda-Werke, Pilsen

Remarks: Basically Flak 18 barrel on a new lightweight 2-wheel carriage. Introduced 1936 and widely used by all services during WW II; deployed on various fixed and mobile mounts. Flak 36 generally used Flakvisier 35 or 36; Flak 37 indicated same gun with Zeiss clockwork-type Flakvisier 37.

3.7 cm Flak 36 on Sonderanhänger 52

3.7 cm Flugabwehrkanone M 42

German designation 3.7 cm Flak M 42
Calibre 37 mm
Length of barrel (L/69): 2560 mm
Length of rifling 2267 mm
Weight of gun 300 kg
Weight installed with shield 1350 kg
Traverse 360°
Elevation −10° to +90°
Type of feed 8 round linked clips
Muzzle velocity (M-Schoss): 925 m/sec; (HE): 850 m/sec; (AP/HE): 815 m/sec
Shell weight (M-Schoss): 0.565 kg; (HE): 0.612 kg; (AP/HE): 0.7 kg
Rate of fire 160–180 rpm
Barrel life 7000 rounds
Manufacturer Rheinmetall-Borsig AG, Düsseldorf

Remarks: Designed for naval use. Development began 1939, in Kriegsmarine service from 1942. Number deployed for shore AA defence. Production terminated late 1944 in favour of 3.7 cm Flak 43.

3.7 cm Flugabwehrkanone 43

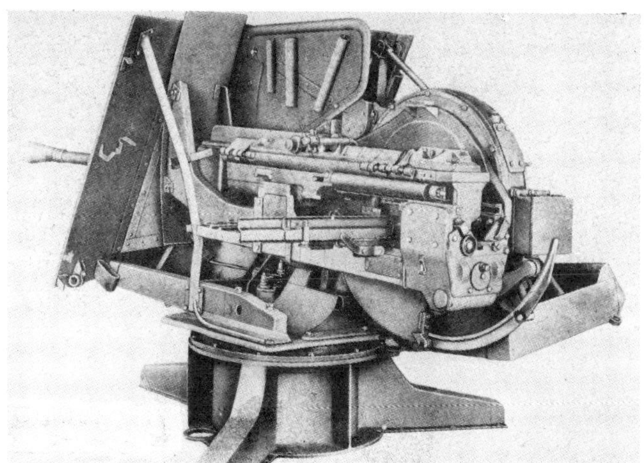

3.7 cm Flak 43. This equipment was transportable either on a single-axle Sonderanhänger 58 *or the twin-axle Sd Ah 104 or Sd Ah 206 trailers*

3.7 cm Flak 43 in firing position, with attached ammunition trays and the wire mesh cage for the ejected empty cartridges.

German designation 3.7 cm Flak 43
Design designation Gerät 338
Calibre 37 mm
Length of piece 3300 mm
Length of bore (L/57): 2130 mm
Length of rifling 1826 mm
Weight travelling 2059 kg
Weight in action 1392 kg
Traverse 360°
Elevation −7°30′ to +90°
Type of feed 8 round linked clips
Muzzle velocity (HE): 820 m/sec; (AP): 770 m/sec; (AP40): 1150 m/sec
Shot weight (AP40): 0.485 kg; (AP): 0.685 kg
Shell weight (HE): 0.625 kg
Maximum effective ceiling 4200 m
Rate of fire (cyclic): 230–250 rpm; (practical): 150 rpm
Barrel life 8000 rounds
Manufacturers Dürkopp-Werke, Bielefeld; Weserhütte, Bad Oeynhausen

Remarks: Original design with new gas-operated breech mechanism and incorporating many features for speedy large-scale production. In service from late 1943. Great improvement over existing 37 mm AA guns.

3.7 cm Flugabwehrkanonezwilling 43

3.7 cm Flakzwilling 43 ready for action, France, 1944. Note the sloping armour plating to protect the crew from low-level strafing attacks

German designation 3.7 cm Flakzwilling 43
Calibre 37 mm
Length of piece 3300 mm
Length of bore (L/57): 2130 mm
Length of rifling 1826 mm
Weight travelling 4290 kg
Weight in action 2780 kg
Traverse 360°
Elevation −8° to +90°
Type of feed 8 round linked clips
Muzzle velocity (HE): 820 m/sec; (AP): 770 m/sec
Shot weight 0.685 kg
Shell weight 0.625 kg
Maximum effective ceiling 4200 m
Rate of fire (cyclic): 500 rpm; (practical): 300–360 rpm
Barrel life 8000 rounds
Manufacturers Metallwerke Creussen, Creussen; DWM, Berlin-Borsigwalde; Gustav Appel, Spandau; Eustergerling Masch.-und-Apparatenbau, Bielefeld; Neumärk Gerätebau GmbH, Landsberg; Dollberg, Berlin-Rudow

Remarks: Two superimposed Flak 43 barrels for greater fire-power on modified mount. Extremely powerful weapon against low-level aircraft. Only limited numbers in service from late 1944 onwards.

Gerät 339

German designation Gerät 339 B.Kp
Calibre 37 mm
Weight of gun 300 kg
Type of feed 8 round linked clips
Muzzle velocity 840 m/sec
Rate of fire 240 rpm
Manufacturers Joint Krupp/Dürkopp project

Remarks: Unsuccessful contender for Flak 43 programme. Only a few prototypes made and tested 1942–43.

Gerät 341

German designation Gerät 341
Calibre 37 mm
Length of piece 4315 mm
Length of barrel (L/77): 2860 mm
Length of rifling 2400 rpm
Weight of gun 470 kg
Muzzle velocity 1040 m/sec
Shell weight 0.725 kg
Rate of fire 250 rpm
Manufacturer Rheinmetall-Borsig AG, Düsseldorf

Remarks: Late 1944 development of 3.7 cm Flak 43 to fire new ammunition. Only four prototypes completed but early results not encouraging. Trials still in progress when war ended. Designation 3.7 cm Flak 45 may refer to this development.

3 cm Flugabwehrkanone M 44

German designation 3 cm Flak M 44
Calibre 30 mm
Length of piece 2500 mm
Length of barrel (L/53.3): 1600 mm
Weight of gun 180 kg
Weight of mounting 240 kg
Weight installed for action 690 kg
Traverse 360°
Elevation −10° to +75°
Muzzle velocity (M-Schoss): 900–920 m/sec (est); (HE): 800 m/sec (est); (AP/HE): 725 m/sec (est)
Shell weight (M-Schoss): 0.33 kg; (HE): 0.44 kg; (AP/HE): 0.5 kg
Rate of fire (est): 400–420 rpm
Manufacturer (gun): Rheinmetall-Borsig AG, Düsseldorf; (mounting): Dr Böhme & Co., Minden, Westphalia

Remarks: Designed for naval use but war ended before production could begin.

3 cm Flugabwehrkanone M 44

German designation 3 cm Flak M 44 (300 M)
Calibre 30 mm
Length of piece 2500 mm
Length of barrel (L/53.3): 1600 mm
Weight of gun 165 kg
Muzzle velocity (M-Schoss): 900–920 m/sec (est); (HE): 800 m/sec (est); (AP): 725 m/sec (est)
Shell weight (M-Schoss): 0.33 kg; (HE): 0.44 kg
Shot weight 0.5 kg
Rate of fire (est): 400 rpm
Manufacturer Mauser-Werke AG, Obern dorf-am-Neckar

Remarks: Late 1944 project for naval AA gun; not proceeded with. Intended to use same naval mount as 3.7 cm Flak 45.

3 cm Maschinenkanone 303 (Br)

German designation 3 cm MK 303 (Br)
Calibre 30 mm
Length of gun 3145 mm
Length of barrel (L/73) 2200 mm
Weight of gun 170 or 185 kg
Muzzle velocity (M-Schoss): 1060-1100 m/sec (est); (HE): 900 m/sec (est); (AP/HE): 950 m/sec (est)
Shell weight (M-Schoss): 0.33 kg; (HE): 0.44 kg; (AP/HE): 0.5 kg
Rate of fire (est): 400 rpm
Manufacturer Brünner-Waffenwerke, Brünn

Remarks: Projected naval AA gun for U-boats and light surface craft, but also intended for shore installations. Designed to use same naval mount as 3.7 cm Flak 37. Development not completed before May 1945.

Fledermaus

German designation 'Fledermaus'
Calibre 37 mm
Length of piece 4430 mm
Length of barrel (L/77): 2846 mm
Weight of gun (est): 400 kg
Weight of barrel 120 kg
Muzzle velocity (est): 1000 m/sec
Shell weight 0.725 kg
Rate of fire (est): 250 rpm
Manufacturer Stübgen, Erfurt

Remarks: Very little information has survived regarding this weapon but it does not appear to have been more than a design project for mounting on anti-aircraft AFVs.

2 cm Flugabwehrkanone 28 and 29

2 cm Flak 28 with its wheels attached to the gun mount for travelling

2 cm Flak 29, the 20 mm Oerlikon on a pedestal mount

German designation 2 cm Flak 28 und 29
Original designation Oerlikon Type FF
Calibre 20 mm
Length of bore (L/60): 1200 mm
Weight travelling 259 kg
Weight in action 227 kg
Traverse 360°
Elevation −15° to +85°
Type of feed 15 round box magazine; 60 round drum magazine
Muzzle velocity (AP): 810 m/sec; (HE): 850 m/sec
Shot weight 0.124 kg
Shell weight 0.128 kg
Maximum effective ceiling 1500 m
Rate of fire (cyclic): 280 rpm; (practical): 120 rpm

An Oerlikon MG FF/B German aircraft cannon on an improvised ground mount. Eastern Front, 1944

Original manufacturer Werkzeug-u. Maschinenfabrik Oerlikon, Oerlikon, Switzerland

Remarks: Evolved from an original 1914 German design by Reinhold Becker. First produced by Oerlikon 1921, and subsequently widely exported and also licence-built abroad. Oerlikons in German service comprised early purchases from Switzerland (used mainly by Kriegsmarine), augmented with guns acquired in Czechoslovakia, Poland, France (where built under licence), former Baltic States and Italy. A certain number also captured from British troops. Same basic cannon also used on various Luftwaffe aircraft.

---

2 cm Flugabwehrkanone Madsen

German designation 2 cm Flak Madsen
Original designation Madsen Model 1933 and 1935; Canon Mitrailleur CHM de 20 mm mle 1935
Calibre 20 mm
Length of piece 2250 mm
Length of barrel (L/60): 1200 mm
Weight of gun 52 kg
Weight complete 307 kg
Weight in action 260 kg
Traverse 360°
Elevation −5° to +85°
Type of feed round magazine
Muzzle velocity (AP): 730 m/sec; (HE): 890 m/sec
Shot weight 0.128 kg
Shell weight 0.136 kg
Maximum effective ceiling 2120 m

2 cm Flak Madsen on a pedestal mount

Rate of fire (cyclic): 400 rpm; (practical): 200–250 rpm
Manufacturer Dansk Industrie Syndikat Madsen A/S, Copenhagen or Herlev

2 cm Flak Madsen on a mobile mount

Remarks: One of commercially most successful light AA guns, sold to or licence-built in 20 countries before WW II. German forces acquired 20 mm Madsen guns from Czechoslovakia, Poland, Norway, Belgium and France.

---

2 cm Flugabwehrkanone Breda and 2 cm Maschinengewehr 282(i)

German designations 2 cm Flak Breda or 2 cm Breda(i); 2 cm MG 282(i)
Original designation Cannone-mitragliera da 20/65 modello 35 (Breda)
Calibre 20 mm
Length of piece 2031 mm
Length of bore (L/65): 1300 mm
Length of rifling 1159 mm
Weight of gun 68.5 kg
Weight complete 370 kg
Weight in action 330 kg
Traverse 360°
Elevation −10° to +80°
Type of feed 12 round clips
Muzzle velocity 830–850 m/sec
Shot weight 0.16 kg
Shell weight 0.31 kg
Maximum effective ceiling 2500 m
Rate of fire (cyclic): 220–240 rpm; (practical): 150 rpm
Manufactuter Societa Italiano Ernesto Breda, Brescia

Remarks: Designed as light dual-purpose gun. Adopted by Italian Army 1935 and used throughout WW II. Similar modello 39 combined same barrel with improved mountings. Used by German troops in North Africa and Italy.

2 cm Flak Breda. This was a dual-purpose AA/AT weapon carried in four pack loads or MT drawn

2 cm Flugabwehrkanone Scotti

German designation 2 cm Flak Scotti oder 2 cm Scotti(i)
Original designation Cannone-mitragliera da 20/77 (Scotti)
Calibre 20 mm
Length of piece 2275 mm
Length of bore (L/77): 1540 mm
Length of rifling 1400 mm
Weight of gun 73 kg
Weight in action 227.5 kg
Traverse 360°
Elevation −10° to +85°
Type of feed 12 round clips
Muzzle velocity 830–850 m/sec
Projectile weight 0.125 kg
Maximum effective ceiling 2135 m
Rate of fire (cyclic): 230–250 rpm; (practical): 120 rpm
Manufacturer Isotta Fraschini SpA, Torino

Remarks: An original Italian design, first produced by Oerlikon in Switzerland 1932. Adopted by Italian Army in 1937, but not manufactured in large numbers until 1942. Later series used optional strip or belt feed. Only limited German service use in Italy.

2 cm Flak Scotti was generally similar to the 2 cm Flak Breda

2.5 cm Flugabwehrkanone Hotchkiss, or 2.5 cm Hotchkiss 38 and 39

2 cm Flak Hotchkiss 38

2 cm Flak Hotchkiss 39. This version had a heavier platform mounting

German designations 2.5 cm Flak Hotchkiss; 2.5 cm Flak Hotchkiss 38 und 39
Original designation Mitrailleuse de 25 mm sur affut universel Hotchkiss
Calibre 25 mm
Length of piece 2997 mm
Length of bore (L/60): 1500 mm
Weight complete 1234 kg
Weight in action 850 kg
Traverse 360°
Elevation −5° to +80°
Type of feed 10 round magazine
Muzzle velocity (HE): 900 m/sec; (AP): 875 m/sec
Shot weight 0.324 kg
Shell weight 0.29 kg
Maximum vertical ceiling 5000 m
Rate of fire (cyclic): 350 rpm; (practical): 175 rpm
Manufacturer Société de la Fabrication des Armes Hotchkiss et Cie, St. Denis

Remarks: Designed as light dual-purpose weapon. First 25 mm Hotchkiss adopted by French Army 1930. Over 1100 guns of this type in service in May 1940, of which some retained by Vichy France. Mle 38 had light, mle 39 – heavier mounting. Used by German occupation troops in France.

3.7 cm Flugabwehrkanone Breda

German designation 3.7 cm Flak Breda or 3.7 cm Breda(i)
Original designation Cannone-mitragliera da 37/54 modello 39
Calibre 37 mm
Length of piece 3280 mm
Length of bore (L/54): 1998 mm
Weight complete 2975 kg
Traverse 360°
Elevation −10° to +90°
Type of feed 6 round trays
Muzzle velocity 800 m/sec
Projectile weight 0.8 or 1.25 kg
Maximum effective ceiling 4000 m
Rate of fire (cyclic): 200 rpm; (practical): 140 rpm
Manufacturer Societa Italiano Ernesto Breda, Brescia

Remarks: Entered Italian service in 1939 and originally intended for static Home defence. From 1942 also used on mobile mounts; numbers deployed on Eastern Front. Was classed as 'medium' by Italian forces.

3.7 cm Flugabwehrkanone M 39a(r)

German designation 3.7 cm Flak M 39a(r)
Original designation *137 mm Zenitnaya Pushka obr. 1939 g*
Calibre 37 mm
Length of piece 2738 mm
Length of bore (L/66.7): 2468 mm
Weight in action 2100 kg
Traverse 360°
Elevation −5° to +85°
Type of feed 5 round clips
Muzzle velocity 880 m/sec
Shell weight 0.785 kg
Maximum effective ceiling 3000 m
Rate of fire (cyclic): 160–180 rpm; (practical): 80 rpm
Manufacturers Various Soviet State arsenals

3.7 cm Flak M 39a(r) in travelling position

3.7 cm Flak M 39a(r) in firing position

Remarks: Based to large extent on 25 mm Bofors design. Entered Red Army AA service in autumn 1939, and remained in use in slightly modified form until well after WW II. Robust and simple to maintain; often employed in dual-purpose role. Many captured guns converted to take German fire control equipment.

4 cm Flugabwehrkanone 28

German designation 4 cm Flak 28
Original designations Bofors 40 mm L/60; (e): Ordnance Q.F. 40 mm Mks 1,1* and 3; (p): 40 mm armata przeciwlotnicza wz. 36
Calibre 40 mm
Length of piece (L/56.2): 2250 mm
Length of rifling 1932.5 mm
Weight travelling 2320–2500 kg
Weight in action 1730 kg
Traverse 360°
Elevation −5° to +90°
Type of feed 4 round clips
Muzzle velocity 823–875 m/sec
Shell weight 0.955 kg
Maximum effective ceiling 5000 m
Rate of fire (cyclic): 180 rpm; (practical): 120 rpm
Barrel life 2500 rounds
Original manufacturer AB Bofors, Bofors, Sweden
Licensed manufacturers MAVAG, Hungary; Starachowice, Poland; FN, Herstal, Belgium; Various British arsenals; Kongsberg, Norway, various plants in Austria, Denmark and France

4 cm Flak 28 in action

4 cm Flak 28 in travelling position

Remarks: Most widely used and one of most successful light AA guns of WW II, in service with both Allied and Axis forces. First produced by Bofors 1930 based on original 1928 design, and subsequently widely exported and licence-built abroad. Guns in German service came mostly from Poland. Additional Bofors were captured by German forces from Dutch and British troops, and a small number from Red Army in former Baltic States in June–July 1941. During German occupation of Norway limited production also undertaken by Kongsberg, mostly for Kriegsmarine.

Medium anti-aircraft guns

Medium anti-aircraft guns, for the purpose of this book, are taken to be those with calibres between 40 mm and 75 mm. These calibres were supposed to cover the altitude 'band' between 1500 and just over 3000 metres, but as will be seen, this was one area where the Germans were unable to design a suitable weapon. But here they were not alone for none of the Allies produced a service weapon capable of the same task.

In 1935 Rheinmetall were asked to commence work on a 50 mm gun for trial purposes. Krupp, Mauser and Gustloff were also requested to submit proposals, but it would appear that the priority for this was not high. In 1936, a Rheinmetall prototype was ready – this was the *Gerät 56 V1*. Progress after that date was slow, mainly due to the uncertainty as to the gun's precise role, but in 1940 work had progressed to the stage where a new version, the *Gerät 56 V2*, was selected for troop trials as the *5 cm Flak 41*. Sixty were made and the first were issued in November 1941. The best that could be said for this gun was that it was not a success. In service it proved cumbersome and slow to traverse, the rate of fire was too low, and the shell fired was really too light and underdeveloped for its task. However, of the sixty produced, twenty-four were still in service when the war ended, and by that time a battery had been converted for remote fire control. One of the worst problems involved with the *Flak 41* was the ammunition. When the gun fired it gave rise to a great deal of smoke and flash which did not help the gun layer.

The Krupp-designed prototype, the *Gerät 56 K*, was completed during late 1939. By that time, the need for a 5 cm was being seriously questioned and it seemed likely that the batch of 60 from Rheinmetall would be the sum total required. Nevertheless, a great deal of experimental work was carried out with the Krupp gun, especially with regard to new ammunition.

Despite the lack of success shown by the *Flak 41* it did reveal the need for some form of 'intermediate' Flak weapon with a calibre between 50 and 60 mm. Then, in March 1942, all further work on such a weapon was forbidden as the result of an economising campaign conducted by Reichsminster Speer and, not withstanding entreaties from GL/Flak and OKL, Speer refused to admit the need for such a weapon. Eventually even Hitler was called in to provide support for the decision. But by 1943 Allied air raids had reached the point where it was realised that there *was* a need for an 'intermediate' gun and work restarted on a new design with a calibre of 55 mm. Four firms (Rheinmetall, Krupp, Dürkopp and Gustloff) became involved in the programme and many of the resultant details showed the influence of the Krupp *Gerät 56 K*. The important difference was that the new gun was not just another anti-aircraft gun, but a part of what is now known as a 'weapons system' in which the gun is fully integrated with a central fire control radar, a predictor and all the other fire control systems needed. Development of such an extensive programme led to frequent delays and many time-consuming trials to the extent that by May 1945 only a few prototypes had been completed. The end result was the *Gerät 58* which had many novel features, including the 'soft recoil' system where the gun was sprung forward to fire when the resultant forces pushed back the barrel ready for the next shot.

By 1944 it was painfully obvious that the delays inflicted on the 'intermediate' programme would mean that no suitable gun was going to be ready for service for some time, and once again the stage was set for a stop-gap solution. In this case the new weapon was one of the oddest conversions that the Germans produced for the starting point was an anti-tank gun, the *5 cm Pak 38*. A version of this anti-tank gun was the *5 cm KwK 39*. When the need for some form of heavy aircraft cannon arose, the tank gun was taken by Rheinmetall and converted into the *5 cm BK (Bordkanone)*. Further development by Mauser resulted in the *MK 214 A*, an aircraft gun intended for use against heavy bomber formations by such advanced aircraft as the Me 262A-1a. By this time the original anti-tank gun was considerably lighter and had been fitted with an automatic feed and 50-round drum magazine. All indications were that this aircraft gun would become quite an effective anti-aircraft gun and plans were made to mount it on modified carriages intended for the *Gerät 58*. The task was given to Dürkopp at Bielefeld but none were produced before the war ended. By late 1944 the need for any form of Flak gun was becoming so desperate that at one point the basic *5 cm KwK 39* was considered for fitting on a simple tubular steel mounting.

The only other gun that came within the 40 to 75 mm range was a project by Krupp to produce a 4.5 cm Flak weapon which was obviously based on the Bofors 40 mm design. Exactly when this project was put forward is not known, but it appears to have got no further than the wooden model stage.

Foreign equipment was not considered for German service after 1940 for the simple reason that there was nothing in this calibre range used by any other combatant. The nearest that came to it was a Czech gun, the Škoda 47 mm M 37. Small numbers of these were taken into German service after 1939 but the calibre was an offshoot of an anti-tank gun programme rather than an attempt at an 'intermediate' anti-aircraft gun.

At this point we have to mention the German attempts to evolve an exotic AA weapon, a form of electro-magnetic gun. Considerable experimental work was put into a device which could throw an anti-aircraft shell to a useful height and by May 1945 an experimental projector about 2 metres long had fired a 10 gram projectile with a 5 mm diameter at a muzzle velocity of 1080 metres/second. A result of these experiments was a specification for a gun firing a 40 mm shell weighing 6.5 kg at a muzzle velocity of 2000 metres/second. The projector for this shell was to be about 10 metres long and was to be mounted on a *12.8 cm Flak 40* carriage. Of all the many hare-brained schemes put forward by German research teams this one was perhaps the most bizarre. The electro-magnetic gun does work in principle but to propel a useful shell the entire output of a large power station would be needed. This project was only beaten for uselessness by the Wind Gun found at Hillersleben in 1945 or the Zippermayer Explosive Vortice projector.

Appendix 1

Numbers of 5 cm Flak 41 in Luftwaffe service Sept 1942 – Feb 1945

1942		**1944**	
September	39	January	58
October	43	February	58
November	42	March	50
December	42	April	57
1943	May		57
January	39	June	57
February	40	July	47
March	40	August	32
April	39	September	25
May	39	October	30
June	35	November	30
July	46		
August	48	December	30
September	49	**1945**	
October	...	January	29
November	49	February	24
December	...		

5 cm Flugabwehrkanone 41

German designation 5 cm Flak 41
Prototype designation Gerät 56 V2
Calibre 50 mm
Length of piece 4686 mm
Length of barrel (L/67): 3342 mm
Length of rifling 2979 mm
Weight travelling 5720 kg
Weight in action 3100 kg
Weight of gun 550 kg
Traverse 360°
Elevation −10° to +90°
Type of feed 5 round clips
Muzzle velocity (HE): 840 m/sec; (AP): 830 m/sec
Shell weight 2.2 kg
Shot weight 2.23 kg
Maximum effective ceiling 9400 m
Rate of fire (practical): 130 rpm
Barrel life 5–7000 rounds
Manufacturer Rheinmetall-Borsig

Remarks: An unsatisfactory weapon. First prototypes completed 1936; after prolonged trials ordered into limited production 1940, issued for troops trials in November 1941. Of only 60 guns completed 24 still in service late in 1944.

Gerät 56 V1a

German designation Gerät 56 V 1a
Calibre 50 mm
Weight travelling 4300 kg
Weight in action 2500 kg
Weight of gun 650 kg
Type of feed 5 round clips
Muzzle velocity 840–860 m/sec
Projectile weight (approx): 2.1 kg
Rate of fire 80–100 rpm
Manufacturer Rheinmetall-Borsig, Düsseldorf

Remarks: Development begun 1935, first prototype completed during 1936. Not considered suitable for service but used for trials which led to 5 cm Flak 41.

Gerät 56 G

German designation Gerät 56G
Calibre 50 mm
Weight of gun 600 kg
Traverse 360°
Elevation −15° to +105°
Type of feed 4 round clips
Muzzle velocity 840 m/sec
Rate of fire (cyclic): 180 rpm
Manufacturer Gustloff-Werke, Suhl

Remarks: Prototype only, not accepted for service. Development ceased 1939.

Gerät 56 M

German designation Gerät 56 M
Calibre 50 mm
Weight of gun 500 kg
Traverse 360°
Elevation −15° to +105°
Type of feed 6 round clips
Muzzle velocity 840 m/sec
Rate of fire (cyclic): 150 rpm
Manufacturer Mauser-Werke AG, Oberndorf-am-Neckar

Remarks: Prototype only; development ceased 1939.

Gerät 56 K

German designation Gerät 56 K
Calibre 50 mm
Weight of gun 550 kg
Traverse 360°
Elevation −15° to +105°
Type of feed 6 round clips
Muzzle velocity 840 m/sec
Rate of fire (cyclic): 135 rpm
Manufacturer Friedr. Krupp AG, Essen

Remarks: Prototype weapon completed 1939. Not accepted for service but used for extensive ammunition and carriage trials.

Gerät 58

German designation Gerät 58
Calibre 55 mm
Length of piece 6150 mm
Length of barrel (L/76.5): 4211 mm
Weight travelling 5490 kg
Weight in action 2990 kg
Weight of gun 650 kg
Traverse 360°
Elevation −10° to −90°
Type of feed 5 round clips
Muzzle velocity 1020–1050 m/sec
Shell weight 2.03 kg
Rate of fire (cyclic): 140 rpm
Manufacturer Rheinmetall-Borsig, Düsseldorf; Dürkopp, Bielefeld

Remarks: Designed as part of an integrated weapon system which included radar and complex fire control equipment. Development commenced 1943 but not concluded before end of WW II. An unknown number of completed carriages converted to take 5 cm Flak 214 guns early in 1945.

Gerät 58 K

German designation Gerät 58 K
Calibre 55 mm
Length of piece 5800 mm
Length of barrel (L/76): 4220 mm
Traverse 360°
Elevation −10° to +90°
Muzzle velocity (est): 1060–1070 m/sec
Shell weight 1.92 kg
Rate of fire (cyclic): 130–140 rpm
Manufacturer Friedr. Krupp AG, Essen

Remarks: Counterpart to Rheinmetall Gerät 58. Development commenced during 1943 but prototype not completed. Dürkopp of Bielefeld and Gustloff-Werke at Suhl both prepared similar projects but it is not known if any prototypes were actually built.

5 cm Flugabwehrkanone 214

German designation 5 cm Flak 214
Calibre 50 mm
Length of weapon 4160 mm
Length of barrel 2825 mm
Weight of weapon 480 kg
Weight of barrel 201 kg
Muzzle velocity (est): 920–930 m/sec
Shell weight 1.54 kg
Rate of fire (est): 140–150 rpm
Gun designers Mauser-Werke AG, Oberndorf-am-Neckar
Carriage designers Dürkopp, Bielefeld

Remarks: Emergency adaptation of MK 214A aircraft cannon for mounting on Gerät 58 carriage. A batch of 50 ordered early in 1945 but none completed..

5 cm automatische Flugabwehrkanone

German designation Not known
Calibre 55 mm
Length of piece 6550 mm
Length of barrel (L/85.5): 4702 mm
Weight of gun 850 kg
Type of feed 4 round clips
Muzzle velocity (est): 1120 m/sec
Shot weight 2.25 kg
Rate of fire (est): 140 rpm
Manufacturer Skoda-Werke, Pilsen

Remarks: A project undertaken by Skoda engineers on German request. Prototype apparently not completed before end of WW II.

4.7 cm Flugabwehrkanone 37(t)

German designation 4.7 cm Flak 37(t)
Original designation 4.7 cm kanon PL vz. 37
Calibre 47 mm
Weight in action 1670 kg
Traverse 360°
Elevation −7° to +85°
Muzzle velocity 800 m/sec
Shell weight 1.5 kg
Maximum effective ceiling 5185 m
Rate of fire (cyclic): 25 rpm; (practical): 15 rpm
Manufacturer Skoda-Werke, Pilsen

Remarks: Designed as dual-purpose weapon. Adopted by Czech Army 1937, and in same year offered to, but rejected after tests by German Army. All available guns taken over after German occupation of Czechoslovakia. Limited production undertaken 1939–40. Only small numbers in service.

Heavy anti-aircraft guns

Heavy anti-aircraft guns are usually regarded as those with calibres of 75 mm and upwards. In Germany there was no official division of anti-aircraft guns into light, medium or heavy and they were all known under the general term of *Flugabwehrkanonen* – Flak.

By the end of WW I Germany, like most other large combatant nations, had already produced specialised anti-aircraft guns with calibres of 77 mm, 80 mm, 88 mm and 105 mm. All these guns were scrapped after 1918 and by the terms of the Versailles Treaty the two main German armament firms of Krupp and Rheinmetall were restricted in the types and calibres of weapons they could produce. Krupp was forbidden to design and produce any guns with calibres below 17 cm and Rheinmetall were restricted to guns below 17 cm. In addition, production totals were to be closely monitored and confined within restricted limits. Of the two firms, Krupp was the hardest hit as the firm had been virtually built on weapon production and design. To remain in being some form of 'foreign' outlet was sought.

A solution to the problem was found in 1921 when Krupp negotiated an agreement with the Swedish firm of Bofors. In return for sales rights to Krupp designs Bofors agreed to allow a small Krupp design team to work in Sweden and carry out development work. By 1922 this work was being financed by the *Reichswehrministerium* via the 'front' office of Koch and Kienzle in Berlin. The Krupp-Bofors association proved very fruitful to both concerns and there was much cross-fertilisation of ideas and technical knowledge. One of the first major projects was a design study for a 75 mm anti-aircraft gun sufficiently advanced to counter the estimated increases in aircraft size and performance. By 1930 both the Krupp team and the Rheinmetall design team, still in being in Germany, had produced prototypes. Neither was adopted for use by the military authorities, who asked for more powerful weapons to counter what were seen to be great strides in aircraft performance expected in the near future. After about 1933 the Krupp design was offered commercially and a number of guns were sold to Spain and some South American states. In 1939 the few still on the production lines were delivered to the German Navy and used for the air defence of the North German coastal strips.

The easiest way to produce the extra power desired by the German staff planners was to increase the calibre and hence projectile weight. As 75 mm had been considered too light the next 'preferred' calibre would have been 105 mm. But a projectile of that calibre was seen as being too large and heavy for hand-loading and as the resultant gun was needed for field defence of the land forces as well as home defence, hand-loading was required to avoid the complexity and weight of power-loading. A compromise was sought and found with the choice of 88 mm, almost mid-way between the two extremes of 75 and 105 mm.

The Krupp design team went to work as soon as the new calibre was fixed. In Sweden the 75 mm project was altered and enlarged and many new features were added so that by 1931 the new design was taken back to Essen and work began on constructing a prototype. Under conditions of great secrecy this prototype gun was completed during 1932 and almost at once work also began on setting up a production line. The years of research and development with Bofors produced an excellent design and the Krupp gun was accepted for service almost immediately afterwards as the *8.8 cm Flak 18*.

Within a few months the new gun was in large scale production. By that time the NSDAP was in power and all pretensions of adhering to the Versailles Treaty were discarded. By 1934 the *Flak 18* was in service in some numbers and various trials and exercises had been carried out. Among the results of these exercises was the realisation that the service life of a *Flak 18* barrel would be about 900 rounds, which was considered to be too short. The German military planners were working on the assumption that the next major war would be of short but intense duration. During a short war the barrels of many *Flak 18*s would soon become useless and the cost of stockpiling large numbers of spare barrels was considered prohibitive. To find a possible solution a great deal of research work was carried out and Rheinmetall eventually came up with the idea of a three-piece barrel liner in the shape of the *Rohr Aufbau 9 (RA9)*. As most of the barrel wear took place at the forcing cone between the chamber and the first section of rifling, the *RA9* was designed so that this section could be easily removed and replaced in the field. Only the vulnerable portions of the barrel needed to be stockpiled, but this saving had to be considered against the more involved and costly production methods needed. In 1936 such expenditure was acceptable and the new barrel went into production for fitting on to the *8.8 cm Flak 36*. Later anti-aircraft guns adopted the principle of the *RA9* but after about 1941 the idea rebounded on the Germans in an unforeseen manner. Most of the barrel wear on the early *Flak 18* was caused by the copper driving bands on the projectiles then in use. After 1940 the copper scarcity, then just becoming apparent, enforced a change from copper to sintered iron (*Sintereisen*) for the driving bands and this change soon showed that the barrel life was considerably increased using these bands. Thus there was no longer any need for the expensive multi-part barrels and a change back to monobloc barrels was put under way. But by 1942 such a change involved a major disruption as most machines were geared to machining the shorter lengths of barrel used on the multi-part liners. It was a good example of German military planners overreaching themselves and imposing on German industry a major task at a time when it could least be afforded.

Almost as soon as the *Flak 36* entered service the *Flak 37* appeared. This was basically a *Flak 36* with revised data transmission which was simpler and quicker to use. Most *Flak 37*s were used for home defence of the Reich after 1941.

The *Flak 18, 36* and *37* became the backbone of the Luftwaffe anti-aircraft defences, but such was the soundness of their design that they were also used in other roles. Early field experiences during the Spanish Civil War showed they were very powerful anti-tank weapons and from 1940 they were increasingly used in that role in support of Wehrmacht formations. Despite their height and bulk the *8.8 cm Flak* guns were often very effective against armoured targets and such was their power that they were used in that role right up to 1945. Many sources refer to the '88' as the most famous gun of WW II, such was the success of the gun and the aura that gradually surrounded it as the war continued. Special armour-piercing ammunition was developed and used but the *Flak 18, 36* and *37* were never really intended as anti-tank guns and the role was imposed upon them by the use of heavier armour on Allied tanks. Other tasks the *8.8 cm Flak* series were called upon to perform were long-range field artillery and coastal defence. Many were mounted on special Flak trains. The standard 8.8 cm guns were fitted on mobile cruciform platforms but as numbers were gradually diverted from field use to defence of the Reich the need for mobile carriages disappeared and they were mounted directly into concrete. These versions were suffixed '*/2*', eg *8.8 cm Flak 36/2*. The various types of barrels were all interchangeable with the various types of carriage so it was possible to see a *Flak 18* barrel fitted to a *Flak 37* mounting and other combinations.

Back in 1933 it had been forecast that the 8.8 cm Flak series would not be the ultimate answer to the air defence problem and that an even heavier calibre gun would still be needed. It was decided that a new 105 mm gun was required but as the new gun

did not need to be as mobile as the 88 mm series the loading problems were not so acute. In 1933 specifications were issued for a *Gerät 38* and Krupp and Rheinmetall were invited to submit prototypes. Both prototypes, and a further short series of guns for troop trials, were ready in 1935 and two different systems of power control were incorporated on each batch. By 1936 the trials were completed and Rheinmetall were awarded the production contract for the *10.5 cm Flak 38*. In 1939 the design was improved and a multi-barrel construction adopted – this then became the *10.5 cm Flak 39*.

With the *8.8 cm* and *10.5 cm Flak* guns in production and service the HWA felt that they had a considerable ascendancy over any target likely to show itself, but they were not content to rest on their laurels. As early as 1936 two new specifications were put out, one for a 12.8 cm gun, the *Gerät 40*, and the other for a 15 cm gun, the *Gerät 50*. This increase in calibres was felt necessary as at that time it was not thought that the copper driving bands then in use could take an increase in muzzle velocity without unacceptable barrel wear taking place, so little work was devoted to improvements in the performance of existing weapons. The 12.8 cm project was given to Rheinmetall and the 15 cm went to Krupps. As ever the Krupp approach to the *Gerät 50* was painstaking and thorough but time-consuming to the extent that Rheinmetall was given a 'cover' project as the *Gerät 55*.

From this point on the 12.8 and 15 cm projects become intertwined. Much was expected of the 15 cm projects and by late 1938 the two prototypes were ready for trials. By that time the *12.8 cm Gerät 40* had undergone considerable firing trials as it had been ready in 1937. It had proved itself an excellent gun and limited production had already begun as the *12.8 cm Flak 40*. These early production examples were mobile and carried on a special transporter in one load. But production was very limited pending the results of the 15 cm trials. By November 1938 it was apparent that the performance of both 15 cm projects was no better than that of the *12.8 cm Flak 40*, despite many novel features. As well as the lack of performance both guns would have to be transported in several loads on special transporters which was considered a disadvantage. Attempts to try to improve performance on both projects went ahead until January 1940 by when all work was stopped as it was seen that no useful improvement could be made. Production of the *12.8 cm Flak 40* was gradually stepped up but it was not until 1942 that it reached the necessary level, and by that time the requirement for being mobile had been dropped and all production was for either static or railway mountings.

Thus the 12.8 cm gun became a weapon for home defence but the need for an improved heavy gun for the defence of field units remained unfulfilled. By the late 1930s it was apparent that the existing 8.8 cm guns would soon be no match for the next generation of aircraft which would have greatly increased performances in speed and altitude. The 10.5 cm guns were really too unwieldy for field use and in any case their performance was little better than that of the 8.8 cm guns. The 12.8 cm was too bulky and immobile for the field role so the answer was considered to be an entirely new gun.

Once again, 88 mm was selected as the calibre. Improvements in propellants and the use of sintered iron driving bands made an increase in muzzle velocity possible, and by the end of 1939 a new specification was issued to Rheinmetall for the *Gerät 37*. As a result of tactical experience in Spain it was decided that the new gun would have to be a dual-purpose weapon capable of fire against ground targets as well as anti-aircraft use and this extra requirement considerably increased the complexity of the resultant design. Rheinmetall approached the project in their

12.8 cm Flak 40/2 in a concealed gun pit

usual brisk manner and introduced a number of new and untried features that were to cause problems in the field later. The early prototypes soon showed that the expected performance could be achieved but the innovations caused constant troubles, not the least of which was the extraction of steel cartridge cases from the multi-section barrel. The use of steel cases had been enforced by the chronic raw materials shortage imposed by the Allied blockade, and copper was one of the very early casualties. The designers had followed the German convention of using a horizontal sliding block breech which dictated the use of a metal case for obturation, and the production of these cases remained a constant headache until the end of the war. Steel was used from about 1940 onwards in place of brass in case manufacture and was found to be generally satisfactory, but when used with multi-section barrel liners such steel cartridge cases proved to have unfortunate expansion characteristics which caused constant blockages.

The case extraction was only one problem with the new gun, which was soon designated *8.8 cm Flak 41*. The dual purpose role involved many new features that went wrong due to lack of development. The need for a back-up design was realised resulting in a specification for a new *Gerät 42* which was given to Krupp. In their usual thorough manner Krupp went ahead and worked on a family of weapons based on the *Gerät 42* specification that included an anti-tank gun and tank gun. But the Krupp development proved too time-consuming so that when the first full-scale wooden mock-ups had finally been made a new and much improved specification was put forward. That was in February 1943, by which time it was felt that the Krupp *Gerät 42* would be unable to meet the new specification without a great deal of extra development work. The *Gerät 42* was thus dropped and work went ahead with the tank and anti-tank gun projects. These were later to emerge as the *8.8 cm KwK 43* and the *8.8 cm Pak 43*.

With the *Gerät 42* out of the way all development had to be concentrated on the Flak 41 and after considerable efforts it emerged as an excellent weapon that could out-perform any other German gun apart from the *12.8 cm Flak 40*. The first pre-production batch was sent to North Africa during the last part of that protracted campaign but combat experience only highlighted the numerous troubles still inherent in the design. Thereafter the *Flak 41* was usually confined to defence of the Reich which negated its original purpose and rendered the many dual-purpose additions unnecessary. Production of the *Flak 41* was slow and difficult to the extent that by February 1945 there were only 289 in service.

Increasing aircraft performance by the end of 1940 again highlighted the need for a powerful Flak gun, especially around targets of high political or economic importance. Special Flak towers had already been built around such cities as Berlin to take the expected 15 cm guns but as none were to be ready before a target date of 1943 (if then) these towers had to be equipped with alternative weapons. A single *12.8 cm Flak 40* was thought to be a waste of the potential of these towers, which commanded a considerable field of fire over and around the target area to be protected, so the decision was made to 'double-up' on the common carriage – the guns were 'mirror-imaged' to aid loading. The new weapon, known as *12.8 cm Flakzwilling 40*, went straight into production without trials and the first were ready for installation near Berlin in August 1942. Production of these massive and complex guns was very slow and expensive – each combination cost RM202,000. By February 1945 there were only 34 in service, usually grouped in batteries of four.

Although the *Geräte 50* and *55* had been dropped in early 1940 the idea of a 15 cm equipment was still in being. In January 1940, almost as soon as work on the *Geräte 50* and *55* had ceased, a new specification was issued. Again Krupp and Rheinmetall were involved, the Krupp project becoming the *Gerät 60* and the Rheinmetall counterpart the *Gerät 65*. Work continued on these projects until the spring of 1941 when it was appreciated that the work then in progress on jet and rocket aircraft would further increase the demands made upon anti-aircraft defences. Then followed an odd interlude in the already complicated Flak defences story with the investigation into a form of really super-heavy anti-aircraft weapon. At the same time as the Luftwaffe began to appreciate the need to counter the jet aircraft, the German Navy also became involved with the same problem. As ever, the Navy approach differed from that of the Luftwaffe. The Navy wanted twin guns in turrets with calibres of 20.3 or 24 cm, and these guns would have had all the attendant naval ammunition supply and handling systems. The Luftwaffe wanted single guns mounted in concrete pits with calibres of 21 or 24 cm. To simplify things the ammunition supply system was to be manual wherever possible. For once, the Navy and Luftwaffe got together to work out a common specification but it took until the summer of 1942 to finalise it. As a compromise the Luftwaffe 24 cm gun was adopted for placing on a carriage of Navy design. New specifications were issued and this time the Krupp project was allocated the code name *Gerät 80* and the Rheinmetall project became *Gerät 85*. Little work was carried out by either for in late 1943 it was decided to see what transpired with the *Geräte 60* and *65*.

Originally the *Geräte 60* and *65* were supposed to be transported in one load but in October 1942 an order was made prohibiting the production of mobile carriages for all Flak guns over 10.5 cm. The designation of both 15 cm projects was altered to *Geräte 60 F* and *65F* respectively, but that was not the only change. An experimental barrel with the multi-piece liner was ready by early 1942 but it showed all the usual problems of case extraction to the extent that the barrel design was altered to monobloc construction. Very advanced sights were designed for the *Geräte 60F* and *65F* along with remote control equipment. During December 1942 new and more demanding specifications were issued which involved almost complete redesigns but in view of the expected improvements this was accepted. The first of these improved guns was expected to be ready during June 1944 but in the middle of September 1943 an order was made cancelling all design and development work on guns with calibres greater than 12.8 cm. The intention was to give priority to jet and rocket fighter production, but this was not accepted by GL/Flak who campaigned vigorously to reverse the order. Later, in October 1943, a partial reversal was effected, but only purely experimental work was to be allowed for research. Thus, in one move all work on the *Geräte 60, 65,* and *80* and *85* came to an abrupt end. It is difficult to see what use the 24 cm project would have been. The guns would have been very complex and expensive in money, materials and facilities, and produced to meet a threat to which the only viable answer was some kind of guided missile. Considerable field research work was carried out by the Luftwaffe Weapon Test Centre at Tarnewitz which showed that each bomber shot down by gunfire cost the Reich at least RM250,000. Unless the 24 cm project was able to provide a virtual 100 per cent certainty of a hit the only alternative was the guided missile, work on which was in its early stages.

So after late 1943 the defence of the Reich and the various field formations was carried out by the old stalwarts in the shape of the *8.8 cm Flak 18, 36* and *37*, small numbers of the *8.8 cm Flak 41*, the *10.5 cm Flak 38* and *39* and the *12.8 cm Flak* and *Flakzwilling 40*. The only way to improve this situation, once all the various projects had been dropped, was to develop the existing weapons, their ammunition and their tactical use. To improve the 10.5 cm guns it was proposed that they should be adapted to take the *Flak 41* barrel, but this idea was dropped when it was realised that the *Flak 41* extraction problems would remain with the new combination, and in any case production of Flak 41 barrels was still very slow. To improve the *8.8 cm Flak 36* and *37* it was proposed that the barrels be lengthened from L/56 to L/66 and rechambered to take *Flak 41* ammunition (an idea simply to fit *Flak 41* barrels was not proceeded with as the barrels would be too

heavy for the carriage). Trials with this combination were successful to the extent that the barrels were lengthened to L/74 calibres and a muzzle brake fitted. The result was the *8.8 cm Flak 37/41*, but very few were made, perhaps as few as thirteen. As always the usual case extraction troubles were carried over.

Work on the 10.5 cm series was not proceeded with as the 12.8 cm showed much more development potential. New ammunition and a longer barrel and muzzle brake were developed. The carriage of the *Flak 40* was retained in a modified form and the new gun would have been the *12.8 cm Flak 45*, but the war ended just as the prototype was nearing completion.

The use of anti-aircraft guns in defence of the Reich was an ever-changing scene to meet an ever-increasing threat. When the Allied bomber fleets started to grow in number and range, many Flak guns were gradually withdrawn from field units to home defence. This gradual change was reflected in the growing manufacturing practice of producing guns for static concrete emplacement only and increasing numbers were delivered without platforms, cruciform carriages or tractors, with a considerable saving in manufacturing potential. But static emplacements were of little use when the Allied bombers began to concentrate on such targets as Hamburg. Thousand-bomber raids completely overwhelmed the defence while other guns statically emplaced near by were unable to help. One possible answer was seen as the mobile railway battery. This led to 8.8, 10.5 and 12.8 cm batteries being formed and towed around the Reich to wherever they were needed and then parked in railway sidings.

For static batteries a considerable increase in results could be obtained by the formation and use of '*Gross-Batterien*'. These were combinations of up to three Flak batteries all firing on one designated target. Visible targets were tracked by the three *Kommandogeräte* assigned to the three batteries and the fire of all three six-gun batteries could thus be concentrated. At night a single *Würzburg* radar fire control system was used. The results were very encouraging until a high-level decision was made to concentrate twenty-four guns under one *Kommandogerät* in the '*Mammoth-Batterie*'. Various technical problems made this combination a much less viable proposition but it was persisted with and the potential of the '*Gross-Batterie*' was never fully realised.

As the war ended the 8.8, 10.5 and 12.8 cm Flak guns of 1939 vintage were still in use along with the later additions of the *Flak 41* and the *Flakzwilling 40*. Much of the very involved and potential-consuming research and development had been wasted in the 15 and 24 cm projects, but at the end of the war the Flak arm was nevertheless one of the most efficient and cohesive components of the German forces. At one point (July 1943) as many as 2500 Allied aircraft were shot down in a single month, but that freak 'high' was never repeated. By late 1944 increased numbers of static sites were being overrun and their precious guns lost in ever-increasing numbers. Despite frantic appeals for higher production totals of Flak guns the war ended with the Allied air fleets having absolute air supremacy over all the remaining German territories.

During the German campaigns of 1939 and 1940 large numbers of foreign heavy anti-aircraft guns were taken over from the defeated armies, many of them intact and with stockpiles of ammunition ready to hand. The statically emplaced guns were usually sited around important production centres and towns which the Germans were anxious to use and protect, so the simplest and most economic solution was to take over the existing guns and installations for their own use. In this way numerous different types of anti-aircraft guns came into German service, many of them of ancient origin and doubtful performance.

Czechoslovakia was the first unwilling provider of modern Škoda guns of 7.65 and 8 cm calibre. The old 8.35 cm kanon PL vz. 22/24 was taken over in some numbers and as most of the Czech guns were mobile they were used by many German field formations. The static guns were left in place to protect the numerous important production centres, many of which were diverted to produce modern German Flak designs.

France yielded a number of anti-aircraft guns based on the old 75 mm mle 1897 field gun barrel. During WW I numbers of these barrels had been placed on rudimentary anti-aircraft carriages around the more important French cities and arsenals and they were still there in 1940. The Germans took them over and kept them in use, along with as many other French anti-aircraft guns as possible. The 1940 campaign also produced numbers of useful guns from Holland and Belgium – these were usually 75 mm Vickers guns. Norway was forced to yield the few guns produced at the Kongsberg Arsenal, but the main 1940 windfall came from the aftermath of the Dunkirk evacuation. Large and very useful numbers of British 3.7-in mobile anti-aircraft guns fell into German hands from the wreckage of the BEF and the Belgian Army, which had been given a few to bolster their defences. These 3.7-in guns were very highly thought of by the Germans who carefully sited them just inland from the Channel coast. Some were emplaced as coastal defence guns, as at Walcheren, but most were used in the AA role. So highly did the Germans think of these British guns that in early 1943 they went as far as making special 9.4 cm ammunition in Germany for them. 100,000 rounds were produced, after which it was thought that the gun barrels would be worn out.

The invasion of Soviet soil in 1941 bought massive amounts of all varieties of Russian equipment into the German fold. Among this booty were large numbers of 76.2 mm obr. 1931 and 1938 anti-aircraft guns and 85 mm obr. 39 guns. These were all good, sound and modern designs which were eagerly taken into German use. Starting in 1943 obr. 1938 and 1939 guns were transported to Northern Italy and there bored out to take 8.8 cm German ammunition. On return to Germany they were initially used by various Home Defence Flak units, and then issued to some front-line formations.

Units in Italy often found themselves using various types of Italian anti-aircraft guns, some of them of Škoda origin. The most useful Italian design was the Cannone da 90/53 which had a performance not far removed from that of the German 8.8 cm Flak 18-37 series. Many of the other Italian guns were less capable but after 1943 the Cannone da 75/46 modello 34 was a useful prize and many were used on the Eastern Front.

The German Navy also had a considerable part to play in defence of the German territories as many of the coastal defences were its responsibility. Usually the Kriegsmarine took over the standard types of weapon from the Luftwaffe, but they also produced their own weapon designs. One of these, used in some numbers, was the *10.5 cm SKC/32nL in 8.8 cm MPL C/30,* a Rheinmetall dual-purpose design mounted on the *8.8 cm Flak 18-37* cruciform platform. It had been intended that the *10.5 cm SKC/32* would act also as a coastal defence gun but it had an indifferent performance. A better showing was expected from the *12.8 cm SKC/40* which was a navalised *Flak 40* designed for fitting into a twin turret with all the attendant power laying, power feed and ramming. Since this gun was intended for land use it can only be seen as a typical example of the usual lack of co-ordination regarding German weapon projects. For the purpose for which the Navy gun was intended the *12.8 cm Flakzwilling 40* would have sufficed with a much reduced cost outlay and complexity. As it was, the first Navy *12.8 cm SKC/40* was not ready until late 1944 and the large number of troubles with the installation were still under investigation when the war ended. Why such a programme ever got under way is even more difficult to understand when a version of the 12.8 cm gun was already in production for the German Navy as the *12.8 cm Flak 40M.*

Appendix 1

Production centres for Flak guns (August 1944)

8.8 cm Flak 18, 36, 37
A. O. Hering, Neustadt; Gebr. Böhler & Co. AG, Kapfeberg-Deuchendorf/Steiermark; J. M. Voith, Heidenheim/Brenz; F. Werleim & Co., Wien; Skoda-Werke, Pilsen und Dubnica; Ost. Maschinenbau GmbH., Keuwerk Eintrachthütte; Friedr. Krupp Grusonwerk, Magdenburg-Bückau; Masch.-Fabr. Augsburg-Nürnberg (MAN) AG, Augsburg; Berlin-Erf. Masch.-Fabr. AG, Erfurt; Masch.-Fabr. Andritz AG, Graz-Andritz; Ost. Maschinenbau GmbH, Sosnowitz.

8.8 cm Flak 41
Rheinmetall-Borsig AG, Düsseldorf; Škoda-Werke, Dubnica.

10.5 cm Flak 38, 39
Berlin-Erfurter Maschinenfabrik AG, Erfurt; Benteler Werke, Bielefeld; Krupp Grusonwerk AG, Magdeburg-Buckau; Eisenwerke Weserhütte, Bad Oeynhausen; Hann. Maschinenfabrik AG vorm Georg Egerstorf, Hannover; Mitteldeutsche Stahlwerke, Gröditz; Röchling'sche Eisen-und-Stahlwerke GmbH., Völkingen/Saar; Masch.-Fabr. Augsburg-Nürnberg (MAN) AG, Augsburg.

12.8 cm Flak 40
Friedr. Krupp AG, Essen; Skoda-Werke, Pilsen; Hann. Maschinenfabrik AG vorm Georg Egerstorf, Hannover; Oberschul. Gerätbau GmbH., Laurahütte, Kattowitz.

12.8 cm Flakzwilling 40
Hann. Maschinenfabrik AG vorm Georg Egerstorf, Hannover; Oberschul. Gerätbau GmbH., Laurahütte, Kattowitz.

Appendix 2

Experimental projectiles for standard barrels

DS = discarding sabot. Skirted = used muzzle squeeze bore attachment.
Inc = incendiary

Gun	Rifling twist	Proj. calibre (cm)	Type	Weight (kg)	MV (m/sec)
8.8 cm Flak 18–37	1 in 30	8.8 cm	Normal (HE)	9	841
		8.8/7.0	DS HE	5/4.4	1085
		8.8/7.0	Skirted (HE)	4.4	1195
8.8 cm Flak 41,37/41	1 in 30	8.8	Normal (HE)	9.4	1021
		8.8/7.0	DS HE	5/4.4	1290
		8.8/7.0	Skirted (HE)	4.4	1360
10.5 cm Flak 38, 39	1 in 35	10.5	Normal (HE)	15.1	900
		10.5/8.8	DS HE	10.8/9	1067
		10.5/8.8	DS Inc	11/9.2	1060
		10.5/8.8	Skirted (HE)	9.08	1130
		10.5/8.8	Skirted (Inc)	9.08	1130
12.8 cm Flak 40	1 in 33	12.8	Normal (HE)	26	900
		12.8/10.5	DS Inc or HE	17.4/14.4	1109
		12.8/10.5	Skirted (Inc or HE	16.7	1118
12.8 cm Flak 45	1 in 30	12.8	Normal (HE)	31	930
		12.8/10.5	DS Inc or HE	17.4/14.4	1234
		12.8/10.5	Skirted (Inc or HE)	16.7	1201

Appendix 2A

Experimental shells for experimental barrels

DS = discarding sabot. Skirted = used muzzle squeeze bore attachment.
Inc = incendiary

Gun	Rifling twist	Proj. calibre (cm)	Type	Weight (kg)	MV (m/sec)
10.5 cm Flak 38, 39	1 in 35	10.5	Normal (HE)	15.1	900
	1 in 20	10.5/7.0	DS (HE)	6/4.4	1350
	1 in 22.5	10.5/8.0	Skirted (HE)	8.7	1155
	Smooth bore	10.5/4.5	Fin stab. (HE)	9.3/7.13	951
12.8 cm Flak 40	1 in 33	12.8	Normal (HE)	26	900
	1 in 18	12.8/7.5	DS (HE)	8.5/5.6	1423
12.8 cm Flak 45	1 in 30	12.8	Normal (HE)	31	930
	1 in 21	12.8/9.6	Skirted (HE)	14.2	1289
		12.8/9.6	Skirted (Inc)	14.5	1271
	1 in 18	12.8/7.5	DS (HE)	8.5/5.6	1539
	Smooth bore	12.8/7.3	Fin stab. (HE)	7.7/6.3	1500
10.5 cm Flak 40/39	1 in 25.5	10.5	Normal (HE)	15.1	900
		10.5/8.0	Skirted (HE)	8	1405
		10.5/8.8	DS (HE or Inc)	10.2/8.8	1344

Appendix 3

Heavy anti-aircraft guns in service September 1942 – February 1945

	8.8 cm Flak 18, 36, 37	8.8 cm Flak 41	10.5 cm Flak 38, 39	12.8 cm Flak 40	12.8 cm Flak-zwilling 40
1942					
September	5184	—	500	16	7
October	5265	4	488	22	7
November	5413	18	517	26	8
December	6148	24	580	32	10
1943					
January	6183	36	658	45	10
February	6508	43	681	52	12
March	6673	62	755	72	14
April	6670	67	797	83	14
May	6379	41	878	89	14
June	6448	46	949	93	14
July	6617	52	1055	105	14
August	7024	61	1141	123	16
September	7269	67	1222	140	18
October	7641	56	1270	143	20
November	7809	69	1307	159	20
December	8214	75	1392	179	20
1944					
January	8658	78	1490	197	20
February	8870	91	1530	220	20
March	9010	110	1611	260	24
April	9333	116	1711	289	24
May	9787	117	1784	309	24
June	10,107	117	1868	357	24
July	10,286	149	1919	401	25
August	10,704	157	1969	471	27
September	9125	156	1758	492	32
October	9639	158	1835	501	30
November	9734	191	1867	514	30
December	9878	252	1911	525	31
1945					
January	9442	318	1902	570	33
February	8769	289	1850	534	33

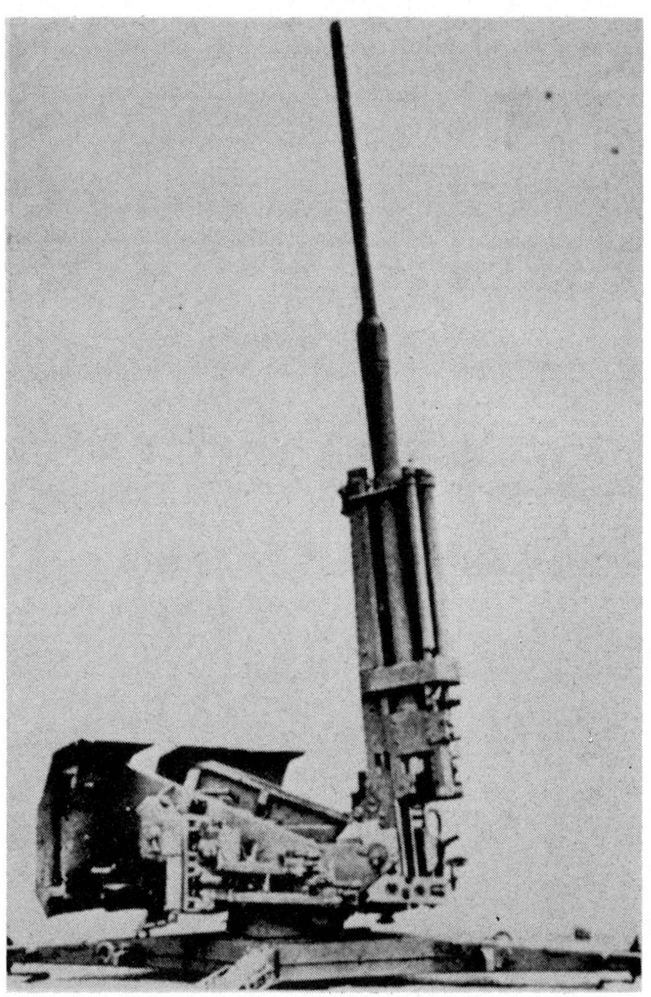

8.8 cm Flak 41 at maximum elevation

Appendix 4 Part 1

French heavy anti-aircraft guns impressed in German service
Availability August 1943 – October 1944

	7.5 cm Auto-kanone (f)	7.5 cm Flak M 17/34(f)	7.5 cm Flak M 30(f)	7.5 cm Flak M 33(f)	7.5 cm Flak M 36(f)	9 cm Flak M 39(f)
1943						
August	33	43	58	143	293	16
September	33	41	55	150	288	16
October	33	41	73	133	297	20
November	33	48	69	160	311	35
December	33	48	72	145	267	36
1944						
January	33	96	61	134	222	36
February	33	75	70	100	266	21
March	33	93	77	141	228	18
April	33	56	71	146	240	18
May	45	56	73	150	281	19
June	45	50	74	153	303	19
July	45	3	43	76	186	19
August	12	1	40	78	193	19
September	—	1	1	1	16	—
October	—	—	—	—	8	—

Appendix 4 Part 2

Soviet heavy anti-aircraft guns impressed in German service
Availability August 1943 –February 1945

	7.62 cm Flak M 38(r)	7.62/8.8 cm Flak M 38(r)	7.62/8.8 cm Flak M 31(r)	7.62 cm Flak M 31(r)
1943				
August	39	28	94	196
September	33	37	126	146
October	7	3	243	74
November	7	6	266	74
December	6	35	323	70
1944				
January	6	29	358	70
February	6	89	448	69
March	6	75	520	69
April	6	134	546	68
May	6	128	587	68
June	2	117	626	19
July	2	153	699	7
August	1	163	723	—
September	1	119	709	—
October	—	119	687	—
November	—	116	656	—
December	—	686*		—
1945				
January	—	584*		—
February	—	414*		—

* Combined totals of 7.62/8.8 Flak M 38(r) and M 31(r)
Totals for 8.5 cm Flak M 39(r) not available

The Soviet 85 mm M1939 was a modern weapon with good ballistics and captured guns were a welcome addition to the German arsenal ►

Appendix 4 Part 3

Czech, British and Italian heavy anti-aircraft guns impressed in German service
Availability August 1943–February 1945

	7.5 cm Flak Skoda	8.35 cm Flak M 22(t)	9 cm Flak M 12	9.4 cm Flak Vickers m 39(e)	9 cm Flak 41(i)	10.2 cm Flak (i)
1943						
August	—	107	12	62	—	—
September	—	106	12	62	—	—
October	—	106	12	53	—	—
November	—	106	12	55	—	—
December	—	106	12	52	—	—
1944						
January	—	106	12	50	72	—
February	—	96	12	39	147	4
March	—	96	12	33	213	4
April	12	57	12	33	213	4
May	12	57	12	33	231	4
June	12	19	12	37	266	4
July	12	—	—	33	244	4
August	12	—	—	33	264	4
September	12	—	—	33	245	4
October	—	20	—	33	256	—
November	—	—	—	33	298	—
December	—	—	—	33	315	—
1945						
January	—	—	—	33	303	—
February	—	—	—	1	302	—

8.8 cm Flugabwehrkanone 18, 36 and 37

German designation 8.8 cm Flak 18, 36 oder 37
Calibre 88 mm
Length of piece (L/56): 4930 mm
Length of barrel 4686 mm
Length of rifling 4126 mm
Weight travelling (Flak 18): 6861 kg; (Flak 36, 37): 8200 kg
Weight in action (mobile): 5150 kg; (static): 3710 kg
Weight of gun (Flak 18): 1440 kg; (Flak 36, 37): 1450 kg
Traverse 2 × 360°
Elevation −3° to +85°
Muzzle velocity 820–840 m/sec
Shell weight 9 kg
Shot weight 9.6 kg
Maximum ceiling 10,600 m
Rate of fire 15–20 rpm
Barrel life (copper bands): 2000–2500 rounds; (iron bands): 6000 rounds
Manufacturers *See* Appendix 1

Remarks: (Flak 18): First of the famous 8.8 cm Flak series. Original design evolved by Krupp engineers in Sweden 1931, prototype completed 1932, and production commenced in Germany 1933. First combat use in Spain 1936–39, often in dual-purpose role. Remained in service until end of WW II. (Flak 36): Introduced 3-sectioned barrel liner. In service from 1936. (Flak 37): Featured improved fire control equipment and other detail changes. All three types had many interchangeable parts, including barrels and carriages. Also fitted on various experimental self-propelled mounts.

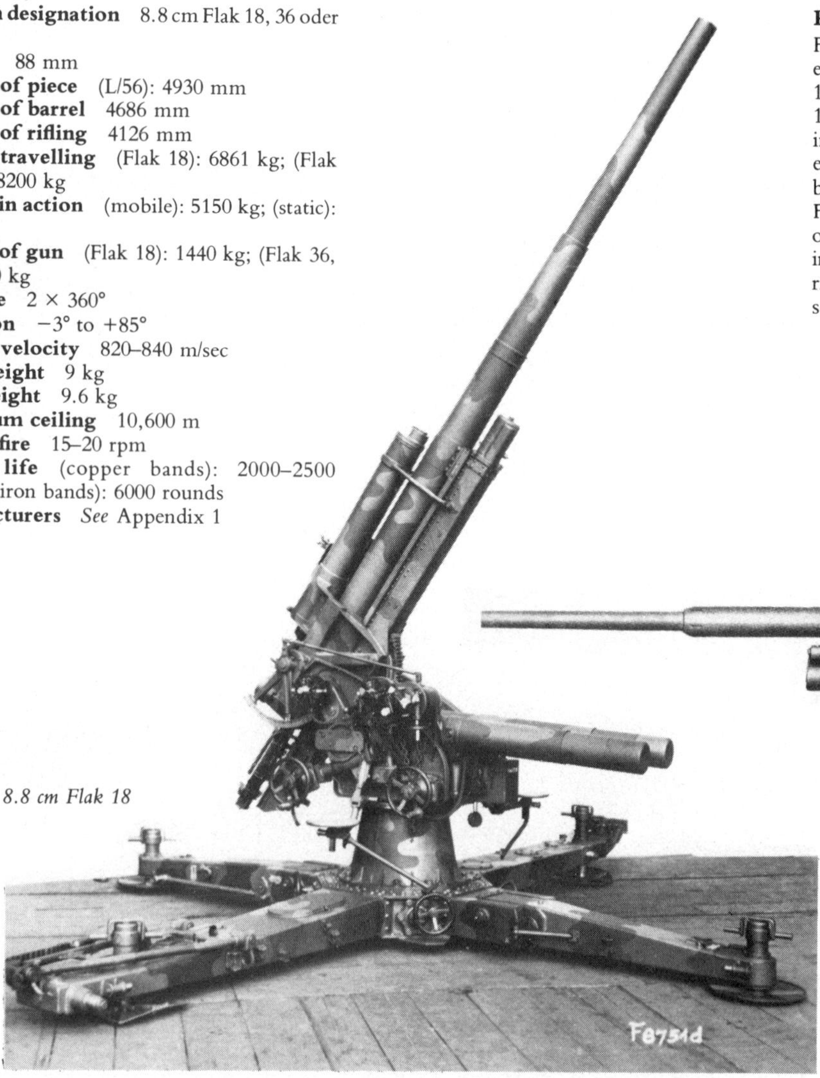

8.8 cm Flak 18

Static mount for 8.8 cm Flak 18/2, 36/2 and 37/2

8.8 cm Flak 18 on Sonderanhänger 201

8.8 cm Flak 36 firing artillery support

8.8 cm Flak 37$41, a more powerful interim version with the barrel rechambered to take Flak 41 ammunition, lengthened to L/74 calibres and fitted with a muzzle brake

8.8 cm Flugabwehrkanone 41

German designation 8.8 cm Flak 41
Design designation Gerät 37
Calibre 88 mm
Length of piece (L/74): 6548 mm
Length of barrel 6293 mm
Length of rifling 5411 mm
Weight complete 11,240 kg
Weight in action 7840 kg
Weight of gun 2130 kg
Traverse 360°
Elevation −3° to +90°
Muzzle velocity (HE): 1000 m/sec; (AP): 980 m/sec
Shell weight 9.4 kg
Shot weight 10 kg; (AP40): 7.5 kg
Maximum vertical range 14,700 m
Rate of fire 22–25 rpm
Barrel life 1500 rounds
Manufacturers *See* Appendix 1

Remarks: More powerful version of the basic weapon designed to 1939 specifications. Prototype known as Gerät 37. Development completed 1941 but due to many technical problems first issued to troops only early in 1943. After first combat experience in Tunisia nearly all production reserved for Home Defence purposes. An excellent AA gun when carefully maintained.

8.8 cm Flak 41. Static mount for this model was designated Flak 41/2

8.8 cm Flak 41 on Sonderanhänger 202. *This trailer consisted of two interchangeable front and rear wheel units, each with double tyres. The same trailer was also used with the 8.8 cm Flak 18.36 and 37*

10.5 cm Flugabwehrkanone 38 and 39

German designation 10.5 cm Flak 38 und 39
Prototype designation Gerät 38
Calibre 105 mm
Length of piece (L/63.3): 6648 mm
Length of rifling 5531 mm
Weight complete 14,600 kg
Weight in action 10,240 kg
Weight of gun 2510 kg
Traverse 360°
Elevation −3° to +85°
Muzzle velocity (HE): 880 m/sec; (AP): 860 m/sec
Shell weight 15.1 kg
Shot weight 15.6 kg
Maximum vertical ceiling 12,800 m
Rate of fire 12–15 rpm
Barrel life (copper bands): 1500 rounds; (iron bands): 3500 rounds
Manufacturer *See* Appendix 1

Remarks: Designed to specifications formulated in 1933. Rheinmetall Gerät 38 selected in 1936 as Flak 38. Flak 39 differed in having improved electrical and data transmission system and sectioned barrel. First delivered during 1940. Both versions built and used in some numbers throughout WW II. A 1943 project designated 10.5 cm Flak 40 did not advance beyond drawing board stage. It would have fired a 17 kg shell at a muzzle velocity of 1050 m/sec.

10.5 cm Flak 39 (3rd version)

10.5 cm Flak 38

10.5 cm Flak 39/2. These statically mounted guns were moved to their sites on the Sd Ah 203

12.8 cm Flugabwehrkanone 40

German designation 12.8 cm Flak 40
Prototype designation Gerät 40
Calibre 128 mm
Length of piece (L/61): 7835 mm
Length of barrel 7490 mm
Length of rifling 6477 mm
Weight travelling 27,000 kg
Weight in action (mobile): 17,000 kg; (static): 13,000 kg
Weight of gun 4828 kg
Traverse 360°
Elevation −3° to +87°
Muzzle velocity (HE): 880 m/sec
Shell weight 26 kg
Maximum vertical ceiling 14,800 m
Rate of fire 12–14 rpm
Barrel life 1000–2000 rounds
Manufacturers *See* Appendix 1

Remarks: Designed to 1936 specifications and originally known as Gerät 40. Prototype guns tested 1937, ordered into limited production 1939. Only six mobile examples completed before production switched to static version in 1942.

12.8 cm Flak 40/1 mobile equipment

12.8 cm Flak 40/1 on Sonderanhänger 220 *(Sd Ah 220)*

12.8 cm Flak 40/2 in an open concrete gun pit. Note the armoured ammunition lockers

12.8 cm Flugabwehrkanonezwilling 40

German designation 12.8 cm Flakzwilling 40
Calibre 128 mm
Length of piece (L/61): 7835 mm
Length of barrel 7490 mm
Length of rifling 6477 mm
Weight emplaced 26,000 kg
Weight of each gun 4820 kg
Traverse 360°
Elevation 0° to +87°
Muzzle velocity 880 m/sec
Shell weight (each firing): 2 × 26 kg = 52 kg
Maximum vertical ceiling 14,800 m
Rate of fire (each barrel): 12–14 rpm
Barrel life (each barrel): 1000–2000 rounds
Manufacturers *See* Appendix 1

Remarks: Originally known as Gerät 44. Work on first prototype commenced during late 1940, type ordered late in 1941. First operational use in defence of Berlin in spring 1942. Proved rather complex and costly to produce, and only limited numbers delivered.

12.8 cm Flakzwilling 40/2

12.8 cm Flugabwehrkanone 45

German designation 12.8 cm Flak 45
Calibre 128 mm
Length of piece with m/b (L/78): 9984 mm
Length of piece (L/75): 9600 mm
Length of rifling 8064 mm
Muzzle velocity 940 m/sec
Shell weight 28 kg
Manufacturer Rheinmetall-Borsig AG

Remarks: Revised 12.8 cm Flak 40 barrel on Flak 40 carriage. Development began during 1943 and a prototype was completed just before end of WW II. Allocated same wartime code name as 12.8 cm Flak 40. Numerous sub-calibre and sabot rounds were under development for this gun (*see* Appendix 3).

7.5 cm Flugabwehrkanone L/60

German designation 7.5 cm Flak L/60
Calibre 75 mm
Length of piece (L/60): 4500 mm
Weight in action 3140 kg
Weight of gun 735 kg
Traverse 360°
Elevation −5° to +85°
Muzzle velocity 850 m/sec
Shell weight 6.5 kg
Maximum ceiling 11,300 m
Rate of fire 20–25 rpm
Manufacturer Friedr. Krupp AG, Essen

Remarks: Designed by Krupp engineers in Sweden 1925–30 and pioneered many features later incorporated in 8.8 cm Flak 18. Rejected by German military authorities and released for export. After 1939 all undelivered guns impressed in German service and used mainly by naval personnel in coastal AA role.

7.5 cm Flugabwehrkanone L/59

German designation 7.5 cm Flak L/59
Prototype designation 7.5 cm Flak P L/65
Calibre 75 mm
Length of piece (L/65): 4875 mm
Length of barrel (L/59): 4425 mm
Weight travelling 6492 kg
Weight in action 4200 kg
Traverse 360°
Elevation −5° to +85°
Muzzle velocity 800 m/sec
Shell weight 6.62 kg
Maximum ceiling 9000 m
Rate of fire 20–25 rpm
Manufacturer Rheinmetall-Borsig AG, Düsseldorf

Remarks: Development commenced in 1925, completed 1930. Trials with prototypes showed unsuitability for service and design abandoned.

8.8 cm Flugabwehrkanone 37/41

German designation 8.8 cm Flak 37/41
Calibre 88 mm
Length of piece with m/b (L/88): 7744 mm
Length of piece (L/74): 6548 mm
Length of rifling 5411 mm
Weight complete 8450 kg
Weight in action 5250 kg
Weight of gun 2130 kg
Traverse 360°
Elevation −3° to +85°
Muzzle velocity 1000 m/sec
Shell weight 9.4 kg
Maximum vertical range 14,700 m
Rate of fire 15–20 rpm
Manufacturers Joint Rheinmetall-Krupp venture

Remarks: In an attempt to improve Flak 18/37 range a Flak 41 barrel was placed on Flak 37 carriage. Trials indicated excess recoil forces resulting in Flak 41 barrel being replaced by lengthened Flak 18 barrel fitted with muzzle brake and converted to take Flak 41 ammunition. Combination also featured power loaders. Development work commenced in 1942 but not completed until 1944. Total only some 13 guns built and delivered.

8.8 cm Flugabwehrkanone 39/41

German designation 8.8 cm Flak 39/41
Calibre 88 mm
Length of piece (L/74): 6548 mm
Length of barrel 6293 mm
Length of rifling 5411 mm
Weight in action 9460 kg
Weight of gun 2130 kg
Traverse 360°
Elevation −3° to +85°
Muzzle velocity 1000 m/sec
Shell weight 9.4 kg
Maximum vertical ceiling 14,700 m
Rate of fire 12–15 rpm
Manufacturer Rheinmetall-Borsig AG, Düsseldorf

Remarks: Development started 1942 and ceased in 1944 as this combination of an 8.8 cm Flak 41 gun and 10.5 cm Flak 39 carriage was considered unsuitable for service.

Gerät 42

German designation Gerät 42
Calibre 88 mm
Length of piece with m/b (L/75): 6600 mm
Length of piece (L/71.5): 6292 mm
Length of rifling 5192 mm
Weight in action (approx): 7700 kg
Weight of gun 1840 kg
Traverse 360°
Elevation −3° to +90°
Muzzle velocity 1000 m/sec
Shell weight 10 kg
Rate of fire 22–25 rpm
Manufacturer Friedr. Krupp AG, Essen

Remarks: Development commenced in 1939 as 'back-up' project to Rheinmetall 8.8 cm Flak 41. Reached wooden mock-up stage by early 1942 when further work abandoned due to new and more exacting specifications which Gerät 42 could not meet. Results of development work later incorporated into 8.8 cm KwK 43 and 8.8 cm Pak 43 designs.

Gerät 50

German designation Gerät 50
Calibre 149.1 mm
Length of piece (L/52): 7753 mm
Length of rifling 6113 mm
Weight of four travelling loads 44,600 kg
Weight in action 32,000 kg
Weight of gun 5680 kg
Traverse 360°
Elevation −1°30′ to +90°
Type of feed 10 round hopper magazine
Muzzle velocity 890 m/sec
Shell weight 40 kg
Maximum vertical ceiling 16,300 m
Rate of fire 10 rpm
Manufacturer Friedr. Krupp AG, Essen

Remarks: Projected heavy AA gun according to specifications issued in 1936. Parallel development Gerät 55 by Rheinmetall. All work ceased in January 1940.

Gerät 50; *loading the automatic magazine*

Gerät 55

German designation Gerät 55
Calibre 149.1 mm
Length of piece (L/55): 8200 mm
Length of barrel 7753 mm
Weight of three loads 38,900 kg
Weight in action 22,200 kg
Weight of gun 6350 kg
Traverse 360°
Elevation −3° to +88°
Muzzle velocity 870 m/sec
Shell weight 40 kg
Maximum vertical ceiling 16,300 m
Rate of fire 9 rpm
Manufacturer Rheinmetall-Borsig AG, Düsseldorf

Remarks: Parallel development to Gerät 50, based on same specifications. Virtually an enlarged version of 12.8 cm Flak 40. All work ceased in January 1940.

Gerät 55

Gerät 60

German designation Gerät 60
Calibre 149.1 mm
Length of piece Not recorded
Length of rifling 8946 mm
Weight in action 37,000 kg
Weight of gun 6800 kg
Traverse 360°
Elevation −3° to +90°
Muzzle velocity 980 m/sec
Shell weight 42 kg
Maximum vertical ceiling (approx): 18,000 m
Rate of fire 9 rpm
Manufacturer Friedr. Krupp AG, Essen

Remarks: Specification issued January 1940, with back-up project Gerät 65 by Rheinmetall. Development work continued until spring 1941 when slowed down and finally abandoned late 1942 in favour of static version Gerät 60F.

Gerät 65

German designation Gerät 65
Calibre 149.1 mm
Length of piece (L/102.5): 15375 mm
Length of barrel 15000 mm
Length of rifling 13438 mm
Weight travelling 37,000 kg
Weight in action 26,000 kg
Traverse 360°
Elevation −3° to +90°
Muzzle velocity 950 m/sec
Shell weight 42 kg
Maximum vertical ceiling (est): 18,000 m
Rate of fire 9 rpm
Manufacturer Rheinmetall-Borsig AG, Düsseldorf

Remarks: Parallel development to Gerät 60. Project work commenced early in 1940, slowed down after spring 1941, and finally abandoned late in 1942 in favour of static version Gerät 65F.

Gerät 60 Fest

German designation Gerät 60F
Calibre 149.1 mm
Length of piece (L/94): 14100 mm
Length of rifling 12450 mm
Weight in action 79,850 kg
Muzzle velocity (est): 1230 m/sec
Shell weight 42 kg
Maximum vertical range (est): 25,000 m
Manufacturer Friedr. Krupp AG, Essen

Remarks: Static version of Gerät 60. Development started late in 1942 but cancelled in September 1943.

Gerät 65 Fest

German designation Gerät 65F
Calibre 149.1 mm
Length of piece (L/104): 15600 mm
Muzzle velocity (est): 1200 m/sec
Shell weight 45 kg
Maximum vertical ceiling (est): 25,000 m
Manufacturer Rheinmetall-Borsig AG, Düsseldorf

Remarks: Static version of Gerät 65. Development started late in 1942 but cancelled in September 1943.

Gerät 80

German designation Gerät 80
Calibre 238 mm
Length of piece (L/64): 15232 mm
Length of rifling 14756 mm
Weight emplaced 152,000 kg; (with armour): 450,000 kg
Weight of gun 57,000 kg
Traverse 360°
Elevation −10° to +85°
Muzzle velocity 1040 m/sec
Shell weight 200 kg
Maximum vertical ceiling 30,000 m
Rate of fire 7 rpm
Manufacturer Friedr. Krupp AG, Essen

Remarks: 1942 project combining proposed Luftwaffe 24 cm Flak gun with Navy-designed carriage in competition with Rheinmetall Gerät 85. Cancelled October 1943 while still in early stages.

Gerät 85

German designation Gerät 85
Calibre 238 mm
Length of piece (L/72.5): 17400 mm
Length of barrel 16800 mm
Length of rifling 15745 mm
Weight of gun 38,000 kg
Traverse 360°
Elevation 0° to +90°
Muzzle velocity (est): 1000 m/sec
Shell weight 180 kg
Maximum vertical ceiling (est): 23,000 m
Manufacturer Rheinmetall-Borsig, Düsseldorf

Remarks: 1942 Rheinmetall project combining proposed 24 cm Luftwaffe Flak gun with Navy-designed carriage in competition with Krupp Gerät 80. Development cancelled October 1943.

7.5 cm Flugabwehrkanone (b)

German designation 7.5 cm Flak (b)
Original designation 75 mm FRC mle 27
Calibre 75 mm
Length of piece (L/52): 3900 mm
Weight in action 8027 kg
Traverse 360°
Elevation 0° to +70°

Muzzle velocity 700 m/sec
Shell weight 6.5 kg
Maximum vertical ceiling 7500 m
Rate of fire 15 rpm
Manufacturer Fonderie Royale des Canons, Liège

Remarks: Most of these guns were mounted on mobile platforms stabilised by outriggers. Only small number in service 1940 and taken over by German forces in Belgium.

7.5 cm Feldkanone 97(f)

7.5 cm Flak 97(f) on a concrete base for permanent installation

7.5 cm Flak 97(f) on a pedestal mount

German designation 7.5 cm FK 97(f)
Original designation Canon de 75 mm Anti-Arien sur plâte-forme Schneider
Calibre 75 mm
Length of piece (L/33): 2721 mm
Weight emplaced 3000 kg
Traverse 360°
Elevation +10° to +70°

Muzzle velocity 575 m/sec
Shell weight 6.25 kg
Maximum ceiling 6500 m
Rate of fire 12–15 rpm
Manufacturers (gun): Fonderie de Bourges, Bourges; (carriage): Schneider et Cie, Le Creusot; De Dion Bouton

Remarks: Despite their age (design dated from 1915) French Army still had 913 of these guns in service in May 1940. Most were retained for a while by German forces but gradually dismantled and diverted to Atlantic coast defences. Note that original field gun designation was retained.

7.5 cm Flugabwehrkanone M 17/34(f)

German designation 7.5 cm Flak M 17/34(f)
Original designation Canon de 75 mm contre aeronefs sur remorque Schneider
Calibre 75 mm
Length of piece (L/53): 4000 mm
Length of rifling 3250 mm
Weight complete 4940 kg
Weight in action 3460 kg
Traverse 360°
Elevation 0° to +70°
Muzzle velocity 685–715 m/sec
Shell weight 6.1 kg
Maximum ceiling 8200 m
Rate of fire 20–30 rpm

Remarks: Basically modernised mle 1917 with modified carriage. Substantial number captured by German forces in 1940 and impressed in service, mainly in France.

Battery of 7.5 cm Flak M17/34(f)

7.5 cm Flugabwehrkanone M 30(f)

7.5 cm Flak M 30(f) in travelling position

7.5 cm Flak M 30(f) in action position

German designation 7.5 cm Flak M 30(f)
Original designation Canon de 75 mm contre aeronefs mle 1930
Calibre 75 mm
Length of piece (L/53): 4000 mm
Weight complete 4200 kg
Traverse 360°
Elevation 0° to +78°
Muzzle velocity 685–715 m/sec
Maximum ceiling 8200 m
Rate of fire 20–30 rpm

Remarks: Similar to mle 1917/34 but minus complicated fire control equipment. After 1940 impressed in German service and used mainly in France.

7.5 cm Flugabwehrkanone M 33(f)

German designation 7.5 cm Flak M 33(f)
Original designation Canon de 75 mm contre aeronefs mle 1933
Calibre 75 mm
Length of piece (L/53): 4005 mm
Weight in action 4200 kg
Traverse 360°
Elevation 0° to +70°
Muzzle velocity 685–715 m/sec
Shell weight 9 kg
Maximum ceiling 7200 m
Rate of fire 20–30 rpm

Remarks: Modernised mle 1917/34. Improved barrel, revised mounting. Only limited number built and taken over by German forces after 1940.

7.5 cm Flak M 33(f) in a gun pit

7.5 cm Flugabwehrkanone M 36(f)

German designation 7.5 cm Flak M 36(f)
Original designation Canon de 75 mm contre aeronefs mle 1936 (Schneider)
Calibre 75 mm
Length of piece (L/54): 4050 mm
Weight travelling 5560 kg
Weight in action 4100 kg
Traverse 360°
Elevation −5° to +70°
Muzzle velocity 700 m/sec
Shell weight 6.44 kg
Maximum vertical ceiling 8200 m
Rate of fire 20–25 rpm

7.5 cm Flak M 36(f) in travelling position

Remarks: Probably the best heavy French AA gun in 1939–40, but available only in small numbers. All captured weapons impressed in German service in France. Also known as 7.5 cm Flak 36 (Schneider).

7.5 cm Flak M 36(f) in action position

7.5 cm Flugabwehrkanone Vickers M.35(h) and 7.5 cm Flugabwehrkanone (d)

German designations 7.5 cm Flak Vickers M.35(h); 7.5 cm Flak (d)
Original designation Vickers Model 1931
Calibre 75 mm
Length of piece (L/43): 3225 mm
Weight travelling 3325 kg
Weight in action 2825 kg
Traverse 360°
Elevation 0° to +90°
Muzzle velocity 750 m/sec
Shell weight 6.5 kg
Maximum vertical ceiling 10,000 m
Rate of fire 12–15 rpm
Manufacturer Vickers-Armstrong Ltd., Crayford and Elswick

Remarks: Ordered during 1920s, but design rejected by British War Office and released for export. Guns in German service included captured weapons from Holland, Belgium, Denmark and Soviet Union (ex-Lithuanian). Used only by special Home Defence 'barrage units' 1940–43.

7.5 cm Flugabwehrkanone Vickers(e)

German designation 7.5 cm Flak Vickers(e)
Original designation Ordnance, QF, 3.in. 20 cwt
Calibre 76.2 mm
Length of piece (L/45): 3556 mm
Length of barrel 3429 mm
Length of rifling 2981.5 mm
Weight complete 7983 kg
Weight of gun 1017 kg
Traverse 360°
Elevation −10° to +90°
Muzzle velocity 610 m/sec
Shell weight 7.26 kg
Maximum vertical ceiling 7160 m
Rate of fire 15–20 rpm
Manufacturer Vickers-Armstrong Ltd., Crayford and Elswick

Remarks: Vickers 3-in AA guns of slightly differing variants recovered by German forces after BEF evacuation from France in June 1940. Later some also captured in Balkan area.

7.5 cm Flugabwehrkanone M 37(t) and 7.5 cm Flugabwehrkanone Skoda

German designation 7.5 cm Flak M 37(t) oder Flak Skoda
Original designation 7.5 cm kanon PL vz. 37
Calibre 75 mm
Length of piece (L/48.7): 3650 mm
Weight travelling 4150 kg
Weight in action 2800 kg
Traverse 360°
Elevation 0° to +85°
Muzzle velocity 775 m/sec
Shell weight 6.5 kg
Maximum vertical ceiling 9200 m
Rate of fire 15–20 rpm
Manufacturer Škoda, Pilsen

Remarks: Czech AA guns taken over by German forces 1938–39 and subsequently delivered to Italian Army. After September 1943 available guns impressed by German troops in Italy.

7.5 cm Flak M 37(t) in travelling position

7.5 cm Flugabwehrkanone 264/3(i)

German designation 7.5 cm Flak 264/3(i)
Original designation Cannone da 75/46 modello 34
Calibre 75 mm
Length of piece (L/46): 3450 mm
Length of rifling 2844 mm
Weight travelling 4405 kg
Weight in action 3300 kg
Weight of gun 747 kg
Traverse 360°
Elevation 0° to +90°
Muzzle velocity 750 m/sec
Shell weight 6.5 kg
Maximum vertical ceiling 8200 m
Rate of fire 20–25 rpm
Manufacturer Ansaldo, Turin

Remarks: Ansaldo design of late 1920s, with many modern features. Adopted as standard Italian Army AA gun 1934, but only 226 completed by late 1942. Used in action by both Italian and German troops in North Africa, Italy and on Eastern Front.

7.5 cm Flugabwehrkanone 264/4(i)

German designation 7.5 cm Flak 264/4(i)
Original designation Cannone da 75/46 modello 40
Calibre 75 mm
Length of piece (L/46): 3450 mm
Length of rifling 2844 mm
Weight emplaced 2450 kg
Weight of gun 686 kg
Traverse 2 × 360°
Elevation −1° to +90°
Muzzle velocity 750 m/sec
Shell weight 6.5 kg
Maximum vertical ceiling 8200 m
Rate of fire 20 rpm
Manufacturer Ansaldo, Turin

Remarks: Same basic weapon as 75/46 modello 34 but designed for static use only. Entered service 1940 but by late 1942 only 45 of 232 guns ordered had been delivered. All available guns impressed into German service after 1943.

7.62 cm Flugabwehrkanone 266/1(i)

German designation 7.62 cm Flak 266/1(i)
Original designation Cannone da 76/40 CA
Calibre 76.2 mm
Length of piece (L/41.2): 3139 mm
Length of barrel 3042.86 mm
Length of rifling 2580.58 mm
Weight emplaced 2676 kg
Weight of gun 600 kg
Traverse 360°
Elevation −5° to +75°
Muzzle velocity 690 m/sec
Shell weight 6.05 kg
Maximum vertical ceiling 5900 m
Rate of fire 20 rpm

Remarks: Dual purpose AA/coastal defence gun of WW I origin. Most available guns modernised 1935 (*see* 7.62 cm Flak 266/2(i); only a few taken into German service after 1943.

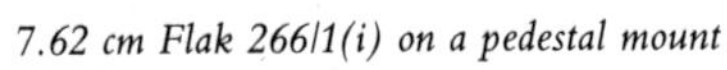
7.62 cm Flak 266/1(i) on a pedestal mount

7.62 cm Flugabwehrkanone 266/2(i)

German designation 7.62 cm Flak 266/2(i)
Original designation Cannone da 76/40 modificata 35
Calibre 76.2 mm
Length of piece (L/41.2): 3139 mm
Length of barrel 3042.86 mm
Length of rifling 2590.58 mm
Weight emplaced 5243 kg
Weight of gun 627 kg
Traverse 360°
Elevation −6° to +81°
Muzzle velocity 690 m/sec
Shell weight 6 kg
Maximum vertical ceiling 5900 m
Rate of fire 20 rpm

Remarks: Modernised version of Cannone da 76/40 mod. 33. After September 1943 available guns impressed into German service into Italy.

7.62 cm Flak 266/2(i), static mount

7.62 cm Flugabwehrkanone 266/3(i)

German designation 7.62 cm Flak 266/3(i)
Original designation Cannone da 76/45 CA
Calibre 76.2 mm
Length of piece (L/46.9): 3573 mm
Length of barrel 3438.8 mm
Length of rifling 2769.5 mm
Weight emplaced 2204 kg
Weight of gun 704 kg
Traverse 360°
Elevation −5° to +80°
Muzzle velocity 756 m/sec
Shell weight 6 kg
Maximum vertical ceiling 7800 m
Rate of fire 20 rpm

Remarks: Modernised version of Cannone da 76/40, built in small numbers for static defence. Some impressed by German forces in Italy after September 1943.

7.62 cm Flak 266/3(i), static mount

7.62 cm Flugabwehrkanone M 31(r)

7.62 cm Flak M 31(r) on its two wheel trailer

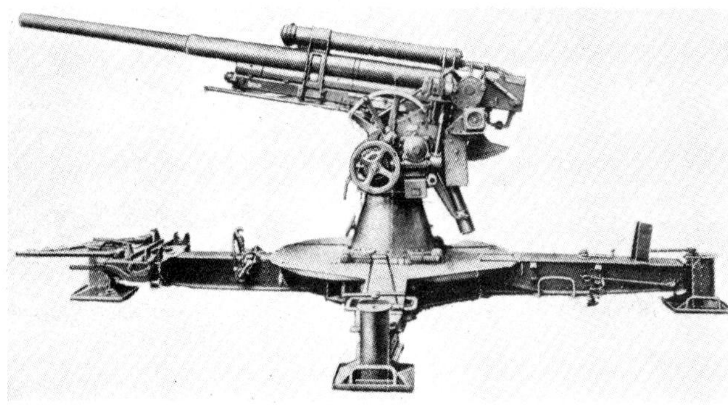

7.62 cm Flak M 31(r) in firing position

German designation 7.62 cm Flak M 31(r)
Original designation *76.2 mm Zenitnaya Pushka obr. 1931 g*
Calibre 76.2 mm
Length of piece (L/55): 4191 mm
Length of rifling 3372.5 mm
Weight travelling 4820 kg
Weight in action 3650 kg
Traverse 2 × 360°
Elevation −2° to +82°
Muzzle velocity 813 m/sec
Shell weight 6.61 kg
Maximum vertical ceiling 9300 m
Rate of fire 15–20 rpm
Original manufacturer Various Soviet State arsenals

Remarks: First standard Soviet heavy AA gun. Design incorporated many Vickers features. Substantial numbers in service 1941 and many captured by German troops. From 1942 most used with available captured ammunition by various German Home Defence AA units; remaining guns later rebored to 8.8 cm calibre.

7.62/8.8 cm Flugabwehrkanone M 31(r)

German designation 7.62/8.8 cm Flak M 31(r)
Original designation *76.2 mm ZP obr. 1931 g*
Calibre 88 mm
Length of piece 4191 mm
Weight in action 3650 kg
Weight of gun 830 kg
Traverse 2 × 360°
Elevation −2° to ×82°
Muzzle velocity Not recorded
Shell weight 9 kg
Maximum vertical ceiling (est): 9000 m
Rate of fire 15–20 rpm

Remarks: Rebored Soviet *76.2 mm ZP1931* AA guns. Used by various German Home Defence AA units, later also in dual role by front-line formations.

7.62/8.8 cm Flak M 31(r) in ground support role on the Eastern Front

7.62 cm Flugabwehrkanone M 38(r)

German designation 7.62 cm Flak M 38(r)
Original designation *76.2 mm ZP obr. 1938 g*
Calibre 76.2 mm
Length of piece (L/55): 4191 mm
Length of rifling 3606 mm
Weight travelling 4210 kg
Weight in action 3047 kg
Traverse 2 × 360°
Elevation −3° to +82°
Muzzle velocity 813 m/sec
Shell weight 6.61 kg
Maximum vertical ceiling 9300 m
Rate of fire 15–20 rpm
Original manufacturer Various Soviet State arsenals

Remarks: Modernised *76.2 mm obr. 1931* AA gun with twin axle carriage. Large numbers in service 1941 and many captured and later used by German troops. By late 1944 most remaining guns rebored to 8.8 cm.

7.62/8.8 cm Flugabwehrkanone M 38(r)

German designation 7.62/8.8 cm Flak M 38(r)
Original designation *76.2 mm ZP obr. 1938 g*
Calibre 88 mm
Length of piece 4191 mm
Weight in action 3047 kg
Weight of gun 920 kg
Traverse 2 × 360°
Elevation −3° to +82°
Muzzle velocity Not recorded
Shell weight 9 kg
Maximum vertical ceiling (est): 9000 m
Rate of fire 15–20 rpm

Remarks: Rebored Soviet *76.2 mm ZP 1938* AA guns. Initially used by various German Home Defence AA units, from late 1944, also by front-line formations.

7.65 cm Flugabwehrkanone 33(t)

German designation 7.65 cm Flak 33(t)
Original designation 8 cm kanon PL vz. 33
Calibre 76.5 mm
Length of piece (L/50): 3825 mm
Weight in action 2440 kg
Traverse 360°
Elevation 0° to +85°
Muzzle velocity 808 m/sec
Shell weight 6.765 kg
Maximum vertical ceiling 8390 m
Rate of fire 15–20 rpm
Manufacturer Škoda, Pilsen

Remarks: Adopted by Czech Army and also sold to Romania, Yugoslavia and Lithuania. Most guns in German service came from Czech stocks.

7.65 cm Flugabwehrkanone 37(t)

German designation 7.65 cm Flak 37(t)
Original designation 8 cm kanon PL vz. 37
Calibre 76.5 mm
Length of piece (L/52.8): 4040 mm
Weight travelling 5100 kg
Weight in action 3800 kg
Traverse 360°
Elevation −1° to +85°
Muzzle velocity 800 m/sec
Shell weight 8 kg
Maximum vertical ceiling 11,470 m
Rate of fire 15–20 rpm
Manufacturer Škoda, Pilsen

Remarks: Improved vz. 33 guns, with longer barrel. Adopted by Czech Army 1937. All available guns taken over by German forces 1938–39.

8.35 cm Flugabwehrkanone 22(t)

German designation 8.35 cm Flak 22(t)
Original designation 8.35 cm kanon PL vz. 37
Calibre 83.5 mm
Length of piece (L/55): 4600 mm
Weight travelling 8800 kg
Weight in action 8800 kg
Traverse 360°
Elevation −5° to +85°
Muzzle velocity 800 m/sec
Shell weight 10 kg
Maximum vertical ceiling 11,300 m
Rate of fire 15 rpm
Manufacturer Škoda, Pilsen

Remarks: Modified Škoda vz. 22 guns. Some requisitioned by German forces in occupied Czechoslovakia 1938–39, additional examples captured in Yugoslavia 1941.

8.5 cm Flugabwehrkanone M 39(r)

German designation 8.5 cm Flak M 39(r)
Original designation *85 mm ZP obr. 1939 g*
Calibre 85 mm
Length of piece (L/55.2): 4693 mm
Length of rifling 3493.5 mm
Weight travelling 4300 kg
Weight in action 3057 kg
Traverse 2 × 360°
Elevation −3° to +82°
Muzzle velocity 800 m/sec
Shell weight 9.2 kg
Maximum vertical ceiling 10,500 m
Rate of fire 15–20 rpm
Manufacturer Various Soviet State arsenals

Remarks: Dual-purpose weapon. Basically scaled-up version of *76.2 mm ZP obr. 1938 g*; also known as *KS-12*. One of the most successful Soviet gun designs of that period.

8.5/8.8 cm Flugabwehrkanone M 39(r)

German designation 8.5 cm Flak M 39(r)
Original designation *85 mm ZP obr. 1939 g*
Calibre 88 mm
Length 4693 mm
Weight in action 3057 kg
Weight of gun 920 kg
Traverse 2 ×360°
Elevation −3° to +82°
Muzzle velocity Not recorded
Shell weight 9 kg
Maximum vertical ceiling (est): 10,500 m
Rate of fire 15–20 rpm

Remarks: Rebored Soviet *85 mm ZP 1939* AA guns.

9 cm Flugabwehrkanone M 39(f)

German designation 9 cm Flak M 39(f)
Original designation Canon de 90 mm de DCA à traction mécanique mle 1926
Calibre 90 mm
Length of piece (L/50): 4500 mm
Length of barrel 3780 mm
Weight travelling 8570 kg
Weight in action 5760 kg
Traverse 360°
Elevation −4° to +80°
Muzzle velocity 810 m/sec
Shell weight 9.615 kg
Maximum ceiling 11,600 m
Rate of fire 15 rpm
Original manufacturer Ateliers de Havre, Le Havre

Remarks: Heaviest modern French AA gun, but only 17 in service 1940. Afterwards all remaining guns impressed by German forces in France. Kept in small-scale production after 1940.

9 cm Flak M 39(f) in travelling position

9 cm Flak M 39(f) in firing position

9 cm Flugabwehrkanone M 12(t)

German designation 9 cm Flak M 12(t)
Original designation 9 cm kanon PL vz. 12/20
Calibre 90 mm
Length of piece (L/45): 4050 mm
Weight emplaced 6500 kg
Traverse 360°
Elevation −5° to +90°
Muzzle velocity 775 m/sec
Shell weight 10.2 kg
Maximum vertical ceiling 6000 m
Rate of fire 10–15 rpm
Manufacturer Škoda, Pilsen

Remarks: Notable as one of earliest heavy AA guns. Most exported to Yugoslavia, Romania, China and Soviet Union, but some still left in Czechoslovakia 1938–39. Guns impressed in German service comprised small number of ex-Czech and captured Soviet weapons.

9 cm Flak M 12(t), improved version

9 cm Flak M 12(t) pedestal mount static equipment (original version)

9 cm Flugabwehrkanone 41(i) and 309/1(i)

German designation 9 cm Flak 41(i) oder 309/1(i)
Original designation Cannone da 90/53 CA
Calibre 90 mm
Length of piece 5300 mm
Length of barrel (L/53): 4770 mm
Length of rifling 4045 mm
Weight travelling 8950 kg
Weight in action 6240 kg
Traverse 360°
Elevation −2° to +85°
Muzzle velocity 830 m/sec
Shell weight (HE): 10.33 kg
Shot weight 11.25 kg
Maximum effective ceiling 11,400 m
Rate of fire 20 rpm
Manufacturer Ansaldo, Turin

Remarks: First produced late in 1939. Built in three versions of which 90/53 static gun numerically most important. Best Italian heavy AA gun of WW II years; also successfully used against ground targets, tanks and in coastal defence.

9.4 cm Flugabwehrkanone Vickers M 39(e)

9.4 cm Flak Vickers M 39(e) mobile equipment

German designations 9.4 cm Flak Vickers M 39(e)
Original designation Ordnance QF, 3.7-inch Mks I, II and III on Mountings Mks I, IA and IIIA; Ordnance QF, 3.7-inch Mks II and IIA on Mountings Mks II, IIA and IIB
Calibre 94 mm
Length of piece 4957 mm
Length of bore (L/50): 4699 mm
Length of rifling 3987.4 mm
Weight complete 9326 kg
Weight in action 8900 kg
Weight of gun 1740 kg
Traverse 360°
Elevation −5° to +80°
Muzzle velocity 792 m/sec
Shell weight 12.96 kg
Maximum effective ceiling 9760 m
Rate of fire 10 rpm
Manufacturer Vickers-Armstrong Ltd, Crayford and Elswick

Remarks: Standard British heavy AA gun of WW II years. Designed to 1934 specifications, first delivered in January 1938. Guns in German service recovered in France after BEF evacuation in May–June 1940; some also captured from France and Belgium. Highly regarded by German Flak troops. Some mounted horizontally as coastal defence guns, as at Walcheren, Holland.

10.2 cm Flugabwehrkanone (i)

German designation 10.2 cm Flak(i)
Original designation Cannone da 102/35
Calibre 102 mm
Length of piece (L/35): 3570 mm
Traverse 360°
Elevation −10° to +70°
Muzzle velocity 755 m/sec
Shell weight 13.1 kg
Maximum vertical ceiling 9450 m
Rate of fire 10–12 rpm
Manufacturer Ansaldo, Torino (not confirmed)

Remarks: Designed and used as dual-purpose AA/coastal defence gun. Featured electrical data transmission. Only small number (4) taken over by German troops after 1943.

10.2 cm Flak(i) static AA guns in position

Light field artillery

This section will deal with artillery up to 105 mm calibre.

When WW I ended in 1918 the German Army possessed two main types of field piece, namely the *7.7 cm FK 16* and the *10.5 cm leFH 16*. Both used the same carriage, which was a development of a design that entered service in 1896. When the war ended there were large numbers of these two pieces still coming off the production lines but the terms of the Versailles Treaty restricted the number of guns in the new German Army. In addition, large numbers were confiscated and issued as war reparations to such nations as Belgium, or else sold on the open market. But a good many were carefully hidden away for possible future use and throughout the 1920s and the 1930s the *7.7 cm FK 16* and the *10.5 cm leFH 16* were the mainstays of the German field artillery. It was on these two pieces that the Wehrmacht trained and kept operationally up-to-date. In 1939 many were still in use and were kept in service for training and equipping many second-line formations. Inevitably, many were later incorporated in the 'Atlantic Wall' defences, but these were nearly all the *10.5 cm leFH 16* as the *7.7 cm FK 16* had been updated by that time.

The updating of the FK 16 consisted of placing new 7.5 cm barrels on the old carriage. Originally it was intended that this new barrel/carriage combination would be used by cavalry artillery units only but its use spread to all arms of the German Army. It was redesigned *7.5 cm FK 16nA* (nA = neuer Art = new model) and almost all the old 7.7 cm guns were involved in the modification programme. After 1940 the *FK 16nA* was gradually withdrawn from front-line service and most were issued to second-line and occupation units.

The *7.5 cm FK 16nA* was regarded as an interim model, as by the late 1920s the analysis of WW I combat reports and a close watch on technical developments abroad were combined to make a forecast of trends in field artillery. By late 1928 it had been decided by the German military planners that in future the main weight of light field artillery would be invested in 105 mm howitzers with only limited backing to be provided by 75 mm guns. Specifications were issued for the new 75 mm gun and Krupp and Rheinmetall both built prototypes. The Rheinmetall prototype was ready in 1931 but the Krupp entry was the one selected for service. Development was so slow that it was not until 1938 that the first examples were issued to the troops as the *7.5 cm leFK 18*. The *leFK 18* used a good carriage but its performance was little different from that of the old *FK 16* and as a result relatively few were produced. By 1938 the main emphasis was on the 105 mm howitzers. The only other 'German' 75 mm gun of the early war years was the *7.5 cm FK 38* which was built by Krupp during the late 1930s to a Brazilian order. An order for 64 was fulfilled and delivered, but the production line was kept in being and the rest of the output went to the German Army. About 80 were delivered and the last examples came off the lines during 1942. The *FK 38* was an updated *FK 18* but the later versions used pressed steel wheels to make them more suitable for powered traction.

As stated above, a decision made during the late 1920s placed an emphasis on the 105 mm howitzer as the mainstay of the German field artillery. The old *10.5 cm leFH 16* was used until new equipment became available, but after 1940 most were relegated to second-line and occupation units. Many ended up as part of the 'Atlantic Wall' defences and it would be safe to say that nearly every piece mentioned in this section was at some time or other emplaced as part of the sea defences of German territories. In this way the *10.5 cm leFH 16* remained in service until 1945, and after 1940 the numbers were increased by some of the old 1918 war reparations that had been delivered to Belgium. The Belgian howitzers had been modernised by the addition of pneumatic tyres and other refinements and were quickly taken into German service.

The contract for the new 10.5 cm howitzer went to Rheinmetall-Borsig. This firm had started design work during the late 1920s on a 7.5 cm long range gun known as the *7.5 cm WFK L/42*, but the trend to the heavier calibre of 105 mm meant the end of the work carried out on that promising design and the carriage was adapted to take a 10.5 cm barrel. The new model was ready in 1935 and went into series production soon after, so that by 1939 the new gun, the *10.5 cm leFH 18*, was the 'standard' field piece for the Wehrmacht. A wide range of ammunition was developed for this weapon, including a leaflet shell, and the *leFH 18* proved to be a reliable, sturdy field howitzer. But it tended towards being rather heavy for its calibre and in action proved to be rather low on range when pitted against the Soviet artillery after 1941. In an attempt to improve performance it was decided to increase the propellant charges and provide for the increased recoil loads by fitting a muzzle brake. In this form the piece was known as the *10.5 cm leFH 18M*. The increased range was obtained, but when sabot ammunition was used the sabot rings

10.5 cm le FH 18/40 in action

tended to foul the muzzle brake vanes, necessitating a redesigned brake for later models.

Just before the war started in 1939 Krupp of Essen produced a commercial version of the *leFH 18* and sold a number to Holland. The events of 1940 brought these pieces back into the German fold, but as they used slightly different ammunition these guns could not be used as front line equipment. As the number of pieces involved was about 80 the Germans decided to remove the 'Dutch' barrels and replace them with *leFH 18M* barrels. The result was the *10.5 cm leFH 18/39* which remained in service until 1945.

With the 10.5 cm howitzers in use in large numbers during the early war years the German artillery had the edge in fire weight over nearly all their opponents, most of whom used obsolete WW I equipment of about 75 mm calibre. But as the war went on more modern opponents in the shape of the British 25 pr (87.6 mm) and the many Soviet 76.2 mm guns showed that the 10.5 cm howitzers could be outranged without undue difficulty and the events of the Russian winter campaign of 1941–42 showed the other major shortcoming of the German piece. The alternate cycle of frost/thaw during that winter reduced the rough dirt roads that supplied the Eastern Front into quagmires through which the heavy German howitzers could make only slow progress. Despite the fact that the sturdy carriage provided a good steady fire platform for the *leFH 18* series, it was really too heavy for the job, and many were lost during that winter of 1941–42 simply because they got bogged down in frozen mud during the retreat and could not be moved. In March 1942 a specification was issued for a new and lighter carriage for the 10.5 cm family but the specification laid stress on the fact that the new carriage was needed quickly and was to be capable of easy mass production. Once again, everything was ready for another typical German feat of improvisation. The carriage of the *7.5 cm Pak 40* was adapted to take a *leFH 18M* barrel fitted with a new design of muzzle brake that allowed sabot ammunition to be fired, and the new combination became the *10.5 cm leFH 18/40.* Only about 30 kg in weight was saved by this gun/carriage combination, but it had the advantage that it offered the chance of increased output totals and thus the *leFH 18/40* entered service. The main production centres were Schichau at Elbing, Menck und Hambrock at Hamburg and the Krupp facilities in Markstädt. Krupp also carried out some development work on the *leFH 18/40* specification which involved fitting a longer barrel to produce the *10.5 cm leFH 18/42,* but it did not enter service.

It was intended that the *leFH 18/40* was to be a stop-gap equipment only but it was kept in production until 1945, despite the many troubles caused by the 10.5 cm recoil forces on a carriage that was really too light to absorb them. It was intended that a completely new design would be produced and this, in time, emerged as the *10.5 cm leFH 42*. Despite its being a sound sturdy design, the *leFH 42* was not adopted for service, as by the time it was ready the specifications for such a weapon had been changed as the result of combat experience on the Eastern Front. The new specifications called for 360° traverse, a high angle of elevation to provide the high angle fire needed in the Russian forests, a range of at least 13,000 metres and the weight no more than that of the *leFH 18/40*. For once, Rheinmetall did not enter a design and it was left to Krupp and the Skoda-Werke at Pilsen to put forward their submissions. Krupp submitted two proposals and Skoda only one, but it was the Skoda design that actually reached the prototype stage first. The designation of the new howitzer was *10.5 cm leFH 43*, and all three designs used some novel and original features. But they arrived too late on the scene and only the Skoda gun was actually test-fired before the war ended. Both Krupp models remained in the wooden mock-up stage.

By 1944 there began a gradual drift back to the use of 75 mm as a field calibre in the German Army. Field artillery units began to be issued with numbers of *7.5 cm Pak 40* equipments, giving them a useful dual-purpose weapon which however had range limitations due to the lack of elevation available. The Pak guns were redesignated *7.5 cm FK 40*, and their numbers were increased by the issue of *8.8 cm Pak 43* guns as well; these became the *8.8 cm K 43*. The limitations of the *Pak 40* were recognised and one of the strangest gun combinations of WW II then emerged to fulfil the need. The 7.5 cm barrel was placed on the carriage of the *10.5 cm leFH 18/40*, which was itself a modified *Pak 40* carriage. This new weapon was designated the *7.5 cm FK 7M85*. Relatively few of these oddities actually reached the front before the war ended. By that time many *Pak 40*s had been modified by giving them an increased elevation arc which would seem to have made the 7M85 programme unnecessary; these guns were the *7.5 cm FK 7M59*.

In the calibre range up to 105 mm came other German guns that were usually used by Corps rather than field formations. The oldest of these was the *10 cm K 17,* or *17/04*. This WW I veteran was still in use in 1939, many having been delivered to Austria after 1918, but few seem to have been actually used in the front line. Most ended up as part of the 'Atlantic Wall' defences. A slightly later design, the *s 10 cm K 18*, was a result of the design work carried out during the late 1920s. Both Krupp and Rheinmetall submitted proposals but for once the end result was a

s 10 cm K 18 in action on the Eastern Front

combination of both designs and the production line was set up at the Spreewerk at Berlin-Spandau. The *s 10 cm K 18* was really too big and numbers were subsequently withdrawn from field use and deployed as coast defence weapons. In 1940 Krupp and Rheinmetall were asked to increase the range of the *s 10 cm K 18*, but as early as 1937 there had been reports that the *K 18* was too heavy for the shell it fired. Both Krupp and Rheinmetall produced almost identical designs which involved the use of a longer barrel placed on a modified *K 18* carriage. The result would have been the *10.5 cm s K 40* but it did not go into production. One reason was that the new gun was even heavier than the original but the increase in range was only marginal. Another reason that prevented production was that by the time it was ready in 1941 all production facilities were operating at full stretch to meet the demands of the Eastern Front. It was not until 1942 that a small batch was actually produced and issued, but these used a modified version of the *s 10 cm K 18* carriage. The original designation of this field gun was *10.5 cm s K 18/40*, later changed to *10.5 cm s K 42*. Production was carried out at the Spreewerk, Berlin-Spandau.

One other gun remains to be mentioned, the *10.5 cm leK 41*. It was an attempt by Rheinmetall and Krupp to provide a field gun with a better range than the 10.5 cm howitzers, at the same time keeping the weight to that of the howitzer. Work continued on both projects, and prototypes were built, despite the fact that by the time they were ready the emphasis had been changed to decreased weight. Both the Krupp and Rheinmetall models were again too heavy for their intended role and during 1941 the programme was terminated.

A characteristic of the German war production story is that military planners had underestimated the demands of a major European war. The over-rapid expansion of the German armed forces after 1939 made such demands on the German war economy that it was soon unable to cope with the massive flood of demands for all forms of equipment. Light field artillery was high on the list of demands and it was only the use of captured or impressed equipment that enabled the Wehrmacht to carry out its multitude of tasks. As a general rule all German-designed and produced material was issued to the front line formations and everything else was relegated to training, garrison and occupation units. Artillery detachments all over Europe often found themselves using all sorts of equipment, some of it very good, some of it very old, and some of it from some very obscure corners of Europe. The range of light artillery pieces in German service up to 1945 was vast. Much of the equipment mentioned below was used during the early war years by various field formations, but after about 1942 a great deal of equipment was shifted to the Atlantic coast where it was installed to cover a variety of roles. Some were actually sited in bunkers for beach defences, while others were used for battery defences to cover some of the larger and specialised installations. Many of these guns were situated with all available ammunition close at hand, and there they remained until 1945 or until overrun.

Czechoslovakia was first on the list of those forced to contribute to the German armour. In 1938 the Czech Army had a magnificient gun park and it was equipped with many good modern guns. These were taken over by the German Army en bloc, and during the campaigns of 1939 and 1940 whole divisions of the Wehrmacht were equipped with Czech material. All the Czech guns came from the Škoda Armament Works at Pilsen, and after 1939 this large and important arsenal came under German control as the Skoda-Werke. Initially the fact that the Czech guns had different calibres seemed to pose a problem, but so much ammunition and manufacturing facilities also came under German control that this never materialised. More Škoda guns were taken over from Poland, Yugoslavia, Greece and eventually Italy. All were good sound designs that gave good service and many remained in use until 1945.

In 1939 it was the turn of Poland to give up her gun park, or what was left of it after the devastating effects of the 'Blitzkrieg'. Poland had never produced any original designs, despite some attempts, and her guns were all imports which were often adapted to suit the needs of her army. Thus the Germans were able to take over numbers of a 75 mm field gun, the wz. 02/26, which was originally a Russian design, while the 105 mm armata wz. 29 was a Polish modification of a French Schneider design, the mle 1913. Both types were used by the Germans, usually by units based in Poland.

10.5 cm le FH 16 in action

France had a massive gun park in 1940, much of it rather ancient equipment, and nearly all of it fell into German hands after the lightning campaign of May and June 1940. The Germans took over every artillery item they could find and render serviceable. Large numbers of the venerable mle 1897, the famous 'soixante-quinze', were eagerly taken into use and more came from Poland, and later Greece. A later version of the mle 1897, the 97/33, was added to the total, but just as important were the vast stocks of ammunition that were captured. French 75 and 105 mm shells were frequently fired from German guns as they were found to be completely compatible, as was a great deal of other ancillary equipment, including tractors. The German occupation forces also took over stocks of 105 mm howitzers as well and many of these were used by various artillery formations until they were assigned to the coastal defence role. Among these howitzers were numbers of the mle 1913 which were also acquired in Poland, Belgium and Yugoslavia. In 1943 Italy also added the few still left in service.

Belgium and Holland contributed to the German artillery park in a rather unusual way. After 1918 the Belgians received a number of *7.7 cm FK 16* guns and these were gradually modified in a number of respects so that they were almost identical to the German *7.5 cm FK 16nA*. Assimilating these pieces into the German Army was no problem, and the same applied to the Dutch guns, which were a Krupp commercial design, the Model 1903. After 1920, Krupp took over the Dutch Siderius firm and ran it as a virtual 'shadow' concern. However, Krupp did not use Siderius for weapon development as was the general case elsewhere but merely concentrated on modernising the Model 1903. By 1940 there were no fewer than three different versions in service and all of course came under German control. Holland was also forced to hand over a number of Bofors 105 mm Model 1927 guns which were used for coastal defence by their new masters.

Denmark contributed a French 105 mm Schneider gun, the M. 30, which was also used for coast defence by the Germans. Norway had bought a number of Erhardt field guns during the first decade of the century and in 1940 they reverted to German control. The guns were the Model 01 and despite their age they were used to equip some artillery units based in Norway. The Erhardt concern later formed a large part of what was to become the Rheinmetall-Borsig conglomerate.

After Dunkirk large stocks of British field guns fell into German hands, despite the attempts of their crews to spike many of them. The bulk of the captured British guns were 18 prs and 25 pr Mark 1s, some with split and some with box trails. Many of these guns were used for screening coastal defence batteries but most appear to have been withdrawn after about 1943. Another British gun the Germans were always glad to use, especially in North Africa, was the 25 pr Mark 2.

But it was in the Soviet Union that the Wehrmacht gained its greatest hoard of artillery. From the campaigns of 1941 and 1942 came vast numbers of field guns, nearly all of them sturdy, serviceable weapons that could easily be turned against their former owners. All used the same 76.2 mm ammunition and vast stocks of these rounds were captured and were likely to be captured in the future. As a result on the Eastern Front many front-line artillery units found themselves using Soviet guns against the Soviets. Such large numbers were captured that it became commonplace to find Russian guns being used on other fronts as well as many being incorporated into the 'Atlantic Wall'. Apart from the 76.2 mm guns, numbers of 107 mm guns (slightly outside the confines of this section but included here to simplify matters) were taken over and numbers of these ended up along the Channel coast.

September 1943 brought the Italian surrender and in one fell swoop the German divisions in Italy took over as much of the Italian equipment as they could. The Italian material was largely obsolescent but by autumn 1943 any additional weapons were most welcome to the German forces.

7.5 cm Feldkanone 16 neuer Art

7.5 cm FK nA mounted on a concrete pivot for coastal defence. France, 1944

German designation 7.5 cm FK 16nA
Calibre 75 mm
Length of piece (L/36): 2700 mm
Length of rifling 2036 mm
Weight travelling 2415 kg
Weight in action 1524 kg
Traverse 4°
Elevation −9° to +44°
Muzzle velocity 662 m/sec
Shell weight 5.83 kg
Maximum range 12,875 m
Rate of fire 10–12 rpm
Manufacturer Rheinmetall-Borsig AG, Düsseldorf

Remarks: Modernised version of WW I gun – new barrel on 7.7 cm FK 16 carriage. In service from 1934, but after 1939 increasingly used for training purposes. Later many deployed along 'Atlantic Wall'.

7.5 cm leichte Feldkanone 18

Horse-drawn leFK 18 on the Eastern Front

German designation 7.5 cm leFK 18
Calibre 75 mm
Length of piece (L/26): 1940 mm
Length of barrel 1660 mm
Length of rifling 1412 mm
Weight travelling 2010 kg
Weight in action 1120 kg
Traverse 60°
Elevation −5° to +45°
Muzzle velocity 425 m/sec
Shell weight 5.83 kg
Maximum range 9425 m
Rate of fire 8–10 rpm
Barrel life 8000–10,000 rounds
Manufacturer Friedr. Krupp AG, Essen

Remarks: Development commenced 1930–31 simultaneously by Krupp and Rheinmetall. Final design combined Rheinmetall gun with Krupp carriage, and introduced divided recoil system, later a characteristic of most German guns. In production from 1938; in service throughout WW II.

7.5 cm Feldkanone 38

German designation 7.5 cm FK 38
Calibre 75 mm
Length of piece (L/34): 2550 mm
Length of barrel 2335 mm
Length of rifling 1914 mm
Weight travelling 1860 kg
Weight in action 1366 kg
Traverse 50°
Elevation −5° to +45°
Muzzle velocity 605 m/sec
Shell weight 5.85 kg
Maximum range 11,500 m
Rate of fire 8–10 rpm
Manufacturer Friedr. Krupp AG, Essen

Remarks: General refinement of FK 18 for export. Brazilian order completed just prior to September 1939. By early 1942 another batch of 80 guns with some changes produced for German artillery troops.

7.5 cm Feldkanone 7M85

German designation 7.5 cm FK 7M85
Calibre 75 mm
Length of piece (L/46): 3700 mm
Length of barrel 3450 mm
Length of rifling 2461 mm
Weight in action 1778 kg
Traverse 30°30′
Elevation −5° to +42°
Muzzle velocity 550 m/sec
Shell weight (HE): 5.74 kg
Maximum range 10,275 m
Rate of fire 12–15 rpm
Barrel life 6000 rounds
Manufacturer Not confirmed

Remarks: 7M85 was designed to 1944 specification requesting an easily produced light dual-purpose field/anti-tank gun. Completed gun combined 7.5 cm Pak 40 barrel with 10.5 cm leFH 18/40 carriage to save production facilities. On trials late 1944 proved overweight and only small series produced. One of first weapons to use new designation system intended for all new equipment, introduced in September 1944. In this case 7 = calibre group; M = ammunition group ballistically suited to this weapon, and 85 = last two digits of weapon drawing number.

7.5 cm Feldkanone 7M59

German designation 7.5 cm FK 7M59
Calibre 75 mm
Length of piece (L/46): 3700 mm
Length of barrel 3450 mm
Length of rifling 2461 mm
Weight in action 1453 kg
Traverse 65°
Elevation −5° to +35°
Shell weight 5.74 kg
Muzzle velocity (HE): 550 m/sec
Maximum range 13,300 m
Manufacturer Not confirmed

Remarks: 1945 improvisation to provide easily produced dual purpose field/anti-tank gun. Essentially 7.5 cm Pak 40 with extended elevation arc for increased range. Only small series completed and used.

7.5 cm Feldkanone L/42

German designation 7.5 cm FK L/42
Design designation 7.5 cm WFK L/42
Calibre 75 mm
Length of piece (L/42): 3150 mm
Weight in action 1625 kg
Traverse 60°
Elevation −5° to +42°
Muzzle velocity 701 m/sec
Shell weight 6.56 kg
Maximum range 13,480 m
Rate of fire 10 rpm
Manufacturer Rheinmetall-Borsig AG, Düsseldorf

Remarks: Designed as light field gun 1930 but only prototype completed. Development abandoned following decision to standardise on 105 mm as field piece calibre.

10.5 cm leichte Feldhaubitze 16 and 327(b)

German designations 10.5 cm leFH 16; 10.5 cm leFH 327(b)
Original designation (b): Obusier de 105 GP
Calibre 105 mm
Length of piece (L/22): 2310 mm
Length of barrel 1878 mm
Length of rifling 1634 mm
Weight travelling 2300 kg
Weight in action 1525 kg
Weight of gun 509 kg
Traverse 4°
Elevation −9° to +40°
Muzzle velocity 395 m/sec
Shell weight 14.81 kg
Maximum range 9225 m
Rate of fire 4–5 rpm
Manufacturer Friedr. Krupp AG, Essen

Remarks: WW I howitzer on same carriage as 7.7 cm FK 16. Served as first standard field howitzer with new Wehrmacht artillery regiments. After 1939 mostly relegated to reserve units; subsequently number deployed along 'Atlantic Wall' defences.

10.5 cm leichte Feldhaubitze 18

German designation 10.5 cm leFH 18
Calibre 105 mm
Length of piece (L/28): 2941 mm
Length of barrel 2612 mm
Length of rifling 2392 mm
Weight travelling 3490 kg
Weight in action (horse towed): 2040 kg; (mot): 2065 kg
Traverse 56°
Elevation −5° to +42°
Muzzle velocity 470 m/sec
Shell weight 14.81 kg
Maximum range 10,675 m
Rate of fire 4–6 rpm
Barrel life 10–12,000 rounds
Manufacturer (original): Rheinmetall-Borsig AG, Düsseldorf

Remarks: Based on 1928–29 design studies; prototype construction began 1933. Adopted as standard field howitzer by Wehrmacht 1935 and ordered in large-scale production at several centres soon afterwards. Reliable and stable weapon. Remained in service until end of WW II.

10.5 cm leFH 18 in action on the Eastern Front. Note that this gun has cast steel wheels with solid rubber tyres

10.5 cm leichte Feldhaubitze 18 (Mündungbremse)

German designation 10.5 cm leFH 18M
Calibre 105 mm
Length of piece with m/b 3308 mm
Length of piece (L/28): 2941 mm
Length of barrel 2612 mm
Length of rifling 2392 mm
Weight in action 2065 kg
Traverse 56°
Elevation −5° to +42°
Muzzle velocity 540 m/sec
Shell weight 14.81 kg
Maximum range 12,325 m
Rate of fire 4–6 rpm
Barrel life 10–12,000 rounds
Original manufacturer Rheinmetall-Borsig AG, Düsseldorf

Remarks: 1940 modification of 10.5 cm leFH 18 to take more powerful charge for longer range. Extra forces countered by modified recoil system and muzzle brake. In service until end of WW II.

10.5 cm leFH 18M on the Eastern Front, winter 1943–44

10.5 cm leichte Feldhaubitze 18/39

German designation 10.5 cm leFH 18/39
Calibre 105 mm
Length of piece with m/b 3308 mm
Length of piece (L/28): 2941 mm
Length of barrel 2612 mm
Length of rifling 2392 mm
Weight in action 1950 kg
Traverse 60°
Elevation −5° to +45°
Muzzle velocity 540 m/sec
Shell weight 14.81 kg
Maximum range 12,325 m
Rate of fire 4–6 rpm
Barrel life 10–12,000 rounds
Manufacturer (gun): Rheinmetall-Borsig AG, Düsseldorf; (carriage): Friedr. Krupp AG, Essen

Remarks: Field howitzers produced to Dutch order, delivered 1939. Due to different ammunition all captured guns replaced on same carriages by 10.5 cm leFH 18M barrels. About 80 such conversions completed 1940–41.

10.5 cm leichte Feldhaubitze 18/40

German designation 10.5 cm leFH 18/40
Calibre 105 mm
Length of piece (L/28): 2941 mm
Length of barrel 2612 mm
Length of rifling 2392 mm
Weight travelling 2300 kg
Weight in action 1900 kg
Traverse 60°
Elevation −5° to +42°
Muzzle velocity 540 m/sec
Shell weight 14.81 kg
Maximum range 12,325 m
Rate of fire 6–8 rpm
Barrel life 10,000 rounds
Manufacturers: Schichau, Elbing; Menck u. Hambrock, Hamburg; Friedr. Krupp, Markstädt

Remarks: Designed to meet a new specification issued in March 1942 requesting reduction in weight for same performance and easier series manufacturer. To speed production solution combined 10.5 cm leFH 18M barrel with 7.5 cm Pak 40 carriage.

10.5 cm leichte Feldhaubitze 18/42

German designation 10.5 cm leFH 18/42
Calibre 105 mm
Length of piece (L/31): 3255 mm
Weight in action 2035 kg
Traverse 60°
Elevation −5° to +45°
Muzzle velocity 585 m/sec
Shell weight 14.81 kg
Maximum range 12,700 m
Manufacturer Friedr. Krupp AG, Essen

Remarks: Unsuccessful 1942 attempt to improve 10.5 cm leFH series performance. Only one prototype built and tested, but proved too heavy. Development abandoned in favour of 10.5 cm FH 43 project.

10.5 cm leichte Feldhaubitze 42

German designation 10.5 cm leFH 42
Calibre 105 mm
Length of piece (L/28): 2941 mm
Weight in action 1630 kg
Traverse 70°
Elevation −5° to +45°
Muzzle velocity 595 m/sec
Shell weight 14.81 kg
Maximum range 13,000 m
Manufacturer Rheinmetall-Borsig AG, Düsseldorf

Remarks: Designed in 1942 as ultimate development of leFH 18 series. Only one prototype built and tested. Development abandoned in favour of more promising 10.5 cm leFH 43 project.

10.5 cm leichte Feldhaubitze 43 Skoda

10.5 cm leFH 43 Skoda with the new type of gun carriage comprising four trails
On the move two of the trails were attached to the gun muzzle while the other two were clamped together for towing

German designation 10.5 cm leFH 43 Skoda
Design designation 10.5 cm leFH 14.81/610
Calibre 105 mm
Length of piece (L/35): 3456 mm
Length of barrel 2470 mm
Weight in action 2200 kg
Traverse 360°
Elevation −5° to +75°
Muzzle velocity 610 m/sec
Shell weight 14.81 kg
Maximum range 13,000 m
Manufacturer Škoda-Werke, Pilsen

Remarks: An ingenious solution to a most exacting late 1943 specification accepted only by Skoda and Krupp. Prototype gun completed 1945 featured novel four-leg carriage with hydraulic leverage for 360° traverse.

10.5 cm leichte Feldhaubitze 43 Krupp

German designation 10.5 cm leFH 43 Krupp
Calibre 105 mm
Length of piece (L/28): 2941 mm
Weight in action 2400 kg
Traverse 360°
Elevation −4° to +70°
Muzzle velocity (est): 595 m/sec
Shell weight 14.81 kg
Maximum range (est): 13,000 m
Manufacturer Friedr. Krupp AG, Essen

Remarks: First of two Krupp submissions to leFH 43 project. This version featured standard 10.5 cm howitzer barrel on a new four-trail carriage. Only wooden mock-up completed.

10.5 cm leichte Feldhaubitze 43 Krupp

German designation 10.5 cm leFH 43 Krupp
Calibre 105 mm
Length of piece (L/35): 3675 mm
Weight in action (approx): 2450 kg
Traverse 360°
Elevation −10° to +70°
Muzzle velocity 655 m/sec
Shell weight 14.81 kg
Maximum range (est): 14,200 m
Manufacturer Friedr. Krupp AG, Essen

Remarks: Second of two Krupp submissions to leFH 43 project. This version combined 10.5 cm leFH 18/42 barrel with a modified 8.8 cm Pak 43 carriage. Only wooden mock-up completed.

10 cm Kanone 17, 17/04 and 17/04(ö)

German designation 10 cm K 17 oder 17/04
Calibre 105 mm
Length of piece (L/45): 4725 mm
Weight in action 3300 kg
Traverse 6°
Elevation −2° to +45°
Muzzle velocity 650 m/sec
Shell weight 18.5 kg
Maximum range 16,500 m
Manufacturer Friedr. Krupp AG, Essen

Remarks: WW I veteran, still in service 1939. Later used mainly by reserve artillery detachments and in coastal defence.

10 cm leichte Kanone 41

10 cm, leK 41 (Rheinmetall version)

German designation 10 cm leK 41
Calibre 105 mm
Length of piece (L/40): 4200 mm
Weight in action 2640 kg
Traverse 60°
Elevation −5° to +45°
Muzzle velocity 665 m/sec
Shell weight 15 kg
Maximum range 15,000 m
Manufacturer Rheinmetall-Borsig AG, Düsseldorf

10 cm leK 41 (Krupp version)

Remarks: Development of this gun to an OKH requirement commenced 1938 and ceased in 1941. Prototype met specifications but proved to have too light shelling power. Krupp also submitted a design to same specifications but no data can be discovered for this model. It was based on 10 cm K 17.

schwere 10 cm Kanone 18

German designation s 10 cm K 18
Calibre 105 mm
Length of piece (L/52): 5460 mm
Length of barrel 5173 mm
Length of rifling 4252 mm
Weight travelling 6434 kg
Weight in action 5642 kg
Traverse 60°
Elevation 0° to +45°
Muzzle velocity 835 m/sec
Shell weight (HE): 15.14 kg; (AP/HE): 15.56 kg
Maximum range 19,015 m
Rate of fire 6 rpm
Barrel life 6000–10,000 rounds
Designers (gun): Rheinmetall-Borsig AG, Düsseldorf; (carriage): Friedr. Krupp AG, Essen
Manufacturer Spreewerk, Berlin-Spandau

Remarks: Project specifications given to both Krupp and Rheinmetall and final model combined both designs. Developed 1926–29, accepted for service after trials 1933–34, kept in limited production until late 1943. From 1942 also deployed along Atlantic coast.

Schwere 10 cm Kanone 18 with aluminium wheels for horse traction

s 10 cm K 18 with solid rubber tyres for motor traction

s 10 cm K 18 on horse-drawn limber. The gun and carriage were transported in two sections

s 10 cm K 18 on limber for motor traction in one section

s 10 cm K 18 in an open gun emplacement on a platform for coastal defence. France, 1943

schwere 10 cm Kanone 18/40 and 42

German designation (initial): s 10 cm K 18/40; (in service): s 10 cm K 42
Calibre 105 mm
Length of piece (L/60): 6300 mm
Length of rifling 4849 mm
Weight travelling 6449 kg
Weight in action (steel wheels): 5620 kg; (alloy wheels): 5430 kg
Traverse 56°
Elevation 0° to +45°
Muzzle velocity 905 m/sec
Shell weight 15.14 kg
Maximum range 21,150 m
Rate of fire 6 rpm
Barrel life 4000–5000 rounds
Designers Joint Rheinmetall/Krupp project
Manufacturer Spreewerk, Berlin-Spandau

Remarks: Development commenced 1937, first two prototypes completed by 1941. Design combined s 10 cm K 40 barrel with modified s 10 cm K 18 carriage and featured variable recoil system. Built in limited production from 1942. Original designation s 10 cm K 18/40 later changed to s 10 cm K 42.

7.5 cm Feldkanone 02/26(p)

German designation 7.5 cm FK 02/26(p)
Original designation 75 mm armata polowa wz. 02/26
Calibre 75 mm
Length of piece (L/30): 2286 mm
Weight travelling 1940 kg
Weight in action 1190 kg
Traverse 10°
Elevation −11° to +16°
Muzzle velocity 600 m/sec
Shell weight 7.24 kg
Maximum range 10,700 m
Rate of fire 10 rpm
Original manufacturer Putilov Arsenal, St. Petersburg/Petrograd

Remarks: Former Tsarist Russian guns relined in Poland to 75 mm calibre. Deliveries commenced in 1926. Total of 446 in Polish Army service in August 1939. In German hands used mainly by reserve artillery units and later along 'Atlantic Wall' defences.

7.5 cm Feldkanone 97(f), 231(f) and 97(p)

German designations 7.5 cm FK 97(f) oder 231(f); 7.5 cm FK 97(p) – more usual designation
Original designations (f): Canon de 75 mle 1897; (p): 75 mm armata polowa wz. 1897
Calibre 75 mm
Length of piece (L/36): 2720 mm
Length of barrel 2587 kg
Weight travelling 1970 kg
Weight in action 1140 kg
Weight of piece 461.5 kg
Traverse 6°
Elevation −11° to +18°
Muzzle velocity 550–675 m/sec
Shell weight 6.195 kg
Maximum range 11,100 m
Rate of fire 12 rpm
Manufacturers (gun): Fonderie de Bourges, Bourges; (carriage): Schneider et Cie, Le Creusot

Remarks: Notable as the most widely used light field gun ever produced. Introduced first successful hydraulic recoil system. Very large numbers captured by German forces in 1940; later some also acquired in Greece. In German service used mainly for coastal defence in France. In 1942 600+ '75' barrels successfully converted to 7.5 cm Pak 97/38 and 97/40 to counter heavier-armoured Soviet tanks.

7.5 cm FK 97(f) during a training shoot

7.5 cm FK 97(f) on a concrete mount for coastal defence

7.5 cm Kanone 232(f)

German designation 7.5 cm K 232(f)
Original designation Canon de 75 mle 97/33
Calibre 75 mm
Length of piece (L/36): 2720 mm
Length of barrel 2587 mm
Weight travelling 1550 kg
Weight in action 1500 kg
Weight of piece 461.5 kg
Traverse 58°
Elevation −6° to +50°
Muzzle velocity 575 m/sec
Shell weight 6.195 kg
Maximum range 11,100 m
Rate of fire 12 rpm
Manufacturer Schneider et Cie, Le Creusot

Remarks: Slightly modernised mle 1897. New split-trail carriage and other minor changes. Proved not very successful; most guns exported. Only limited number remained in French service 1939–40 and later in German hands. Used mainly for coastal defence in France.

7.5 cm Feldkanone 234(b)

German designation 7.5 cm FK 234(b)
Original designation Canon de 75 mle GP II
Calibre 75 mm
Length of piece (L/37.4): 2806 mm
Length of rifling 2227.4 mm
Weight travelling 2477 kg
Weight in action 1510 kg
Weight of gun 375 kg
Traverse 8°
Elevation −7° to +43°
Muzzle velocity 579 m/sec
Shell weight 6.125 kg
Maximum range 11,000 m
Rate of fire 12 rpm
Manufacturer Société anonyme John Cockerill, Liège

Remarks: Belgian conversions of 7.7 cm FK 16 war reparations to take new and longer 75 mm barrel. Substantial numbers in Belgian Army service 1940. Captured guns used locally by German occupation troops; also for coastal defence.

7.5 cm Feldkanone 235(b)

German designation 7.5 cm FK 235(b)
Original designation Canon de 75 mle TR
Calibre 75 mm
Length of piece (L/30): 2250 mm
Length of rifling 1744.5 mm
Weight travelling 1835 kg
Weight in action 1190 kg
Weight of gun 351 kg
Traverse 6°32′
Elevation (normal): −10° to +15°; (trail bent): up to +21°
Muzzle velocity 540 m/sec
Shell weight 6.52 kg
Maximum range 9900 m
Rate of fire (possible): 12 rpm; (normal): 6 rpm
Manufacturer Fonderie Royale des Canons, Liège

Remarks: Licence-built Krupp Mod. 1905, in production until 1914. Slightly modified post-WW I. Remaining guns surrendered to German forces 1940 and later deployed for local (and coastal) defences and training.

7.5 cm FK 235(b) in an open coastal defence emplacement. France, 1944

7.5 cm Feldkanone 236(b)

German designation 7.5 cm FK 236(b)
Original designation Canon de 75 mle GP III
Calibre 75 mm
Length of piece (L/37.3): 2800 mm
Length of rifling 2227.4 mm
Weight travelling 2337 kg
Weight in action 1390 kg
Weight of gun 367 kg
Traverse 3°24′
Elevation −8° to +35°
Muzzle velocity 579 m/sec
Shell weight 6.125 kg

Maximum range 11,000 m
Rate of fire (possible): 12 rpm; (normal): 6 rpm
Manufacturer Société anonyme John Cockerill, Liège

Remarks: Another Belgian conversion of ex-German 7.7 cm FK 16 by inserting 75 mm sleeve in original barrel. Produced in substantial numbers. After 1940 saw widespread German use, mainly because almost identical with 7.5 cm FK 16 nA.

7.5 cm Feldkanone 237(i)

German designation 7.5 cm FK 237(i)
Original designation Cannone da 75/27 modello 06
Calibre 75 mm
Length of piece (L/30): 2250 mm
Length of barrel 2030 mm
Length of rifling 1744.5 mm
Weight travelling (horse-drawn): 1080 kg; (tractor): 1700 kg
Weight in action (horse-drawn): 1015 kg; (tractor): 1080 kg
Weight of gun 345 kg
Traverse 7°
Elevation −10° to +16°
Muzzle velocity 502 m/sec
Shell weight 6.35 kg
Maximum range 10,240 m
Rate of fire 4–6 rpm

Remarks: Initially licence-built Krupp Mod. 1906. Despite obsolescence many guns of this type in Italian service 1940. After 1943 substantial numbers taken over by German forces in Italy.

7.5 cm FK 237(i) in action in North Africa

7.5 cm Feldkanone 243(h)

German designation 7.5 cm FK 243(h)
Original designation L/30 or M 02/04 vd
Calibre 75 mm
Length of piece (L/30): 2250.8 mm
Length of barrel 2030.7 mm
Length of rifling 1745.7 mm
Weight in action 1299 kg
Weight of piece 350 kg
Traverse 9°
Elevation −8° to +40°
Muzzle velocity 500 m/sec
Shell weight 6.5 kg
Maximum range 10,600 m
Rate of fire 8 rpm
Manufacturer Siderius, Holland

Remarks: One of three almost identical Dutch modifications of Krupp Mod. 1903 field gun (others: OM 04 and NM 10), and only one taken into German service after invasion of Holland. In German service used mainly for local coastal defence.

7.5 cm Feldkanone 244(i)

German designation 7.5 cm FK 244(i)
Original designation Cannone da 75/27 modello 11
Calibre 75 mm
Length of piece (L/28.4): 2132 mm
Length of barrel 2030 mm
Length of rifling 1748 mm
Weight travelling 1900 kg
Weight in action 1076 kg
Weight of piece 305 kg
Traverse 52°9′
Elevation −15° to +65°
Muzzle velocity 502 m/sec
Shell weight 6.35 kg
Maximum range 10,240 m
Rate of fire 4–6 rpm

Remarks: Designed by Deport in France; intended as cavalry gun but also issued to field batteries. After 1943 limited number taken into German service for local use in Italy.

7.5 cm leichte Feldkanone 245(i)

German designation 7.5 cm leFK 245(i)
Original designation Cannone da 75/27 modello 12
Calibre 75 mm
Length of piece (L/30): 2250 mm
Length of barrel 2030 mm
Length of rifling 1744.5 mm
Weight travelling 1700 kg
Weight in action 900 kg
Weight of piece 395 kg
Traverse 7°
Elevation −12° to +18°30′
Muzzle velocity 500 m/sec
Shell weight 6.35 kg
Maximum range 10,240 m
Rate of fire 4–6 rpm
Manufacturer Vickers-Terni, Turin

Remarks: Revised horse-drawn version of modello 06 gun. Only limited numbers built and in service. After September 1943 remaining guns used by German forces in Italy.

7.5 cm Feldkanone 246(n) and 01(n)

German designation 7.5 cm FK 246(n) oder 01(n)
Original designation (n): 7,5 cm feltkanon L/31 M/01
Calibre 75 mm
Length of piece (L/31): 2325 mm
Length of barrel 2167 mm
Weight travelling 1773 kg
Weight in action 1037 kg
Weight of piece 330 kg
Traverse 7°
Elevation −7° to + 15°30′
Muzzle velocity 500 m/sec
Shell weight 6.5 kg
Maximum range 10,600 m
Rate of fire 8 rpm
Manufacturer Erhardt, Düsseldorf

Remarks: Pre-WW I German export to Norway. Captured remaining guns used by German occupation troops in Norway, usually as infantry guns.

7.5 cm Feldkanone 248(i)

German designation 7.5 cm FK 248(i)
Original designation Cannone da 75/32 modello 37
Calibre 75 mm
Length of piece with m/b (L/34): 2574 mm
Length of barrel 2207 mm
Length of rifling 1862.5 mm
Weight travelling 1250 kg
Weight in action 1200 kg
Weight of piece 347 kg
Traverse 50°
Elevation −10° to +45°
Muzzle velocity 624 m/sec
Shell weight 6.3 kg˙
Maximum range 12,500 m
Rate of fire 6–8 rpm
Manufacturer O.T.O.-Terni, Turin

Remarks: Designed by Ansaldo as first modern post-WW I Italian field gun. Despite many good points only small numbers built. After 1943 all available guns taken into German service.

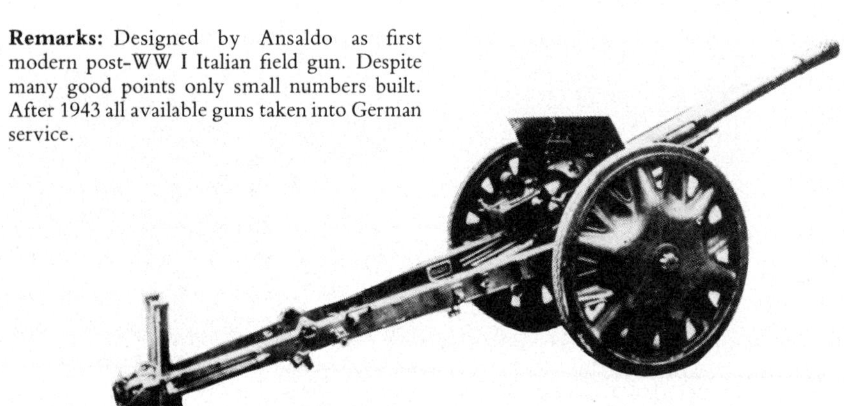

7.5 cm leichte Feldhaubitze 255(i)

German designation 7.5 cm leFH 255(i)
Original designation Obice da 75/18 modello 35
Calibre 75 mm
Length of piece (L/20.7): 1557 mm
Length of barrel 1374.6 mm
Length of rifling 1133.5 mm
Weight travelling 1850 kg
Weight in action 1050 kg
Weight of piece 172 kg
Traverse 48°
Elevation −10° to +45°
Muzzle velocity 425 m/sec
Shell weight 6.4 kg
Maximum range 9560 m
Rate of fire 6–8 rpm
Manufacturer Ansaldo, Turin

Remarks: Designed as field howitzer by combining Obice da 75/18 mod. 34 barrel with new split-trail field carriage. Generally successful weapon but produced only in limited numbers. After 1943 all available guns taken into German service.

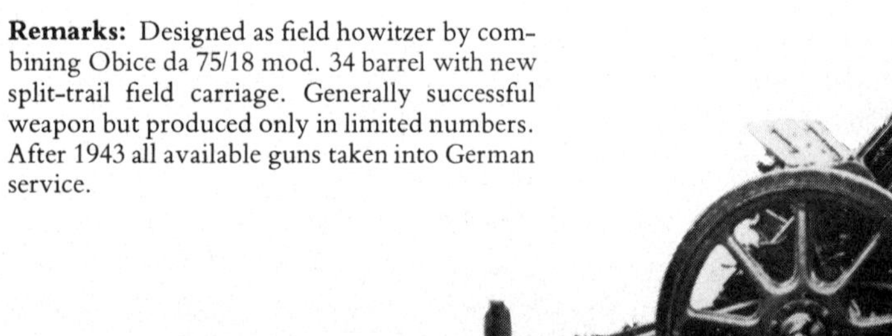

7.62 cm Feldkanone 288(r)

German designation 7.62 cm FK 288(r)
Original designation *76.2 mm Pushka obr. 1942 g/ZiS-3* (76–42)
Calibre 76.2 mm
Length of piece (L/42): 3200 mm
Length of barrel 2994 mm
Length of rifling 2588 mm
Weight travelling 1120 kg
Weight in action 1120 kg
Weight of piece 400 kg
Traverse 50°
Elevation −5° to +37°
Muzzle velocity 680 m/sec
Shell weight 6.21 kg
Maximum range 13,000 m
Rate of fire Up to 25 rpm
Manufacturer Various Soviet State arsenals

Remarks: Very successful dual-purpose weapon evolved under wartime conditions. Combined obr. 1939 gun with muzzle brake and a new split-pole trail carriage. Entered Red Army service late in 1942 and produced in very large numbers. Captured examples much appreciated by German troops.

7.62 cm Feldkanone 288/1(r)

German designation 7.62 cm FK 288/1(r)
Original designation *76.2 mm Pushka obr. 1941 g/ZiS-3* (76–41)
Calibre 76.2 mm
Length of piece (L/42): 3194.3 mm
Length of barrel 2985.6 mm
Length of rifling 2585.6 mm
Weight travelling 1110 kg
Weight in action 1110 kg
Weight of piece 400 kg
Traverse 56°
Elevation −10° to +18°
Muzzle velocity 680 m/sec
Shell weight 6.21 kg
Maximum range 11,000 m
Rate of fire Up to 25 rpm
Manufacturer Various Soviet State arsenals

Remarks: Late 1941 war emergency solution combining 76.2 mm field gun barrel with muzzle brake and *57 mm PTP obr. 1971/ZiS-3* carriage. Limited production; only small number in German service.

7.62 cm Feldkanone 295/1(r)

German designation 7.62 cm FK 295/1(r)
Original designation *76.2 mm Pushka obr. 1902/30 g* L/30
Calibre 76.2 mm
Length of piece (L/30): 2286 mm
Length of barrel 2196 mm
Length of rifling 1790 mm
Weight travelling 2350 kg
Weight in action 1320 kg
Weight of piece 389 kg
Traverse 5°20′
Elevation −5° to +37°
Muzzle velocity (small charge): 455 m/sec; (large charge): 635 m/sec
Shell weight 6.4 kg

Maximum range (small charge): 9500 m; (large charge): 12,400 m
Rate of fire 8 rpm
Original manufacturer Putilov Arsenal, St. Petersburg/Petrograd; 1930s conversions at various Soviet State arsenals

Remarks: Soviet 1930 modernisation of old Tsarist field guns. Produced in two versions (*see* 7.62 cm FK 259/2). In large-scale service with Red Army 1941. Many captured guns used by German artillery detachments all over Europe. A version of this gun, 7.62 cm Flak 295(E), was fitted on captured Soviet armoured trains.

7.62 cm Feldkanone 295/2(r)

German designation 7.62 cm FK 295/2(r)
Original designation *76.2 mm Pushka obr. 1902/30 g* L/40
Calibre 76.2 mm
Length of piece (L/40): 3046 mm
Length of barrel 2550 mm
Length of rifling 1905 mm
Weight travelling 2380 kg
Weight in action 1350 kg
Weight of piece 419 kg
Traverse 5°20′
Elevation −5° to +37°
Muzzle velocity (small charge): 475 m/sec; (large charge): 680 m/sec
Shell weight 6.4 kg
Maximum range (small charge): 9800 m, (large charge): 13,000 m
Rate of fire 8 rpm
Original manufacturer Putilov Arsenal, St. Petersburg/Petrograd; 1930s conversions at various Soviet State arsenals

Remarks: Longer-barrelled 1930 Soviet conversion of old Tsarist guns. In Soviet and German service as its first version.

7.62 cm FK 295/2(r) field guns which formed part of the artillery park of the Luftwaffe 'Hermann Goering' Division, captured by the 1st Canadian Army in Holland in late spring 1945

7.62 cm Feldkanone 296(r) and 36(r)

German designation 7.62 cm FK 296(r) oder 36(r)
Original designation *76.2 mm Pushka obr. 1936 g* (76–36)
Calibre 76.2 mm
Length of piece (L/51.1): 3895 mm
Length of barrel 3270 mm
Length of rifling 1905 mm
Weight travelling 2400 kg
Weight in action 1350 kg
Weight of piece 439 kg
Traverse 60°
Elevation −5° to +75°
Muzzle velocity 706 m/sec
Shell weight 6.4 kg
Maximum range 13,580 m
Rate of fire Up to 25 rpm
Manufacturer Various Soviet State arsenals

Remarks: Very efficient long-barrelled field gun with obvious anti-tank capabilities. Introduced in Red Army service 1939, first used during Soviet-Finnish war same year. From 1941 many captured guns used by German artilllery detachments; special ammunition also provided in Germany for these guns. Initially an emergency solution, 7.62 cm Pak 36(r) conversion proved one of most effective AT guns for some time. Large numbers so modified 1942–43.

7.62 cm FK 296(r) modified with pneumatic tyres in service with the Afrika Korps

7.62 cm Feldkanone 297(r), 39(r) and 7.62 cm Panzerabwehrkanone 39(r)

German designations 7.62 cm FK 297(r) oder 39(r); 7.62 cm Pak 39(r)
Original designation *76.2 mm Pushka obr. 1939 g* (76–39)
Calibre 76.2 mm
Length of piece (L/42): 3200 mm
Length of barrel 2588 mm
Length of rifling 1905 mm
Weight travelling 2350 kg
Weight in action 1570 kg
Weight of piece 425 kg
Traverse 57°
Elevation −6° to +45°
Muzzle velocity 680 m/sec
Shell weight 6.4 kg
Maximum range 13,290 m
Rate of fire Up to 25 rpm
Manufacturer Various Soviet State arsenals

Remarks: Soviet dual-purpose field gun, smaller, lighter and more mobile than obr. 76–36. Large numbers in service; many used by German artillery detachments. 7.62 cm Pak 39(r) used by German troops virtually same as original Soviet gun.

7.62 cm FK 297(r) with muzzle brake, modified by the German Army

A Soviet 76-39 field gun immediately after capture by German troops

7.65 cm Feldkanone 5/8(ö), (t) and 300(j)

German designations 7.65 cm FK 5/8(ö) oder (t); 7.65 cm FK 300(j)
Original designations (t): 8 cm kanon vz. 05/08; (j): 80 mm M 5/8
Calibre 76.5 mm
Length of piece (L/30): 2285 mm
Length of barrel 2077 mm
Length of rifling 1784 mm
Weight travelling 2447 kg
Weight in action 1065 kg
Weight of gun 355 kg
Traverse 7°52′
Elevation −7°30′ to +18°
Muzzle velocity 433 m/sec
Shell weight 7 or 8 kg
Maximum range 9300 m
Rate of fire 8–10 rpm
Manufacturer Škoda, Pilsen

Remarks: Originally designed as dual-purpose field/mountain gun. Obsolete in 1939 but many still in service. Used by various local German artillery detachments; also against partisans in Yugoslavia.

7.65 cm Feldkanone 17(ö), (t) and 303(j)

German designations 7.65 cm FK 17(ö) oder (t); 7.65 cm FK 303(j)
Original designations (t): 8 cm kanon vz. 17; (j): 80 mm M 17
Calibre 76.5 mm
Length of piece (L/30): 2297 mm
Length of barrel 2078 mm
Length of rifling 1915 mm
Weight travelling 2089 kg
Weight in action 1319 kg
Weight of gun 379 kg
Traverse 8°
Elevation −10° to +45°
Muzzle velocity 554 m/sec
Shell weight 8 kg
Maximum range 11,400 m
Rate of fire 10–12 rpm
Manufacturer Škoda, Pilsen

Remarks: Dual-purpose field/mountain gun, and used as such by Yugoslav Army. Captured guns used by local German artillery detachments; also against partisans in Yugoslavia and for training.

7.65 cm Feldkanone 304(j)

German designation 7.65 cm FK 304(j)
Original designation (j): 80 mm M 28
Calibre 76.5 mm
Length of piece (L/40): 3060 mm
Weight travelling 2977 kg
Weight in action 1816 kg
Weight of gun 508 kg
Traverse (on carriage): 7°30′; (on platform): 360°
Elevation −8° to +80°
Muzzle velocity 600 m/sec
Shell weight 8 kg
Maximum range 13,100 m
Rate of fire 10 rpm
Manufacturer Škoda, Pilsen

Remarks: Initially designed as a field/mountain gun with AA capabilities. Export model only. All guns in German service acquired from Yugoslavia.

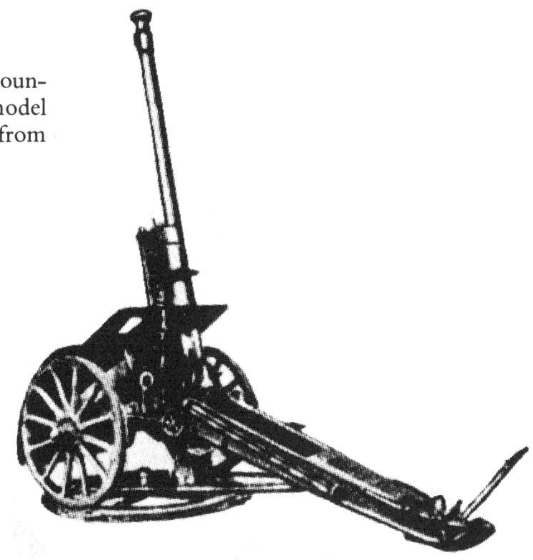

8 cm Feldkanone 30(t)

German designation 8 cm FK 30(t)
Original designation 8 cm kanon vz. 30
Calibre 76.5 mm
Length of piece (L/40): 3060 mm
Weight travelling 2977 kg
Weight in action 1816 kg
Traverse 8°
Elevation −8° to +80°
Muzzle velocity 600 m/sec
Shell weight 8 kg
Maximum range 13,500 m
Rate of fire 10–12 rpm
Manufacturer Škoda, Pilsen

Remarks: Designed to include theoretical AA capability. Used same carriage as 100 mm vz. 30 (NPH) field howitzer. Large numbers in Czech service in 1938 and subsequently in German hands.

8.38 cm Feldkanone 271(e)

German designation 8.38 cm FK 271(e)
Original designation QF 18 pr Mks I-II* on Carriage Mk IIPA
Calibre 83.8 mm
Length of piece (L/29.38): 2463 mm
Length of barrel 2454 mm
Length of rifling 2038 mm
Weight travelling 2724 kg
Weight in action 1518.3 kg
Weight of piece 462.6 kg
Traverse 8°
Elevation −5° to +16°
Muzzle velocity 495 m/sec
Shell weight 8.39 kg
Maximum range 10,150 m
Rate of fire 10–12 rpm
Manufacturer Numerous British arsenals and commercial concerns

Remarks: British field gun of pre-WW I design, subsequently modernised, but earlier versions remained in service 1939–40. Most BEF guns left behind after Dunkirk evacuation impressed by German artillery troops for coastal defence. Some additional guns of same type captured summer 1941 in former Baltic States.

8.76 cm Feldkanone 280(e)

German designation 8.76 cm FK 280(e)
Original designation Ordnance QF 25 pr Mks II and III on Carriage, 25 pr Mk I
Calibre 87.6 mm
Length of piece (L/28.25): 2476 mm
Length of barrel 2350 mm
Length of rifling 1885.6 mm
Weight travelling with limber 3327 kg
Weight in action 1800 kg
Weight of piece 453.6 kg
Traverse (on platform): 360°; (on carriage): 8°
Elevation −5° to +40°
Muzzle velocity (normal charge): 453 m/sec; (super charge): 532 m/sec
Shell weight 11.34 kg
Maximum range (normal range): 10,790 m; (super charge): 12,253 m
Rate of fire 12–14 rpm
Manufacturer Various British State arsenals

Remarks: British gun-howitzer designed to 1936 specifications. First used in action by British Army in Norway 1940. Over 12,000 delivered by 1945. Much appreciated by German artillery troops where captured 25 pr Mk II guns sometimes equipped whole regiments.

8.76 cm FK 280(e) with its German crew. At one period in Libya the divisional artillery of the German 90th Light Division consisted entirely of captured Soviet 7.62 cm guns and British 25 prs

8.76 cm Feldkanone 281(e)

German designation 8.76 cm FK 281(e)
Original designation QF 25 pr Mk I on Carriage 25/18 pr Mk IVP
Calibre 87.6 mm
Length of piece (L/28): 2457 mm
Length of barrel 2346 mm
Length of rifling 1885.3 mm
Weight travelling with limber 2516 kg
Weight in action 1600 kg
Weight of piece 450 kg
Traverse 9°
Elevation −5° to +37°30°
Muzzle velocity 453 m/sec
Shell weight 11.34 kg
Maximum range 10,790 m
Rate of fire 12–14 rpm
Manufacturer Royal Ordnance Factory, Leeds

Remarks: British gun-howitzer designed to 1935 specifications. For financial reasons basically rebored old 18 pr barrels on modified 18 pr carriages. Main equipment of British artillery regiments 1939–40. Guns acquired by German forces in France and elsewhere mainly used for coastal defence purposes.

8.76 cm Feldkanone 282(e)

German designation 8.76 cm FK 282(e)
Original designation QF 25 pr Mk I on Carriage 25/18 pr Mk VP
Calibre 87.6 mm
Length of piece (L/28): 2457 mm
Length of barrel 2346 mm
Length of rifling 1885.3 mm
Weight travelling with limber 2521 kg
Weight in action 1605 kg
Weight of piece 450 kg
Traverse 50°
Elevation −4°40′ to +37°50′
Muzzle velocity 453 m/sec
Shell weight 11.34 kg
Maximum range 10,790 m
Rate of fire 12–14 rpm
Manufacturer Royal Ordnance Factory, Leeds

Remarks: Rebored 18 pr barrels on split-trail carriages originally intended for 25 pr Mk II guns. Number captured by German forces in France and impressed into service.

10 cm leichte Feldhaubitze 30(t)

10 cm leFH 30(t) in action with the Waffen-SS Division 'Totenkopf'

German designation 10 cm leFH 30(t)
Original designation (t): 10 cm houfnice vz. 30
Calibre 100 mm
Length of piece (L/25): 2500 mm
Weight travelling 3077 kg
Weight in action 1766 kg
Traverse 8°
Elevation −8° to +80°
Muzzle velocity 430 m/sec
Shell weight 16 kg
Maximum range 10,600 m
Rate of fire 6–8 rpm
Manufacturer Škoda, Pilsen

Remarks: Czech Army version of Mod. 28 high-angle field gun supplied to Yugoslavia. Total of 158 in service 1938, many of which impressed by regular German artillery units.

10 cm leichte Feldhaubitze 315(i) and 14(ö)

German designation 10 cm leFH 315(i); 10 cm leFH 14(ö)
Original designation (i): Obice da 100/17 modello 14
Calibre 100 mm
Length of piece (L/19): 1930 mm
Length of rifling 1500 mm
Weight travelling 2367 kg
Weight in action 1417 kg
Weight of gun 403 kg
Traverse 5°21′
Elevation −8° to +48°
Muzzle velocity 407 m/sec
Shell weight 13.375 kg
Maximum range 9280 m
Rate of fire 6–8 rpm
Manufacturer Škoda, Pilsen

Remarks: Originally produced for Austro-Hungarian Empire armies. After 1918 large numbers taken over by Italy; some delivered to Poland and number remained in Austrian service. Basic weapon used M 16/19 mountain howitzer barrel on a field carriage.

10 cm leichte Feldhaubitze 14/19(t) and (p), 316(j) and 318(g)

◄ *10 cm leFH 14/19(t) in action*

10 cm leFH 14/19(t) in a Canadian captured weapons dump. Holland, May 1945 ►

German designations 10 cm leFH 14/19(t) und (p); 10 cm leFH 316(j) oder 318(g)
Original designations (t): 10 cm houfnice vz. 14/19; (p): 100 mm haubica wz. 1914/1918 P i A; (j): 100 mm M 14/19; (g): 100–14/19
Calibre 100 mm
Length of piece (L/24): 2400 mm
Length of barrel 2175 mm
Length of rifling 1899 mm
Weight travelling 2855 kg
Weight in action 1490 kg
Weight of gun 430 kg
Traverse 5°
Elevation −7°30′ to +48°
Muzzle velocity 395 m/sec
Shell weight 16 kg
Maximum range 9800 m
Rate of fire 8 rpm
Manufacturer Škoda, Pilsen

Remarks: Essentially modernised M. 14 gun. Large numbers in Czech service 1938. Subsequently used by various regular German artillery detachments 1939–41; additional pieces captured in Poland, Yugoslavia and Greece. Many used by Italian Army were WWI reparations.

10 cm leichte Feldhaubitze 317(j)

German designation 10 cm leFH 317(j)
Original designation (j): 100 mm M 28
Calibre 100 mm
Length of piece (L/25): 2500 mm
Weight travelling 3509 kg
Weight in action 1798 kg
Weight of piece 490 kg
Traverse 11°
Elevation −8° to +80°
Muzzle velocity 449 m/sec
Shell weight 14 kg
Maximum range 10,700 m
Rate of fire 5–6 rpm
Manufacturer Škoda, Pilsen

Remarks: Designed to Yugoslav specifications but only some 20 delivered. Originally designated Škoda 100 mm Mod. 28 (FE). Captured remaining guns impressed in German service and used locally, in part against partisans.

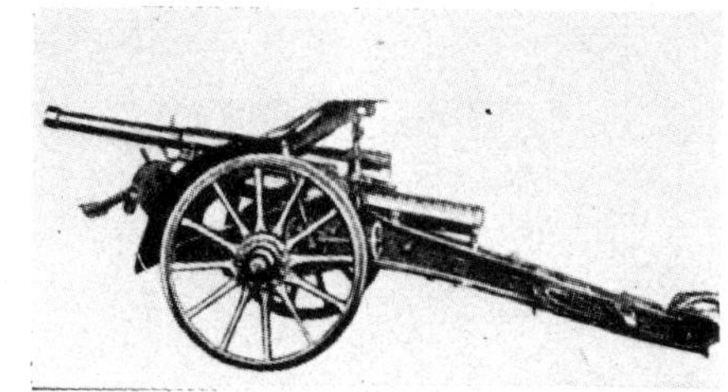

10.5 cm Kanone 29(p)

German designation 10.5 cm K 29(p)
Original designation 105 mm armata wz. 29
Calibre 105 mm
Length of piece (L/28): 2987 mm
Weight in action 2880 kg
Traverse 63°
Elevation 0° to +43°
Muzzle velocity 550 m/sec
Shell weight 15.4 or 15.7 kg
Maximum range 15,200 m
Rate of fire 6 rpm
Manufacturer State Arsenal, Starachowice

Remarks: Polish conversion of French Schneider L 13 S mounted on a new split-trail carriage. All serviceable captured guns impressed by German artillery units.

10.5 cm Kanone 320(i)

10.5 cm K 320(i) in action on the Eastern Front

German designation 10.5 cm K 320(i)
Original designation (i): Cannone da 105/32
Calibre 105 mm
Length of piece (L/35): 3640 mm
Length of barrel 3360 mm
Length of rifling 3708 mm
Weight travelling 3770 kg
Weight in action 3030 kg
Weight of gun 1270 kg
Traverse 6°
Elevation −10° to +30°
Muzzle velocity 668 m/sec
Shell weight 16.1 kg
Maximum range 16,200 m
Rate of fire 3–4 rpm
Manufacturer Škoda, Pilsen

Remarks: Originally manufactured for Austro-Hungarian Empire armies as Škoda 104 mm Mod. 1915. After 1918 most remaining guns taken over by Italian Army. Some still in service 1940, and a few taken over by German troops in Italy after September 1943.

10.5 cm leichte Feldhaubitze 324(f)

German designation 10.5 cm leFH 324(f)
Original designation Canon de 105 court mle 1934 Schneider
Calibre 105 mm
Length of piece (L/20): 2090 mm
Length of barrel 1948 mm
Weight in action 1722 kg
Weight of piece 346 kg
Traverse 45°
Elevation −8° to +43°
Muzzle velocity 465 m/sec
Shell weight 15.7 kg
Maximum range 10,700 m
Rate of fire 5 rpm
Manufacturer Schneider et Cie, Le Creusot

Remarks: Substantial number in French service 1939–40 and many captured by German forces. Subsequently used mainly by local garrison troops and for coastal defence.

10.5 cm leichte Feldhaubitze 325(f)

German designation 10.5 cm leFH 325(f)
Original designation Canon de 105 court mle 1935 B
Calibre 105 mm
Length of piece (L/16.7): 1760 mm
Length of barrel 1505 mm
Weight travelling 1700 kg
Weight in action 1627 kg
Weight of piece 470 kg
Traverse 58°
Elevation −6° to +50°
Muzzle velocity 442 m/sec
Shell weight 15.7 kg
Maximum range 10,300 m
Rate of fire 5 rpm
Manufacturer Atelier de Bourges, Bourges

Remarks: Designed by State Arsenal at Bourges. By 1939 total of 410 delivered. Captured guns used by German occupation troops in France for coastal defence and training.

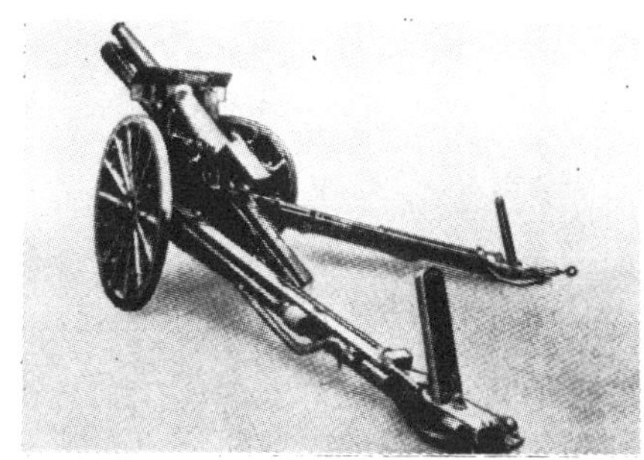

10.5 cm leichte Feldhaubitze 326(i)

German designation 10.5 cm leFH 326(i)
Original designation Obice da 105/14
Calibre 105 mm
Length of piece (L/14): 1470 mm
Length of barrel 1372.8 mm
Length of rifling 1181.9 mm
Weight travelling 1740 kg
Weight in action 1400 kg
Weight of piece 368 kg
Traverse 5°
Elevation −5° to +70°
Muzzle velocity 330 m/sec
Shell weight 16.3 kg
Maximum range 8160 m
Rate of fire 4–6 rpm
Manufacturer Ansaldo, Turin

Remarks: Accepted for service in 1937 but manufactured only in small numbers. After 1943 some impressed into German service in Italy.

schwere 10.5 cm Kanone 332(f)

German designation s 10.5 cm K 332(f)
Original designation Canon de 105 L mle 1936 Schneider
Calibre 105 mm
Length of piece (L/37.6): 3905 mm
Length of barrel 3802 mm
Weight travelling (tractor towed): 4800 kg; (horse towed): 4090 kg
Weight in action (tractor): 3920 kg; (horse): 3540 kg
Weight of piece 1105 kg
Traverse 50°
Elevation 0° to +47°48′
Muzzle velocity 725 m/sec
Shell weight 15.7 kg
Maximum range 16,000 m
Rate of fire 4 rpm
Manufacturer Schneider et Cie, Le Creusot

Remarks: Most modern 105 mm French gun before WW II. Produced in two versions; total of 159 in service 1939. All captured guns impressed by German coastal artillery detachments.

schwere 10.5 cm Kanone 35(t) and 10.5 cm Kanone 339(j)

German designations s 10.5 cm K 30(t); 10.5 cm K 339(j)
Original designations (t): 10.5 cm hruby kanon vz. 35; (j): 105 mm M 36
Calibre 105 mm
Length of piece (L/42): 4400 mm
Weight travelling 4600 kg
Weight in action 4200 kg
Weight of piece 1458 kg
Traverse 50°
Elevation −6° to +42°
Muzzle velocity 730 m/sec
Shell weight 18 kg
Maximum range 18,100 m
Rate of fire 8 rpm
Manufacturer Škoda, Pilsen

Remarks: Advanced Czech design produced in some numbers for Czech and Yugoslav armies. Used by Germans during their campaigns in Balkans and Greece 1941, thereafter relegated to coastal defence and reserve artillery units. Kept in production for German forces until 1941.

Battery of s 10 cm K 35(t) at firing practice

s 10.5 cm K 35(t) emplaced as part of the 'Atlantic Wall' defences ➤

10.5 cm Kanone 331(f), 333(b), 338(i) and (j), 13(p)

German designations 10.5 cm K 331(f), 333(b), 338(i) oder (j), und 13(p)
Original designations (f)(b): Canon de 105 mle 1913 Schneider (L 13 S); (i): Cannone da 105/28; (j): 105 mm M 13; (p): 105 mm armata wz. 13
Calibre 105 mm
Length of piece (L/28.4): 2987 mm
Length of barrel 2360.5 mm
Weight travelling 2650 kg
Weight in action 2300 kg
Weight of piece 891 kg
Traverse 6°
Elevation −5° to +37°
Muzzle velocity 550 m/sec
Shell weight 15.74 kg
Maximum range 12,000 m
Rate of fire 4 rpm
Manufacturer Schneider et Cie, Le Creusot

Remarks: Sturdy French gun but obsolescent by 1939. Most L 13 S sold to Poland where later modernised. Was also built under licence in Italy. In German service used mostly by coastal artillery detachments and for training, but also as regimental artillery by some units based in France.

10.5 cm K 331(f); modified version with pneumatic tyres ➤

schwere 10.5 cm Kanone 335(h)

German designation s 10.5 cm K 335(h)
Original designation (h) 27 Bofors
Calibre 105 mm
Length of piece (L/42): 4410 mm
Weight travelling 4100 kg
Weight in action 3650 kg
Weight of piece 1183 kg
Traverse 60°
Elevation −3° to +45°
Muzzle velocity 750 m/sec
Shell weight 16 kg
Maximum range 16,500 m
Rate of fire 5 rpm
Manufacturer AB Bofors, Bofors, Sweden

Remarks: Swedish gun designed for export as Model 27. After 1940 captured guns employed by German occupation troops for coastal defence.

10.7 cm Kanone 352(r)

German designation 10.7 cm K 352(r)
Original designation *107 mm Pushka obr. 1910/30 g* (107–10/30)
Calibre 106.7 mm
Length of piece (L/38): 4054 mm
Length of barrel 3314 mm
Length of rifling 2667.5 mm
Weight travelling 2580 kg
Weight in action 2380 kg
Weight of piece 1041 kg
Traverse 6°
Elevation −5° to +37°
Muzzle velocity 670 m/sec
Shell weight 17.18 kg
Maximum range 16,350 m
Rate of fire 5–6 rpm
Original manufacturer Putilov Arsenal, St. Petersburg/Petrograd (Schneider licence); 1930s modifications made at various Soviet State arsenals

Remarks: Soviet-modernised Schneider guns built under licence in Tsarist Russia. Modifications included new longer barrel. Large numbers in service with Soviet artillery at Corps level 1941. Many captured by German troops and subsequently used by regular artillery detachments and for coastal defence.

Medium and heavy artillery

Whereas the role of the light field artillery was to provide direct fire support to units at divisional level and below, the role of medium and heavy artillery was rather more complex. In general it was to back up the light artillery and provide long range fire at specific targets such as concentration areas and fortifications, but very often other roles were tackled. Medium artillery is generally taken to cover the calibres from 105 mm up to about 155 mm, and heavy artillery generally covers all calibres above 155 mm, but there were several overlapping exceptions. One has already been seen in the last section where the *s 10 cm K 18* and *10.5 cm sK 18/40* were included, even though they were regarded by the Germans as medium guns.

The terms of the Versailles Treaty of 1919 insisted on the destruction or dispersal of nearly all the huge artillery machine that had been so carefully built up and used during WW I. Only a very limited number of medium and heavy guns were allowed to remain for the Reichswehr to train with and retain some semblance of efficiency and technical competence. There were two main types of 15 cm weapon in use, the *15 cm sFH 13* and the *15 cm K 16,* and in a larger calibre the *lg 21 cm Mrs.* All were veterans of WW I but all three types survived through the 1920s and 1930s to remain in use when WW II began. They saw action during the early war years but thereafter were gradually relegated to training and second-line units, and many were later incorporated in the 'Atlantic Wall'.

All through the 1920s the German armament designers were busy working out the shape and form of the next generation of artillery weapons. High on the list of future requirements was a 15 cm howitzer on which both Krupp and Rheinmetall spent considerable time from about 1926 onwards. It was felt that both firms' submissions had good points and the outcome was that Rheinmetall's gun was selected for fitting onto the Krupp carriage, the same type as used by the *s 10 cm K 18*. Production of this combination, the *15 cm sFH 18*, began in 1934, and it went on to become the backbone of the German medium artillery. The bulk of the production was carried out at four main centres, the Spreewerk at Berlin-Spandau, MAN at Augsburg, Dörries-Füllner at Bad Warmbrunn, and a Škoda factory at Dubnica/Slovakia. No record survives of the actual number produced but it must have been substantial for the *sFH 18* remained in service right up to 1945, and many were delivered to Finland and Italy.

Despite the use of eight different charges the maximum reach of the *sFH 18* was only 13,325 metres, and early combat experience revealed that it was frequently outranged by the opposition, especially after 1941 when the German artillery encountered the formidable Soviet 152 mm guns and howitzers. This shortcoming had been recognised as early as 1938 and both Krupp and Rheinmetall were given contracts to develop a new design. Both produced prototypes known as the *15 cm sFH 40* but neither was accepted for service, mainly because by the time the prototypes were ready there was no spare production capacity and during late 1941 the project was dropped. However, during 1942 a small number of barrels were made under the control of Krupps at Essen and fitted to *sFH 18* carriages. The result was the *15 cm sFH 18/40*, later to be redesignated the *15 cm sFH 42*, but it was not a success. Accuracy proved to be poor at short and medium ranges even though the maximum range was increased to 15,100 metres, and only 46 were made.

Another 15 cm piece that was no great success was the *15 cm sFH 36*. This howitzer was the result of a 1935 request for a lightened version of the *sFH 18* to be towed by horses. The first models were ready by 1938 and again, both Krupp and Rheinmetall were involved. To keep the weight down a shorter barrel with a muzzle brake was fitted and much use was made of light alloys in the carriage and wheels. This use of light alloys was the downfall of the *sFH 36* for by 1942 such materials were in short supply and what there was was earmarked for aircraft production. Also, by that time the German policy was to concentrate on the use of tractors rather than horses for towing artillery, so the *sFH 36* was discontinued.

It was the general German practice to use metal cartridges for almost every calibre. This made possible the nearly universal use of the horizontal sliding block breech mechanism, which was almost a trademark of German artillery designs, but by about 1942 the raw material situation in Germany was becoming increasingly precarious to the extent that the use of metal cartridge cases was seen as a luxury that could be ill afforded. Changes of metal from the usual brass to steel made little difference and only the use of combustible bagged charges seemed to offer any real savings. But German designers knew so little about this method and the mechanical difficulties involved that they had to virtually re-learn the art of sealing the breech of a gun using bagged charges. As an experiment an *sFH 18* was converted for bagged charges and thus became the *15 cm sFH 18/43*. A novel and quite ingenious sealing system was used on this piece but the experimental work was not completed before the war ended.

One other 15 cm howitzer project that was also never completed was the *15 cm sFH 43*. It was the counterpart to the *10.5 cm leFH 43* project mentioned in the last section and Krupp, Rheinmetall and Škoda all carried out design work. Most of the projects were scaled-up versions of the *leFH 43* models but none were finished by the time the war ended. Krupp approached the project in their usual thorough manner and at one point put forward a plan to mount the new piece on the cruciform carriage of their *12.8 cm K 44*. All these *sFH 43* projects would have used the breech obturation developed on the *15 cm sFH 18/43*.

While the main weight of the German medium artillery rested with the *15 cm sFH 18*, combat experience on the Eastern Front showed it to lack range and be rather heavy to move. By the middle of 1942 a specification for a new 12.8 cm gun was put out to Krupp and Škoda. To save raw materials the gun was to use bagged charges and it had to have all the other features specified for the *10.5 cm leFH 43* and *15 cm sFH 43*, ie 360° traverse, high angle fire, etc. The Škoda model did not advance beyond the drawing board (their design designation was 25/940/S), but the Krupp version reached the wooden model stage before it was stopped as a much more promising design was in the offing.

The new design was the *12.8 cm K 44* and both Krupp and Rheinmetall produced prototypes. Of the two the Krupp version was chosen and it was one of the most remarkable gun designs of the war. It was intended from the start as a dual purpose field and anti-tank weapon and thus was low and mounted on a 360° traverse cruciform platform. A well-sloped shield provided crew protection. As a field piece or as an anti-tank gun the *12.8 cm K 44* would have been a formidable weapon but it arrived too late for any large scale production to get under way. About fifty barrels were produced but as there were no carriages they had to be placed on captured French and Soviet carriages as the *12.8 cm K 81/1* and *81/2*.

There were two other German 15 cm artillery pieces in service, both of them guns. One was the Rheinmetall *15 cm K 18* which entered service in 1938 as a replacement for the ageing *15 cm K 16*. Development work on this gun had been in progress for some time before 1938 and the design was already approved in 1935. The bulk of the weapon meant it had to be transported in two loads and overall it can only be described as a cumbersome piece of artillery, but it remained in use until the end of WW II. The other 15 cm gun was the *15 cm K 39* which was originally a Krupp commercial design sold to Turkey. The *K 39* was

24 cm K 3 in action

designed for use as a field gun or a coastal gun, and for the latter role it used a specially-evolved turntable platform which could be towed in sections with the gun. This 360° platform could be erected at any convenient site and as a result the *K 39* was used as a coastal defence weapon.

The general performance of the German 15 cm artillery was such that by the late 1930s it was decided by OKH that a new and heavier gun would be needed. Krupp were given the development contract and proposed the *17 cm K 18 in Mrs Laf.* The *17 cm K 18* was 'partnered' by an even heavier weapon, the *21 cm Mrs 18*, which shared the same carriage. This carriage was a very advanced design and had several unusual features. One was that it used a dual-recoil system in that, not only did the barrel recoil on its usual mechanism, but the whole platform also recoiled along carriage rails. When the first of these carriages were ready the barrel production was held up and to allow the units that would be equipped with the new gun to get some experience with the carriage, eight *15 cm SK C/28* and a small number of *15 cm K 16* barrels were fitted to the available carriages and issued for training. When the first 17 cm barrels started to come off the Hanomag production lines at Hannover during 1941 they replaced the temporary barrels, none of which were kept in service longer than was necessary. Hanomag also produced 21 cm equipments along with Krupp at Essen, but during 1942 it was realised that although the *21 cm Mrs 18* was a very good weapon it had only about half the range of the *17 cm K 18*, so its production was stopped to concentrate on the 17 cm gun.

The introduction of a 21 cm howitzer also brought a demand for a 21 cm gun. This time Krupp were given the contract as they had already built a 21 cm gun for possible commercial exploitation in about 1936. This gun, the *21 cm K L/50*, was built along the same lines as the *15 cm K 39* and used a very similar portable turntable for the secondary coastal defence role. The time taken to bring this gun into action was inordinately long and it seems to have had only a few buyers, but the experience gained from it was used in the design of the *21 cm K 38*. An order for fifteen of these guns was made in 1938 and delivery was to have been by 1940, but by 1943 only seven had been produced and production was then terminated (one of the seven was sent to Japan). As a technical design the *21 cm K 38* was rated as one of the best artillery pieces to emerge from WW II, but with so few actually being produced they had little impact at the time.
1940, but by 1943 only seven had been produced and prduction was then terminated (one of the sevan was sent to Japan). As a technical design the *21 cm K 38* was rated as one of the best artillery pieces to emerge from WW II, but with so few actually being produced they had little impact at the time.

Before the German take-over of Czechoslovakia the Pilsen-based Škoda concern had negotiated sales to Turkey of two of their export models, the 21 cm kanon 'V' vz. 39, sometimes known as the K52, and the 24 cm houfnice vz. 39(U) or vz. 166/600. Two guns and two howitzers were actually delivered before the start of the war prevented any further deliveries. As the Škoda concern had by then become the Skoda-Werke under German control the assembly lines were kept going and the output went to the German forces instead. Only ten howitzers were delivered as the *24 cm H 39*, but production of the gun (the *21 cm K 39*) was greater and it was built in three models, all basically similar. The initial batch was the *21 cm K 39*, but a further twenty were ordered with some production changes as the *21 cm K 39/40*', and another batch as the *21 cm K 39/41* – some of these were fitted with muzzle brakes making them the largest guns in German service to be so equipped. Some slight production changes on the howitzer produced the *24 cm H 39/40.*

In 1937 Krupp produced a 24 cm gun, the *24 cm K L/46,* another commercial model similar to the *15 cm K 39*. This gun had several unusual features, one of which was the use of data transmission from a central position for gun-laying. Another feature was the general size and weight of the gun, which had to be transported in three or four loads. Only one unit was equipped with the *24 cm K L/46* – 1 Batt/Art. Rgt 84.

Rheinmetall were also involved in a 24 cm gun design on which they started work during 1935. The prototype was completed in 1937 and the gun entered service in 1938 as the *24 cm K 3*. It was a huge piece with dual-recoil carriage and the unusual refinement that the gun could be brought in and out of action using only the winches and ramps built into the carriage. Even so it took 25 men about 1½ hours of hard labour to prepare the *K 3* for action. Krupp of Essen were responsible for the assembly of the small batch built, probably only about ten.

Krupp's involvement with the *K 3* convinced them they could do better and probably at less cost (each *K 3* cost RM500,000. The number of loads required to transport the *K 3* was also felt to be excessive. OKH also changed their ideas on super-heavy guns and issued a revised specification for a gun to fire a 160 kg projectile to a range of 48–49,000 metres. Again, both Krupp and Rheinmetall produced designs, but Krupp was the only one to begin building a prototype as the *24 cm K 4*. The carriage was designed to take either a 24 cm gun or a 30.5 cm howitzer and the whole weapon would have been moved slung between two turretless Tiger 1 tanks. A self-propelled version was also under consideration. Work was at the stage of assembling the prototype of a two-load version which had the barrel and cradle on a separate transporter when an Allied air raid destroyed all available material. Work on the *K 4* was then abandoned.

The next largest German weapon was the *28 cm H L/12*, a Krupp design that dated back to well before WW I. The *H L/12*

was a squat and massive piece that seems to have been kept in service because it was too large to move away from the various defensive positions where it was usually sited. It took about three to four days to emplace one of these monsters as the howitzer carriage was embedded in a heavy platform. Despite these difficulties the *H L/12* was in action on various occasions such as the bombardment of Sevastopol and on other campaigns when there was time to move this veteran into position.

After the obsolete *H L/12* it comes as a distinct contrast to consider the next piece, which was the thoroughly modern *35.5 cm H M1*. Development of this howitzer began in about 1936 and all the work was carried out by Rheinmetall-Borsig at their Düsseldorf plant. The first example entered service in 1939 and by 1941 there was one battery equipped with them (1 Batterie der Art. Abt. (mot.) 641). The actual number produced is still uncertain and reports vary from three to seven. On the move the *H M1* was carried in six loads and time into action was about two hours. The general design was virtually an enlarged version of the 24 cm *K 3*.

The largest of all the German heavy artillery pieces was the *42 cm Gamma Haubitze*, which also originated in pre-WW I days. The *Gamma* had been designed and built by Krupp as one of a series of super-heavy howitzers intended to smash the ring of concrete forts in Belgium and France that seemed to prevent the manoeuvres planned by the General Staff in any future war. That war began in 1914 and a mobile development of the *Gamma*, the *42 cm 'Dicke Bertha'*, was instrumental in the destruction of the Liège forts, although less successful when used against the Verdun forts in 1917. After 1918 the *Gamma* somehow escaped the attentions of the various Treaty commissions and the sole example was re-assembled at Krupp's training ground at Meppen, where it was used in various experiments involving concrete-piercing shells. There it might have remained, but in 1942 the *Gamma* was moved to the Crimea where it took part in the bombardment of Sevastopol along with the rest of the German 'siege train'. About 80 shells were fired on that occasion and pictorial evidence would seem to suggest that it also took part in the bombardment of Warsaw in 1944.

The lack of a properly balanced and equipped air force was one reason why the Germans were forced to resort to such odd weapons as the *15 cm Hochdruckpumpe*. After 1941 the Allied air fleets began to seriously disrupt the economic and production output of the Third Reich to such an extent that some form of retaliation was sought. The Luftwaffe bomber units were equipped with aircraft designed for the tactical support of armies in the field and, apart from the increasingly efficient Allied defences, were quite unable to carry on a prolonged strategic bomber attack. For these reasons some form of radical device had to be sought. Two such weapons actually saw action, the *V-1* flying bomb and the *V-2* rocket, but the third was never used. This third weapon, the *V-3*, was the *15 cm HDP* which had a rather prolonged development history. The *HDP* was based on an old ballistic idea which involved the use of several propellant chambers set alongside the length of a barrel. After a projectile was set in motion by an initial charge, other charges were fired as the projectile passed each chamber in turn. In theory each explosion would add to the final projectile velocity to the extent that very long ranges could be achieved. In practice the detonation of the charges in the side chambers was very difficult to time and achieve and up to WW II various such trial guns proved to be unsuccessful in several countries. After 1918 the principle came to the attention of one Herr Cönders who was the chief engineer of the Röchling Stahlwerke AG of Saarbrücken. (Cönders later evolved the Röchling anti-concrete shell which was so successful that it was rarely, if ever, used in action as the German staffs feared that a 'dud' would give away the secret of the shell.) After much theoretical paperwork, Cönders decided that his shells should have some form of long-range delivery system and went back to the idea of the multi-chamber gun. By May 1943 his company had built a 2 cm working model which seemed to show that the idea was feasible, but any further work had to have the approval of the Minister for Armaments, Speer. He showed the Cönders submission to Hitler who enthusiastically took over the idea and at once ordered full-scale production and the erection of a massive concrete firing installation at Mimoyèques, near Calais. This installation was to have 50 barrels, each one 150 metres long, and the target was London. By the end of September 1943 a short version of the *HDP* had been assembled on the Hillersleben range and the first few firings revealed a multitude of problems both in the projectile and the gun chambers. The first Hillersleben gun was joined by a second, but one was destroyed in a major accident. By that time a larger scale gun had been built on the island of Misdroy in the Baltic. Continuous development work with these guns gradually reduced the number of problems but it was never possible to prevent odd explosions inside the barrel and the only answer seemed to be to build any future guns in easily-replaced sections. The Misdroy firings, which began in March 1944, showed that the shell design was basically unstable and after much political argument the HWA were bought into the project, which was not of their choice and to which they were violently opposed – only 'the Führer's Will' compelled their involvement. New designs of shell were produced by a variety of contractors and at last it seemed that the majority of troubles were past and the projected range of 150 kilometres was possible.

At this point the Allied intelligence services and air forces once more weighed in to affect the course of events. For some time they had been watching the extensive excavations at Mimoyèques, and although they had no clear idea of the purpose of the site they correctly deduced it might have something to do with the rumoured 'new weapons', and in one heavy raid destroyed the site and much of the gun barrel installation work. This did not mark the end of the *HDP*, for in December 1944 two shortened versions were set up on a hillside near Hermeskeil and used to bombard Antwerp and Luxembourg during the Ardennes offensive. With the failure of that short campaign the two guns and their records were destroyed, and the only 'hard' evidence the Allies were able to find after the war was the wreckage of the two trial guns that had been set up on the Hillersleben range.

But the *HDP* was not the only scheme that diverted industrial and trial potential away from more viable projects. Until the end of the war various Krupp and Skoda design teams were busy devising super-heavy artillery. Skoda were busy on a 42 cm howitzer which, like the Krupp *K 4*, would have been transported slung between two turretless Tiger 1 tanks. Krupp were working on their *38 cm R 2* howitzer, and a 42 cm howitzer, the *K 5*. Even more unlikely were two more Krupp projects for 52 cm howitzers, both with a range of 25,000 metres; they were known as the *R 1* and *'Siegfried'*.

When the war ended the old stalwarts in the shape of the *sFH 18*, the *17 cm K 18* and the *21 cm Mrs 18* were still in service despite the huge amount of development work carried out on the '43' series and other similar projects. Compared with the relatively few types of equipment taken into the field by many of the Allied armies, the size and scope of the German medium and heavy artillery formations were too diverse and varied for logistical safety. Only the training and skill of the German gun teams made the German artillery such a formidable opponent.

Contemplating the use of 'foreign' equipment in the medium and heavy artillery bracket by the Germans during WW II one is somewhat overwhelmed by the size of the field to cover. Wherever the triumphant German forces went through Europe they captured amounts of medium and heavy artillery pieces, many of them ancient, some modern and efficient, but most of them in good condition and ready for further use. Overall, the accent was on the ancient because many artillery pieces were developed and produced to meet the needs of the tactical concepts prevalent before 1914 and just after. The top end of the scale was where the

35.5 cm M 1 being prepared for action

preponderance of age tended to be accentuated for the size of the really heavy guns was prodigious indeed, and the heavier artillery is the longer it tends to survive. Artillery pieces of 30.5 cm and upwards, if not common before 1914, were not unusual as the states of Europe drifted towards their supposed situations of isolation behind rings of concrete. To invade other nations – and the aggressive nations of 1914 saw no alternative – these rings of concrete had to be breached, but to do so by head-on attack involved the use of really heavy artillery, and howitzers with calibres of 38 cm and up to 42 cm were developed and built.

The cost of these massive pieces and the supply of their very costly ammunition meant that they had to be used as efficiently as possible, and by 1918 most of the heavy pieces that were ready for action in 1914 were still in the field. After 1918 they were little used except for the ritual firings to impress visiting dignitaries and many were still in service all over Europe ready for the holocaust that followed the events of 1939. By then the heavy artillery piece was an anachronism as the aircraft, and particularly the dive bomber, had rendered it obsolete. Only when air supremacy was complete and assured could the heavy howitzer be employed and transported, and the days of leisurely moves with ponderous artillery were over. This did not prevent the victorious Wehrmacht from taking into their own service any medium and heavy artillery they could capture, for in 1939 and 1940 the Luftwaffe was in the ascendancy and the future seemed assured.

Events prior to 1939 resulted in the Czechoslovak gun park coming under German control. Before 1918 that territory had been a part of the old Austro-Hungarian Empire, which in its heyday was only a little behind Germany as the military head of Central Europe. As Krupp built larger and larger guns, the Škoda works followed, so that by the end of 1918 the Austro-Hungarian armies had a substantial number of 24 cm guns, 30.5 and 42 cm howitzers on hand. When the old Empire was split up into its constituent parts this artillery was divided to equip the armies of the resultant nations, but the lion's share went to the new composite state of Czechoslovakia, and what was more important that country also retained the Škoda plant and its design staff. Thus by 1938 the Czech medium and heavy artillery was second to none. When the Czech state fell under the German wing in 1938 and 1939 the acquisition of the Škoda concern was a major prize that cannot be overestimated, and the Czech Army artillery immediately became a major part of the German military machine.

At the bottom end of the scale were the 149 mm guns and howitzers used in the field. These included the vz. 14, 14/16 and 15 from WW I and the more modern vz. 25, 33 and 37. The M.28 was a heavy fixed-position gun sold to Yugoslavia. Numbers of the M.15/16 heavy guns remained in use in Italy. The next 'preferred' Škoda calibre was 21 cm and it was here that howitzers began to become predominant – the M.18 and 18/19 howitzers were widely used by the Germans as were the 22 cm versions sold to Poland and Yugoslavia. The *21 cm K 52* gun has already been mentioned but under German control the Czechs also produced the M.1939 210 mm gun and 305 mm howitzer for delivery to the Soviet Union in exchange for wheat and raw materials essential to the expanding German war economy.

After Czechoslovakia, France was the next major unwilling contributor to the German armoury, for in 1940 nearly all the French gun park fell into German hands. Much of it was intact, although rather elderly, having been intended for use during WW I. Not much of this huge amount was used as field artillery by the Germans apart from a few of the more efficient guns, and the heavy weapons mentioned above. It is known that numbers of 155 mm GPF and GPF-T guns were taken into first-line service and used widely. The same applied to the 220 mle 1916 howitzer which was used on the Eastern Front in 1941 and 1942. The 155 mm C 17 S came into German hands from a number of sources apart from France and this too became an important German weapon, but most of the other guns were kept on hand until they too were incorporated in the 'Atlantic Wall' defences.

The invasion of the Soviet Union in 1941 revealed many unpleasant truths to the German medium artillery arm for, until then, two years of war had seemed to indicate that their weapons were second to none. As the winter of 1941 turned into the thaw of 1942 it was painfully obvious that the German artillery pieces were overweight and lacked range as opposed to their Soviet equivalents. This led to the formulation of the specification for the '43' series of guns and howitzers. In the meantime the gunners in the front line had to do something to counter their opponents, and the easiest solution was to simply use as much captured Soviet material as possible. So many Soviet guns and howitzers were captured that the Germans were later able to afford the luxury of transporting large numbers across Europe to bolster the 'Atlantic Wall' defences and the artillery formations based in France and Italy.

Apart from the usual incorporation of captured artillery into the German ranks, mention must be made of the careful examination and analysis of technical points to which each weapon type was subjected. Any good points were incorporated into the next generation of German guns and carriages. The same applied to captured ammunition, and wherever possible captured stocks were fired by German artillery. Any odd sizeable stocks of shells, that could not be used by the artillery for reasons of non-standard calibres or insufficient numbers of pieces to hand to utilise them, were passed to the engineers, who used them for coastal and beach defences in controlled minefields, and for the planned demolition of piers and bridges. Other artillery equipment that helped the German war effort was the use, often after a minimum of adaptation, of captured fuses, cartridge cases, explosives, fire control equipment, and even such basics as communication equipment.

15 cm schwere Feldhaubitze 13, 406(h) or 409(b)

German designation 15 cm sFH 13; 15 cm sFH 406(h) oder 409(b)
Original designations (h): 15 cm sFH; (b): Obusier de 150 L/17
Calibre 149.7 mm
Length of piece (L/17): 2550 mm
Length of barrel 2266 mm
Weight travelling 2332 kg
Weight in action 2270 kg
Weight of piece 382 kg
Traverse 7°
Elevation 0° to +45°
Muzzle velocity 390 m/sec
Shell weight 39.165 kg
Maximum range 8900 m
Rate of fire 3 rpm
Manufacturer Friedr. Krupp AG, Essen

Remarks: First issued to German artillery detachments 1917 and generally acknowledged as an excellent heavy howitzer for its day. Obsolescent by 1939 and afterwards used mainly in coastal defence role. Belgian and Dutch guns represented post-WW I German reparations.

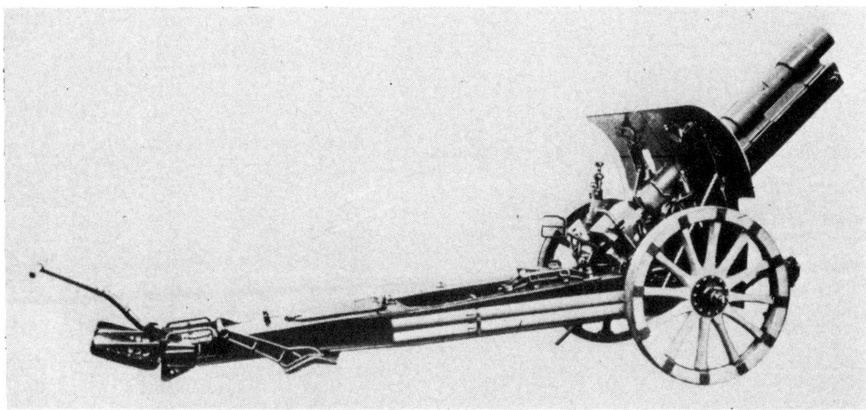

15 cm s FH 13. Introduced early in 1917 this gun was mounted on a box trail carriage designed for horse traction in one load

---

15 cm schwere Feldhaubitze 18 and 18M

The horse-traction 15 cm s FH 18 was moved in two sections. This view shows the limber and wheeled carriage for the gun barrel

For motorised traction the 15 cm s FH 18 was fitted with solid rubber tyres and moved in one section, the gun trails being carried on a two-wheel limber

15 cm s FH 18M in action on the Eastern Front, 1943

15 cm s FH 18 horse limber and gun carriage

15 cm s FH 18M. In 1942 a replaceable chamber liner was introduced to overcome excessive erosion caused by the increased powder charges, and a muzzle brake fitted to reduce the stress on the carriage

15 cm s FH 18 in action with the Waffen-SS artillery.

15 cm s FH 18 with steel shod aluminium wheels for horse traction

German designation 15 cm sFH 18
Calibre 149 mm
Length of piece (L/29.5): 4440 mm
Length of barrel 3985 mm
Length of rifling 3623 mm
Weight travelling 6304 kg
Weight in action 5512 kg
Traverse 60°
Elevation −3° to +45°
Muzzle velocity 520 m/sec

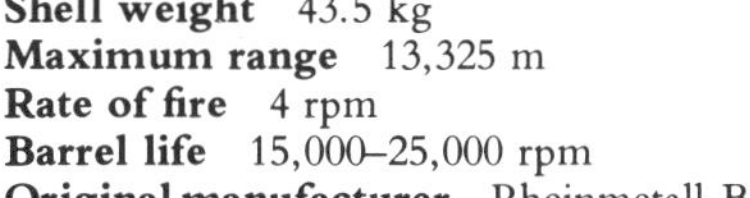

Shell weight 43.5 kg
Maximum range 13,325 m
Rate of fire 4 rpm
Barrel life 15,000–25,000 rpm
Original manufacturer Rheinmetall-Borsig AG, Düsseldorf
Manufacturer Spreewerk, Berlin-Spandau; MAN, Augsburg; Dörries-Füllner, Bad Warmbrunn; Škoda, Dubnica/Slovakia

Remarks: Joint Krupp-Rheinmetall design worked on independently during 1926–30. Best features of both combined 1933 to become standard Wehrmacht heavy field howitzer. In production from late 1933, in service from early 1934. During WW II some also supplied to Italian Army and used as Obice da 149/28. sFH 18M represented 1942 modification with muzzle brake and renewable barrel liner.

15 cm schwere Feldhaubitze 36

German designation 15 cm sFH 36
Calibre 149 mm
Length of piece (L/24): 3555 mm
Length of barrel 2965 mm
Length of rifling 2475 mm
Weight travelling 3500 kg
Weight in action 3280 kg
Traverse 56°
Elevation −1° to +43°
Muzzle velocity 485 m/sec
Shell weight 43.5 kg
Maximum range 12,300 m
Rate of fire 4 rpm
Manufacturer Rheinmetall-Borsig AG, Düsseldorf

Remarks: Development started 1935 by both Krupp and Rheinmetall in response to an OKH specification calling for a horse-drawn one-load equipment. Rheinmetall proposal approved 1938 and put into limited production late in 1939. Only small series built; production terminated late 1941.

15 cm s FH 36, Krupp prototype

15 cm s FH 36. This Rheinmetall prototype was designed for horse traction

15 cm s FH 36, Krupp version, on limber for horse traction

15 cm schwere Feldhaubitze 40

15 cm s FH 40 was developed to achieve increased performance over the s FH 18 but was not accepted into service. Illustration shows the Rheinmetall prototype

German designation 15 cm sFH 40
Calibre 149 mm
Length of piece (L/32.5): 4875 mm
Length of barrel 3297 mm
Weight travelling 6200 kg
Weight in action 5402 kg
Traverse 60°
Elevation 0° to +70°
Muzzle velocity 595 m/sec
Shell weight 43.5 kg
Maximum range 15,400 m
Rate of fire 4 rpm
Manufacturers Rheinmetall-Borsig AG, Düsseldorf; Friedr. Krupp AG, Essen

Remarks: Development commenced 1938, first prototypes completed 1938. Not accepted for service but during 1942 barrel placed on sFH 18 carriage to produce sFH 18/40 or 42.

---

15 cm schwere Feldhaubitze 18/40 or 42

German designation 15 cm sFH 18/40, later sFH 42
Calibre 149 mm
Length of piece with m/b (L/36): 5388 mm
Length of piece less m/b (L/32.5): 4875 mm
Length of barrel 3297 mm
Weight travelling 6480 kg
Weight in action 5660 kg
Traverse 56°
Elevation 0° to +45°
Muzzle velocity 595 m/sec
Shell weight 43.5 kg
Maximum range 15,100 m
Rate of fire 4 rpm
Barrel life 10,000 rounds
Manufacturers Spreewerk, Berlin-Spandau; MAN, Augsburg; Dörries-Füllner, Bad Warmbrunn; Škoda, Dubnica/Slovakia

Remarks: Initial sFH 40 resulted from unsuccessful Krupp and Rheinmetall developments to 1938 specification for sFH 36. During 1942 46 sFH 40 barrels combined with sFH 18 carriages to become sFH 18/40, later known as sFH 42. Proved not very successful in service.

15 cm s FG 18/40(42) was basically the s FH 40 barrel and breech mechanism on the s FH 18 carriage

---

15 cm schwere Feldhaubitze 43 Krupp

German designation 15 cm sFH 43 Kp
Calibre 149 mm
Length of piece (L/41): 6158 mm
Weight travelling 8400–8975 kg
Weight in action 7400–7900 kg
Traverse 360°
Elevation −5° to +70°
Muzzle velocity (est): 660 m/sec
Maximum range (est): 18,000 m
Shell weight 43.6 kg
Designers Friedr. Krupp AG, Essen

Remarks: Designed to specification issued late in 1943 but only wooden mock-up completed before end of WW II. Final model would have used same carriage as 12.8 cm K 44.

15 cm s FH 43 full scale wooden mock-up

15 cm s FH 43 wooden mock-up in travelling position

15 cm schwere Feldhaubitze 43 Skoda

German designation 15 cm sFH 43 Sk
Design designation 15 cm sH 43.5/600
Calibre 149 mm
Weight travelling 6650 kg
Weight in action 5950 kg
Traverse 360°
Elevation −5° to +65°
Muzzle velocity 600 m/sec
Shell weight 43.5 kg
Maximum range (est): 15,000 m
Designers Skoda-Werke, Pilsen

Remarks: Project only.

12.8 cm Panzerabwehrkanone 44, 80 and 12.8 cm Kanone 44

German designations 12.8 cm Pak 44, K 44 or Pak 80; Also referred to as 12.8 cm Pak 43, oder Pjk 44
Calibre 128 mm
Length of piece (L/55): 7023 mm
Length of bore 6623 mm
Length of rifling 5538 mm
Weight of gun 3353 kg
Weight of gun and carriage 10,160 kg
Traverse 360°
Elevation −7°51′ to +45°27′
Muzzle velocity (AP): 950 m/sec; (HE): 750 m/sec
Shot weight 28.3 kg
Shell weight 28 kg
Armour penetration 219 mm at 500 m (0°); 202 mm at 1000 m; 187 mm at 1500 m
Maximum range (HE): 24,410 m
Barrel life 1000–2000 rounds
Manufacturer Friedr. Krupp, Bertha-Werke, Breslau

Remarks: A series of weapons evolved late 1944 around 12.8 cm K 44, an advanced gun design. Prototypes built by both Krupp and Rheinmetall, but only Krupp design accepted. Despite excellent performance production difficulties resulted in only small numbers reaching action. Krupp Pak 80 had elevation limited from −7° to + 15°.

12.8 cm Kanone 81/1

German designation 12.8 cm K 81/1
Calibre 128 mm
Length of piece (L/55): 7023 mm
Length of bore 6623 mm
Length of rifling 5538 mm
Weight in action 12,150 kg
Traverse 60°
Elevation −4° to +45°
Muzzle velocity (AP): 950 m/sec; (HE): 750 m/sec
Shot weight 28.3 kg
Shell weight 28 kg
Maximum range (HE) approx: 24,000 m

Remarks: Dual-purpose weapon combining new German 12.8 cm K 81 barrel (version of K 44 intended for tank use) with captured French Canon de 155 GPF-T carriages. Reportedly about 50 completed late 1944 and used in action.

12.8 cm Kanone 81/2

German designation 12.8 cm K 81/2
Calibre 128 mm
Length of piece (L/55): 7023 mm
Length of bore 6623 mm
Length of rifling 5538 mm
Weight in action (approx): 8200 kg
Traverse 40°
Elevation −4° to +45°
Muzzle velocity (AP): 950 m/sec; (HE): 750 m/sec
Shot weight 28.3 kg
Shell weight 28 kg
Maximum range (HE) approx: 24,000 m

Remarks: Dual-purpose weapon combining new German 12.8 cm K 81 barrel with captured Soviet 152 mm Gun-Howitzer obr. 1937 (ML-20) carriage. Only small number completed late 1944 and used in action.

15 cm Kanone 16 or 429(b)

German designation 15 cm K 16 oder 15 cm K 429(b)
Original designation (b): Canon de 150 L/43
Calibre 149.3 mm
Length of piece (L/43): 6410 mm
Length of barrel 6020 mm
Weight travelling (2 loads): 17,372 kg
Weight in action 10,870 kg
Weight of piece 4090 kg
Traverse 8°
Elevation −3° to +42°
Muzzle velocity 757 m/sec
Shell weight 51.4 kg
Maximum range 22,000 m
Rate of fire 3 rpm
Barrel life 3000–4000 rounds
Manufacturer Friedr. Krupp AG, Essen

Remarks: Produced during WW I in two almost identical versions. First issued 1916, numbers still in service 1939 but mainly for training. During 1941 as an emergency production measure small number of K 16 barrels placed on 21 cm Mrs 18 carriages to become 15 cm K 16 in Mrs Laf. Belgian guns comprised 1919 reparations, recaptured by German forces 1940.

15 cm Kanone 16 being manhandled into position. Introduced in 1917, this gun saw only limited service during World War II

15 cm Kanone 18

German designation 15 cm K 18
Calibre 149.1 mm
Length of piece (L/55): 8200 mm
Length of barrel 6432 mm
Weight travelling (2 loads): 18,700 kg
Weight in action 12,460 kg
Traverse (on platform): 360°; (on carriage): 11°
Elevation −2° to +43°
Muzzle velocity 865 m/sec
Shell weight 43 kg
Maximum range 24,825 m
Rate of fire 2 rpm
Barrel life 3000–5000 rounds
Manufacturer Rheinmetall-Borsig AG, Düsseldorf

Remarks: Development commenced 1933, entered service 1938. Intended as standard Wehrmacht heavy gun. Design incorporated several unusual features, including special two-piece platform for 360° traverse. Remained in service until end of WW II.

15 cm Kanone 18 was provided with a ground traversing platform which was anchored by means of securing pickets in the firing position. The equipment was transported in two loads

15 cm K 18 barrel on the transporter

15 cm K 18 carriage in the travelling position. Note the traversing platform under the gun carriage

15 cm Kanone 39

15 cm K 39 was designed to fire the same shells as the K 18 and was almost identical ballistically. It was transported in three loads

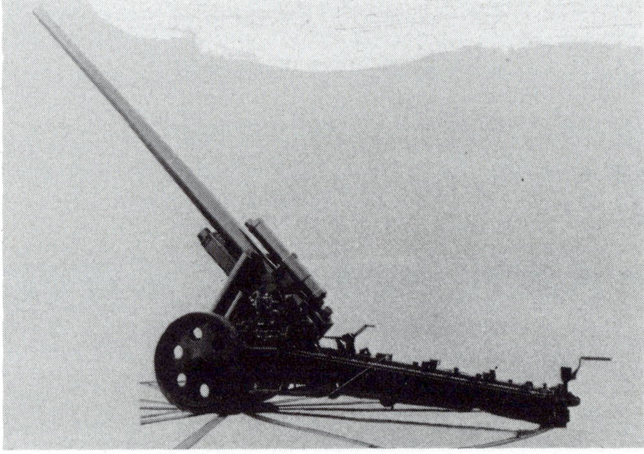

15 cm K 39 on firing platform for the coastal defence role

German designation 15 cm K 39
Calibre 149.1 mm
Length of piece (L/55): 8250 mm
Length of barrel 7868 mm
Length of rifling 6505 mm
Weight travelling (2 loads): 18,282 kg
Weight in action 12,186 kg
Traverse (on turntable): 360°; (on carriage): 60°
Muzzle velocity 865 m/sec
Shell weight 43 kg
Maximum range 24,825 m
Rate of fire 2 rpm
Barrel life 3000–5000 rounds
Manufacturer Friedr. Krupp AG, Essen

Remarks: Designed to Turkish order as dual-purpose heavy field gun/coastal defence weapon on split-trail carriage or portable turntable. Transported in 3 loads. After September 1939 production diverted to German war effort. First equipment issued 1940. Remained in service throughout WW II years. Usually diverted to coastal role by 1945.

15 cm K 39 gun barrel on its transporter
15 cm K 39 gun carriage in travelling position ►
◄ *K 39 firing platform on its transporter*

15 cm Schiffskanone C/28 in Mörserlafette

German designation 15 cm SKC/28 in Mrs Laf
Calibre 149.1 mm
Length of piece (L/55): 8291 mm
Length of barrel 7815 mm
Length of rifling 6584 mm
Weight travelling (2 loads): 22,735 kg
Weight in action 16,870 kg
Traverse (on platform): 360°; (on carriage): 16°
Elevation 0° to +50°
Muzzle velocity 890 m/sec
Shell weight 43 kg
Maximum range 23,700 m
Rate of fire 2 rpm
Manufacturer (gun): Rheinmetall-Borsig AG, Düsseldorf; (carriage): Hanomag, Hannover

Remarks: An urgent interim measure to provide heavy guns. Combination of 15 cm SKC/28 naval barrels adapted to take Army ammunition with 21 cm Mrs 18 carriages. Only 8 completed during 1941 and used in action on Eastern Front and elsewhere.

15 cm SK C/28 in Mrs Laf. *This equipment was moved in two loads, the barrel transporter of the 21 cm Mrs being used for this purpose*

15 cm Schiffskanone *C/28* in Mörserlafette

17 cm Kanone 18 in Mörserlafette

German designation 17 cm K 18 in Mrs Laf
Calibre 172.5 mm
Length of piece (L/50): 8529 mm
Length of barrel 8103 mm
Length of rifling 6464 mm
Weight travelling (2 loads): 23,375 kg
Weight in action 17,510 kg
Traverse (on platform): 360°; (on carriage): 16°
Elevation 0° to +50°
Muzzle velocity (HE): 860 m/sec; (AP): 830 m/sec
Shell weight (HE): 68 kg; (AP): 71 kg
Maximum range 28,000 m
Rate of fire 1–2 rpm (40 per hour)
Barrel life 1500 rounds
Original designers Friedr. Krupp AG, Essen
Manufacturer Hanomag, Hannover

Remarks: Modern design including carriage with two separate recoil systems. In action gun could be easily rotated through 360° by one man. Introduced in service 1941 and proved best German heavy gun; allocated production priority. Captured examples often used by Allied artillery troops in Europe.

Battery of 17 cm K 18 guns prepare for action

17 cm K 18 in an open gun emplacement for coastal defence. The gun trails have been fitted with a device to run on the circular-shaped rail as an aid to traverse

17 cm Kanone 18 in Mörserlafette *was designed for motorised transport in two loads, but over short distances the gun and carriage could be moved as one load*

langer 21 cm Mörser

German designation lg 21 cm Mrs
Calibre 211 mm
Length of piece (L/14.6): 3063 mm
Length of barrel 2675 mm
Length of rifling 2296 mm
Weight in action 9220 kg
Traverse 4°
Elevation +6° to +70°
Muzzle velocity 393 m/sec
Shell weight (HE): 113 kg; (anti-concrete): 121.4 kg
Maximum range 11,100 m
Rate of fire 1–2 rpm
Manufacturer Friedr. Krupp AG, Essen

Langer 21 cm Mörser was introduced in 1916 but only limited numbers were in service during World War II

Remarks: Introduced in service 1916 as two-load equipment, modernised during mid-1930s for single-load transport. Used in action until 1942 when progressively replaced by 21 cm Mrs 18. Later deployed to static defence positions.

21 cm Mörser 18

German designation 21 cm Mrs 18
Calibre 210.9 mm
Length of piece (L/31): 6510 mm
Length of barrel 6070 mm
Length of rifling 5274 mm
Weight travelling (2 loads): 22,700 kg
Weight in action 16,700 kg
Traverse (on platform): 360°; (on carriage): 16°
Elevation 0° to +70°
Muzzle velocity 565 m/sec
Shell weight 113 or 121.4 kg
Maximum range 18,700 m
Rate of fire 1 rpm
Barrel life 8000–10,000 rounds
Manufacturers Friedr. Krupp AG, Essen; Hanomag, Hannover

21 cm Mrs 18 in firing position

Remarks: Development began 1933, first complete equipment entered service 1939. Used in two variants; same carriage as 17 cm K 18. Production ceased 1942 in favour of 17 cm K 18. Used operationally on all fronts and fired wide range of ammunition specially developed for this weapon.

21 cm Mörser 18 was normally transported in two sections but over short distances could be moved as one load

A modified 21 cm Mrs 18 with four pneumatic tyres. This version did not go into service

21 cm Kanone L/50

German designation 21 cm K L/50 Kp
Calibre 209.3 mm
Length of piece (L/50): 10,500 mm
Weight emplaced 35,100 kg
Traverse on platform 360°
Elevation −4° to +45°
Muzzle velocity 875 m/sec
Shell weight 120 kg
Maximum range 34,000 m
Manufacturer Friedr. Krupp AG, Essen

Remarks: Krupp commercial model built just prior to 1939. Small numbers used by German Army but considered too heavy for general issue.

21 cm Kanone 38

German designation 21 cm K 38
Calibre 210.9 mm
Length of piece (L/55.5): 11620 mm
Length of barrel 11075 mm
Length of rifling 8717 mm
Weight travelling (2 loads): 34,825 kg
Weight in action 25,435 kg
Traverse (on platform): 360°; (on carriage): 18°
Elevation 0° to +50°
Muzzle velocity 905 m/sec
Shell weight 120 kg
Maximum range 33,900 m
Rate of fire 1 rpm
Barrel life 2000 rounds
Manufacturer Friedr. Krupp AG, Essen

Remarks: Development commenced 1938 as possible replacement for 21 cm Mrs 18. Total of 15 on order 1940 but only seven delivered by 1943 when production stopped. Technically one of best guns of WW II years. One complete equipment sent to Japan.

21 cm Kanone 38, designed for motorised transport in two loads

21 cm K 38 barrel on its transporter

Rear view of the 21 cm K 38 in firing position ➤

21 cm K 38 carriage on its transporter

21 cm Kanone 52, 39, 39/40 and 39/41

German designations 21 cm K 52, 39, 39/40, 39/41
Design designation 21 cm kanonu VX
Calibre 210 mm
Length of piece with m/b (L/52): 11462 mm
Length of piece less m/b 10766 mm
Length of barrel 9530 mm
Weight travelling (3 loads): 59,100 kg
Weight in action 39,800 kg
Traverse 360°
Elevation −4° to +45°
Muzzle velocity (K 52 and 39): 800 m/sec; (K39/40 and 39/41): 860 m/sec
Shell weight 135 kg
Maximum range 33,000 m
Rate of fire 3 rounds in 2 mins
Barrel life 1200 rounds
Manufacturer Škoda-Werke, Pilsen

21 cm Kanone 39 and 39/40 were moved in three sections

Remarks: Designed to 1938 Turkish order but only two K 52 guns delivered before production came under German military control. Total of 10 built as K 52 or 39. Later K 39/40 (20 built) had improved performance while K 39/41 (40 ordered 1944, 16 delivered before April 1945) introduced changes to simplify production. Last two versions also featured muzzle brakes. Used operationally on Eastern Front.

24 cm Haubitze 39 and 39/40

German designation 24 cm H 39 und 39/40
Design designation Škoda vz. 166/600
Calibre 240 mm
Length of piece (L/28): 6765 mm
Weight travelling (3 loads): 42,900 kg
Weight in action 29,000 kg
Weight of piece 8100 kg
Traverse 360°
Elevation −4° to +70°
Muzzle velocity 597 m/sec
Shell weight 166 kg
Maximum range 18,150 m
Rate of fire 1 round every 2 mins
Barrel life 2000 rounds
Manufacturer Skoda-Werke, Pilsen

Remarks: Designed by Škoda engineers as parallel development to 21 cm K 39. Built to Turkish order but only two delivered before production came under German military control after July 1939. Some guns used in France 1940. Wartime H 39/40 introduced certain changes to simplify production.

24 cm Kanone L/46

German designation 24 cm K L/46 Kp
Calibre 238 mm
Length of piece (L/46): 10948 mm
Weight travelling (3 loads): 57,200 kg
Weight in action (with platform): 29,600 kg; (less platform): 15,600 kg
Traverse 360°
Elevation −4° to +45°
Muzzle velocity 850 m/sec
Shell weight 180 kg
Maximum range 32,000 m
Rate of fire 2 rpm
Manufacturer Friedr. Krupp AG, Essen

Remarks: Essentially scaled-up 15 cm K 39 with different carriage. Introduced into service 1937 but only in small numbers due to cumbersome nature of design. In action used by only 1.Batt/Art.Abt 84.

24 cm Kanone 3

German designation 24 cm K 3
Calibre 238 mm
Length of piece (L/54.6): 13104 mm
Length of barrel 12480 mm
Length of rifling 10177 mm (64 grooves)
Weight travelling (6 loads): 84,636 kg
Weight in action 54,000 kg
Traverse (on platform): 360°; (on carriage): 6°
Elevation −1° to +56°
Muzzle velocity 870 m/sec
Shell weight 152.3 kg
Max range 37,500 m
Rate of fire 1 round every 3–4 mins
Barrel life 500 rounds
Designers Rheinmetall-Borsig AG, Düsseldorf
Manufacturer Friedr. Krupp AG, Essen

Remarks: Development started 1935 and weapon introduced in service 1938. Design featured dual recoil carriage system, travelled in 6 loads, and could be assembled without cranes. Served with schw. Art. Abt(mot.) 83 which had three batteries of two guns each. Among the many experiments carried out with this equipment were pre-rifled shells fired from a barrel with only 8 rifling grooves, use of a 'squeeze-bore' muzzle attachment, and discarding-sabot shells.

24 cm Kanone 3. This fully mobile equipment was transported in five loads. An electric generator accompanied the sections as the sixth load

K 3 carriage and saddle

K 3 breech ring and mechanism

Rear view of the 24 cm K 3

K 3 firing platform

K 3 gun barrel on its transporter

24 cm Kanone 4

German designation 24 cm K 4
Calibre 238 mm
Length of piece (L/72): 17280 mm
Weight travelling (2 loads): 65,500 kg
Weight in action 55,000 kg
Traverse (on platform): 360°; (on carriage): 16°
Elevation 0° to +55°
Muzzle velocity 1100 m/sec
Shell weight 160 kg
Maximum range 49,000 m
Manufacturer Friedr. Krupp AG, Essen

Remarks: An advanced gun evolved in answer to an OKH request for possible 24 cm K3 replacement. Krupp engineers designed two prototypes, two- and single-load versions, but only experimental two-load version completed. This only prototype was severely damaged during an air raid on Essen in 1943 and development abandoned. Another Krupp proposal to same specifications represented an alternative self-propelled tracked carriage designated 24 cm K 4 St.

28 cm Haubitze L/12

28 cm H L/12 barrel on its transporter

28 cm H L/12 cradle on its transporter

28 cm H L/12 turntable on its transporter

German designation 28 cm H L/12
Calibre 283 mm
Length of piece (L/12): 3396 mm
Weight emplaced 37,000 kg
Traverse 360°
Elevation 0° to +65°
Muzzle velocity 379 m/sec
Shell weight 350 kg
Maximum range 11,400 m
Rate of fire 1 rpm
Manufacturer Friedr. Krupp AG, Essen

Remarks: Pre-WW I static design. A few survived until 1939. Notable as only bag-charge German piece. Despite their age and time needed for emplacement ready for action (3–4 days) reportedly used at Sevastopol in 1942.

28 cm H L/12 wooden firing platform

35.5 cm Haubitze M. 1

German designation 35.5 cm H M. 1
Design designation 35 cm Mörser L/27 M2
Calibre 355.6 mm
Length of piece (L/28.9): 10265 mm
Length of barrel 9585 mm
Length of rifling 8050 mm
Weight travelling (7 loads): 123,500 kg
Weight in action 78,000 kg
Traverse (on platform): 360°; (on carriage): 6°

Elevation +45° to +75°
Muzzle velocity 570 m/sec
Shell weight 575 kg
Maximum range 20,850 m
Rate of fire 1 round every 4 mins
Barrel life 2000 rounds
Manufacturer Rheinmetall-Borsig AG, Düsseldorf

An electrically-operated travelling gantry was used to unload and assemble the 35.5 cm M 1 sections.

Remarks: Development in response to an Army request dated 1935. Design incorporated dual recoil system and two-part platform for all-round traverse. Transported in 6 loads. First complete weapon issued 1939 and used by schw. Art. Abt(mot.) 641. Between 3–7 produced.

35.5 cm Haubitze M 1. This equipment was transported in six sections, each of which was carried on a six-wheeled trailer similar to that used for K 3

Gamma Mörser

Rear view of the 'Gamma' Mörser showing the ammunition hoist and breech

42 cm 'Gamma' Mörser. At the rear of it can be seen the travelling gantry used for assembling the equipment

German designation Gamma Mrs
Calibre 420 mm
Length of piece (L/16): 6723 mm
Length of rifling 5323 mm
Weight emplaced 140,000 kg
Traverse 46°
Elevation +43° to +75°
Muzzle velocity 452 m/sec
Shell weight 1003 or 1020 kg
Maximum range 14,200 m
Rate of fire 1 round every 8 mins
Barrel life 1000 rounds
Manufacturer Friedr. Krupp AG, Essen

The 'Gamma' Mörser with an added armour-protected superstructure shelling Warsaw in 1944

Remarks: Built during 1906 as part of development programme leading to 42 cm 'Big Bertha' guns. Not used during WW I, left at Meppen Proving Ground after 1918 and reassembled 1936–37 for experimental work on concrete-piercing shells. Used to bombard Sevastopol defences 1942. Moving Gamma involved use of 10 special railway flat cars; emplacing it took 1½–2½ days.

15 cm Hochdruckpumpe

German designation 15 cm HDP oder 'Der Tausendfüssler'
Calibre 150 mm
Length of gun Hillersleben: 75000 mm; Mimoyèques 150000 mm
Length of side chambers 1500 mm
Length from breech to 1st chamber 6000 mm
Spacing of subsequent chambers 3200 mm
Angle of chambers to bore 45°
Traverse Fixed
Elevation of guns Hillersleben: 6° fixed; Mimoyèques 55° fixed
Muzzle velocity (est): 1463 m/sec
Shell weight 83 kg
Intended range 88,500 m

Remarks: An unusual smooth-bore multi-chambered weapon evolved by Eng. Cönders. Much of design work carried out by Röchling Eisenwerke but Friedr. Krupp AG, Essen, were responsible for ballistic development. Final design was to have 31 chambers. Main installations at Mimoyèques/France destroyed in Allied air raid, but two shorter-barrelled weapons of same type briefly used during German Ardennes offensive winter 1944–45 before destruction.
NB: Above data should be regarded as provisional.

One breech of the Hochdruckpumpe

15 cm Hochdruckpumpe *(HDP) showing one partly destroyed barrel with its attached gas pressure feed pipes*

15cm Spr Gr 4481 high-explosive projectile for the HDP long-range gun

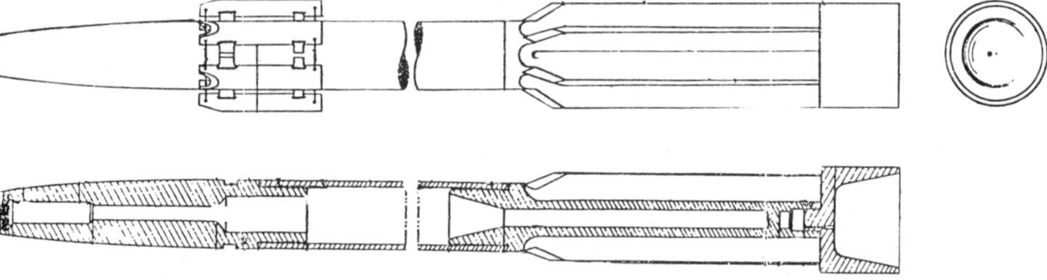

11.4 cm leichte Feldhaubitze 361(e)

German designation 11.4 cm leFH 361(e)
Original designation Q.F. 4.5-in Howitzer Mk 2
Calibre 114.3 mm
Length of piece (L/15.55): 1777.3 mm
Length of barrel 1526.8 mm
Length of rifling 1343 mm
Weight travelling 2222 kg
Weight in action 1494 kg
Weight of piece 463 kg
Traverse 6°
Elevation −5° to +45°
Muzzle velocity 308 m/sec
Maximum range 6040 m
Rate of fire 6–8 rpm
Original manufacturer Coventry Ordnance Works, Coventry

Remarks: Developed during WW I from 4.5-in Howitzer Mk 1. Large numbers of updated Mk 2s with rubber tyres in service 1939–40. Total of 96 captured by German forces in France 1940 and later impressed in coastal defence role. Small number of other WW I howitzers of same type also in Polish and Soviet service. Captured Soviet pieces allocated German designation 11.5 cm leFH 362(r).

11.4 cm le FH 361(e)

11.4 cm Kanone 365(e)

German designation 11.4 cm K 365(e)
Original designation 4.5-in Gun Mk I on Carriage 60 pr Mk IVP
Calibre 114.3 mm
Length of piece (L/42.8): 4881 mm
Length of barrel 4748.2 mm
Length of rifling 3977.8 mm
Weight travelling 7250 kg
Weight in action 5730 kg
Weight of piece 1340 kg
Traverse 7°
Elevation 0° to +42°
Muzzle velocity 686 m/sec
Shell weight 24.95 kg
Maximum range 19,200 m
Rate of fire 2 rpm

Remarks: New 4.5-in barrel combined with modified 60 pr carriage. Total of 76 howitzers made during 1937–38. Most of these left behind in France by BEF 1940 later emplaced by German troops for coastal defence.

12 cm Kanone 370(b)

German designation 12 cm K 370(b)
Original designation Canon de 120 L mle 1931
Calibre 120 mm
Length of piece (L/37): 4426 mm
Length of rifling 3562 mm
Weight travelling 5800 kg
Weight in action 5450 kg
Weight of piece 1824 kg
Traverse 60°
Elevation 0° to +38°30′
Muzzle velocity 760 m/sec
Shell weight 21.93 kg
Maximum range 17,500 m
Rate of fire 1 rpm
Manufacturer Société anonyme John Cockerill, Liège

Remarks: Entered Belgian Army service 1934; total of 24 in line 1939. Noted as stable weapon with good range. Small number of captured guns impressed into German service, mainly for coastal defence.

12 cm leichte Feldhaubitze 373(h)

German designation 12 cm leFH 373(h)
Original designation (h): Lichte Houwitze 12 cm L 14
Calibre 120 mm
Length of piece (L/14): 1725 mm
Length of barrel 1420 mm
Weight travelling 2520 kg
Weight in action 1610 kg
Weight of piece 505 kg
Traverse 6°
Elevation −4°48′ to +43°
Muzzle velocity 317 m/sec
Shell weight 16.5 kg
Maximum range 6050 m
Manufacturer AB Bofors, Bofors, Sweden

Remarks: Purchased from Sweden during WW I. Total of 40 still with Dutch artillery in 1940. German designation allocated, but no record of service use.

12.2 cm leichte Feldhaubitze 386(r)

German designation 12.2 cm leFH 386(r)
Original designation *122 mm Gaubitsa obr. 1909/37 g* (122–09/37)
Calibre 121.92 mm
Length of piece (L/14): 1690 mm
Length of rifling 1132.68 mm
Weight travelling 2480 kg
Weight in action 1450 kg
Weight of piece 475 kg
Traverse 4°
Elevation −5° to +43°
Muzzle velocity 364 m/sec
Shell weight 21.76 kg
Maximum range 8940 m
Original manufacturer Friedr. Krupp, Essen; Licence-built at Putilov Arsenal, St. Petersburg. 1937 modernisation: various Soviet State arsenals

Remarks: Soviet-modernised ex-Tsarist Army howitzers, modified in 1937. Numbers captured by German forces later used on Eastern Front, Balkans and France.

12.2 cm leichte Feldhaubitze 388(r)

German designation 12.2 cm leFH 388(r)
Original designation *122 mm Gaubitsa obr. 1910/30 g*
Calibre 121.92 mm
Length of piece (L/12.8): 1561.6 mm
Length of rifling 1140 mm
Weight travelling with limber 2530 kg
Weight in action 1466 kg
Weight of piece 421.8 kg
Traverse 4°41′
Elevation −3° to +43°
Muzzle velocity 364 m/sec
Shell weight 21.76 kg
Maximum range 8940 m
Rate of fire 6–7 rpm
Original designers Schneider et Cie, Le Creusot
Manufacturers Putilov Arsenal, Leningrad

Remarks: Originally French Schneider 10 S howitzer imported by Tsarist Russia before WW I. During early 1930s modernised in Soviet arsenals. Large numbers captured by German forces during 1941–42 and despite obsolescence used on Eastern Front and elsewhere.

12.2 cm Kanone 390/1(r)

German designation 12.2 cm K 390/1(r)
Original designation *122 mm Pushka obr. 1931 g* (122–31)
Calibre 121.92 mm
Length of piece (L/46.3): 5650 mm
Length of barrel 5483 mm
Length of rifling 4600 mm
Weight travelling 7800 kg
Weight in action 7100 kg
Weight of piece 2340 kg
Traverse 56°
Elevation −4° to +45°
Muzzle velocity 800 m/sec
Shell weight 25 kg
Maximum range 20,870 m
Rate of fire 5–6 rpm
Manufacturer Various Soviet State arsenals

Remarks: An original Soviet design combining new barrel with 152 mm Gun-Howitzer obr. 1934 carriage. A robust and effective gun, used in substantial numbers during WW II. Captured examples served with German artillery detachments on Eastern Front and incorporated into Atlantic coast defences.

12.2 cm Kanone 390/2(r)

German designation 12.2 cm K 390/2(r)
Original designation *122 mm Pushka obr. 1931/37 g* (A-19)
Calibre 121.92 mm
Length of piece (L/46.34): 5650 mm
Length of barrel 5483 mm
Length of rifling 4600 mm
Weight travelling 7907 kg
Weight in action 7117 kg
Weight of piece 2340 kg
Traverse 58°
Elevation −2° to +65°
Muzzle velocity 800 m/sec
Shell weight 25 kg
Maximum range 20,400 m
Rate of fire 5–6 rpm
Manufacturer Various Soviet State arsenals

12.2 cm Kanone 390/2(r) in travelling position

Remarks: Combination of 122 mm obr. 1931 Field Gun barrel and 152 mm Gun-Howitzer obr. 1937 carriage. Captured guns mostly deployed along French coastline apart from use on Eastern Front.

12.2 cm schwere Feldhaubitze 396(r)

German designation 12.2 cm sFH 396(r)
Original designation *122 mm Gaubitsa obr. 1938 g* (122–38)
Calibre 121.92 mm
Length of piece (L/22.7): 2800 mm
Length of barrel 2668 mm
Length of rifling 2263 mm
Weight travelling 2800 kg
Weight in action 2250 kg
Weight of piece 722.5 kg
Traverse 50°
Elevation −3° to +65°
Muzzle velocity 515 m/sec
Shell weight 21.76 kg
Maximum range 12,100 m
Rate of fire 5–6 rpm
Manufacturer Various Soviet State arsenals

Remarks: One of the most successful and reliable Soviet artillery pieces of WW II. In Red Army service from late 1938. Produced and used in very large numbers. Many captured examples in service with regular German artillery detachments; also deployed for coastal defence in France.

Whenever possible German artillery was transported by rail on the Eastern Front to avoid the poor and congested roads. Illustration shows a 12.2 cm K 390/2(r) being loaded on a railway flatcar

15 cm schwere Feldhaubitze 15(t) and (ö)

German designation 15 cm sFH 15(t)
Original designation (t): 15 cm hruba houfnice vz. 15
Calibre 149.1 mm
Length of piece (L/20): 2990 mm
Weight in action 5560 kg
Traverse 8°
Elevation −5° to +65°
Muzzle velocity 508 m/sec
Shell weight 42 kg
Maximum range 11,500 m
Manufacturer Škoda, Pilsen

Remarks: Mobile piece derived from fortification howitzer, transported in four loads. Only 57 built during WW I. Despite obsolescence available pieces impressed in German service and used 1939–41; subsequently gradually withdrawn from use.

15 cm schwere Feldhaubitze 25(t)

German designation 15 cm sFH 25(t)
Original designation 15 cm hruba houfnice vz. 25
Calibre 149.1 mm
Length of piece (L/18): 2700 mm
Weight travelling (2 loads): 6050 kg
Weight in action 3800 kg
Traverse 7°
Elevation −5° to +70°
Muzzle velocity 450 m/sec
Shell weight 42 kg
Maximum range 11,800 m
Manufacturer Škoda, Pilsen

Remarks: First gun evolved by Škoda for new post-WW I Czech State. Entered service 1925, in production until 1933. Following German occupation all available vz. 25 pieces widely used by Wehrmacht artillery during 1939–42.

15 cm schwere Feldhaubitze 37(t)

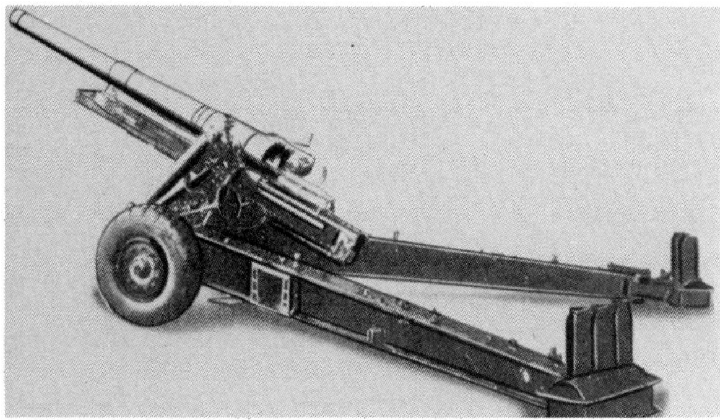

German designation 15 cm sFH 37(t)
Original designation 15 cm hruba houfnice vz. 37 (K4)
Calibre 149.1 mm
Length of piece (L/24): 3600 mm
Weight travelling 5730 kg
Weight in action 5200 kg
Traverse 45°
Elevation −5° to +70°
Muzzle velocity 580 m/sec
Shell weight 42 kg
Maximum range 15,100 m
Manufacturer Škoda, Pilsen

Remarks: Last Škoda heavy howitzer design for Czech Army before German occupation 1938–39. Kept in production for German forces during WW II; some also supplied to Slovak Army.

15 cm schwere Feldhaubitze 400(i)

German designation 15 cm sFH 400(i)
Original designation (i): Obice da 149/12 modello 14
Calibre 149.1 mm
Length of piece (L/14): 2090 mm
Length of barrel 1806 mm
Length of rifling 1644.5 mm
Weight travelling 3070 kg
Weight in action 2344 kg
Weight of piece 870 kg
Traverse 5°
Elevation −5° to +43°
Muzzle velocity 300 m/sec
Shell weight 41 kg
Maximum range 10,000 m
Rate of fire 1–2 rpm
Manufacturer Škoda, Pilsen

Remarks: Originally Austro-Hungarian Škoda 149 mm Model 14. Post-WW I used by Czech, Austrian and Hungarian armies. Italian howitzers comprised weapons captured during WW I and postwar reparations; total of 490 in service 1939. After 1943 substantial numbers impressed by various regular German artillery detachments.

---

15 cm schwere Feldhaubitze 401(i)

German designation 15 cm sFH 401(i)
Original designation (i): Obice da 149/13
Calibre 149.1 mm
Length of piece (L/14.1): 2120 mm
Length of barrel 1835.6 mm
Length of rifling 1542.1 mm
Weight travelling 3340 kg
Weight in action 2765 kg
Weight of piece 870 kg
Traverse 6°
Elevation −5° to +70°
Muzzle velocity 336 m/sec
Maximum range 8790 m
Rate of fire 1–2 rpm
Manufacturer Škoda, Pilsen

Remarks: Originally known as Škoda 149 mm Model 14/16. Numbers captured by Italian forces from Austro-Hungarian armies during WW I remained in service in modified form 1940. After September 1943 many taken over by German forces in Italy.

---

15 cm Kanone 403(j)

German designation 15 cm K 403(j)
Original designation (j): 150 mm M 28
Calibre 149.1 mm
Length of piece (L/46.5): 7025 mm
Weight travelling (3 loads): 24,000 kg
Weight in action 15,000 kg
Weight of piece 5088 kg
Traverse 360°
Elevation −4° to +45°
Muzzle velocity 760 m/sec
Shell weight 56 kg
Maximum range 23,800 m
Rate of fire 1 rpm
Manufacturer Škoda, Pilsen

Remarks: Dual-purpose siege/coastal defence gun, originally known as Škoda 149 mm Model 1928(NOa). Produced for export only and sold to Yugoslavia and Romania. In 1941 less than 20 captured by German forces in Yugoslavia and later used on Eastern Front and in coastal defence role.

15 cm Kanone 403(j) emplaced for static defence

15 cm schwere Feldhaubitze 404(i)

German designation 15 cm sFH 404(i)
Original designation Obice da 149/19
Calibre 149.1 mm
Length of piece (L/20.4): 3034 mm
Length of barrel 2897 mm
Length of rifling 2431.5 mm
Weight travelling (2 loads): 6700 kg
Weight in action 5500 kg
Weight of piece 1610 kg
Traverse 50°
Elevation +5° to +60°
Muzzle velocity 597 m/sec
Shell weight 42.55 kg
Maximum range 14,250 m
Rate of fire 2 rpm
Manufacturer Ansaldo, Pozuoli

Remarks: One of a series of modern and efficient Italian medium howitzers evolved during mid-1930s. Variants included 149/19 mod. 37, 41 and 42. After 1943 kept in production for German forces; used by several regular German artillery detachments.

14.5 cm Kanone 405(f)

German designation 14.5 cm K 405(f)
Original designation Canon de 145 L mle 1916 St. Chamond
Calibre 145 mm
Length of piece (L/50.8): 7362 mm
Length of barrel 7362 mm
Length of rifling 6112.8 mm
Weight travelling 14,060 kg
Weight in action 13,210 kg
Weight of piece 5330 kg
Traverse 6°
Elevation 0° to +38°
Muzzle velocity 784 m/sec
Shell weight 36.2 kg
Maximum range 20,200 m
Rate of fire 1 rpm
Manufacturer (gun): Fonderie de Ruelle; (carriage): St. Chamond

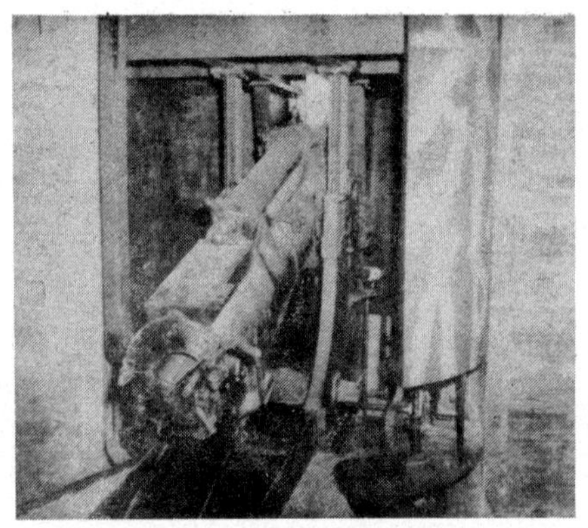

Interior view of an 14.5 cm K 405(f) in a casemate on the French coast. Note the curved lateral armour plates traversing together with the gun

Remarks: Generally known as L 16 St Ch. Evolved 1916 to provide French counterpart to German heavy guns. Essentially 145 mm naval gun barrels on land carriages. Post-WW I some sold to Romania. In 1939 total of 215 still in service, mainly emplaced for costal defence. As such, also taken over by German forces after 1940.

15 cm Kanone 408(i)

German designation 15 cm K 408(i)
Original designation Cannone da 149/40
Calibre 149.1 mm
Length of piece (L/40.5): 6036 mm
Length of barrel 5964 mm
Length of rifling 4965.6 mm
Weight travelling (2 loads): 15,673 kg; (mountain regions, 4 loads): 13,809 kg
Weight in action 11,340 kg
Weight of piece 3855 kg
Traverse 60°
Elevation 0° to +45°
Muzzle velocity 800 m/sec
Shell weight 46 kg
Maximum range 23,700 m
Rate of fire 1 rpm
Manufacturer Ansaldo, Turin

Remarks: Kept in production for German Army after Italian surrender in 1943. By April 1944 twelve had been delivered.

15.2 cm Kanone 15/16(t) or 410(i)

German designations 15.2 cm K 410(i); 15.2 cm K 15/16(t)
Original designation (i): Cannone da 152/37
Calibre 152.4 mm
Length of piece (L/39.5): 6000 mm
Length of barrel 5581 mm
Length of rifling 4502.6 mm
Weight travelling (2 loads): 16,415 kg
Weight in action 11,900 kg
Weight of piece 4870 kg
Traverse 6°
Elevation −6° to +45°
Muzzle velocity 692 m/sec
Shell weight 54 kg
Maximum range 21,840 m
Rate of fire 1 rpm
Manufacturer Škoda, Pilsen

Battery of K 15/16(t) guns in semi-concealed positions just inland from the Channel coast in France, 1944 ▲

Remarks: Designed by Škoda in 1915 to replace M.15. Post-WW I standard heavy gun of Czech and Austrian armies, but by 1938 only with reserve units. Italian guns comprised M 15/16s captured during WW I. Only small numbers involved; little German service use. Some ex-Czech guns used by German artillery in North Africa, later for coastal defence.

15 cm K 15/16(t) in the coastal defence role firing against enemy shipping ►

15.2 cm schwere Feldhaubitze 407(h), 410(b), 412(e) and 412(i)

German designations 15.2 cm sFH 407(h), 410(b), 412(e) und (i)
Original designations (h): Houwitzer 6″; (b): Obusier de 6″; (e): BL 6-inch 26 cwt Howitzer Mk I on Carriages 1P or 1R; (i): Obice da 152/13
Calibre 152.4 mm
Length of piece (L/14.6): 2223 mm
Length of barrel 2027 mm
Length of rifling 1637.6 mm
Weight travelling 4471 kg
Weight in action 4201 kg
Weight of piece 1245.5 kg
Traverse 8°
Elevation 0° to +45°
Muzzle velocity 429 m/sec
Shell weight 45.48 kg
Maximum range 10,430 m
Rate of fire 2–3 rpm

Remarks: First used in action 1916 and produced in large numbers. Later also sold abroad. In British service during late 1930s fitted with pneumatic tyres for motor traction. Most of 220 howitzers with BEF in France captured by German troops after Dunkirk 1940. Used by several regular German artillery detachments.

15.5 cm schwere Feldhaubitze 413(b), 414(f) and (i), 17(p)

German designations 15.5 cm sFH 413(b); 15.5 cm sFH 414(f) und (i); 15.5 cm sFH 17(p)
Original designations (b): Obusier de 155; (f): Canon de 155 C mle 1917 Schneider; (i): Obice da 155/14 PB; (p): 155 mm haubica wz. 1917
Calibre 155 mm
Length of piece (L/15.3): 2332 mm
Length of barrel 2176 mm
Length of rifling 1737 mm
Weight travelling 3720 kg
Weight in action 3300 kg
Weight of piece 1245 kg
Traverse 6°
Elevation 0° to +42°20′
Muzzle velocity 450 m/sec
Shell weight 43.61 kg
Maximum range 11,300 m
Rate of fire 3 rpm
Manufacturer Schneider et Cie, Le Creusot

Remarks: Entered service 1917 and soon proved a most efficient howitzer. Also sold to Tsarist Russia and adopted by US Army in France; widely exported post-WW I. Generally known as C 17 S. In 1939 total of 2043 still with French artillery regiments. After 1940 captured guns used by various German regular artillery detachments and for coastal defence.

15.5 cm s FH 414(f) on a pedestal mount for coastal defence

◀ *15.5 cm s FH 414(f)*

15.5 cm schwere Feldhaubitze 415(f)

German designation 15.5 cm sFH 415(f)
Original designation Canon de 155 C mle 15 St. Chamond
Calibre 155 mm
Length of piece (L/17.8): 2764 mm
Length of barrel 2517 mm
Length of rifling 2255 mm
Weight travelling 3860 kg
Weight in action 3040 kg
Weight of piece 1275 kg
Traverse 5°40′
Elevation −5° to + 40°
Muzzle velocity 367 m/sec
Shell weight 43.5 kg
Maximum range 10,600 m
Rate of fire 2–3 rpm
Manufacturer St. Chamond

Remarks: First used in action 1915; total of 390 built. Many still in service 1939–40 and subsequently captured by German troops. Used mainly by German coastal artillery units in France.

15.5 cm Kanone 416(f), (b) and 431(b)

15.5 cm K 416(f) on a traversing platform for coastal defence

German designations 15.5 cm K 416(f) und (b); 15.5 cm K 431(b)
Original designation (f), (b): Canon de 155 L mle 1917 Schneider
Calibre 155 mm
Length of piece (L/31.91): 4950 mm
Length of barrel 4680 mm
Length of rifling 3691 mm
Weight travelling (1 load): 9900 kg; (2 loads): 12,170 kg
Weight in action 8956 kg
Weight of piece 3800 kg
Traverse 4°30′
Elevation −5° to +40°
Muzzle velocity 665 m/sec
Shell weight 43 kg
Maximum range 17,300 m
Rate of fire 1 rpm
Manufacturer Schneider et Cie, Le Creusot

Remarks: Extemporised French WW I long-range gun combining new barrel with existing mle 1917 carriage. Total of 410 such conversions made, of which some exported to Belgium during 1920s. Modernised for motor traction still in service 1939. In German hands largely emplaced for coastal defence.

15.5 cm K 416(f) in action on the Eastern Front

15.5 cm Kanone 417(f)

15.5 cm K 417(f) in travelling position

15.5 cm K 417(f) in the Western Desert, 1942

German designation 15.5 cm K 417(f)
Original designation Canon de 155 GPF-CA (Grand Puissance Filloux-CA)
Calibre 155 mm
Length of piece (L/38.7): 5915 mm
Length of barrel 5725 mm
Length of rifling 4283 mm
Weight travelling 11,700 kg
Weight in action 10,750 kg
Weight of piece 3870 kg
Traverse 60°
Elevation 0° to +35°
Muzzle velocity 721 m/sec
Shell weight 44.85 kg
Maximum range 16,500 m
Rate of fire 1 rpm

Remarks: Identical to basic Canon 155 GPF but chambered for different ammunition. Only few in service 1939. Captured examples impressed by German coastal artillery units in France.

15.5 cm Kanone 418(f)

15.5 cm K 418(f) on an elaborate traversing platform in a coastal defence position

German designation 15.5 cm K 418(f)
Original designation Canon de 155 GPF (Grand Puissance Filloux)
Calibre 155 mm
Length of piece (L/38.2): 5915 mm
Length of barrel 5725 mm
Length of rifling 4583 mm
Weight travelling 11,700 kg
Weight in action 10,750 kg
Weight of piece 3870 kg
Traverse 60°
Elevation 0° to +35°
Muzzle velocity 735 m/sec
Shell weight 43 kg
Maximum range 19,500 m
Rate of fire 1 rpm

Remarks: Successful WW I design first used in action 1917. Also adopted by US Army 1918 and remained in US service until WW II as 155 mm Gun M1918 M1. Total of 449 in French service 1939, many of which captured by German forces. Initially impressed with regular German artillery detachments, later in coastal defence.

15.5 cm K 418(f) in an 'Atlantic Wall' casemate mounting

15.5 cm Kanone 419(f)

German designation 15.5 cm K 419(f)
Original designation Canon de 155 GPF-T (Grand Puissance Filloux-Touzard)
Calibre 155 mm
Length of piece (L/38.2): 5915 mm
Length of barrel 5725 mm
Length of rifling 4583 mm
Weight travelling 11,700 kg
Weight in action 10,800 kg
Weight of piece 3870 kg
Traverse 60°
Elevation 0° to +39°
Muzzle velocity 735 m/sec
Shell weight 43 kg
Maximum range 19,500 m
Rate of fire 1 rpm

Remarks: Modernised version of basic Canon 155 GPF differing mainly in new 6-wheeled carriage. After 1940 became highly prized German acquisition. Some used in North Africa 1941. Same French carriage later adopted for 12.8 cm K 81/1.

15.5 cm K 419(f) being moved into position in North Africa, 1942

15.5 cm K 419(f) in action in North Africa

15.5 cm Kanone 420(f)

German designation 15.5 cm K 420(f)
Original designation Canon de 155 L mle 16 St. Chamond
Calibre 155 mm
Length of piece (L/47.5): 7362 mm
Length of barrel 7362 mm
Length of rifling 5871.5 mm
Weight travelling 14,000 kg
Weight in action 13,150 kg
Weight of piece 5270 kg
Traverse 6°
Elevation 0° to +38°
Muzzle velocity 790 m/sec
Shell weight 43 kg
Maximum range 21,300 m
Rate of fire 1 rpm
Manufacturer St. Chamond

Remarks: Rebored version of 145 mm mle 1916 St. Chamond naval guns. Some sold to Italy during 1930s. In 1939 deployed mainly as fortification weapons. Captured guns impressed by German coastal defence units. Italian guns taken over by German forces after September 1943 designated 15.5 cm K 420(i); these fired heavier shell than French version.

15.5 cm K 420(f) on a traversing platform for coastal defence. France, 1944

15.5 cm Kanone 420(f)

15.5 cm Kanone 422(f)

German designation 15.5 cm K 422(f)
Original designation Canon de 155 mle 1877–1914 Schneider
Calibre 155 mm
Length of piece (L/27): 4200 mm
Length of barrel 4110 mm
Length of rifling 3171 mm
Weight travelling (1 load): 6353 kg; (2 loads): 7946 kg
Weight of piece 2437 kg
Traverse 4°40′
Elevation −5° to +42°
Muzzle velocity 561 m/sec
Shell weight 42.9 kg
Maximum range 13,900 m
Rate of fire 1–2 rpm
Manufacturer Schneider et Cie, Le Creusot

Remarks: Another extemporised French WW I conversion combining obsolete mle 1877 barrels with new recoil system and Schneider mle 1914 152 mm carriages. By 1939 relegated to fortress artillery role. In German service used mainly for coastal defence; some also by reserve artillery units in occupied France.

15.5 cm Kanone 425(f)

German designation 15.5 cm K 425(f)
Original designation Canon de 155 L mle 1918 Schneider
Calibre 155 mm
Length of piece (L/26.4): 4089 mm
Length of barrel 4089 mm
Length of rifling 3171 mm
Weight travelling 5530 kg
Weight in action 5050 kg
Weight of piece 2400 kg
Traverse 6°
Elevation +1°15′ to +43°35′
Muzzle velocity 561 m/sec
Shell weight 43.1 kg
Maximum range 13,600 m
Rate of fire 4–5 rpm
Manufacturer Schneider et Cie, Le Creusot

Remarks: An extemporised WW I design comprising old mle 1877/1914 barrels on improved carriage minus shield. Some still in service 1939–40. Captured pieces deployed by German troops for coastal defence.

15.5 cm Kanone 432(b)

German designation 15.5 cm K 432(b)
Original designation Canon de 155 L mle 1924
Calibre 155 mm
Length of piece (L/30.48): 4721 mm
Length of rifling 3280.5 mm
Weight travelling (3 loads): 19,234 kg
Weight in action 7840 kg
Traverse 4°
Elevation +5° to +26°
Muzzle velocity 665 m/sec
Shell weight 43 kg
Maximum range 17,000 m
Rate of fire 1 rpm
Manufacturer Société anonyme John Cockerill, Liège

Remarks: Largest artillery piece built in Belgium before WW II. Introduced into service during late 1920s, but only small numbers built. All remaining pieces impressed into German service in Belgium after 1940.

15.2 cm Kanonenhaubitze 433/1(r)

15.2 cm KH 433/1(r) in the coastal defence role. France, 1944

15.2 cm Kanonenhaubitze 433/1(r) at maximum elevation

German designation 15.2 cm KH 433/1(r)
Original designation *152 mm Gaubitsa-Pushka obr. 1937 g*(ML-20)
Calibre 152.4 mm
Length of piece with m/b 4925 mm
Length of piece less m/b (L/29): 4405 mm
Length of rifling 3467 mm
Weight travelling with limber 7930 kg
Weight in action 7128 kg
Weight of piece 2363 kg
Traverse 58°
Elevation −2° to +65°
Muzzle velocity (HE): 655 m/sec; (concrete-piercing): 670 m/sec
Shell weight (HE): 43.5 kg; (concrete-piercing): 40 kg
Maximum range (HE): 17,265 m
Rate of fire 1 rpm

Remarks: One of the most modern Soviet heavy guns. Evolved during mid-1930s for counter-battery fire. Design combined 152 mm Gun-Howitzer obr. 1910/34 g barrel with 122 mm Field Gun obr. 1931/37 carriage. Produced and used on a very large scale throughout WW II. Substantial numbers in German service, both with heavy artillery detachments and for coastal defence.

15.2 cm Kanone 433/2(r)

German designation 15.2 cm K 433/2(r)
Original designation *152 mm Pushka obr. 1910/34 g*
Calibre 152.4 mm
Length of piece with m/b 4922 mm
Length of piece less m/b 4404 mm
Length of rifling 3464 mm
Weight travelling 7820 kg
Weight in action 7100 kg
Weight of piece 2320 kg
Traverse 56°
Elevation −4° to +45°
Muzzle velocity 650 m/sec
Shell weight 43.5 kg
Maximum range 17,600 m
Rate of fire 2–3 rpm

Remarks: An interim Soviet measure to provide more modern heavy guns. Essentially 152 mm Howitzer obr. 1937 g barrel on 122 mm Field Gun obr. 1931 carriage. First used during Soviet-Finnish war 1939–40. Substantial numbers captured by German forces 1941–42 and impressed into service.

15.2 cm Kanone 433/2(r) in travelling position

15.2 cm Kanone 438(r)

German designation 15.2 cm K 438(r)
Original designation *152 mm Pushka obr. 1910/30 g*
Calibre 152.4 mm
Length of piece (L/32): 4855 mm
Length of barrel 4260 mm
Length of rifling 3304 mm
Weight travelling (iron tyres): 4 loads, 19,260 kg; (rubber tyres): 4 loads, 19,307 kg
Weight in action (iron tyres): 6777 kg; (rubber tyres): 6700 kg
Weight of piece 2570 kg
Traverse 4°30′
Elevation −7° to +37°
Muzzle velocity 650 m/sec
Shell weight 43.56 kg
Maximum range 16,800 m
Rate of fire 2–3 rpm
Original manufacturer Schneider et Cie, Le Creusot; 1930s modifications made at various Soviet State arsenals

Remarks: Pre-WW I French guns exported to Tsarist Russia, extensively modernised as interim equipment by Soviets during 1930s. By 1941-42 large numbers captured by German forces and impressed into service.

15.2 cm schwere Feldhaubitze 443(r)

German designation 15.2 cm sFH 443(r)
Original designation *152 mm Gaubitsa obr. 1938 g* (152-38); M-10
Calibre 152.4 mm
Length of piece (L/24.3): 3700 mm
Length of barrel 3528 mm
Length of rifling 3120 mm
Weight travelling 4550 kg
Weight in action 4100 kg
Weight of piece 1400 kg
Traverse 50°
Elevation −1° to +65°
Muzzle velocity (HE): 432 m/sec; (AP): 508 m/sec
Shell weight 51.1 kg
Shot weight 40 kg
Maximum range 12,400 m
Rate of fire 2–4 rpm
Manufacturer Various Soviet State arsenals

Remarks: Entered Red Army service late in 1938. Soon proved a very sturdy and effective artillery piece, with designed anti-tank capability. Produced and used on very large scale. All serviceable captured pieces impressed by German heavy artillery detachments. A wartime development evolved by eng.F.Petrov entered service as *Gaubitsa obr.1943 g.*

15.2 cm schwere Feldhaubitze 445(r)

German designation 15.2 cm sFH 445(r)
Original designation *152 mm Gaubitsa obr. 1909/30 g*
Calibre 152.4 mm
Length of piece (L/14.2): 2160 mm
Length of barrel 1995 mm
Length of rifling 1657 mm
Weight travelling 3050 kg
Weight in action 2725 kg
Weight of piece 1106 kg
Traverse 5°40′
Elevation 0° to +41°
Muzzle velocity 391 m/sec
Shell weight 40 kg
Maximum range 9854 m
Rate of fire 4 rpm
Original manufacturer Putilov Arsenal, St. Petersburg/Petrograd; 1930s modifications at various Soviet State arsenals

Remarks: Original pre-WW I Russian design, modernised to some extent during 1930s. By 1941 mostly with Red Army reserve artillery and training detachments. Captured guns used by German forces only in coastal defence role.

15.2 cm schwere Feldhaubitze 446(r)

German designation 15.2 cm sFH 446(r)
Original designation *152 mm Gaubitsa obr. 1910/30 g* (152-10/30)
Calibre 152.4 mm
Length of piece (L/12): 1830 mm
Length of barrel 1672.8 mm
Length of rifling 1354.8 mm
Weight travelling 3230 kg
Weight in action 2580 kg
Weight of piece 880 kg
Traverse 4°50′
Elevation −6°40′ to +39°45′
Muzzle velocity 391 m/sec
Shell weight 40 kg
Maximum range 9854 m
Rate of fire 3 rpm
Original manufacturer Schneider et Cie, Le Creusot; 1930s modifications made in various Soviet State arsenals

Remarks: Pre-WW I French howitzers exported to Tsarist Russia. During 1930s Soviet modernisation programme completely renovated, including new rubber-tyred carriages. Many still in use 1941, and numbers captured by German forces.

20.3 cm Haubitze 503/3(r), 503/4(r) and 503/5(r)

German designations 20.3 cm H 503/3(r), 503/4(r) und 503/5(r)
Original designation *203 mm Gaubitsa obr. 1931 g* (B-4 and B-4 Srs. II)
Calibre 203.2 mm
Length of piece (L/25): 5087 mm
Length of barrel 4915 mm
Length of rifling 3981 mm
Weight travelling H 503/3(r): 47,590 kg – 6 loads; H 503/4(r): 48,330 kg – 5 loads; H 503/5(r): 52,890 – 5 loads
Weight in action (all models): 17,700 kg
Weight of piece 5200 kg
Traverse 8°
Elevation 0° to +60°

Muzzle velocity 607 m/sec
Shell weight 100 kg
Maximum range 18,025 m
Rate of fire 1 round every 4 mins
Manufacturer Several Soviet State artillery arsenals

20.3 cm H 503/3(r) found abandoned intact bythe American forces in France in late autumn 1944

Remarks: An unusual but efficient Soviet heavy artillery piece evolved under direction of eng. Magdeyev. Introduced into Red Army service 1932 and produced in six slightly different versions until 1937–38. Only three later versions impressed by German artillery troops. These differed solely in method of barrel transport: 503/3 used large artillery wheels with tyres, 503/4 small wheeled limbers, and 503/5 used all-tracked suspension.

21 cm Mörser 18/19(t) or kurzer 21 cm Mörser(t)

Kurzer 21 cm Mörser(t) static equipment was transported in two sections. Illustration shows the barrel and carriage on their transporter

German designation 21 cm Mrs 18/19(t) oder kz 21 cm Mrs(t)
Original designation 21 cm mozdir vz. 18
Calibre 210 mm
Length of piece (L/16): 3360 mm
Weight travelling (4 loads): 25,050 kg
Weight in action 9460 kg
Traverse 360°
Elevation +40° to +71°30′
Muzzle velocity 380 m/sec
Shell weight 135 kg
Maximum range 10,100 m
Manufacturer Škoda, Pilsen

Remarks: Evolved for Austro-Hungarian Empire armies during WW I to provide an interim-calibre heavy static mortar (M.18). After 1918 in Czech service, where subsequently developed into mobile version (M.18/19). During WW II used by German siege artillery units.

21 cm Mörser 18/19(t) mobile equipment was transported in three loads. These illustrations show the top cradle and recoil system, barrel and the firing platform on their transporters

21 cm Haubitze 520(i)

German designation 21 cm H 520(i)
Original designation Obice da 210/22 modello 35
Calibre 210 mm
Length of piece (L/23.8): 5000 mm
Length of barrel 4673 mm
Length of rifling 4115.2 mm
Weight travelling (2 loads): 24,030 kg
Weight in action 15,885 kg
Weight of piece 4755 kg
Traverse 75°
Elevation 0° to +70°
Muzzle velocity 560 m/sec
Shell weight 101 or 133 kg
Maximum range 15,407 m
Rate of fire 1 round every 2–4 mins
Manufacturer Ansaldo, Pozzuoli

Remarks: A first-rate howitzer. Design accepted by Italian Army 1938, total of 346 ordered by 1940 but only 20 in service by late 1942. After September 1943 kept in production for German artillery troops.

22 cm Mörser 530(b) and 531(f)

22 cm Mrs 531(f) in action on the Eastern Front

22 cm Mrs 531(f) on a traversing platform in the coastal defence role

German designations 22 cm Mrs 530(b); 22 cm Mrs 531(f)
Original designations (b): Mortier de 220 TR mle 1916 Schneider; (f): Mortier de 220 mle 1916 Schneider
Calibre 220 mm
Length of piece (L/10.34): 2278 mm
Length of rifling 1579 mm
Weight travelling (1 load): 8600 kg; (2 loads): 10,810 kg
Weight of piece 2810 kg
Traverse 6°
Elevation (possible): −1°20′ to +65°; (firing): +20° to +65°
Muzzle velocity 415 m/sec
Shell weight 100.5 kg
Maximum range 10,800 m
Rate of fire 2 rpm
Manufacturer Schneider et Cie, Le Creusot

Remarks: First delivered to French heavy artillery in 1916 and still in service 1940. Small numbers impressed by German forces and used after 1941 on Eastern Front.

22 cm Kanone 532(f)

German designation 22 cm K 532(f)
Original designation Canon de 220 L mle 1917 Schneider
Calibre 220 mm
Length of piece (L/34.87): 7672.5 mm
Length of rifling 6113.5 mm
Weight travelling (1 load): 25,880 kg; (2 loads): 30,120 kg
Weight in action 25,880 kg
Weight of piece 9280 kg
Traverse 20°
Elevation −10° to +37°
Muzzle velocity 766 m/sec
Shell weight 104.75 kg
Maximum range 22,800 m
Rate of fire 1 rpm
Manufacturer Schneider et Cie, Le Creusot

22 cm K 532(f) of 'Batterie Radetzky' *near Mirus in a coastal defence emplacement*

22 cm Kanone 532(f) in a round-defence coastal emplacement

Remarks: Introduced into French service in 1917. Small number in serviceable condition captured by German forces during 1940 and deployed for coastal defence. Four such guns manned by Marine Art. Abt. 604 at Jerbourg, Guernsey, as 'Batterie Strassburg'.
'Batterie Radetzky' was 16./HKAR 1265.

---

22 cm Mörser 538(j) or Mrs(p)

22 cm Mrs(p) at maximum elevation. Norway, 1941

German designation 22 cm Mrs 538(j) oder 22 cm Mrs(p)
Original designations (j): 220 mm M 28; (p): 220 mm mozdzierz wz. 32
Calibre 220 mm
Length of piece (L/15.5): 4340 mm
Weight travelling (3 or 4 loads): 22,700 kg
Weight in action 14,700 kg
Weight of piece 4350 kg
Traverse 350°
Elevation (possible): −4° to +75°; (firing): +40° to +75°
Muzzle velocity 500 m/sec
Shell weight 128 kg
Maximum range 14,200 m
Rate of fire 1 rpm
Manufacturer Skoda, Pilsen

Remarks: Export model produced during 1920s based on experience gained during WW I. Total of 27 sold to Poland, most of which captured undamaged by German forces. Yugoslav guns probably did not number more than 12.

23.4 cm Haubitze 546/2(e) or 545/2(b)

German designation 23.4 cm H 546/2(e) oder 545/2(b)
Original designation Ordnance BL 9.2-in Howitzer Mk II
Calibre 233.6 mm
Length of piece (L/18.5): 4331 mm
Length of barrel 4042.5 mm
Length of rifling 3080 mm
Weight travelling (4 loads): 21,518 kg
Weight in action 12,662 kg
Weight of piece 4294 kg
Traverse 60°
Elevation (possible): 0° to +50°; (firing): +15° to +50°
Muzzle velocity 450 m/sec
Shell weight 131.5 kg
Maximum range 11,900 m
Rate of fire 1 rpm
Manufacturer Vickers-Armstrong Ltd., Crayford and Elswick

Remarks: Total of 27 left in France by BEF in 1940. Only small numbers used by German heavy artillery troops and all scrapped by 1943.

schwere 24 cm Kanone(t)

German designation s 24 cm K(t)
Original designation 24 cm kanon vz. 16
Calibre 240 mm
Length of piece (L/40): 9600 mm
Weight travelling (4 loads): 143,000 kg
Weight emplaced 86,000 kg
Traverse 360°
Elevation −5° to +41°30′
Muzzle velocity 794 m/sec
Shell weight 198 kg
Maximum range 29,875 m
Rate of fire 1 rpm
Barrel life 1000 rounds
Manufacturer Škoda, Pilsen

Remarks: Long-range gun first produced 1916. Design used same carriage as 30.5 cm mozdir vz. 16 (30.5 cm Mrs(t)). Small number kept in Czech service impressed by German artillery and used during Western campaign in France 1940. Thereafter relegated to coastal defence, but some also retained by heavy artillery batteries.

s 24 cm K(t) on the Eastern Front

24 cm Kanone 556(f)

German designation 24 cm K 556(f)
Original designation Canon de 240 L mle 84/17 St. Chamond
Calibre 240 mm
Length of piece (L/29): 7000 mm
Length of barrel 6455 mm
Weight travelling (2 loads): 43,500 kg
Weight in action 31,000 kg
Weight of piece 14,500 kg
Muzzle velocity 575 m/sec
Shell weight 161 kg
Maximum range 17,300 m
Rate of fire 1 round every 3 mins
Manufacturer St. Chamond

Remarks: Originally designed as a fortification and coastal gun but impressed into field service with specially-designed carriage during WW I (1917). When old barrels had worn out they were replaced by new barrels designated mle 1917. In German hands most older model guns became 24 cm K 556(f) and 1917 version – 24 cm K 556/1(f). Both versions were deployed for coastal defence.

24 cm K 556(f) barrel on its transporter

28 cm Mörser 601(f)

German designation 28 cm Mrs 601(f)
Original designation Mortier de 280 mle 14/16 Schneider
Calibre 279.4 mm
Length of piece (L/12): 3353 mm
Weight travelling (4 loads): 23,000 kg
Weight emplaced 16,000 kg
Weight of piece 4100 kg
Traverse 20°
Elevation +10° to +60°
Muzzle velocity 418 m/sec
Shell weight 205 kg
Maximum range 10,950 m
Rate of fire 1 round every 5 mins
Manufacturer Schneider et Cie, Le Creusot

28 cm Mörser 601(f) on the Eastern Front, 1943

Remarks: Originally designed for trench conditions of WW I. Transported in four loads, when emplaced required special pit under breech for high-angle firing. Small number of these mortars used by German artillery at Leningrad 1941–44.

30.5 cm Mörser (t) and 638(j)

30.5 cm Mrs(t) on its transporter captured by the American troops in Germany early in 1945

German designations 30.5 cm Mrs (t); 30.5 cm Mrs 638(j)
Original designations (t): 30.5 cm mozdir vz. 16; (j): 305 mm M 16
Calibre 305 mm
Length of piece (L/12): 3660 mm
Weight travelling (3 loads): 38,500 kg
Weight in action 23,150 kg
Weight of piece 7240 kg
Traverse 360°
Elevation (possible): −4° to +75°; (firing): +40° to +75°
Muzzle velocity 448 m/sec
Shell weight 289 kg
Rate of fire 1 round every 5 mins
Barrel life 2000 rounds
Manufacturer Škoda, Pilsen

Remarks: Originally produced in 1916 and kept in service with Czech and Yugoslav armies after 1918. Transported in two loads. Six of these heavy mortars were used during German siege of Leningrad 1941–44.

30.5 cm Mrs(t) in action outside Sevastopol, June 1942

30.5 cm Mörser 639(j)

German designation 30.5 cm Mrs 639(j)
Original designation (j): 305 mm M 11/30
Calibre 305 mm
Length of piece (L/10): 3050 mm
Weight travelling (3 loads): 32,885 kg
Weight in action 20,830 kg
Weight of piece 5320 kg
Traverse 120°
Elevation (possible): 0° to +75°: (firing): +40° to +75°
Muzzle velocity 330 or 407 m/sec
Shell weight 380 or 287 kg
Maximum range 9600 or 11,000 m
Rate of fire 1 round every 5 mins
Manufacturer Škoda, Pilsen

Remarks: Intended as static howitzer emplaced on heavy mounting bed. Captured examples used by German forces for coastal defence of Adriatic ports. Moving piece in three loads was a major undertaking.

42 cm Haubitze (t)

Loading the 42 cm H(t)

42 cm Haubitze(t) in action on the Eastern Front

German designation 42 cm H (t)
Original designation 42 cm houfnice vz. 17
Calibre 420 mm
Length of piece (L/15): 6290 mm
Length of rifling 4570 mm
Weight travelling (4 loads): 160,000 kg
Weight emplaced 105,000 kg
Traverse 360°
Elevation +40° to +71°
Muzzle velocity 435 m/sec
Shell weight 1020 kg
Maximum range 14,600 m
Rate of fire 1 round every 5 mins
Barrel life 1000 rounds
Manufacturer Škoda, Pilsen

Remarks: Very few remaining huge howitzers of this type in Czech service taken over by German forces after 1938–39. Used during siege of Sevastopol in 1942.

Railway guns

Although the principle of imparting mobility to large calibre artillery pieces by placing them on special railway mountings dated from the last part of the 19th century, it was WW I that gave the impetus to making the railway gun an important part of many European armouries. The use of railway guns enabled artillery tacticians to switch heavy artillery from one sector of a front to another with a facility that was denied to more conventional field pieces. Railway guns could be quickly concentrated and dispersed as necessary, and by rapid changes of position they could deliver long range harassing fire and remain undetected by the means then in use for long periods at a time. Another useful role for the railway gun was in coastal defence. Where a long stretch of coastline had to be defended a few railway guns could be situated at selected central points and moved to pre-prepared sites as and when the need arose. Thus by 1918 the railway gun was in use by nearly all the major combatants and not the least of these was Germany. But after 1918 all the German railway artillery was scrapped by the Treaty commissions.

After the NSDAP came to power in 1933 the German military began a major rearmament programme and on the list of weapons needed were modern railway guns. Before 1933 a great deal of theoretical work had been carried out on future railway guns but it was not until 1934 that the first practical work began on two new designs. In time these were to emerge as the *K 5(E)* and *K 12(E)*, but by the mid-1930s these designs were a very long way from service and in the meantime OKH expressed a need for some equipment to be delivered in a shorter time scale. Krupp of Essen announced that they would be able to deliver railway guns based on mounting designs used during WW I, the drawings for which they still had in their files at Essen, and utilising a number of obsolete naval gun barrels they happened to have still in their stocks. As the time scale for producing these guns would be considerably shorter than that envisaged for the new designs, orders were placed for railway guns with calibres between 15 and 28 cm, the contract being known as the *'Sofort-Programm'*. It began in 1936 and all the equipments were delivered by 1939.

At the lower end of the calibre scale were the *15 cm K(E)* and *17 cm K(E)*, both of which utilised the same design of mounting with a fully traversing turntable and stabilising outriggers. The guns used were old L/40 naval pieces, which were really too light to justify the use of a railway mounting, but four 15 cm and six 17 cm equipments were made during 1937 and 1938. The 15 cm guns were used by Eisenbahnbatterie 655, and the 17 cm guns by Eisenbahnbatterien 717 and 718.

The next largest calibre in the *'Sofort-Programm'* was 24 cm and here there were two types that used the same mounting. One was the *Theodor Bruno K(E)* which was based on the *24 cm SK L/35* originally built pre-WW I for the 'Wittelsbach' class battleships. Six of these combinations were produced between 1937 and 1939. The other type was the *Theodor K(E)* (originally the *Theodor Karl*), which used the *24 cm SK L/40* from the 1904 'Deutschland' class battleships, but only three of these were produced during 1936 and 1937. Most of these 24 cm guns were shared among Eisenbahnbatterien 664, 674 and 722.

Next in size were the three different types of 28 cm 'Bruno' guns, all of which used the same design of mounting representing a scaled-up version of that used by the 24 cm guns. The most numerous version was the *kurze Bruno K(E)* of which eight were made during 1937 and 1938. Another elderly naval gun was used, this time the *24 cm SK L/40*, and the finished equipments were shared by Eisenbahnbatterien 690, 694, 695, 696 and 721.

Only three examples of the *lange Bruno K(E)* with the *28 cm SK L/45* were made between 1937 and 1938. They were used by Eisenbahnbatterie 688. The only other component of the Bruno series from the *'Sofort-Programm'* was the *schwere Bruno K(E)* based on the *28 cm Küst K L/42*. This elderly coastal defence gun had long been out of service but Krupp still had two examples available and by the end of 1938 both were fitted to Bruno mountings and subsequently issued to Eisenbahnbatterie 689.

One result of the *'Sofort-Programm'* was that Krupp became the exclusive designers and builders of railway gun equipment for the German Army. Hanomag of Hannover later assisted in production but, for once, Rheinmetall took no part in this programme. They did produce paper designs for 15 cm and 24 cm railway guns, but they were not built.

Another result of the *'Sofort-Programm'* was an OKH request to improve the range of the 28 cm Bruno guns to match that of the *K 5(E)*. Not surprisingly in view of the age of the barrels used on the Bruno series, Krupp announced that an entirely new gun design would be needed with some modifications to the Bruno mountings in addition. Krupp were told to go ahead with the design work and the result was three new railway guns known as the *Bruno neue K(E)*. One was delivered each year between 1940 and 1942. More production was planned but the *Bruno neue K(E)* was not a great success due to unexplained internal ballistic difficulties, and the hoped-for range was not achieved. Further production was stopped to concentrate on the *K 5(E)*.

Another off-shoot of the *'Sofort-Programm'* was the *20 cm K(E)*. These guns were completed after the end of the programme proper at a time when Krupp were making *20.3 cm SK C/34* naval guns for the 'Admiral Hipper' class of heavy cruisers. Eight of these guns were declared to be surplus to OKM requirements and acquired by the Army. Krupp altered an old WW I railway mounting design to take the barrels and started production during 1939. Almost as soon as the production programme began the HWA realised that the eight guns would be the only ones in Army use with a calibre of 20.3 cm which would greatly complicate the ammunition supply situation. They asked for the barrels to be bored out to 21 cm but by the time the request was made the eight barrels were completed and it would have been uneconomic to build new barrels, so the *20 cm K(E)* was taken into Army use with the last deliveries being made in 1941. Eventually the supply situation for these equipments was simplified by incorporating all of them into the Atlantic coast defences in France. As coastal guns all the necessary ammunition could be concentrated in a few locations. The guns were installed at Brest, Cherbourg and Paimpol. The location of the Paimpol guns was Plounez, where there were two equipped with turntables. Another two were at Cap de la Hague.

As a result of the *'Sofort-Programm'* and its off-shoots the German Army had on hand a quantity of railway guns to meet their immediate needs but they eagerly awaited the new and more modern weapons. The smallest of these was the *21 cm K 12(E)*, but it was smallest in calibre only, as it turned out to be a massive and complex weapon. It was designed as a result of the achievements of the so-called 'Paris Gun' during WW I. That long-range gun had been designed and built for the German Navy and it was manned by naval personnel, much to the chagrin of the German Army. They resolved to do better themselves but had no chance until the *K 12 (E)* project began. Well before the first practical steps were taken a great deal of complex theoretical work had been carried out during the 1920s and early 1930s. One offshoot of this theoretical work was the *Vögele* portable turntable which enabled railway guns to fire over a much larger arc of traverse than was otherwise possible without the use of normal turntables. Other work had been carried out in the area of the very high muzzle velocities needed to reach the stratosphere through which long range shells would have to travel to achieve the ranges needed for a new 'Paris Gun'. One problem that grew to great proportions with the original 'Paris Gun' was that the barrel became worn very quickly to the extent that each shell fired was serially numbered to ensure that barrel wear was compensated for by loading with correspondingly larger calibre shells. Such complexities were thought to be avoidable, if possi-

21 cm Kanone 12(E) in action

ble, and the theoretical research seemed to indicate that shells with pre-machined splines might prevent much of the wear that would otherwise occur. Only eight rifling grooves with corresponding splines were thought necessary but to prove the principle a sub-calibre barrel of 10.5 cm was made. This was the *K 12 M*, but to ensure that any possibility of the normal copper driving bands and rifling being able to stand the stresses was investigated fully, a further sub-calibre barrel, the *K 12 M.Ku*, was built and tried alongside the *K 12 M*. From these trials the splined shell concept emerged the clear leader as the normal rifling/driving band combination soon proved unable to stand up to the stresses involved.

Work then went ahead with the construction of the new weapon and mounting. The gun was a very lengthy affair and had to be externally braced to prevent it sagging under its own weight. The length in calibres of the barrel was L/158 which also led to problems with the placing of the trunnions. On the first version these were placed so far forward to partly counteract the muzzle preponderance that the carriage had to be raised at high angles of elevation to prevent the breech striking the ground when fired. As a result the rate of fire was considerably reduced. Other complexities included the provision to disconnect the barrel from the recoil mechanism to draw it back for travelling. The first full size barrel was proof fired during 1937 (the sub-calibre barrels were fired during 1935) and the first full equipment was fired during 1938. The designation given to this equipment was *K 12 V*, and as a result of the experiences with the gun and mounting after its entry into service during 1939 a redesign of the carriage was started to try to locate the barrel so that more of the gun was situated in front of the trunnions; the extra weight involved was compensated for by stronger balancing hydraulic presses. The end result was the *K 12 N*, but only one example of each type was made. Overall, the complexity and size of the gun and mounting were too much for the German Army, even though it did have a remarkable range. According to various sources it was as much as 120 kilometres, but the most recorded in action was about 88 kilometres – against Rainham in Kent – and the gun appears to have been firing from the Pas de Calais. The battery involved was probably Eisenbahnbatterie 701, a unit that seems to have spent all its time based in Western Europe. One example of this gun was captured in 1945 at Selazette in Holland and it was closely scrutinised by the Allies. At the end of their exhaustive investigations they announced that the design was one of the most remarkable to emerge from WW II, but to that they added the rider that it was also the most useless. The effort invested in its design was largely wasted for it had little military value, and indeed, one high-ranking officer* regarded it as 'little more than a toy'. A toy it might have been, but as an exercise in practical ballistics it was a major achievement. One item of note was the cost of each equipment – RM1,500,000 for the gun and mounting alone.

* General der Artillerie Karl Thoholte.

While the *K 12(E)* project ended after only two examples had been made, the *28 cm K 5(E)* story was much different. As with the *K 12(E)*, two sub-calibre barrels were built for tests with splined shells but for the *K 5(E)* the calibre was 15 cm. The two barrels were the *K 5 M* and *K 5 M.Ku*. As a result of these sub-calibre trials the first full-size barrel was proof-fired during 1936 and the next year saw the first firing of a complete equipment. By February 1940 eight were in service but in that year there began a strange spate of barrels which split when fired. Trials failed to reveal the exact cause but it was thought that reducing the depth of the rifling grooves might help and indeed it did provide a cure. The barrels with the shallower grooves were designated *K 5 Tiefzug 7 mm*, as opposed to the early barrels which became the *K 5 Tiefzug 10 mm*. Another change took place in 1943 when the considerable experience gained using sintered-iron driving bands seemed to offer the possibility of reaching high muzzle velocities with normally-rifled barrels. As a result of trials some barrels and ammunition were produced in this combination and the barrels were designated *K 5 Vz (Vielzug)*, but the numbers involved were small.

Overall, the design of the *K 5(E)* was excellent and it soon earned its title of the best railway gun ever produced. In service it proved to be trouble-free once the 1940 split-barrel episode was closed, and it remained in production at Essen and Hannover until 1945. By then about 25 had been made, and it had gained the nickname of *'schlanke Bertha'*. The range was of the order of 58 kilometres but this could be extended by the use of the *28 cm R Gr 4331*, a rocket-assisted projectile which could increase the range to about 86 kilometres. With this projectile accuracy was poor owing to the difficulties of timing the rocket motor ignition immediately the projectile had reached the top of its trajectory.

The success of the *K 5(E)* did not mean that it was immune to the normal run of trials and experimentation to which all German weapons were seemingly subjected. One *K 5(E)* was used in a series of trials intended to develop a screw breech mechanism to use bagged charges. That came to nothing, as did another series of tests to develop a muzzle brake for the piece. A great deal of experimental work was carried out into the problem of carrying a *K 5(E)* across country when damaged railway lines were encountered – by 1943 the Allied air forces were already causing considerable disruption to the German railway system. A specification was issued for a new equipment, the *K 5 ERF (Eisenbahn Runden Feld)*, which would normally be transported by rail, but between damaged tracks it was to be carried across country in several loads by converted Tiger II tank chassis. To this end the barrel was to be built in two parts. As an extra the mounting was to be able to accommodate a 38 cm howitzer, type unspecified. Both guns were to be lowered on to ground mountings for firing. From 1943 onwards this project attracted a great deal of detailed design work and development but by 1945 nothing tangible had been produced, which was probably just as well. The idea of a large gun being broken down into its component parts and reassembled under the constant threat of air attack, to say nothing of large tanks and their loads lumbering across country under a similar threat makes one wonder exactly what the German designers were thinking of and why they diverted so much attention to the scheme.

At least one other *K 5(E)* was used in research on sub-calibre fin-stabilised projectiles. The barrel in this case was smooth-bored and had a calibre of 31 cm. It was known as the *K 5 Glatt*, and the projectiles used were versions of the *PPF* or *Peenemünder Pfeil Geschoss*. By 1945 the results had reached the stage where troop trials were about to begin; by that time the projectiles were reaching ranges of the order of 150 kilometres. The trials were carried out on the Rügenwalde and Hillersleben ranges. According to one source two guns were actually used in action. One of these guns was made by Krupp and the other by Hanomag, and they were supposed to be the first of a series of seven. The two guns were based at Aarweiler near Bonn, and reportedly shelled Maastricht and Verviers. The projectile used was the *31 cm SpGr 4861*, with a body calibre of 12 cm, and a weight of 136 kilograms.

By the end of 1938 the Krupp drawing office had finished work on the *K 5(E)* and *K 12(E)* and were able to start new projects. One of these was the adaptation of the naval *38 cm SK C/34* to a railway mounting. These guns had already been redesigned for use in the coastal defence role with longer chambers, and it was this version that Krupp used to produce the *38 cm Siegfried K(E)*. In 1939 eight were ordered, but it was not until 1943 that the first was ready, and by 1945 only three had been completed.

At the same time as the 38 cm project began Krupp also started work on a 40.6 cm railway gun using the *40.6 cm SK C/34*. Again there was a coastal defence version with a longer chamber

than the naval gun, and it was this version that was used for the railway mounting. It soon became apparent that designing a carriage for this gun would be no easy matter as its sheer size would make it difficult to fit normal railway dimensions. As a result the 40.6 cm project was postponed and it is not known if any examples were actually completed. Despite reports from several sources that at least one example was made, no pictorial evidence has come to hand and consequently doubts remain.

The largest of the German railway guns deserves more than the usual coverage for it was the *80 cm K(E)* – the largest gun ever built. Over the years this monster has become almost a legend in many ways and, indeed, its sheer scale even now staggers the imagination. The story of the *80 cm K(E)* began in about 1934–35 at a time when the German military planners were considering the possibility of a head-on assault on the Maginot Line defences. They asked if it would be possible to design and built super-heavy calibre guns to smash the large forts involved. Krupp produced paper gun designs with calibres of 70, 80, 84 and 100 cm. These designs were duly filed and forgotten until Hitler visited the Krupp works at Essen in 1936. The head of the firm, Gustav Krupp von Bohlen und Halbach, showed the plans for the super-heavy project to his visitor and from the response shown deduced that the idea had captured Hitler's imagination. With no official sanction Krupp ordered the detail design work on an 80 cm railway gun and in 1937 these plans were submitted to the HWA. During 1939 they ordered three equipments, two for delivery during 1940 or early 1941 and a third during 1944; extra barrels were to be delivered with each gun.

Drawing up paper plans was one thing, but actually building a gun over twice the size of anything ever built before was another. Well over RM10,000,000 were invested in new plant and in diversion of existing equipment in the tooling stage, and a new 15,000 ton steel press was built and installed to forge massive steel ingots into shape for the carriage and barrel components. It was not long before even the Krupp complex was being stretched to its limits to produce the first parts, and as a result a great deal of other war work had to be diverted elsewhere. The 1940 delivery date slipped by and it was not until late in that year that the first barrel was ready for proof firing on the Hillersleben range. Even there a special gantry and concrete butts had to be erected at great expense and diversion of effort, but the proof firings were successful.

By the summer of 1941 the first full equipment was ready for its acceptance trials on the Rügenwalde ranges. During these trials Hitler paid another visit and was delighted to accept the gun as a gift from Alfried Krupp, who thus revived an old Krupp tradition of presenting the first example of any weapon off the Krupp production lines to the Head of State. No doubt this 'gift' was partly paid for when the bill for the second gun was later presented. It came to no less than RM7,000,000 for the gun and carriage alone – all the special rail cars, cranes, ammunition, etc, were extras.

It was not long before the gun acquired a name in keeping with German railway gun traditions. The first *80 cm K(E)* thus became the *'schwere Gustav'* and was often known simply as *'Gustav'*. The *'Gustav'* so honoured was Gustav Krupp who had initiated the gun design, but by 1941 he was a semi-invalid and no longer head of the firm. Eventually, the second gun was named *'Dora'* after the wife of Erich Müller, head of the gun design team.

By early 1942 the gun crews were trained to the state where they were ready for action, but by then the French forts had long been by-passed, as had the Soviet frontier defences. The next most likely operation in which the *80 cm K(E)* was to be involved would have been 'Operation Felix' – the attack on Gibraltar – but Franco frustrated all Hitler's attempts to get Spain involved in the war and *'Gustav'* was deployed eastwards towards Sevastopol.

Before discussing the *80 cm K(E)* in action it would be of interest to look at the design of the piece and its mounting. Apart from its sheer size there was little to note regarding the gun itself, which was L/40.6 calibres long and used a horizontal sliding block breech mechanism. The mounting was unusual in that in action it used two pairs of railway lines, side by side, but on the move the mounting was split into two halves so that they could be towed along a single pair of lines. To serve the gun no fewer than 1420 men were needed, but most of these were used in the gun assembly stage and thereafter manned the two light Flak detachments deployed to defend the gun in position. The direct gun crew in action numbered about 500, most of whom were in the ammunition team. On the move the gun was broken down into 25 separate loads and in addition there were crew carriages, kitchens, workshops, ammunition trucks and so on.

The actual site had to be carefully surveyed and prepared well in advance. Once on site it took about three to six weeks to get the gun assembled and ready for action. The hard work continued in action for the rate of fire was about one round every thirty minutes. There were two main types of projectile. The HE shell weighed 4800 kg and a concrete-piercing shell weighed 7100 kg (ie 4.72 and 7 tons respectively). Maximum range of the HE shell was about 47 km.

28 cm schwere Bruno Kanone (E) in transit

28 cm Bruno neue Kanone (Eisenbahn)

28 cm BrNK(E) in action

28 cm Bruno Neue Kanone(E)

German designation Br NK (E)
Calibre 283 mm
Length of piece (L/58): 16400 mm
Length of barrel 15247 mm
Length of rifling 12401 mm
Weight emplaced 150,000 kg
Weight of piece 55,260 kg
Traverse (turntable): 360°; (carriage): 1°
Elevation 0° to +50°

Muzzle velocity 995 m/sec
Shell weight 255 kg
Maximum range 46,600 m
Rate of fire 1 round every 3 mins
Barrel life 500 rounds
Length of carriage 24880 mm
Manufacturer Friedr. Krupp AG, Essen

Remarks: Three delivered 1940–42.

28 cm BrNK(E) on a turntable

28 cm Kanone 5 (Eisenbahn)

28 cm K 5(E) on a traversing turntable

28 cm Kanone 5(E) in travelling position

German designation K 5 (E)
Calibre 283 mm
Length of piece (L/76): 21538 mm
Length of barrel 20548 mm
Length of rifling 17374 mm
Weight travelling 210,000 kg
Weight emplaced 218,000 kg
Weight of piece 80,545 kg
Traverse (turntable): 360° (carriage): 18′
Elevation 0° to +50°

Muzzle velocity 1120 m/sec
Shell weight 255 kg
Maximum range 62,400 m
Rate of fire 1 round every 3–5 mins
Barrel life 240 rounds
Length of carriage 21234 mm; (plus overhang): 21934 mm
Manufacturers Friedr. Krupp AG, Essen; Hanomag, Hannover

Remarks: Development commenced during 1934 and first example completed 1937. Total of 8 in service by February 1940. During 1942–43 two K 5 (E) guns deployed with Eisb. Art. Rgt. zbV 679 outside Leningrad. Reportedly some 25 K 5 (E) guns completed by end of WW II.

28 cm kurze Bruno Kanone (Eisenbahn)

German designation kzBrK (E)
Calibre 283 mm
Length of piece (L/40): 11200 mm
Weight emplaced 130,000 kg
Weight of piece 45,300 kg
Traverse (turntable): 360°; (carriage): 18′
Elevation 0° to +45°
Muzzle velocity 820 m/sec
Shell weight 240 kg
Maximum range 29,500 m
Rate of fire 1 round every 5–6 mins
Barrel life 850 rounds
Length of carriage 22800 mm
Manufacturers Friedr. Krupp AG, Essen; Hanomag, Hannover

Remarks: Eight produced, 1937–38.

28 cm kzBK(E) in action

28 cm kurze Bruno Kanone(E) on the turntable

28 cm kzBK(E) showing the turntable recoil mechanism

28 cm lange Bruno Kanone (Eisenbahn)

German designation lgBr K (E)
Calibre 283 mm
Length of piece (L/45): 12735 mm
Length of rifling 9698 mm
Weight emplaced 123,000 kg
Weight of piece 39,800 kg
Traverse (turntable): 360°; (carriage): 18′
Elevation 0° to +40°
Muzzle velocity 865 m/sec
Shell weight 302 kg
Maximum range 28,500 m
Rate of fire 1 round every 5 mins
Barrel life 400 rounds
Length of carriage 22800 mm
Manufacturers Friedr. Krupp AG, Essen; Hanomag, Hannover

Remarks: Three delivered 1936–37.

28 cm schwere Bruno Kanone (Eisenbahn)

German designation sBr K (E)
Calibre 283 mm
Length of piece (L/42): 11930 mm
Length of barrel 11084 mm
Length of rifling 8892 mm
Weight emplaced 118,000 kg
Weight of piece 40,850 kg
Traverse (turntable): 360°; (carriage): 18′
Elevation 0° to +45°
Muzzle velocity 745 m/sec
Shell weight 302 kg
Maximum range 29,400 m
Rate of fire 1 round every 5–6 mins
Barrel life 400 rounds
Length of carriage 22800 mm
Manufacturer Friedr. Krupp AG, Essen

Remarks: Comprised pre-WW I 28 cm Küst.K barrels left in Krupp stockpile after 1918, and specially-designed railway mountings. Only two complete equipments produced 1936–38 and used during WW II.

28 cm schwere Bruno Kanone(E) being elevated

24 cm Theodor Kanone (Eisenbahn)

German designation Th K (E)
Calibre 238 mm
Length of piece (L/40): 9550 mm
Length of barrel 8900 mm
Length of rifling 7820 mm
Weight emplaced 94,000 kg
Traverse (turntable): 360°; (carriage): 40′
Elevation (possible): 0° to +45°; (in action): +10° to +45°
Muzzle velocity 810 m/sec
Shell weight 148.5 kg
Maximum range 26,750 m
Rate of fire 1 round every 3 mins
Barrel life 900 rounds
Length of carriage 18450 mm
Manufacturer Friedr. Krupp AG, Essen

Remarks: Three delivered 1937–38.

24 cm Theodor Kanone(E)

◄ *24 cm ThK(E) on a portable turntable for the coastal defence role. The ramp to enable the gun and carriage to move onto the turntable is in the right background*

24 cm Theodor Bruno Kanone (Eisenbahn)

Battery of 24 cm ThBK(E) in firing position

24 cm Theodor Bruno Kanone(E)

German designation Th.Br.K (E)
Calibre 238 mm
Length of piece (L/35): 8400 mm
Length of barrel 7800 mm
Length of rifling 6300 mm
Weight emplaced 95,000 kg
Weight of piece 24,000 kg
Traverse (turntable): 360°; (carriage): 18′
Elevation 0° to +45°
Muzzle velocity 670 m/sec
Shell weight 151 kg
Maximum range 20,200 m
Rate of fire 1 round every 3 mins
Barrel life 1300 rounds
Length of carriage 20700 mm
Manufacturer Friedr. Krupp AG, Essen

Remarks: Six produced, with deliveries commencing in 1937.

24 cm ThBK(E) rear view showing the ammunition hoist

20.3 cm Kanone (Eisenbahn)

20.3 cm K(E) being elevated for action

20.3 cm K(E). A traverse of 360° was obtained by means of a portable turntable composed of a circular track set over the main track. The gun and carriage were rolled on to the turntable by using a ramp. A power traversing unit was attached to the turntable

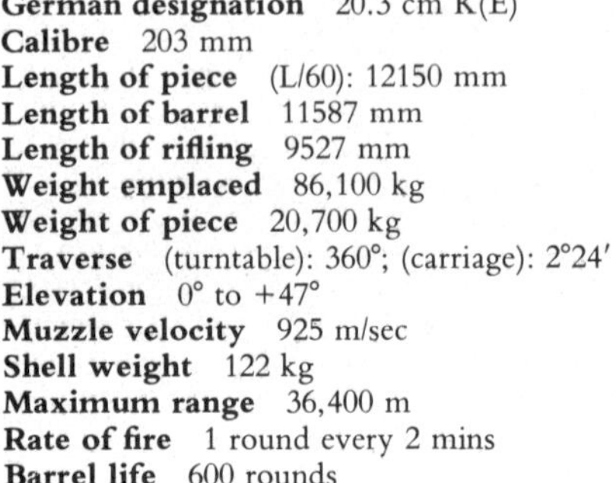

German designation 20.3 cm K(E)
Calibre 203 mm
Length of piece (L/60): 12150 mm
Length of barrel 11587 mm
Length of rifling 9527 mm
Weight emplaced 86,100 kg
Weight of piece 20,700 kg
Traverse (turntable): 360°; (carriage): 2°24′
Elevation 0° to +47°
Muzzle velocity 925 m/sec
Shell weight 122 kg
Maximum range 36,400 m
Rate of fire 1 round every 2 mins
Barrel life 600 rounds
Length of carriage 19445 mm
Manufacturer Friedr. Krupp AG, Essen

Remarks: Comprised 20.3 cm SKC/34 guns originally intended for 'Admiral Hipper' class heavy cruisers on specially-designed railway mountings. First in service 1940; total of 8 delivered. Most used as coastal defence weapons in Cherbourg-Brest region until overrun by US troops in July 1944. At that time replacement 21 cm barrels were under construction at Krupp works. Some were captured by the Allies near Cherbourg, July 1944.

---

21 cm Kanone 12 (Eisenbahn)

German designation K 12 (E)
Calibre 211 mm
Length of piece (L/158): 33300 mm
Length of barrel 32112 mm
Length of rifling 27724 mm
Weight travelling 317,000 kg
Weight emplaced 309,000 kg
Weight of piece 99,700 kg
Traverse 360°
Traverse on carriage 14′
Elevation 0° to +55°
Muzzle velocity (maximum): 1625 m/sec; (normal): 1500 m/sec
Maximum range (approx): 115,000 m
Rate of fire 1 round every 5 mins
Barrel life 90 rounds
Length of carriage 41300 mm; (plus overhang): 47860 mm
Manufacturer Friedr. Krupp AG, Essen

Remarks: Introduced into service during 1939. Used by only one battery, Eisb. Battr. 701; this unit had only one gun in service at any one time.

21 cm Kanone 12V(E) in the final assembly stages

21 cm K 12N(E) shortly after completion

K 12N(E) at maximum elevation

15 cm Kanone (Eisenbahn)

15 cm Kanone (E) showing the hinged outriggers on both sides of the carriage to support the gun in recoil

Battery of 15 cm and 17 cm Kanonen *(E) during a practice shoot*

German designation 15 cm K(E)
Calibre (guns 5, 7, 8): 149.1 mm; (gun 23): 149.3 mm
Length of piece (L/40): 5960 mm
Length of barrel 5571 mm
Length of rifling 4470 mm
Weight emplaced 74,000 kg
Weight of piece 5800 kg
Traverse 360°
Elevation 0° to +45°
Muzzle velocity 805 m/sec
Shell weight 43 or 52.5 kg
Maximum range 22,500 m
Rate of fire 3 rpm
Barrel life 1100 rounds
Length of carriage 20100 mm
Manufacturer Friedr. Krupp AG, Essen

Remarks: Introduced 1937. Used by Eisb. Bttr. 655 and 'Batterie Gneisenau'.

A unit of the 'Batterie Gneisenau' in action. This battery consisted of four railway guns – 15 cm SK L/45 in MPLC 13 auf E-Wagen

---

17 cm Kanone (Eisenbahn)

17 cm Kanone(E) in travelling position

17 cm K(E) in the coastal defence role. Note the concrete bunker for ammunition storage

German designation 17 cm K(E)
Calibre 172.6 mm
Length of piece (L/40): 6900 mm
Length of rifling 4991.5 mm
Weight emplaced 80,000 kg
Traverse 360°
Elevation 0° to +45°
Muzzle velocity 860 m/sec
Shell weight 62.8 kg

Maximum range 26,100 m
Rate of fire 1 rpm
Barrel life 1100 rounds
Length of carriage 20100 mm
Manufacturer Friedr. Krupp AG, Essen

Remarks: Six equipments built between 1937–38. Used by Eisb. Bttr. 717 and 718.

17 cm K(E) with outriggers placed in position

A unit of the 'Batterie Gneisenau' in action. This battery consisted of four railway guns – 15 cm SK L/45 in MPLC 13 auf E-Wagen

Once the *'Gustav'* had been emplaced at Bakhchisaray, about 16 km to the north of Sevastopol, it was ready to take part in the bombardment with the rest of the German siege train. On 5 June 1942 it fired its first round in anger and by the time it had finished some 48 shells had been fired at a variety of targets all of which were completely demolished. Perhaps the most spectacular success was the destruction of an underground ammunition magazine 30 metres under Severnaya Bay. Eight shots were fired at this target, which was completely destroyed and a small sailing ship sunk in the process. By the end of the month Sevastopol was in German hands and *'Gustav'* withdrew westwards for a barrel change, for together with training and proofing shells its barrel had fired about 300 rounds. It was returned to Essen to be fitted with a new liner.

By August 1942 *'Dora'* was ready for action and during that month assembly work began about 16 km from Stalingrad. Exactly what part *'Dora'* took in that battle has not been found recorded, but it was hurriedly dismantled and withdrawn. Both guns reappeared at Rügenwalde during early 1943 and fired a few practice rounds, but thereafter their fate is a mystery. Indications seem to show that *'Gustav'* was sent to the Leningrad Front but that siege was lifted before it could arrive, and there are no records until 1945 when the parts of *'Gustav'* were found at Auerbach in Bavaria by the American 3rd Army. *'Dora'* was found dismantled in many parts, most of them partially destroyed, over a wide area ranging from Oberlichtenau to Leipzig railway yards. Parts of the incomplete third gun were found at Essen and the Meppen ranges, and all three were eventually scrapped.

There were several projects involving the use of the *80 cm K(E)*, one of the most likely proposing the substitution of the 80 cm barrel with a 52 cm barrel. This barrel would have fired not only 52 cm projectiles but also sabot shells and rocket-assisted projectiles, the last of which would have a calculated range of about 190 km. *PPF* shells were under consideration for firing from the 80 cm barrel, but none were ever made. Perhaps the most unlikely project was one to mount an 80 cm barrel on a self-propelled tracked carriage for use in street fighting. It would seem that some design work was actually carried out by Krupp and the vehicle/gun combination would have weighed well over 1500 tons. The project was not cancelled until October 1944.

Of all the various odd German schemes for spectacular weapons none can have been more useless to their cause than the *80 cm K(E)*. Only the peculiar political scene imposed by the personality of Hitler could have produced the situation where such massive guns could even be considered, let alone actually built, and the huge diversion of men, money, manufacturing facilities, talent and time was such that the *80 cm K(E)* project can almost be said to have been a factor in the eventual Allied victory. For all its imposing size and scale the *80 cm K(E)* was a waste of valuable German resources and it was produced at a time when the railway gun was reaching both the peak of its design perfection and its nemesis. As early as 1939 it was obvious to all military thinkers that the aeroplane had taken the place of the railway gun. For all its fire power and mobility the railway gun soon proved vulnerable to air attack, and even the prodigious re-laying of damaged tracks, often at very high speed, could not hide the fact that railway guns were obsolete as early as 1940, if not before.

The only nation to add to the number of railway guns used by the German Army was France. Prior to 1939 the French railway guns were regarded as some of the best in the world, and after 1920 some were even built for export to such nations as Japan. Most of the French designs dated back to WW I, but by 1940 many of them were obsolete. During WW I many elderly gun barrels had been placed on railway mountings, so in 1940 their ages were in some cases considerable. Many of these French railway guns fell into German hands undamaged, but some of the older pieces were scrapped. Most of the rest were simply taken over with no changes and used alongside German railway guns as 'standard' equipment. Many were eventually incorporated into the Atlantic coast defences. These French railway guns ranged in calibre from 24 cm up to a 52 cm railway howitzer.

than the naval gun, and it was this version that was used for the railway mounting. It soon became apparent that designing a carriage for this gun would be no easy matter as its sheer size would make it difficult to fit normal railway dimensions. As a result the 40.6 cm project was postponed and it is not known if any examples were actually completed. Despite reports from several sources that at least one example was made, no pictorial evidence has come to hand and consequently doubts remain.

The largest of the German railway guns deserves more than the usual coverage for it was the *80 cm K(E)* – the largest gun ever built. Over the years this monster has become almost a legend in many ways and, indeed, its sheer scale even now staggers the imagination. The story of the *80 cm K(E)* began in about 1934–35 at a time when the German military planners were considering the possibility of a head-on assault on the Maginot Line defences. They asked if it would be possible to design and built super-heavy calibre guns to smash the large forts involved. Krupp produced paper gun designs with calibres of 70, 80, 84 and 100 cm. These designs were duly filed and forgotten until Hitler visited the Krupp works at Essen in 1936. The head of the firm, Gustav Krupp von Bohlen und Halbach, showed the plans for the super-heavy project to his visitor and from the response shown deduced that the idea had captured Hitler's imagination. With no official sanction Krupp ordered the detail design work on an 80 cm railway gun and in 1937 these plans were submitted to the HWA. During 1939 they ordered three equipments, two for delivery during 1940 or early 1941 and a third during 1944; extra barrels were to be delivered with each gun.

Drawing up paper plans was one thing, but actually building a gun over twice the size of anything ever built before was another. Well over RM10,000,000 were invested in new plant and in diversion of existing equipment in the tooling stage, and a new 15,000 ton steel press was built and installed to forge massive steel ingots into shape for the carriage and barrel components. It was not long before even the Krupp complex was being stretched to its limits to produce the first parts, and as a result a great deal of other war work had to be diverted elsewhere. The 1940 delivery date slipped by and it was not until late in that year that the first barrel was ready for proof firing on the Hillersleben range. Even there a special gantry and concrete butts had to be erected at great expense and diversion of effort, but the proof firings were successful.

By the summer of 1941 the first full equipment was ready for its acceptance trials on the Rügenwalde ranges. During these trials Hitler paid another visit and was delighted to accept the gun as a gift from Alfried Krupp, who thus revived an old Krupp tradition of presenting the first example of any weapon off the Krupp production lines to the Head of State. No doubt this 'gift' was partly paid for when the bill for the second gun was later presented. It came to no less than RM7,000,000 for the gun and carriage alone – all the special rail cars, cranes, ammunition, etc, were extras.

It was not long before the gun acquired a name in keeping with German railway gun traditions. The first *80 cm K(E)* thus became the *'schwere Gustav'* and was often known simply as *'Gustav'*. The *'Gustav'* so honoured was Gustav Krupp who had initiated the gun design, but by 1941 he was a semi-invalid and no longer head of the firm. Eventually, the second gun was named *'Dora'* after the wife of Erich Müller, head of the gun design team.

By early 1942 the gun crews were trained to the state where they were ready for action, but by then the French forts had long been by-passed, as had the Soviet frontier defences. The next most likely operation in which the *80 cm K(E)* was to be involved would have been 'Operation Felix' – the attack on Gibraltar – but Franco frustrated all Hitler's attempts to get Spain involved in the war and *'Gustav'* was deployed eastwards towards Sevastopol.

Before discussing the *80 cm K(E)* in action it would be of interest to look at the design of the piece and its mounting. Apart from its sheer size there was little to note regarding the gun itself, which was L/40.6 calibres long and used a horizontal sliding block breech mechanism. The mounting was unusual in that in action it used two pairs of railway lines, side by side, but on the move the mounting was split into two halves so that they could be towed along a single pair of lines. To serve the gun no fewer than 1420 men were needed, but most of these were used in the gun assembly stage and thereafter manned the two light Flak detachments deployed to defend the gun in position. The direct gun crew in action numbered about 500, most of whom were in the ammunition team. On the move the gun was broken down into 25 separate loads and in addition there were crew carriages, kitchens, workshops, ammunition trucks and so on.

The actual site had to be carefully surveyed and prepared well in advance. Once on site it took about three to six weeks to get the gun assembled and ready for action. The hard work continued in action for the rate of fire was about one round every thirty minutes. There were two main types of projectile. The HE shell weighed 4800 kg and a concrete-piercing shell weighed 7100 kg (ie 4.72 and 7 tons respectively). Maximum range of the HE shell was about 47 km.

28 cm schwere Bruno Kanone (E) in transit

ble, and the theoretical research seemed to indicate that shells with pre-machined splines might prevent much of the wear that would otherwise occur. Only eight rifling grooves with corresponding splines were thought necessary but to prove the principle a sub-calibre barrel of 10.5 cm was made. This was the *K 12 M*, but to ensure that any possibility of the normal copper driving bands and rifling being able to stand the stresses was investigated fully, a further sub-calibre barrel, the *K 12 M.Ku*, was built and tried alongside the *K 12 M*. From these trials the splined shell concept emerged the clear leader as the normal rifling/driving band combination soon proved unable to stand up to the stresses involved.

Work then went ahead with the construction of the new weapon and mounting. The gun was a very lengthy affair and had to be externally braced to prevent it sagging under its own weight. The length in calibres of the barrel was L/158 which also led to problems with the placing of the trunnions. On the first version these were placed so far forward to partly counteract the muzzle preponderance that the carriage had to be raised at high angles of elevation to prevent the breech striking the ground when fired. As a result the rate of fire was considerably reduced. Other complexities included the provision to disconnect the barrel from the recoil mechanism to draw it back for travelling. The first full size barrel was proof fired during 1937 (the sub-calibre barrels were fired during 1935) and the first full equipment was fired during 1938. The designation given to this equipment was *K 12 V*, and as a result of the experiences with the gun and mounting after its entry into service during 1939 a redesign of the carriage was started to try to locate the barrel so that more of the gun was situated in front of the trunnions; the extra weight involved was compensated for by stronger balancing hydraulic presses. The end result was the *K 12 N*, but only one example of each type was made. Overall, the complexity and size of the gun and mounting were too much for the German Army, even though it did have a remarkable range. According to various sources it was as much as 120 kilometres, but the most recorded in action was about 88 kilometres – against Rainham in Kent – and the gun appears to have been firing from the Pas de Calais. The battery involved was probably Eisenbahnbatterie 701, a unit that seems to have spent all its time based in Western Europe. One example of this gun was captured in 1945 at Selazette in Holland and it was closely scrutinised by the Allies. At the end of their exhaustive investigations they announced that the design was one of the most remarkable to emerge from WW II, but to that they added the rider that it was also the most useless. The effort invested in its design was largely wasted for it had little military value, and indeed, one high-ranking officer★ regarded it as 'little more than a toy'. A toy it might have been, but as an exercise in practical ballistics it was a major achievement. One item of note was the cost of each equipment – RM1,500,000 for the gun and mounting alone.

★ General der Artillerie Karl Thoholte.

While the *K 12(E)* project ended after only two examples had been made, the *28 cm K 5(E)* story was much different. As with the *K 12(E)*, two sub-calibre barrels were built for tests with splined shells but for the *K 5(E)* the calibre was 15 cm. The two barrels were the *K 5 M* and *K 5 M.Ku*. As a result of these sub-calibre trials the first full-size barrel was proof-fired during 1936 and the next year saw the first firing of a complete equipment. By February 1940 eight were in service but in that year there began a strange spate of barrels which split when fired. Trials failed to reveal the exact cause but it was thought that reducing the depth of the rifling grooves might help and indeed it did provide a cure. The barrels with the shallower grooves were designated *K 5 Tiefzug 7 mm*, as opposed to the early barrels which became the *K 5 Tiefzug 10 mm*. Another change took place in 1943 when the considerable experience gained using sintered-iron driving bands seemed to offer the possibility of reaching high muzzle velocities with normally-rifled barrels. As a result of trials some barrels and ammunition were produced in this combination and the barrels were designated *K 5 Vz (Vielzug)*, but the numbers involved were small.

Overall, the design of the *K 5(E)* was excellent and it soon earned its title of the best railway gun ever produced. In service it proved to be trouble-free once the 1940 split-barrel episode was closed, and it remained in production at Essen and Hannover until 1945. By then about 25 had been made, and it had gained the nickname of *'schlanke Bertha'*. The range was of the order of 58 kilometres but this could be extended by the use of the *28 cm R Gr 4331*, a rocket-assisted projectile which could increase the range to about 86 kilometres. With this projectile accuracy was poor owing to the difficulties of timing the rocket motor ignition immediately the projectile had reached the top of its trajectory.

The success of the *K 5(E)* did not mean that it was immune to the normal run of trials and experimentation to which all German weapons were seemingly subjected. One *K 5(E)* was used in a series of trials intended to develop a screw breech mechanism to use bagged charges. That came to nothing, as did another series of tests to develop a muzzle brake for the piece. A great deal of experimental work was carried out into the problem of carrying a *K 5(E)* across country when damaged railway lines were encountered – by 1943 the Allied air forces were already causing considerable disruption to the German railway system. A specification was issued for a new equipment, the *K 5 ERF (Eisenbahn Runden Feld)*, which would normally be transported by rail, but between damaged tracks it was to be carried across country in several loads by converted Tiger II tank chassis. To this end the barrel was to be built in two parts. As an extra the mounting was to be able to accommodate a 38 cm howitzer, type unspecified. Both guns were to be lowered on to ground mountings for firing. From 1943 onwards this project attracted a great deal of detailed design work and development but by 1945 nothing tangible had been produced, which was probably just as well. The idea of a large gun being broken down into its component parts and reassembled under the constant threat of air attack, to say nothing of large tanks and their loads lumbering across country under a similar threat makes one wonder exactly what the German designers were thinking of and why they diverted so much attention to the scheme.

At least one other *K 5(E)* was used in research on sub-calibre fin-stabilised projectiles. The barrel in this case was smooth-bored and had a calibre of 31 cm. It was known as the *K 5 Glatt*, and the projectiles used were versions of the *PPF* or *Peenemünder Pfeil Geschoss*. By 1945 the results had reached the stage where troop trials were about to begin; by that time the projectiles were reaching ranges of the order of 150 kilometres. The trials were carried out on the Rügenwalde and Hillersleben ranges. According to one source two guns were actually used in action. One of these guns was made by Krupp and the other by Hanomag, and they were supposed to be the first of a series of seven. The two guns were based at Aarweiler near Bonn, and reportedly shelled Maastricht and Verviers. The projectile used was the *31 cm SpGr 4861*, with a body calibre of 12 cm, and a weight of 136 kilograms.

By the end of 1938 the Krupp drawing office had finished work on the *K 5(E)* and *K 12(E)* and were able to start new projects. One of these was the adaptation of the naval *38 cm SK C/34* to a railway mounting. These guns had already been redesigned for use in the coastal defence role with longer chambers, and it was this version that Krupp used to produce the *38 cm Siegfried K(E)*. In 1939 eight were ordered, but it was not until 1943 that the first was ready, and by 1945 only three had been completed.

At the same time as the 38 cm project began Krupp also started work on a 40.6 cm railway gun using the *40.6 cm SK C/34*. Again there was a coastal defence version with a longer chamber

28 cm K 5(E) in action

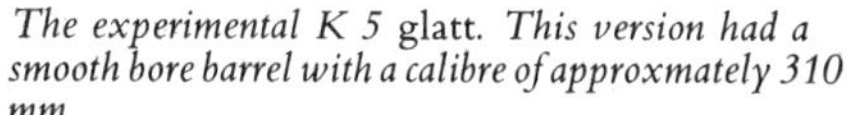

The experimental K 5 glatt. *This version had a smooth bore barrel with a calibre of approxmately 310 mm*

38 cm Siegfried-Kanone (Eisenbahn)

German designation Siegfried K (E)
Calibre 380 mm
Length of piece (L/52): 19630 mm
Length of barrel 18405 mm
Length of rifling 15748 mm
Weight emplaced 294,000 kg
Weight of piece 105,300 kg
Traverse (turntable): 360°; (carriage): 0°
Elevation 0° to +45°
Muzzle velocity 1050 m/sec
Shell weight (light): 495 kg; (heavy): 800 kg
Maximum range (light shell): 55,700 m; (heavy shell): 42,100 m
Rate of fire 1 round every 4–5 mins
Barrel life 240 rounds
Length of carriage 24000 mm
Manufacturer Friedr. Krupp AG, Essen

Remarks: Only three complete equipments delivered, first in 1939 and last during 1943.

38 cm Siegfried-Kanone(E)

38 cm Siegfried-Kanone(E) on a traversing turntable. The recoil mechanism can be seen at the front of the gun carriage

38 cm Siegfried K(E) proof-firing

40.6 cm Kanone (Eisenbahn)

German designation 40.6 cm K (E) oder 'Adolf'
Calibre 406 mm
Length of piece (L/50): 29300 mm
Weight emplaced 323,000 kg
Traverse (turntable): 360°; (carriage): 0°
Muzzle velocity 850 m/sec
Shell weight 960 kg
Maximum range 45,000 m
Designers Friedr. Krupp AG, Essen

Remarks: Development commenced 1938 but was subsequently delayed due to difficulties in fitting such a large equipment on a railway mounting. Most sources state that one such railway-mounted equipment was completed, but there is no pictorial evidence to confirm this.

---

80 cm Kanone (Eisenbahn)

German designations 80 cm K (E), schwere Gustav, Gustav Gerät, Gustav, Dora Gerät, Dora
Calibre 800 mm
Length of piece (L40.6): 32480 mm
Length of barrel 28957 mm
Weight emplaced (approx): 1,350,000 kg
Traverse on carriage 0°
Elevation 0° to +65°
Muzzle velocity (HE): 820 m/sec; (anti-concrete): 700 m/sec
Shell weight (HE): 4800 kg; (anti-concrete): 7100 kg
Maximum range (HE): 47,000 m
Rate of fire (maximum): 1 round every 15 mins
Barrel life (approx): 300 rounds
Length of carriages (approx): 42976 mm
Manufacturer Friedr. Krupp AG, Essen

Remarks: Only two complete equipments delivered, one in 1941 and second during 1942; third gun was never finished. One 'Dora' was deployed outside Leningrad 1942, but reportedly without its special ammunition.

80 cm Kanone schwere Gustav ready for acceptance trials at the Rügenwalde ranges in 1941

Schwere Gustav shelling Sevastopol in summer 1942. Traverse was obtained by moving the gun and carriage on a curved track by two diesel locomotives

Rear view of the schwere Gustav showing the ammunition hoist

Schwere Gustav in position to shell Sevastopol. Note the two parallel sets of rail tracks

19.4 cm Kanone (Eisenbahn) 486(f) or 93(f)

19.4 cm K(E) mounted in an armoured turret on the Atlantic coast of France

19.4 cm K(E) 486(f) moved into a protected coastal defence emplacement. The gun housing has been removed

German designation 19.4 cm K (E) 486(f) or 93(f)
Original designation Matériel de 194 mle 70/93 sur affut-truc tous azimuts; 194 mm 70/93, rayée à gauche
Calibre 194.4 mm
Length of piece (L/30.4): 5886 mm
Length of barrel 5550 mm
Length of rifling 4600 mm
Weight in action 65,000 kg
Weight of piece 10,500 kg
Traverse 360°
Elevation (possible): +3° to +40°; (firing): +10° to +40°
Loading angle +10°
Muzzle velocity 638 m/sec
Shell weight 83 kg
Maximum range 18,300 m
Rate of fire 4–5 rpm
Manufacturer Schneider et Cie, Le Creusot

Remarks: Numbers of these guns were captured in 1940 and later used as coastal guns, some still on their railway carriages. Others were removed and placed on extemporised carriages, such as those on Île de Cezembre.

19.4 cm Kanone(E) 486(f)

24 cm Kanone (Eisenbahn) 557(f) and 557/1(f)

German designation 24 cm K (E) 557(f) und 557/1(f)
Original designations (557(f)): Canon de 240 sur affut-truc mle 84; (557/1(f)): Canon de 240 sur affut-truc mle 17
Calibre 240 mm
Length of piece (L/28): 6700 mm
Length of barrel 6240 mm
Weight in action 90,000 kg
Weight of piece 14,000 kg
Traverse 360°
Elevation −2° to +38°
Muzzle velocity 575 m/sec
Shell weight 159 kg
Maximum range 17,100 m
Rate of fire 1 round every 4–5 mins
Length of carriage 12884 mm
Manufacturer Schneider et Cie, Le Creusot

Remarks: There were two versions of 24 cm K (E) 557(f) differing chamber lengths. 557/1(f) was a later version of same gun.

24 cm Kanone (Eisenbahn) 558(f) or 24 cm Kanone Modell 93/96(f)

German designations 24 cm K (E) 558(f); 24 cm K Mod 93/96(f)
Original designations Canon de 240 sur affut-truc mle 93/96; Canon de 240 T 93/96
Calibre 240 mm
Length of piece (L/41.7): 10055 mm
Length of barrel 9600 mm
Weight of piece 29,000 kg
Traverse 360°
Elevation −1°30′ to +29°
Muzzle velocity 840 m/sec
Shell weight 162 kg
Maximum range 22,700 m
Rate of fire 1 round every 4–5 min
Manufacturer No record, but railway carriages made by St. Chamond

Remarks: After capture by German forces in 1940 most of these guns were removed from their railway mountings and subsequently used exclusively as coastal defence weapons.

27.4 cm Kanone (Eisenbahn) 592(f)

German designation 27.4 cm K (E) 592(f)
Original designations Matériel de 274 mle 17 sur affut-truc a glissement; 274 mle 17 rayée à droite
Calibre 274 mm
Length of piece (L/46.7): 12800 mm
Length of barrel 12330 mm
Weight in action 152,000 kg
Weight of piece 35,000 kg
Traverse 0°
Elevation (possible): 0° to +40°; (firing): +22° to +40°
Muzzle velocity 842 m/sec
Shell weight 237.5 kg
Maximum range 29,100 m
Rate of fire 1 round every 5 mins
Length of carriage (approx): 25900 mm
Manufacturer Schneider et Cie, Le Creusot

28.5 cm Kanone (Eisenbahn) 605(f)

German designation 28.5 cm K (E) 605(f)
Original designations Canon de 285 sur affut-truc mle 17; Canon de 285 T 17
Calibre 285 mm
Length of piece (L/45): 12800 mm
Length of barrel 10830 mm
Weight in action 152,000 kg
Weight of piece 35,000 kg
Traverse 0°
Elevation (possible): 0° to +40° (firing): +20° to +40°
Muzzle velocity 740 m/sec
Shell weight 270 kg
Maximum range 27,100 m
Rate of fire 1 round every 5 mins

32 cm Kanone (Eisenbahn) 651(f) or 651/1(f)

German designation 32 cm K (E) 651(f) oder 651/1(f)
Original designations Matériel de 320 mle 70/30, 70/84 et 70/93 sur affut-truc à glissement; 320 mle 70/73, 70/84 et 70/93 (30 calibres) rayée à gauche
Calibre 320 mm
Length of piece (L/31.6): 10112 mm
Length of barrel 9600 mm
Length of rifling 7860 mm
Weight in action 162,000 kg
Weight of piece (mle 70/84): 48,550 kg; (mle 70/93): 45,736 kg
Traverse 0°
Elevation (possible): +3° to +40°; (firing): +22° to +40°
Loading angle 3°
Muzzle velocity 608 m/sec
Shell weight 387 kg
Maximum range 20,500 m
Rate of fire 1 round every 5 mins

Remarks: 32 cm K (E) 651/1(f) used a longer chamber than other models; there were also other minor changes.

32 cm Kanone (Eisenbahn) 652(f)

German designation 32 cm K (E) 652(f)
Original designations Canon de 320 sur affut-truc mle 17; Canon de 320 T 17
Calibre 320 mm
Length of piece (L/37): 11820 mm
Length of barrel 11200 mm
Weight in action 178,000 kg
Weight of piece 55,000 kg
Traverse 0°
Elevation (possible): +2° to +40°; (firing): +20° to +38°
Loading angle 2°
Muzzle velocity 690 m/sec
Shell weight 392 kg
Maximum range 26,200 m
Rate of fire 1 round every 5 mins

34 cm Kanone -Gl- (Eisenbahn) 673(f)

German designation 34 cm K -Gl- (E) 673(f)
Original designations Matériel de 340 mle 12 sur affut-truc à glissement; 340 mle 1912 rayée à droite 4°
Calibre 340 mm
Length of piece (L/47.3): 16115 mm
Length of barrel 15300 mm
Weight in action 270,000 kg
Weight of piece 98,900 kg
Traverse 0°
Elevation (possible): +3° to +37°; (firing): +23° to +37°
Loading angle +3°
Muzzle velocity 927 m/sec
Shell weight 430 kg
Maximum range 37,600 m
Rate of fire 1 round every 10 mins

Note: Gl = *Gleitlafette* (sliding carriage)

34 cm Kanone -W- (Eisenbahn) 674(f)

34 cm K-W-(E) 674(f). The gun and cradle have been removed from the rail carriage and placed on a pivot mount for coastal defence. France, 1944

34 cm K-W-(E) 674(f) showing the traversing rail

German designation 34 cm K -W- (E) 674(f)
Original designations Matériel de 340 mle 12 sur affut-truc à berecau; 340 mle 12 rayée a 6°
Calibre 340 mm
Length of piece (L/47.3): 16115 mm
Length of barrel 15300 mm
Weight travelling 166,000 kg
Weight in action 164,000 kg
Weight of piece 66,000 kg
Traverse 10°
Elevation (possible): −8° to +42°; (firing): +15° to +42°
Loading angle −8°
Muzzle velocity 930 m/sec
Shell weight 432 kg
Maximum range 44,400 m
Rate of fire 1 round every 10 mins

Note: W = *Wiegenlafette* (cradle mounting).

34 cm Kanone-W-(E) 674(f)

37 cm Haubitze (Eisenbahn) 711(f)

German designation 37 cm H (E) 711(f)
Original designations Matériel de 370 mle 15 sur affut-truc à berceau 370 mle 15 rayée à droite
Calibre 370 mm
Length of piece (L26.6): 9855 mm
Length of barrel 9250 mm
Weight in action 130,000 kg
Weight of piece 38,000 kg
Traverse 12°
Elevation (possible): −5° to +65°; (firing): +45° to +65°
Loading angle −5°
Muzzle velocity (light shell): 535 m/sec; (heavy shell): 475 m/sec
Shell weight (light): 516 kg; (heavy): 710 kg
Maximum range (light shell): 16,400 m; (heavy shell): 14,600 m
Rate of fire 1 round every 5 mins
Manufacturer Schneider et Cie, Le Creusot

Remarks: Three examples used by Batterie 695 (E), two by Batterie 711 (E).

---

40 cm Haubitze (Eisenbahn) 752(f)

German designation 40 cm H (E) 752(f)
Original designations Matériel de 400 mle 15 ou 16 sur affut-truc à berceau 400 mle 15 ou 16 rayée à droite
Calibre 400 mm
Length of piece (L/26.6): 10650 mm
Length of barrel 10000 mm
Weight in action 140,000 kg
Weight of piece 47,000 kg
Traverse 12°
Elevation (possible): −8° to +65°; (firing): +45° to +65°
Loading angle −8°
Muzzle velocity (light shell): 530 m/sec; (heavy shell): 465 m/sec
Shell weight (light): 641 kg; (heavy): 900 kg
Maximum range (light shell): 16,000 m; (heavy shell): 15,000 m
Rate of fire 1 round every 5 mins

Remarks: Total of eight examples were used by German railway artillery. Two were kept as spares and other six divided between Batt. 693 (E) and Batt. 696 (E).

40 rm H(E) 752(f) south of Leningrad, 1943

40 cm Haubitze(E) 752(f)

52 cm Haubitze (Eisenbahn) 871(f)

German designation 52 cm H (E) 871(f)
Original designations Obusier de 520 sur affût-truc à glissement mle 16; Obusier de 520 T Gl 16
Calibre 520 mm
Length of piece (L/16): 8350 mm
Length of barrel 7800 mm
Weight in action 260,000 kg
Weight of piece 44,000 kg
Traverse 0°
Elevation (possible): 0° to +60°; (!ring): 40° to +60°
Muzzle velocity 450 m/sec
Shell weight 1654 kg
Maximum range 14,600 m
Rate of fire 1 round every 6 mins
Length of carriage 30380 mm
Manufacturer Schneider et Cie, Le Creusot

Remarks: One used by German siege artillery to shell Leningrad 1942–43 and subsequently captured during Soviet break-out in January 1944.

Railway anti-aircraft guns

Railways played a part in the anti-aircraft defence of the Reich, and for this purpose special railway waggons were designed and built, usually by Krupp of Essen. As mentioned in the section on heavy anti-aircraft guns the railways became involved with the Luftwaffe Flak defences when the Allied air forces began a policy of concentrating large bomber formations on individual targets, a policy which overwhelmed the local defences while others were unable to intervene due to their static installations. Thus the *Eisenbahn Flak* units were enlarged in number (a few were already in being when the war began) and the numbers of special railway waggons for their use grew rapidly.

Smallest of these special waggons was the *Geschützwagen I(E) leichte Flak* which mounted the *2 cm Flak 30*, the *2 cm Flak 38* or the *2 cm Flakvierling 38*. The crew's quarters were on the same waggon. The *Geschützwagen II(E)* was similar but the gun platform was raised above the level of the crew's quarters. Both forms were used for the defence of freight and troops trains and they were usually situated with one gun a quarter of the length of the train from the engine, another half way to the rear and another some three-quarters of the way along. On especially important trains a further gun was placed in front of the engine, but there were many variations to suit local conditions and the full 'allotment' was seldom deployed. If a train had to stop in one place for prolonged periods the guns were frequently dismounted from their waggons to give them a better field of fire. On occasions the light Flak waggons were used in conjunction with the heavy Flak waggons mentioned below, usually in defence of built-up areas.

The heavy Flak waggon designed for the *8.8 cm Flak 18, 36* and *37* was the *Geschützwagen III(E) schwere Flak*, which could also carry the *10.5 cm Flak 38* and *39*. The *12.8 cm Flak 40* used the *Geschützwagen IV(E) schwere Flak*. Both these waggons featured drop sides to increase the area of the gun platform. Stabilising screw jacks were fitted and ammunition lockers were situated at both ends of the waggon. These heavy Flak waggons were used only in defence of the Reich and even during the fighting in Germany in 1944 and 1945 their movements were restricted to the rear areas. In action they were deployed as nearly as possible as a normal Flak battery, complete with their own radar and command posts.

During 1939 a special undesignated waggon was built to carry either the *15 cm Gerät 50* or *55*. Both guns were mounted on the prototype with no problems, but the demise of the *Geräte 50* and *55* also precluded any production of these special waggons.

Gesch Wg II(E) le Flak with 2 cm Flak 30

Geschützwagen I(E) leichte Flak *with 2 cm Flak 38*

Gesch Wg I(E) le Flak with 2 cm Flakvierling 38

Geschutzwagen II(E) leichte Flak *with 2 cm Flak 30 deployed as part of harbour AA defences*

Gesch Wg II(E) le Flak with 2 cm Flakvierling 38

Gesch Wg II(E) le Flak with 2 cm Flakvierling 38 ready for action

Gun platform of the Gesch Wg II(E) with 3.7 cm Flak 36

Late design light Flak wagons had the gun posts made of concrete and armed with the 2 cm Flakvierling 38 or triple MG 151/20 aircraft cannon, shown here

Concrete railway AA post with 2 cm Flakvierling 38

Geschützwagen III(E) schwere Flak *with 8.8 cm Flak 18 deployed for harbour AA protection.*

A modified version of the Gesch Wg III(E) s Flak with its 8.8 cm Flak and crew protected by light armour superstructure

This version of the Geschützwagen *mounted two 8.8 cm Flak guns with the ammunition stowage situated in the centre section of the flatcar*

A mixed mobile AA defence group near a harbour, comprising several Gesch Wg III(E) units with 10.5 cm Flak and Gesch Wg II(E) with lighter Flak guns

Geschützwagen III(E) s Flak *with 10.5 cm Flak*

An improvised version for use against ground targets. The 8.8 cm Flak 18 was carried by removing its front and rear outrigger arms and bolting the cruciform platform to the railway flatcar floor

Geschützwagen IV(E) s Flak *with 12.8 cm Flak 40 in travelling position*

15 cm Gerät 50 *in action position*

Gesch Wg IV(E) s Flak with its 12.8 cm Flak 40 at maximum elevation

15 cm Gerät 50 *mounted on a railway flatcar, in travelling position*

Armoured trains

Armoured trains (*Eisenbahn Panzerzüge*) were employed quite successfully by the German Army. They were used to some extent in Europe and on a large scale in the Soviet Union where their main function was to patrol and keep open railroads in areas where partisan bands were operating.

These trains were under the direct control of the General Staff and were allotted to Army Groups. Each train carried a train commander (who was usually also the infantry commander), an artillery commander, and a technical officer, responsible for the operation of the train. Various types of armoured trains were put into service by the Wehrmacht, including captured Polish, Czech and Soviet equipment. During 1943 there were 80 such armoured trains in service, numbered *Eisb. Pz. Zug 1* to *Eisb. Pz. Zug 80*. The composition of these trains varied in the types of armoured waggons used and armament carried.

For instance, *Eisb. Pz. Zug 63* consisted of the following units: two armoured gun waggons, each with a turret mounting a Polish-built *10 cm leFH 14/19(p)*; two armoured gun waggons, each with a *2 cm Flakvierling 38*, and a turret-mounted Soviet *7.62 cm FK 295/1(r)* field gun. In addition to the gun waggons there were two armoured infantry/command waggons, each carrying a weapons section comprising two 81 mm mortars, one heavy machine gun, 22 light machine guns and one flamethrower. The armoured steam locomotive was positioned in the middle of these six armoured waggons, and the total personnel carried amounted to 113 men. Placed at the front and rear of the train were railway flat waggons with ramps carrying a light tank, normally a PzKw 38(t) or a French vehicle. These tanks were used as mobile gun positions or, if required, could be run off the flat waggons for action against the enemy.

French Panhard 178 armoured cars, equipped with German grid-type radio aerials and fitted with flanged steel wheels to enable them to run on rails, were used as reconnaissance vehicles in front of the trains, the necessary stores and road wheels being carried on an additional leading flat waggon.

Also put into service were a number of light self-propelled armoured rail units, the first of these being a single improvised vehicle built in a German depot behind the Eastern Front. Called *Panzer Zeppelin*, it was constructed of material salvaged from destroyed tanks, mounted a Soviet BA.10 armoured car turret with a 37 mm gun and carried a small infantry detachment.

In 1944 appeared a number of small armoured rail vehicles built by Steyr in Austria. These units called *Panzersicherungswagen* (armoured security waggons) or *Panzerdraisinen* (armoured trolleys) were equipped with grid-type radio aerials and had a number of weapon portholes. Their crew consisted of a driver, wireless operator, and a machine gun detachment, and the vehicles were used to patrol sections of railway supply lines. Two other types of *Panzerdraisinen* were also put into service, one of which was armed with a PzKw IV turret with *7.5 cm KwK L/24* and a co-axial machine gun, with additional gun ports. A crew of six to eight men were carried. The second type, a larger vehicle, mounted two late-model PzKw IV turrets with *7.5 cm KwK L/48* guns and co-axial machine guns, and retained the armoured skirting as a protection against hollow-charge missiles.

During 1944 eight Italian self-propelled armoured units known as *Littorina Blindate* were built for the German Army by Ansaldo of Genoa. Taken into service as the *Eisenbahn Panzerwagen Littorina Modell 43*, the first carried an armament of two 47 mm guns in two turrets with co-axial machine guns. There were also four Breda 38 machine guns in side mountings, two Breda 38s in the superstructure and two 45 mm mortars firing through a hatch in the roof. The second group of four vehicles were armed with an M35 Breda 20 mm anti-aircraft gun in the roof, which replaced the two mortars, and two Breda 38 machine guns. The weight of these vehicles was about 35 tons and, powered by two diesel engines, they could travel in either direction at 50 km/h. All eight units were equipped with radios and searchlights.

An Eisenbahnpanzerzug *(EisbPzZug) comprising (left to right) an armoured infantry wagon, armoured steam locomotive, an armoured gun wagon with 10 cm leFH 14/19(p), armoured command wagon, an armoured gun wagon with 2 cm Flakvierling 38 and an 7.62 cm FK(r), and a railway flatcar with a PzKpfw 38(t) tank*

Armoured gun wagon with 10 cm le FH 14/19(p)

An armoured flatcar with a PzKpfw 38(t) preceded by another flatcar and a Panhard 178 armoured car on rails

Another version of German armoured train comprising a flatcar with PzKpfw IV turret (7.5 cm L/48), another flatcar with a PzKpfw 38(t), an armoured wagon with 10.5 cm le FH 18/40 in a turret and an armour-protected 2 cm Flakvierling 38, followed by armoured command and infantry wagons, another turreted 10.5 cm le FH 18/40, and the armoured locomotive

An armoured train on the Eastern Front

Another type of armoured train with the 2 cm Flakvierling 38 replaced by 3.7 cm Flak and an self-propelled gun, a Lorraine tractor with 12.2 cm FH(r), instead of a light tank

A Czech armoured train in service with the German Army. This particular train, built in Austro-Hungary and used during World War I, was later handed over to Czechoslovakia and then taken over by the German forces

A Soviet armoured train in German Army service. Various types of captured Soviet armoured trains were impressed by the German forces and used in action. In many cases railway flatcars with light tanks were added for mobile flank defences

Panzerdraisine *with 7.5 cm KwK L/48 in PzKpfw IV turrets. Note the camouflaged concrete Flak posts on railway flatcars*

A closer view of the 10.5 cm le FH 18/40 turret, the protected 2 cm Flakvierling 38 position, and the first armoured personnel wagon

PzKpfw IV turret with 7.5 cm L/48 mounted on a protected railway flatcar

Light reconnaissance train consisting of an unarmoured steam locomotive, two railway flatcars loaded with ballast for exploding mines laid under the track, two flatcars with French Somua tanks, and two armoured open wagons with the train's infantry detachment

'Panzer Zeppelin' with a Soviet B-10 armoured car turret

Infantry detachment detraining from a 'Panzer Zeppelin' to secure position before flank reconnaissance

Panzersicherungswagen *armoured security patrol wagon*

Panzerdraisine *with 7.5 cm KwK L/24 gun in a PzKpfw IV turret*

EisbPzWg 'Littorina'

Coastal artillery

The use of artillery for coastal defence dates back to the very earliest days of artillery itself, and for many centuries any state with even a short stretch of coastline felt compelled to assert its claim to it by the use of extensive fortifications and their attendant weapons. By the late 19th century the art of coastal defence had been brought to a very refined level and the coastlines of Europe were dotted with some very high concentrations of fortifications covering naval establishments, important harbours and the like. The new state of Germany was no exception to this rule and almost as soon as the new German Navy began to expand in numbers and importance during the late 19th century there began a programme of updating old defences and building new ones along the Baltic and North Sea coasts. During WW I these defences were further expanded and modernised to cover the main German Navy base at Kiel and its all-important canal, and most of the artillery used was of naval origin. This was mainly due to the fact that the bulk of the German coastal defences were manned by the Navy, and as the defences were most likely to be attacked by long-range naval weapons the best counter to them would be the same naval weapons. A few specialised coastal defence weapons were developed, usually by Krupp, but much of the artillery used during WW I was naval in origin and consisted of 21 cm, 24 cm, 28 cm and 30.5 cm naval guns of varying lengths.

Surprisingly enough, the bulk of the German coastal defences were unaffected by the terms of the Versailles Treaty of 1919, probably because they were deemed to be defensive in purpose. Thus when the NSDAP came to power in 1933 the coastal defences were largely intact and ready for use. What installations there were required few changes, but at the same time it was felt that more modern and specialised coastal defence guns would be needed in the future and a programme of design and development began in about 1935. The main results of this programme were, in time, the excellent *15 cm SK C/28* and the *30.5 cm SK L/50*. Both these guns were thoroughly modern and efficient weapons. In particular, the *15 cm SK C/28* was such a good gun that it was frequently used as a field gun, in which role its modern mobile carriage was shown to be particularly successful.

As in WW I the bulk of the coastal defences were provided by the German Navy, so most of the weapons used continued to be naval in origin. Starting with the smaller calibres extensive use was made of the *3.7 cm SK C/30*, which was the naval version of the dual-purpose *3.7 cm Flak 18*. To back up these light guns were large numbers of *5 cm KwK 39, 39/1* and *40* tank guns mounted on simple *'Sockellafette'* pivot mounts. Then came the *7.5 cm Pak 40 M* which was the naval version of the *7.5 cm Pak 40* intended for use on light warships and coastal craft. Next was a series of 8.8 cm guns, all of them intended originally for anti-aircraft use on large warships but equally suitable for coastal defences. There were numerous versions of 10.5 and 15 cm guns with a wide variation in ages and an even wider variation in mountings. Some were more modern 15 cm guns designed for use on torpedo boats and U-boats and thus had rather complex waterproof and folding carriages but nevertheless large numbers of these weapons were diverted for coast defence.

The larger-calibre guns showed similar variations in age and origin. Among the more elderly guns was the *17 cm SK L/40* which dated to before WW I, while the *20.3 cm SK C/34* was a modern gun intended for the 'Admiral Hipper' class cruisers. Then came several variants of 24 cm naval guns of varying lengths among which were a small number of 25.4 cm ex-Russian guns which had been captured in 1915 and subsequently installed at Borkum. There were three different lengths of 28 cm gun, all of them dating back to before WW I but still deemed suitable for the coastal defence role (and also for use as railway guns in the 'Bruno' series). Among the really large calibres all the guns were fairly modern. The *30.5 cm SK L/50* has already been mentioned but above this came two naval guns adapted for coastal use by alterations to the chamber dimensions and similar design changes. First of these was the *38 cm SK C/34*, originally intended for use on the 'Bismarck' class battleships, while the largest gun used by the Germans in the coastal defence role was the *40.6 cm SK C/34*, the *'Adolf-Rohr'*.

Considerable investigation was carried out on coastal defence guns by the Kriegsmarine and a number of experimental guns and projects were initiated. At the top end of the scale came projects for 42 cm guns (the *42 cm SK C/34 z* und *C/34 g*), 45 cm guns (*45 cm SK C/34 h*), and even a 53.3 cm gun (*53.3 cm SK C/36 g* or *Gerät 36*). All these guns were intended for use on large warships but coastal versions were also projected. Of the three, only a single example of the 53.3 cm gun was actually completed. It was intended for a super-heavy battleship but as this was never built the guns did not pass the development stage. Another heavy gun intended for coastal use that suffered the same fate was the *30.5 cm SK C/39*, a design based on a WW I naval gun.

As with other weapon developments, the German designers were keen to apply novel and relatively untried techniques to coastal guns, and one result of this was a small series of naval guns with tapered bores. Among these were the *15 cm/11.2 cm Kanone* and the *20.3 cm/17 cm Kanone*, neither of which was accepted for service. Recoilless guns were also considered for the coastal defence role, but the exact reason for this development is uncertain for the advantages over conventional weapons would appear to be few. One project along these lines was the *28 cm Düsen-Kanone*, a Rheinmetall project which was intended for mounting on a motorised carriage. A similar carriage was intended for use with the *28 cm Kanone für R2*, which should have mounted a *28 cm SK C/34* barrel, but neither of these weapons was accepted for service. Another ambitious project that never materialised was the *15 cm Kanone mit Mehrfachladungsraum*, which would have used multiple chambers in place of the conventional single chamber to reduce the size of the breech.

The above outlines the general types of weapons used by the Kriegsmarine in its coastal defence role during WW II, but it must be stressed that it only mentions guns actually approved by the service for use in that role. After 1940 the number of types of weapon was considerably expanded by the sheer size of the task imposed upon the Navy by the conquests of the early war years. Wherever possible, existing coastal defence installations and weapons were taken over and used by the German Navy, but even so it was unable to cover all the occupied coastline of Europe. Consequently the German Army had to take over some of the tasks that would normally have been a Navy responsibility, but it had the disadvantage that it lacked the expertise required for the coastal defence role, and it also lacked the weapons. The only recourse open to the German Army was to employ as many captured weapons as it could lay its hands on, and where Army divisions had a coastal defence responsibility the divisional artillery was used as well. As a result the coastal defences manned by the German Army used a bewildering array of weapons culled from all the corners of Europe and the Operation Order dated 2.1.43 appended to this section can give only a small idea of the range of artillery used.

At this point it would be appropriate to give an outline of the history of the German coastal defences during WW II. In 1939 the bulk of the German coastal defences were where they had been in 1918, ie emplaced along the Baltic and North Sea coasts. The events of 1940 took the German Army to the coasts of Norway and the Atlantic and at the time when the Army was mopping-up the remnants of the Allied armies in France, plans were being hastily made for 'Operation Sea Lion', the invasion of the United Kingdom. To cover the likely invasion route it was decided to transfer four batteries from the Baltic and North Sea coasts to the Pas de Calais. Some of these batteries had already been moved

once to add weight to the North Sea defences during 1939, but the four batteries eventually settled down at Calais, Houlgate, Marcouf and Longues. They were backed up in part by railway guns but for much of 1940 no further reinforcements were possible as there were no more heavy guns available. All around the French coasts existing defences were taken over by German troops wherever possible, and later campaigns into the Mediterranean and the Balkan area resulted in the take-over of other coastal defences, some of them mounting very ancient weapons indeed. There was one further move of a North Sea battery, the transfer of 'Batterie Tirpitz' from Kiel to Constanza in Romania (via Ostvoorne in the Hook of Holland) to cover the oilfields there against a possible Soviet attack. The battery was still there when Romania was taken over by the Soviets in 1944.

By the end of 1940 there was still no substantial increase in the numbers of specialised coastal defence guns ready for use and with the abandoning of 'Operation Sea Lion' the defence of the German-occupied territories in the West became an item of the first priority. By the end of 1940 the invasion of the Soviet Union was well into the planning stage and it was at this time that the first permanent defences were erected along the Atlantic coast. By the middle of 1941 the idea of the 'Atlantic Wall' was firmly established and vast amounts of concrete began to be poured into the construction of what was to become the largest fortification programme of all time. The entire coastline occupied by the German forces was gradually taken over by massive concrete fortifications covering every possible approach to almost every likely invasion point. From the north of Norway to the border between France and Spain the 'Atlantic Wall' consumed colossal amounts of building materials, labour, effort and time. Hitler himself devoted a considerable portion of his time and effort to the design and detail of the fortifications which were built by the services themselves, the Organisation Todt, and the use of enforced, conscripted and, in some cases, contracted labour. Throughout 1941, 1942 and 1943 the 'Atlantic Wall' continued to grow in size and strength until by 1944 it seemed to be almost impregnable.

But as the concrete bunkers and associated structures grew in numbers and variety so did the need to arm them with weapons. As has already been mentioned the German Army had to employ as many captured weapons as possible, and they also had recourse to divisional artillery. But for the Navy there was little alternative but to take over guns intended for cancelled warships or, later, to impress artillery dismantled from other large warships for various reasons taken off the active list and laid up damaged beyond repair in various ports. A typical example of this could be seen on the Island of Fano, off Denmark, where were installed two twin-gun turrets which had been taken from the damaged battleship *Gneisenau*. Each turret had two *15 cm SK C/28* guns which were used to command the approaches to Esbjerg. Also in Denmark was an example of another common German Navy practice which was to salvage as much armament as possible from beached or scuttled warships. The Island of Fano also had a battery of four Bofors Model '06 naval guns which had been salvaged from the Danish coastal defence ship *Peder Skram*, which was scuttled in Copenhagen harbour in August 1943. On a more modest scale the small harbour on Sark in the Channel Islands displayed an 8.8 cm naval gun salvaged from a beached light naval craft. There were numerous other similar cases.

But in 1941 there was a growing need for heavy coastal defence guns to equip the numerous Navy batteries that were planned as part of the 'Atlantic Wall'. Such was the schedule imposed on the equipping of the new batteries that there was no time available for the design and construction of conventional carriages and their complicated accessories, so once again Krupp came to the rescue. They had ready to hand a number of 38 and 40.6 cm naval gun barrels that were intended for fitting to warships that had been cancelled in October 1939. For coastal defence use these guns had to have some form of carriage and Krupp engineers solved the problem. It was proposed that a proof mounting for firing heavy barrels could be easily adapted for the coastal defence role. Fitting such items as ammunition hoists and data transmission systems were relatively minor changes and the new carriage could be made available fairly quickly, so the Krupp proposal was adopted and the *Bettungsschiessgerüst C/39* went into production. In all 37 *C/39* carriages were contracted for. The first four examples were fitted with *38 cm c/34* barrels and were erected on the Channel Coast near Wimereux as part of the Batterie 'Siegfried', later to be renamed 'Batterie Todt'. The guns were initially installed in 360° traverse mountings in open concrete pit mountings. Almost as soon as they were finished Hitler himself decided that the installations were too vulnerable to air attack and insisted that they must be covered in concrete to a depth of 3.5 metres even though this would restrict traverse to 120°. As a result the 'Batterie Todt' and the other Pas de Calais batteries had the protection of not only their armoured turrets but also the thick concrete carapace of their roofs and surrounds.

The next three *C/39* carriages were erected on the island of Hela and fitted with *40.6 cm SK C/34 barrels*. After a short while these combinations were moved to Sangatte in the Pas de Calais. Before the move the battery was named 'Schleswig Holstein' but in France this was changed to 'Batterie Grossdeutschland' and finally to 'Batterie Lindemann'. These guns were roofed in by

15 cm SK L/40, a pre-WW I coastal defence gun retained in service

concrete but their armoured turrets were retained. The 'Siegfried' and 'Lindemann' batteries also had the unusual distinction of taking part in the first coastal battery versus coastal battery artillery duel in history, for in September 1944 the two batteries exchanged fire with the two 15-in guns installed at Wanstone Farm near Dover, and 'Winnie' and 'Pooh', the two 14-in naval guns at St. Margaret's-at-Cliffe, also near Dover. As a result of this duel one of the 'Lindemann' guns received a direct hit and was put out of action but the other guns were kept in action until they were overrun by Allied forces.

The other *C/38* carriages were installed at Hanstedt in Denmark where they were fitted with *38 cm SK C/34* guns. These guns were intended to close the Skagerrak to Allied warships and were thus coupled with the similar installation across the water at Kristiansand-Sud in Norway. Further north, the 'Trondenes' and 'Engeloy' batteries effectively closed the approaches to Narvik in Norway with their *40.6 cm SK C/34* guns, also installed on *C/39* carriages (they are still there in 1978). It was also intended that the *C/39* carriage would be used to take captured French naval guns. Six of these French guns would have been the 38 cm mle 35/36 guns fitted to the battleship *Jean Bart*, and a further eight would have taken the 34 cm mle 12 naval gun which had been intended for the 'Normandie' class battleships of pre-WW I design which were never completed. Examples had fallen into German hands in 1940 and some had already been diverted to coastal defence use by the French, as with the Cap Cepet installation, but as with the *Jean Bart* guns none seem to have been fitted to *C/39* carriages for German service. It would appear that most of the 37 carriages contracted for were finished even if not delivered. One single example was used at Krupp's Meppen range.

Only four examples of a variation of the *C/39* were completed as the *Bettungsschiessgerüst C/40*, but the guns used on these carriages had a most colourful story. They were originally made by the Putilov Arsenal at Reval (Tallinn) in what was then the Russian province Estonia. That was in 1914, and the guns were built to a Schneider-Canet design for installation in the dreadnought *Imperator Aleksandr III*. Under the Kerenski government late in April 1917 she was renamed *Volya*, commissioned without trials into the Black Sea Fleet two months later, came under German control in 1918, under British control in 1919, and was then handed over to the White Russian forces when she was again renamed as *General Alekseyev*. After escape from Bolshevik troops the battleship was interned at Bizerta in 1920, taken over by the French Navy in 1924 but left to deteriorate until finally scrapped during 1935–36. The guns were removed and stored until early 1940 when they were declared surplus and offered as

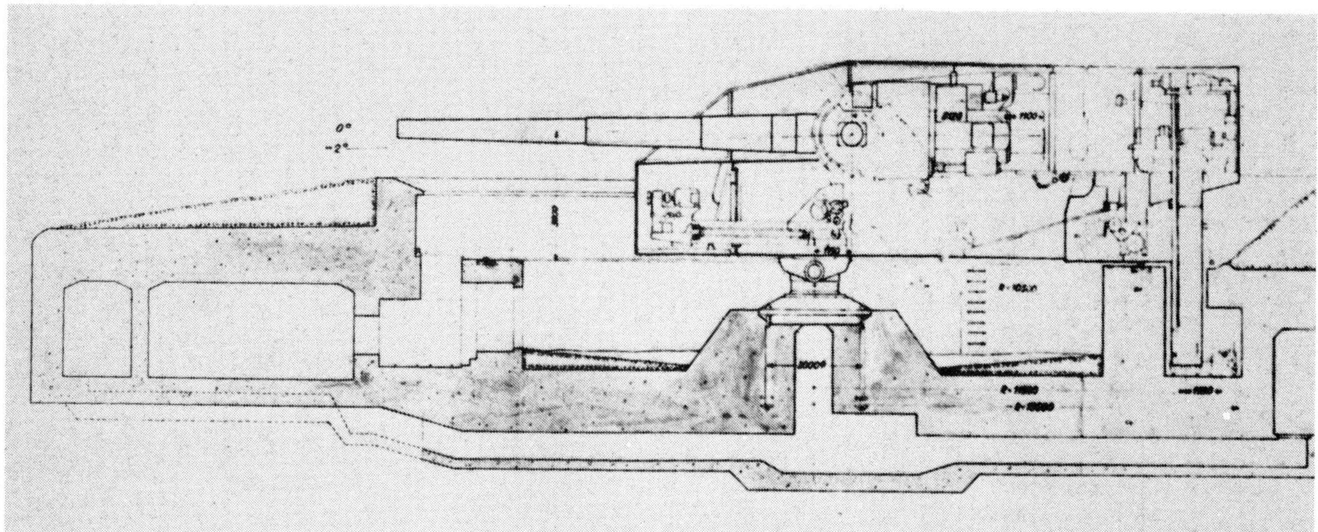

38 cm SK C/34 in B-Gerüst *C 39 with reinforced frontal protection*

38 cm SK L/50 in B-Gerüst *C 39 in a reinforced turret*

part of the very limited Western Allied help to Finland. The guns were loaded in the cargo vessel *Nina*, but she had only reached Norway when the first Soviet-Finnish war came to an end, and only a few weeks later the ship with its cargo fell into German hands when they invaded Norway. From there the guns were sent to Krupp at Essen for refurbishing and were then taken to Guernsey in the Channel Islands where, at La Frie Baton, they were installed in standard concrete pit mountings and set behind armoured turrets. Originally the battery was to be called 'Nina' but this was later changed to 'Batterie Mirus' after a German artillery specialist who was killed during an air attack in transit between the islands. The battery was intended as part of the installations designed to close the Gulf of St. Malo, the other half of which would have been installed near Paimpol on the French mainland. The mainland battery was to have had *38 cm SK C/34* guns mounted in two twin turrets but these were never delivered and their place had to be taken by two *20.3 cm K(E)* guns. But the old Russian guns did not perform as well as expected and indeed, during early proof firing, three of the four guns damaged their trunnions. Then began one of the oddest journeys of WW II for two lorries set out for Tallinn in German-occupied Estonia to find spare parts at the old Putilov Arsenal. The parts were found and three months later the guns were again ready for action, even though the firing charges were thereafter reduced along with a loss in range. Despite all the efforts of the Germans the 'Mirus' guns saw very little action apart from firing on destroyer sweeps in 1944 but one night the guns opened up on what was thought to be a large naval force approaching the Islands. Several rounds were fired as part of a large artillery action but in the morning the 'naval force' turned out to be two barrage balloons that had escaped from their moorings. After the events of June 1944 had by-passed the Channel Islands the 'Mirus' guns had no further part to play and eventually they were scrapped in 1951.

Of all the sectors of the 'Atlantic Wall' none was more heavily defended than the Channel Islands. They were the only British territory to fall under German domination and thus became a valuable propaganda prize to which Hitler himself became deeply involved, down to the siting of individual bunkers and guns. It has been stated that of all the facilities diverted to the Atlantic Wall one tenth went to the Channel Islands of Jersey, Guernsey and Alderney, all of which were designated as 'fortress areas' to be defended 'to the last round and ration tin'. This designation was given to nearly every port on the 'Atlantic Wall' and it must be stated that in nearly every case this order was obeyed.

The use of the *C/39* and *C/40* carriages led to the consideration of similar devices for other captured guns. The *C/41* was intended for 30.5 cm mle 06/10 naval guns captured in France and the type was also intended for use with 34 cm mle 12 French guns. Eight carriages were contracted for use with the 30.5 cm guns and twelve with the 34 cm guns but it is not known for certain how many were actually delivered. It would seem that a further batch of eight for use with elderly ex-Austro-Hungarian Škoda 30.5 cm C/13 naval guns was not completed. The same applies to another carriage, the *C/42*, which was intended for use with French 24 cm mle 02/06 guns. Again, Krupp were responsible for the design of the carriage and actually made part of it but most of the work was carried out by the Ardeltwerken at Eberswalde with the first deliveries intended for the end of 1943. How many were actually made and installed is not known.

To journey along the 'Atlantic Wall' in 1943 or 1944 would have been an experience the like of which we (hopefully) will never see again. The whole length of the Wall was covered with huge concrete structures of all types. Each battery however large or small was equipped not only with casemates or pit mountings for its guns but also with all the other structures needed to serve them. Among these were observation posts, ammunition stores and magazines, troop shelters, command posts, signal installations, radar bunkers and even individual shelters purpose-built for such items as anti-tank guns and searchlights. Every battery was meant to look after its own defences so in addition to the main guns each battery had to have field or anti-tank guns for landward defence and each structure was equipped with machine guns for close defence. In addition most batteries also had some form of anti-aircraft defence. As a result of these additions even a modest battery grew into a major construction work and the number of troops needed to man and defend a battery constituted a major drain on German manpower resources. The rest of the German defences in the South of France and the Mediterranean and Balkans added to the total, as did the batteries on the Black Sea and at Sevastopol, but the bulk were deployed along the English Channel, Atlantic coast and the North Sea.

At no time did Hitler actually know where the intended Allied invasion would take place, or when. The Allies in their turn did everything possible in the deception war to divert attention from their intended invasion points and keep the garrisons in such theatres of war as Norway and the Balkans up to full strength and even to get them reinforced in some cases. Hitler suspected that the invasion would take place in Norway or the Pas de Calais and it was a major part of Allied strategy to keep him thinking that way. Apart from this diversion of forces the 'Atlantic Wall' had one major flaw in that if it was penetrated at any point the rest became almost useless because the German forces in the West in 1944 had few reserves behind the coastal defences. Another flaw was that Hitler's obsession with the Pas de Calais led to the chosen stretch of the Normandy coastline which was the actual invasion area being left relatively undefended and on 6 June 1944 the Allies were able to build up a sizeable beachhead without the expected major battle. Thereafter the 'Atlantic Wall' was turned and all the effort expended on it proved to be largely wasted. In trying to defend everything Hitler in the end could defend virtually nothing, but this did not see any diversion of effort from the rest of the coastal defences. Many German-defended French ports held out for months and the Channel Islands did not capitulate until the war ended in May 1945, even though they were never attacked. The Norwegian batteries stood by to retaliate against an invasion that never came and indeeed, as the war ended a battery equipped with four *38 cm SK C/34* guns in two massive armoured turrets was still under construction at Oxby in Denmark.

Most of this section has been devoted to the efforts of the German Navy, but mention must also be made of the part played by the German Army in coastal defence. As stated above they had the role of coast defence imposed upon them, and had to carry it out with whatever equipment they could find. Training centres for teaching the art of coastal gunnery were set up at Rügenwalde and eventually at Sète in the South of France, both of them under Navy command and guidance – the Navy schools were at Swinemunde on the Baltic and Beziers in the South of France. Like the Navy installations the Army batteries were lavishly equipped with concrete protection, but generally speaking were simpler and more open. Their fire was usually directed by a Navy control post and, as a rule, used in support of a naval fire plan. As stated above their weapons were many and various and the equipment used and carried by the troops manning the Army batteries was frequently made up from captured stocks, including such mundane items as telephone wire and rifles.

For the troops garrisoning the 'Atlantic Wall' the war was a long series of exercises and periods of protracted boredom. Most of the time they manned their guns waiting for an enemy that never came. A few batteries such as those in the Pas de Calais saw some action but most of the rest just stood and waited. For the German forces the 'Atlantic Wall' was a diversion of effort and facilities they could ill afford. In some ways the conquests of the early war years were a major factor in the defeat of Germany in the long run as they required an inordinate proportion of the German war product to keep them under German control. Even without Hitler's obsession for massive fortifications the defence of the West would have been a major task; with it, the result was disaster.

Appendix 1

Listing of German Naval Coastal batteries

Extracted and translated from 'Bettungsschiessgerüste für Marine-Küstenartillerie'. OKM AWa B. D.-Nr.116. Berlin 1943

Gun type	PEACETIME **Battery name**	**Location**	WARTIME **Battery name**	**Location**	**Number of guns**	**Remarks**
21 cm SK L/50	Plantagenbatterie	Swinemünde	Plantagenbatterie	Swinemünde	4	
24 cm SK L/40	Skagerrak	Sylt	Mestersand	North Norway	4	
24 cm SK L/40			Burgas	Bulgaria	2	
24 cm SK L/40	Hamburg	Norderney	Hamburg	Cherbourg	4	
24 cm SK L/40	Oldenburg	Borkum	Oldenburg	Calais	2	25.4 cm Russian gun
28 cm SK L/40	Graf Spee	Wangerooge	Graf Spee	Brest	4	
28 cm SK L/45	Tirpitz	Kiel	Tirpitz	Constanza, Romania	3	
28 cm SK L/45	Goeben	Swinemünde.	Musoen	Drontheim, Norway	4	
28 cm SK L/45	Prinz Heinrich	Fehmarn	Prinz Heinrich	Channel Coast, Leningrad	2	
28 cm SK L/50	Grosser Kurfürst	Pillau	Grosser Kurfürst,	Framzelle,	4	
			Kiberg	North Norway	3	One spare platform
28 cm SK L/50	Coronel	Borkum	Coronel	Borkum	2	Two extra guns on Insel Beer
28 cm SK L/50	Coronel	Borkum	Coronel	Borkum	2	4 guns on Borkum
30.5 cm SK L/50	Friedr. August	Wangerooge	Friedr. August	La Tresorie, France	3	
30.5 cm SK L/50	Friedr. August	Wangerooge	von Schröder	Heligoland	3	
30.5 cm SK L/50(r)			Mirus	Guernsey	4	C/40 carriage
38 cm SK C/34			Todt	Haringzelle	4	C/39 carriage Formerly 'Siegfried'
38 cm SK C/34			Hanstholme 11	Hanstedt, Norway	4	C/39 carriage
38 cm SK C/34			Vara	Kristiansand-Süd, Norway	3	C/39 carriage
40.6 cm SK C/34			Lindemann	Sangatte, France	3	C/39 carriage
40.6 cm SK C/34			Trondenes	Harstadt, Norway	4	C/39 carriage
40.6 cm SK C/34			Engeloy	Narvik	3	
28 cm SK L/50			Grosser Kurfürst	Framzelle	4	L/37 twin mounting
28 cm SK C/34			Rosenburg	Hook of Holland	3	L/37 twin mounting
15 cm SK C/28			Lüderlitz	Borkum	4	C/34 twin mounting
15 cm SK C/28			Zanker	Sylt	4	C/34 twin mounting
15 cm SK C/28				Fanø, Denmark	4	Guns ex-*Gneisenau*
17 cm SK L/40	Ehrhardt Schmidt	Kiel	Ehrhardt Schmidt	Kiel	3	
17 cm SK L/40	Jakobsen	Heligoland	Jakobsen	Heligoland	3	
20.3 cm SK C/34			Ars	Ile de Ré, France	4	Guns ex-*Seydlitz*
20.3 cm SK C/34			Crognon	Ile de Croix, France	4	
28 cm SK C/34			Oerlandet	Dronthheim	3	
28 cm SK C/34			Fjell	Bergen, Norway	3	
19.4 cm K 485(f) SfL				6 in Holland 6 at St. Malo 1 at Swinemünde	13	
20.3 cm SK L/45(r)			von der Goltz	Sevastopol	4	
22 cm K 532(f)			Strassburg	Guernsey	4	
24 cm K(E)558(f)			Prefailles	St. Nazaire	2	Ex-railway guns
24 cm K(E)558(f)			La Bats	St. Nazaire	2	Ex-railway guns
24 cm K(E)558(f)			Ofoten	Narvik	4	
24 cm SK C/97(h)			Brandenburg	Insel Beer, Holland	2	
30.5 cm H Mod. 16			Lödingen	Narvik	4	Bofors howitzers
34 cm K Mod. 12(f)			Plouharnel	Ruiberon, France	4	Ex-railway gun
21 cm K 39/40			Engaloy II	Narvik	3	Army guns
21 cm K 39/40			Trondenes II	Harstadt, Norway	3	Army guns
38 cm SK C/34				Cap de la Hague, France	4	Planned, not built
38 cm SK C/34				Paimpol, France	4	Planned, not built

F 28 cm SK l/45, a naval gun of pre-WW I design

Appendix 2

Translation of construction order for the Atlantic Wall dated 2 January 1943

Supreme Command of the Armed Forces. 2.1.43.

Armed Forces Ops Staff/General i/c Engineers & Fortifications/Branch L (Artillery)
Extract No 39 (Artillery) No 1700/42 Sec
Orders *Secret. Command Orders*

With ref. to: OKW/WF st/Gen i/c Eng & Forts. (LI) Extracts Order Art No. 951/42 Secret and Orders of 16/8/42. (LI) No. 1300/42 Secret Command Order of 10.10.42.
Concerning: Additions for the artillery emplacement construction on the Channel and Atlantic coasts.
To: [Circulation List]

1. The Artillery emplaced in coastal defences is to be grouped as:
   a) close-defence guns against air and sea assault.
   b) coastal batteries (according to Armed Forces Ops Staff Sec. Ord. No. 004688/42★ paragraphs 1 & 3).
   c) fixed (position) batteries of the Army (according to Armed Forces Ops Staff Sec. Ord. No. 004688/42★ paragraph 2).
   d) mobile batteries.
2. The employment of the artillery in paragraph 1a, 1c and 1d is controlled by Supreme Commander, West, and that of the artillery in paragraph 1b by Armed Forces Ops Staff Sec. Ord. No. 004688/42.
3. The Führer has ordered, with regard to the anticipated complete air cover of the attacking enemy, that all batteries which cannot provide adequate AA defence (from Quad 20 mm up to 8.8 cm Flak) should be protected by concrete roofs.
   ★ (only to be supplied to sites marked ★ above.)
   In these cases therefore because of the shortcomings of armoured turrets an all-round field of fire cannot be expected.
4. That since in the near future the planned exchange of the least suitable guns for more modern weapons cannot be guaranteed, the building plan will proceed on the basis of the guns presently to hand.
5. Previous construction plans for the artillery now scrapped.
   a) Artillery Installations I and II.
      Artillery Installation I (Building Regulation 604)
      Provision for:
      1) Three 2 cm Flak 38
      2) Four 2.5 cm Pak 113(f)
      3) Four 3.7 cm Pak
      4) Two 3.7 cm Flak 36
      5) Three 4.7 cm Pak(t)
      6) Two 4.7 Pak 181(f)

7) Two 4.7 Pak 185(b)
8) Two 5 cm Pak 38
9) Four 7.5 cm GebGesch 36
10) Two 7.5 cm FK 231(f)
11) Two 7.5 cm K232(f)
12) Two 7.5 cm Pak 97/38
13) Two 7.5 cm FK 236(b)
14) Two 7.5 cm FK 16 (nA)
15) Two 7.5 cm Pak 40
16) Two 7.5 cm leFK 18
17) One 7.5 cm Pak 231(f)
18) Four 7.5 cm leIG 18
19) One 7.62 cm FK 296(r)
20) One 10.5 cm K 331(f)
21) One 10.5 cm leFH 18
22) One 10.7 cm K 352(r)
23) One 15 cm sFH 18
24) Two 15 cm sIG 33
25) One 15.5 cm K 414(f)
26) One 15.5 cm K 425(f)

Artillery Installation II (Bdg. Reg. 605)
Provision for:
1) Three 2 cm Flak 38
2) Four 2.5 cm Pak 113(f)
3) Four 3.7 cm Pak
4) Two 3.7 cm Flak 36
5) Three 4.7 cm Pak(f)
6) Two 4.7 cm Pak 181(f)
7) Two 4.7 cm Pak 185(b)
8) Two 5 cm Pak 38
9) Four 7.5 cm GebGesch 36
10) Two 7.5 cm FK 231(f)
11) Two 7.5 cm K 232(f)
12) Two 7.5 cm Pak 97/38
13) Two 7.5 cm FK 236(b)
14) Two 7.5 cm FK 16 (nA)
15) Two 7.5 cm Pak 40
16) Two 7.5 cm leFK 18
17) One 7.5 cm Pak 231(f)
18) Four 7.5 cm leIG 18
19) One 7.62 cm FK 296(r)
20) One 10.5 cm K 331(f)
21) One 10.5 cm leFH 18
22) One 10.7 cm K 352(r)
23) One 15 cm sFH 18
24) One 15 cm sIG 33
25) One 15.5 cm K 414(f)
26) One 15.5 cm K 425(f)
27) One 8.8 cm Flak 36
28) One s 10 cm K 18
29) One 12.2 cm K 390(r)
30) One 15.2 cm K 433(r)
31) One 15.2 cm sFH 443(r)
32) One 15.5 cm K 416(f)
33) One 15.5 cm K 418(f)
34) One 15.5 cm K 422(f)

b) Embrasured gun emplacements for field guns with 60° traverse (Bdg. Reg. 611)
Provision for:
1) FK 18
2) le FH 18
3) s 10 cm K 18
4) sIG 33
5) sFH 18
6) 10 cm (8 cm) FK M 30(t)
7) 10 cm leFH 18 M14/19 (t)(p)
8) 15 cm FH M24(t)
9) 10.5 cm K 331(f)
10) 15.5 cm sFH 414(f)
11) 15.5 cm K 422(f)
12) 15.5 cm K 425(f)
13) 12.2 cm K 390(r)
14) 12.2 cm leFH 396(r)
15) 15.2 cm K 433(f)

c) Embrasured gun emplacements with 90° traverse (Bdg. Reg. 649) for 10.5 cm K 331(f) on medium traversing carriage.

d) Embrasured gun emplacement with 120° traverse (Bdg. Reg. 650) for 10.5 cm K 331(f) on medium traversing carriage.

e) Ammunition, Observation and Command Bunkers suitable installations as in reference 2 above, as well as
Command Post for Coastal Battery (Bdg. Reg. 636)
Survey Post for Coastal Battery (Bdg. Reg. 637)

f) Special Constructions for the Navy and for large calibre guns not mentioned in paragraph 7.

6. Construction Plans under further development for artillery.
a) Embrasured emplacements for guns on heavy traversing mounts with 90° traverse.
b) Embrasured emplacements for guns on heavy traversing mounts with 120° traverse.
When the first heavy mount is delivered a further decision will be taken on the introduction of these emplacements.

7. With the raising of traversing speeds, and the reduction in embrased emplacements and open gun positions (360° traverse), traversing mounts are currently in production.
Expected delivery to Commander in Chief West as follows:
a) Medium mounts (16 tons)
February 1943 – 50
March to July 1943 – 150 per month
suitable for:

10.5 cm K 17/04	10.5 cm K 35(t)
10.5 cm K 331(f)	11.4 cm K 365(e)
10.5 cm K 332(f)	12 cm K 370(b)
10.5 cm K 335(h)	15.5 cm sFH 414(f)
10.5 cm K 29(p)	15.5 cm sFH 17(p)

The mounts are distinguished by small differences in pattern for each type of weapon. The first type will be for the 10.5 cm K 331(f). (See paragraph 11(b).)
b) Heavy traversing mounts (32 ton)
from July 1943 – 12 per month
intended for:

12.2 cm K 390/2(r)	15.5 cm K 425(f)
15 cm K 15/16(t)	15.5 cm K 432(b)
14.5 cm K 405(f)	15 cm K 16
15.5 cm K 416(f)	15 cm K 18
15.5 cm K 418(f)	15 cm sFH 25(t)
15.5 cm K 420(f)	15.2 cm sFH 433/1(r)

c) 'Super-heavy' traversing mounts for 17 cm calibre weapon upwards as well as E-Batteries will be ordered later as Special Constructions.

8. For the protection of gun detachments in open gun positions against enemy bombers, an armoured cover is being developed, which will be suitable for guns with traversing mounts as well as those on traversing platforms. The date of the start of construction is still under discussion and will be notified to you later. With these armoured roofs, we shall first of all fit out those which cannot have concrete covers (for technical or other reasons).

9. On the basis of the technical installation arrangements outlined in paragraphs 5 to 8, the following improvements are to be made to battery positions:
a) Installation of traversing mounts in embrasured emplacements as in paragraphs 5(c) (d) and 6.
b) Installation of mobile wheeled carriages in embrasured emplacements as in paragraph 5 (b).
c) Installation of traversing mounts in open battery positions.
d) Retention of the existing types of construction (where fully serviceable).
e) In the cases of (c) and (d) above, fitting with armoured covers within the limits of paragraph 8.

10. We shall have as our goal the provision of the following munitions dumps:
a) Batteries in the Channel Islands – 20 munitions dumps.
b) Batteries in the Defence Zone – 10 ,, ,, .
c) All other Batteries – 5 ,, ,, .
By this means an allotment can be made for the strengthening of the munitions dumps now lacking protection to the extent of providing bombproof storage for the existing ammunition dumps.

11. In examining and selecting the design and order of construction of the above, the following should be taken into consideration:
a) Check the existence of adequate AA protection.
b) Initially *only* medium mounts for the 10.5 cm K 331(f) will be manufactured: positions for these have priority. Those for the other types of gun under construction requiring medium mounts will as soon as possible be segregated as required.

3.7 cm Abkommrohr Kanone

German designation 3.7 cm Abkommrohr K
Calibre 37 mm
Length of piece (L/20): 740 mm
Muzzle velocity 406 m/sec
Shell weight 0.47 kg
Maximum range 4573 m

Remarks: Sub-calibre gun barrel used for training on 8.8 cm SK C/35. Similar 5 cm Abkommrohr K was used as training device with 15 cm SK C/28.

3.7 cm Schiffskanone C/30 in Einheitslafette C/34

German designation 3.7 cm SK C/30 in EhL C/34
Calibre 37 mm
Length of piece (L/83): 3076 mm
Length of barrel 2962 mm
Traverse 360°
Elevation −10° to +80°
Muzzle velocity 1000 m/sec
Shell weight 0.745 kg
Maximum range, horizontal 6600 m
Maximum effective ceiling 2000 m
Manufacturer Rheinmetall-Borsig AG, Düsseldorf

Remarks: Naval dual-purpose AA/coastal defence gun. Emplaced on static single or twin mountings to protect heavy coastal guns or harbours.

3.7 cm Schiffskanone *C/30* in Einheitslafette *C/34 (3.7 cm SK C/30 in EhL C/34)*

Sockellafetten für 5 cm Kampfwagenkanone

German designations Sockellafette 1a für KwK 39/1; Sockellafette 1b für KwK 39; Sockellafette 1c für KwK 39, KwK 39/1 und KwK 40
Data for KwK 39, 39/1
Calibre 50 mm
Length of piece (L/60): 3000 mm
Weight of piece 435 kg
Muzzle velocity (HE): 550 m/sec; (AP): 1190 m/sec
Shell weight 1.82 kg
Shot weight 2.06 kg
Maximum range 6500 m
Rate of fire 15–20 rpm
Barrel life 8000–10,000 rounds
Manufacturer Rheinmetall-Borsig AG, Düsseldorf

Remarks: These mountings were all simple open arrangements, sometimes in casements but more often in open pits. There were no fire controls and all aiming was carried out by layer 'pushing' the gun to correct firing angle. Fitted with various types of shield.

Sockellafette für 5 cm KwK *in a sunken concrete emplacement*

Sockellafetten für 7.5 cm Kampfwagenkanone

Sockellafette IIa auf Kreuzbettung mit 8.8 cm KwK 43, *a provisional test installation. Pivot mounts were also made for the 8.8 cm Pak 43*

German designations Sockellafette 1c für 7.5 cm KwK 51, 67, 68; Sockellafette 1d für 7.5 cm KwK 67; Sockellafette 111 für 7.5 cm KwK 42
Data for 7.5 cm KwK 42
Calibre 75 mm
Length of piece (L/70): 5250 mm
Weight of piece 900 kg
Muzzle velocity (HE): 700 m/sec; (AP): 925 m/sec
Shell weight 5.74 kg
Shot weight 6.8 kg
Maximum range 9850 m
Rate of fire 6 rpm
Barrel life 2000 rounds
Manufacturer Rheinmetall-Borsig AG, Unterlüss

Remarks: Series of simple extemporised pivot mountings similar to that used for 5 cm KwK intended for surplus tank guns in various defensive locations. Some mounts were designed and produced at Skoda-Werke, Pilsen.

7.5 cm Panzerabwehrkanone 40M in Lafette Marine 39/43

German designation 7.5 cm Pak 40M in LM 39/43
Calibre 75 mm
Length of piece (L/46): 3700 mm
Length of barrel 3450 mm
Length of rifling 2461 mm
Weight (total): 2680 kg
Weight of mounting 2020 kg
Traverse 360°
Elevation −10° to +40°
Muzzle velocity (HE): 550 m/sec
Shell weight 5.74 kg
Maximum range 7680 m
Rate of fire 12–15 rpm
Barrel life 6000 rounds
Manufacturer Rheinmetall-Borsig AG, Unterlüss

Remarks: Standard 7.5 cm Pak 40 on naval pedestal mount. Crew protected by 10 mm armour shield. Introduced into service during 1944 and used on light coastal craft or as shore gun, specifically against MTBs.

8.8 cm Schiffskanone C/35 in Unterseebootslafette C/35

German designation 8.8 cm SK C/35 in UbtsL C/35
Calibre 88 mm
Length of piece (L/45): 3990 mm
Length of barrel 3735 mm
Length of rifling (approx): 3344 mm
Weight of piece (approx): 4250 kg
Traverse Up to 360°
Elevation −4° to +30°
Muzzle velocity 700 m/sec
Shell weight 9 kg
Maximum range 12,350 m
Rate of fire 8–10 rpm
Manufacturer Rheinmetall-Borsig AG, Düsseldorf

Remarks: Specially designed naval gun, different from standard 8.8 cm Flak. Introduced in Kriegsmarine service during 1937. Only small number installed in casemates for coastal defence.

10.5 cm Schiffskanone C/32 in 8.8 cm Marine Pivotlafette C/30D

German designation 10.5 cm SK C/32 in 8.8 cm MPL C/30D
Calibre 105 mm
Length of piece (L/45): 4740 mm
Weight complete (B): 13,850 kg; (C): 15,231 kg
Weight of piece 1706 kg
Traverse Up to 360°
Elevation −3° to +79°
Muzzle velocity 785 m/sec
Shell weight 15.06 kg
Maximum range 15,350 m
Rate of fire 6 rpm
Manufacturer Rheinmetall-Borsig AG, Düsseldorf

Remarks: Designed as dual-purpose AA/surface gun for use on warships. Introduced in service during 1932; after 1939 fitted for pedestal mount used for 8.8 cm Flak series for coastal defence/AA use.

10.5 cm SK C/32 in 8.8 cm MPL C/30D armoured cupola for the AA role

---

10.5 cm Schiffskanone L/60 in Einheitslafette

German designation 10.5 cm SK L/60 in EhL
Calibre 105 mm
Length of piece (L/65): 6840 mm
Length of barrel 6300 mm
Weight in action 11,750 kg
Weight of piece 4635 kg
Traverse Up to 360°
Elevation −10° to +80°
Muzzle velocity 900 m/sec
Shell weight 15.1 kg
Maximum range (horizontal): 17,500 m
Maximum vertical ceiling 12,500 m
Rate of fire 15 rpm
Manufacturer Rheinmetall-Borsig AG, Düsseldorf

Remarks: Designed as dual-purpose AA/coastal defence gun; naval version of 10.5 cm Flak 38. First delivered to Kriegsmarine during 1937.

12.7 cm Abkommrohr Kanone

German designation 12.7 cm Abkommrohr K
Calibre 127 mm
Length of piece (L/35): 4445 mm
Muzzle velocity 600 m/sec
Shell weight 28 kg
Maximum range 14,000 m

Remarks: Sub-calibre training gun used with 38 cm SK C/30.

12.8 cm Schiffskanone C/40 in 12.8 cm Doppellafette(PzK) C/40

German designation 12.8 cm SK C/40 in 12.8 cm DoppL(PzK) C/40
Calibre 128 mm
Length of piece (L/61): 7835 m
Traverse 360°
Elevation 0° to +85°
Muzzle velocity 900 m/sec (est)
Shell weight 26 kg
Rate of fire (each barrel): 12–14 rpm (est)

Remarks: Projected coastal version of 12.8 cm Flak 40 in naval turret; not completed.

28 cm Schiffskanone L/40

28 cm SK L/40 in an open gun emplacement with 360° traverse

Loading the 28 cm SK L/40

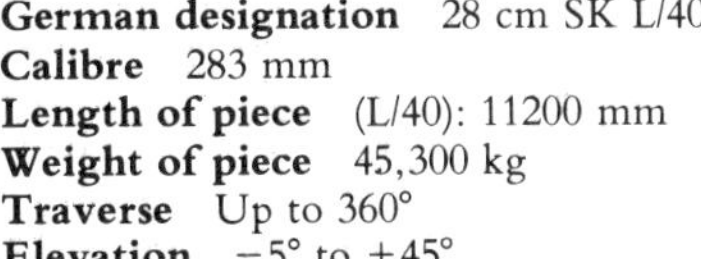

German designation 28 cm SK L/40
Calibre 283 mm
Length of piece (L/40): 11200 mm
Weight of piece 45,300 kg
Traverse Up to 360°
Elevation −5° to +45°

Muzzle velocity 820 m/sec
Shell weight 240 kg
Maximum range 29,500 m
Rate of fire 1 round every 5–6 mins
Manufacturer Friedr. Krupp AG, Essen

Remarks: Designed before WW I for 'Deutschland' class battleships. Many guns used for coastal defence purposes during WW I and WW II on a wide variety of mountings.

---

28 cm Schnellade-Kanone L/45 in Bettungsschiessgerüst

28 cm SK L/45 in an open emplacement. The gun has been covered with an armoured roof

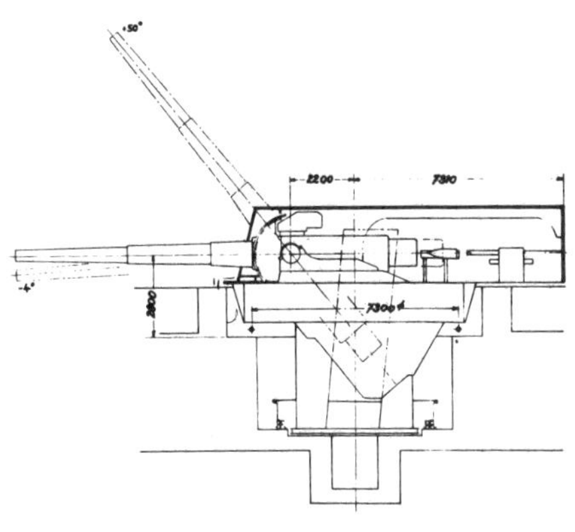

28 cm SK L/45 in Küstenlafette *C/37 showing the principal parameters, lowest depression loading position, highest elevation and recoil*

28 cm SK L/45 in action on the French coast

German designation 28 cm SK L/45 in B-Gerüst
Calibre 283 mm
Length of piece (L/45): 12735 mm
Length of barrel 9698.5 mm
Length of rifling 8490 mm
Weight emplaced 94,697 kg
Weight of piece 39,800 kg

Traverse Up to 360°
Elevation 0° to +37°
Muzzle velocity 875 m/sec
Shell weight 284 kg
Maximum range 36,100 m
Rate of fire 1 round every 5 mins
Manufacturer Friedr. Krupp AG, Essen

Remarks: Designed before WW I as shipboard gun and used as such during that conflict. Number of guns retained for coastal defence and emplaced in several types of mountings.

28 cm Schnellade-Kanone L/50 in Bettungsschiessgerüst

28 cm SK L/50 in B-Gerüst *at maximum elevation*

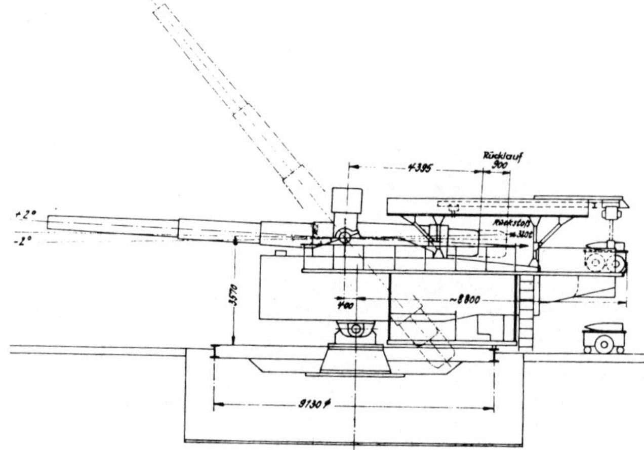

28 cm SK L/50 in Bettungsgerüst *(*B-Gerüst*) as fitted at the 'Batterie Grosser Kurfürst', showing the principal parameters and barrel positions*

German designation 28 cm SK L/50 in B-Gerüst
Calibre 283 mm
Length of piece (L/50): 14150 mm
Length of barrel 13304 mm
Length of rifling 11113.5 mm
Weight of piece 55,050 kg
Weight of mounting 31,670 kg

Traverse Up to 360°
Elevation −2° to +50°
Loading angle +2°
Muzzle velocity 905 m/sec
Shell weight 284 kg
Maximum range 39,100 m
Rate of fire 1 round every 5–6 mins
Manufacturer Friedr. Krupp AG, Essen

Remarks: Designed before WW I for battle-cruisers *Moltke, Goeben* and *Seydlitz*. Retained guns adapted for coastal defence and emplaced in various types of mountings.

28 cm Schnellade-Kanone C/34 in Drehschiesslafette C/28

28 cm SK C/34 in DrhL C/28

Rear view of the 28 cm SK C/34 in DrhL C/28

German designation 28 cm SK C/34 in DrhL C/28
Calibre 283 mm
Length of piece (L/54.5): 15415 mm
Length of barrel 14505 mm
Traverse 180°36′
Elevation −8° to +40°
Loading angle +2°
Muzzle velocity 890 m/sec
Shell weight 330 kg
Maximum range Not recorded
Manufacturer Friedr. Krupp AG, Essen

Remarks: Two main turrets from laid-up battleship *Gneisenau* transferred to coastal defence during 1943. Turret 'B' used by 'Batterie Fjell' and turret 'C' by 'Batterie Oerlandet', both in Norway.

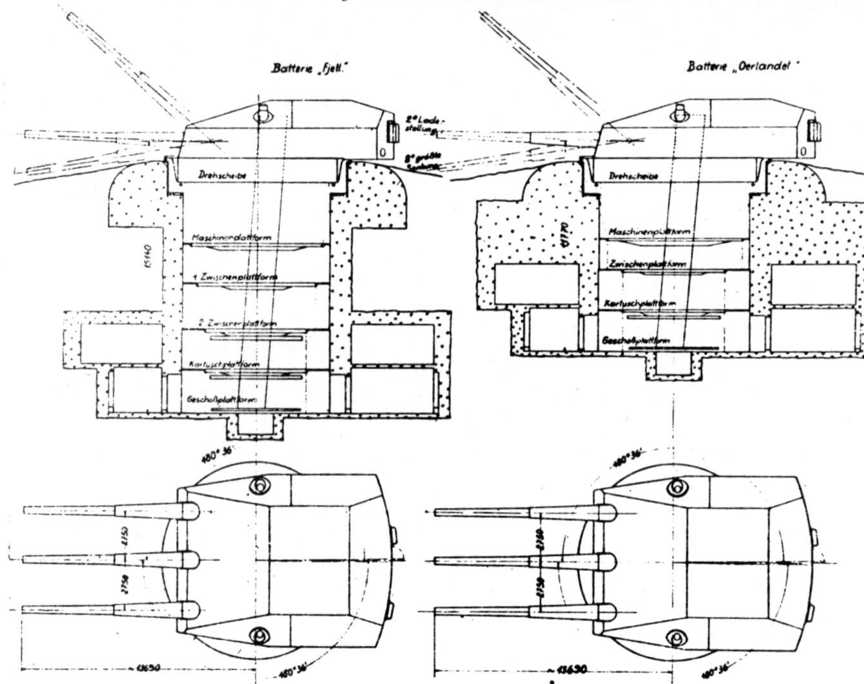

28 cm SK C/34 guns as emplaced with the 'Batterie Fjell' (left) and 'Batterie Oerlandet' (right)

28 cm Küstenhaubitze

28 cm KstH in action on the Eastern Front

28 cm Küsten-Haubitze. *This equipment was similar to the 28 cm L/12 but was of lighter construction*

German designation 28 cm KüstH
Calibre 283 mm
Length of piece (L/12): 3396 mm
Weight emplaced 63,600 kg
Weight of piece 10,800 kg
Traverse 360°
Elevation 0° to +65°
Muzzle velocity 350–379 m/sec
Shell weight 350 kg
Maximum range 11,400 m
Rate of fire 1 rpm
Manufacturer Friedr. Krupp AG, Essen

Remarks: Pre-WW I coastal defence design closely related to 28 cm Haubitze L/12. Notable as only coastal howitzer in German service during WW II.

30.5 cm Schiffskanone L/50

German designation 30.5 cm SK L/50
Calibre 305 mm
Length of piece (L/50): 15250 mm
Length of barrel 14185 mm
Length of rifling 11490 mm
Weight complete less armour 177,000 kg
Traverse Up to 360°
Elevation −4° to +45°
Muzzle velocity (HE): 1050–1120 m/sec; (APHE): 820–855 m/sec
Shell weight (HE): 250 kg; (APHE): 405 kg
Maximum range (HE): 51,000 m; (APHE): 32,500 m
Manufacturers Friedr. Krupp AG, Essen; Škoda, Pilsen (under contract)

Remarks: Designed as main battleship armament during WW I. Retained guns adapted for coastal defence. Emplaced in single and twin turrets.

30.5 cm SK L/50 in B-Gerüst *preparing for action*

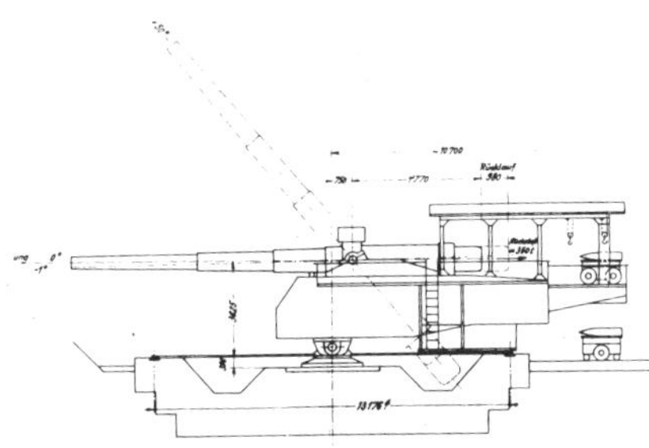

30.5 cm SK L/50 in B-Gerüst *('Batterie Friedrich August')*

30.5 cm SK L/50 in twin turrets

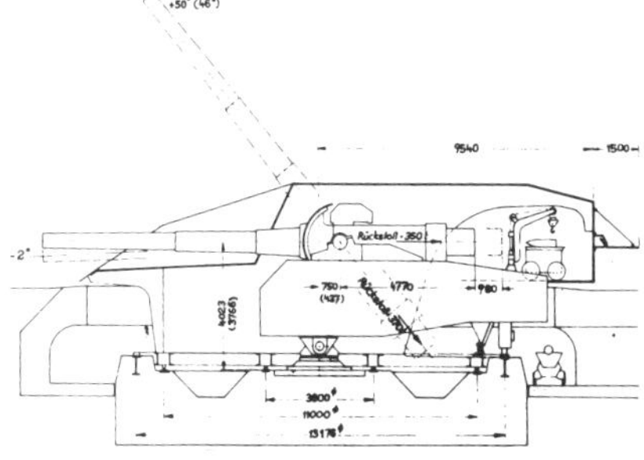

30.5 cm SK L/50 in B-Gerüst mit Schutzschild *('Batterie von Schröder') showing the principal parameters*

38 cm Schiffskanone C/34 'Siegfried'

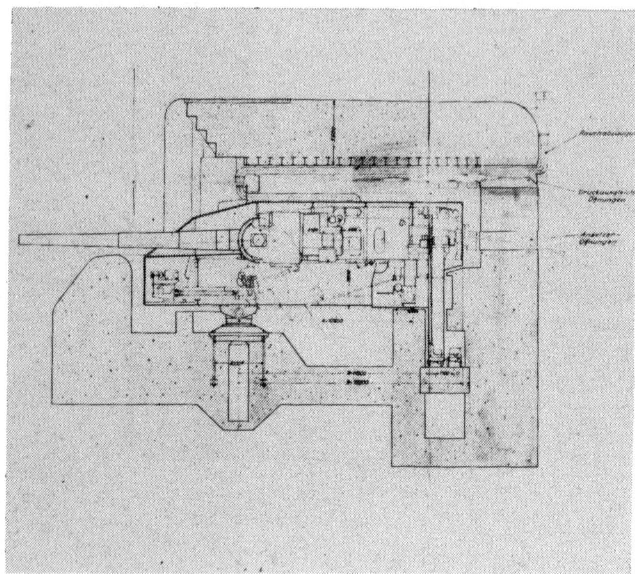

38 cm SK C/34 in B-Gerüst C 39 im Bunker
('Batterie Todt', formerly 'Siegfried')

38 cm SK L/50 in B-Gerüst C 39 *in a reinforced turret*

38 cm SK C/34 of 'Batterie Todt

German designation 38 cm SK C/34 'Siegfried'
Calibre 380 mm
Length of piece (L/52): 19630 mm
Length of barrel 18405 mm
Length of rifling 15748 mm
Weight of piece 105,300 kg
Traverse 120°
Traverse in barbette Up to 360°

Elevation −4° to +60°
Muzzle velocity (light shell): 1050 m/sec; (heavy shell): 820 m/sec
Shell weight (light): 425 kg; (heavy): 800 kg
Maximum range (light shell): 55,700 m; (heavy shell): 42,000 m
Rate of fire 1 round every 5–6 mins
Manufacturer Friedr. Krupp AG, Essen

Remarks: Naval guns intended for cancelled 'Bismarck' class battleships. Adapted for coastal defence purposes during 1940. These guns had longer chamber than original shipboard version. Emplaced on Bettungsschiessgerüst C/39. First four formed 'Batterie Siegfried' (renamed 'Batterie Todt' after 1942) near Wimereux in France.

40.6 cm Schiffskanone C/34 in Schiessgerüst C/39

40.6 cm SK C/34 in B-Gerüst C 39 *with reinforced frontal protection*

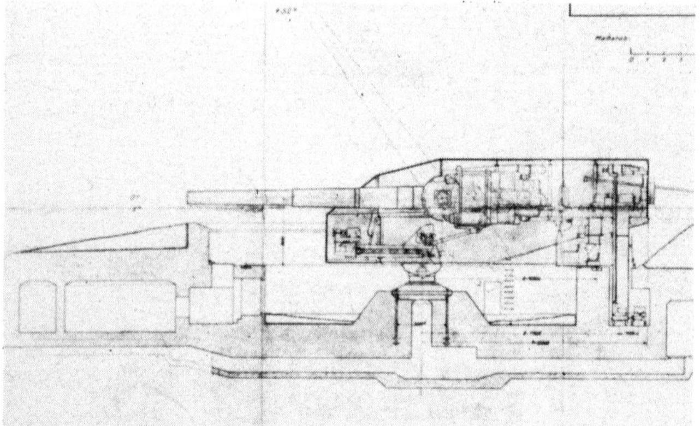

40.6 cm SK C/34 in B-Gerüst C 39 *in a reinforced turret shielded by casemate*

German designation 40.6 cm SK C/34 in SG C/39; 40.6 cm Adolf Rohr
Calibre 406 mm
Length of piece (L/50): 20300 mm
Traverse in casemate 100°
Elevation 0° to +60°
Muzzle velocity (standard shell): 810 m/sec; (long range shell): 1050 m/sec
Shell weight (standard): 1030 kg; (long range): 600 kg
Maximum range (standard shell): 42,800 m; (long range shell): 56,000 m
Rate of fire 1 rpm
Manufacturer Friedr. Krupp AG, Essen

Remarks: Naval guns intended for 'H' class battleships cancelled in October 1939. Adapted for coastal defence purposes during 1942–43. First three guns installed on Hela as 'Batterie Schleswig-Holstein' in 1943, transferred to Pas de Calais 'Batterie Grossdeutschland' and later renamed as 'Batterie Lindemann'.

40.6 cm SK C/34 of the 'Batterie Lindemann'

9.5 cm Küstenkanone(f)

German designation 9.5 cm KstK(f)
Original designation Canon de côte de 95 mle 1893
Calibre 95 mm
Weight emplaced 1800 kg
Traverse 360°
Elevation −10° to +17°
Muzzle velocity 418 m/sec
Shell weight 12.09 kg
Maximum range 8000 m
Rate of fire 10 rpm
Manufacturer de Bange, Le Creusot

Remarks: An obsolete gun of limited performance for WW II conditions. A small number impressed by German artillery troops for harbour defence in occupied France after 1940.

20.3 cm Schiffskanone L/45(r)

German designation 20.3 cm SK L/45(r)
Original designation Not known
Calibre 203.2 mm
Length of piece (L/45): 9144 mm
Traverse 360°
Elevation −5° to +37°30′
Muzzle velocity 720 m/sec
Shell weight 95 kg
Maximum range 25,000 m
Rate of fire 4 rpm

Remarks: Undamaged naval guns of Soviet battery 'Maksim Gorky' near Sevastopol. Impressed into German service during 1943 and renamed 'Batterie von der Goltz'. (Above data should be regarded as provisional.)

24 cm Schnellfeuerkanone C/97(h) in Drehschiesslafette C/97(h)

German designation 24 cm SK C/97(h) in 24 cm DrhL C/97(h)
Calibre 240 mm
Length of piece (L/40): 9600 mm
Length of barrel 8916 mm
Length of rifling 7351 mm
Weight of piece 24,500 kg
Traverse 342°
Elevation −4° to +20°
Muzzle velocity 820–850 m/sec
Shell weight 170 kg
Maximum range Not recorded
Rate of fire 4 rounds every 5 mins

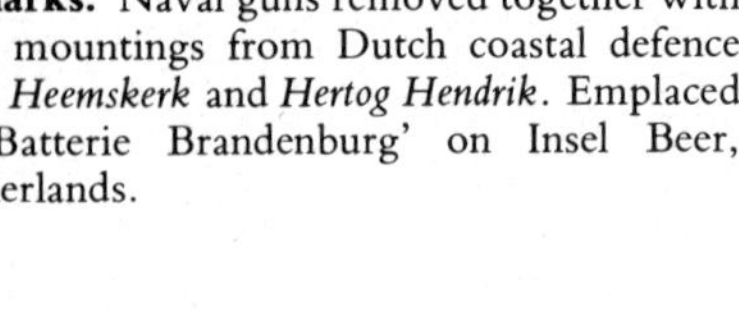

Remarks: Naval guns removed together with their mountings from Dutch coastal defence ships *Heemskerk* and *Hertog Hendrik*. Emplaced as 'Batterie Brandenburg' on Insel Beer, Netherlands.

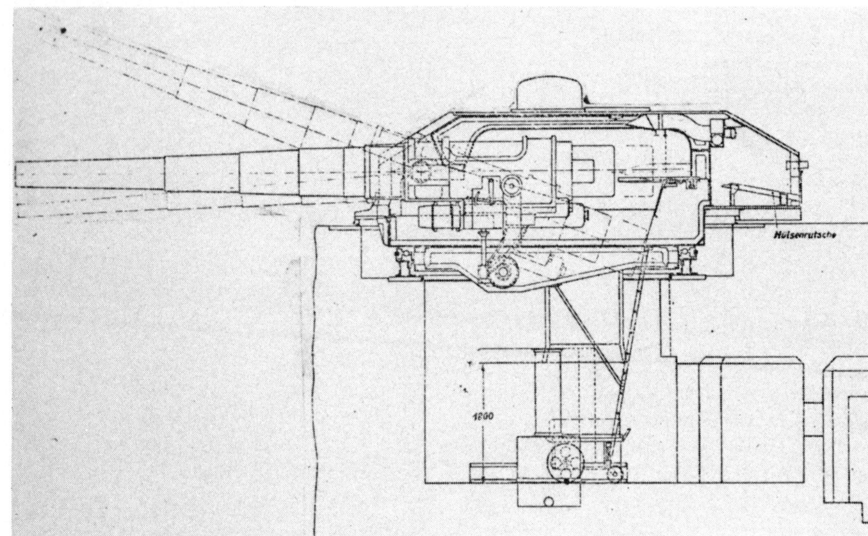

24 cm SK C/97 in 24 cm DrhL C/97(h)

27 cm Küstenmörser 585(f)

German designation 27 cm KstMrs 585(f)
Original designation Mortier de côte de 270 mle 1889
Calibre 270 mm
Length of piece (L/12.4): 3350 mm
Weight travelling (3 loads): 46,200 kg (approx)
Weight emplaced 26,500 kg
Weight of piece 5800 kg
Traverse 300°
Elevation 0° to +60°
Muzzle velocity 420 m/sec
Shell weight 152.2 kg
Maximum range 10,400 m
Rate of fire 1 round every 5 mins
Manufacturer Schneider et Cie, Le Creusot (not confirmed)

Remarks: An obsolete French coastal defence mortar impressed into German service. Available and used only in small numbers.

30.5 cm Schiffskanone C/14 in Schiessgerüst C/40, 30.5 cm Kanone 14(r) or (Eisenbahn) 626(r)

German designations 30.5 cm SK C/14 in S-Gerüst C/14; 30.5 cm K 14(r); 30.5 cm K(E) 626(r)
Original designation 305 mm Naval Gun Model 1914
Calibre 305 mm
Length of piece (L/52): 15,818 mm
Weight of piece 49,000 kg
Traverse 360°
Elevation −2° to +48°
Muzzle velocity (light shell): 1020 m/sec; (heavy shell): 825 m/sec
Shell weight (light (HE)): 250 kg; (heavy (HE or AP)): 405 kg
Maximum range (light shell): 38,000 m; (heavy shell): 28,000 m
Original designers Schneider-Canet, Le Creusot
Manufacturer Putilov Arsenal, Reval (Tallinn), Estonia

Remarks: Naval guns from former Russian battleship *General Alekseyev* (ex-*Imperator Aleksandr III*, ex-*Volya*) captured en route from France to Finland by German forces in Norway during April 1940. Extensively restored and new ammunition made by Friedr. Krupp AG, Essen, during 1941. Emplaced as 'Batterie Mirus' at La Frie Baton, Guernsey.

30.5 cm K 14(r) of the 'Batterie Mirus' in action

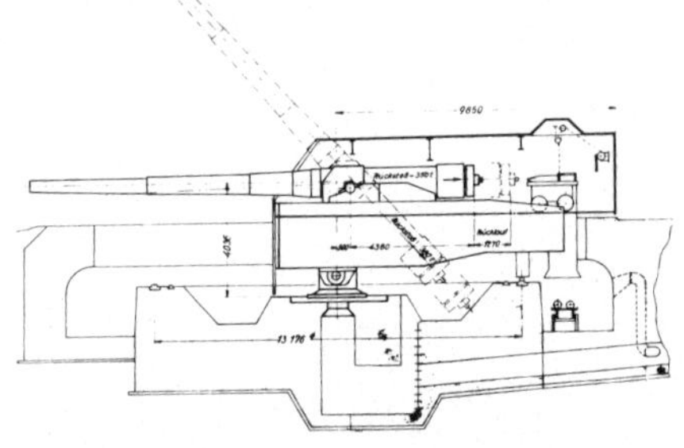

30.5 cm K 14(r) in Bettungsschiessgerüst C 40 *of the 'Batterie Mirus', showing the principal parameters and elevation*

30.5 cm Haubitze Modell 16 (Bofors)

German designation 30.5 cm H Mod 16
Calibre 305 mm
Length of piece (L/30): 9150 mm
Weight emplaced 134,000 kg
Traverse 360°
Elevation 0° to +70°
Muzzle velocity 620 m/sec
Shell weight 385 kg
Maximum range 20,000 m
Rate of fire 1 rpm
Manufacturer AB Bofors, Bofors, Sweden

Remarks: Swedish WW I design. Four pieces emplaced as 'Batterie Lödigen' near Narvik in Norway.

---

34 cm Kanone Modell 12(f)

German designation 34 cm K Mod 12(f)
Original designation Canon de côte 340 mle 1912
Calibre 340 mm
Length of piece (L/47.3): 16115 mm
Length of barrel 15300 mm
Weight emplaced 166,000 kg
Weight of piece 98,900 kg
Traverse Up to 360°
Elevation +15° to +50°
Muzzle velocity 867 m/sec
Shell weight 427 or 540 kg
Maximum range 30,000 m
Rate of fire 1 round every 4 mins

Remarks: Captured French coastal defence guns. Four pieces emplaced in two double turrets at Cap Cépet near Toulon; manned by 4/Marine Art. Abt. 682.

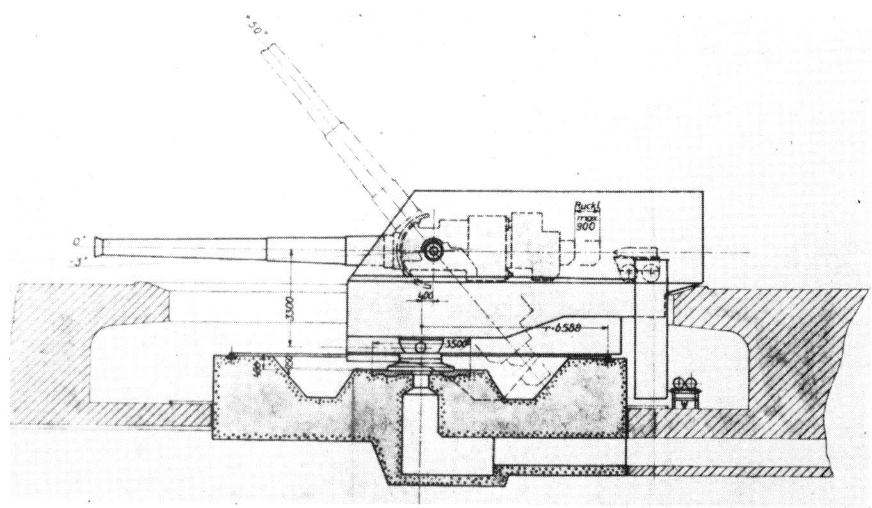

34 cm K Mod 12(f) in Bettungsschiessgerüst C 41

Fortress weapons

Although fortress weapons had been a specialised form of artillery for many years they had gradually fallen from favour during the latter half of the 19th century. This was due not only to their relatively specialised nature but also to their cost which was often very high. Fortress weapons by their very nature tended to be heavy guns protected by massive layers of armour plate which imposed severe limitations on traverse and elevation, so wherever possible the fortifications that gradually spread across Europe between various countries tried to use conventional field weapons behind some form of protection. The fall of the Liège forts in Belgium in 1914 seemed to many to mark the demise of fixed fortifications in modern warfare but the lessons of Fort Douamont and Fort Vaux during the 1917 Verdun battles re-emphasised the importance of the fort as a bastion in defence. Both the French and the Germans noted the Verdun message but characteristically they extracted different content. To the French the lessons of Verdun indicated the overall importance of the fort in the rigid defence line, and the result was the Maginot Line. The Germans used the fort as a form of tactical anchor in a defence line which was largely made up of mobile field formations able to strike at, and exploit, any advance against them. Thus to the Germans the concept of fixed fortress weapons was never a high priority, but that held true only during the period when the General Staff was dominant. With the emergence of Adolf Hitler the emphasis was changed and fortress weapons once more came to the fore with the setting up of the pre-war *Westwall* (called 'Siegfried Line' by the Allies) and after that the massive 'Atlantic Wall'.

With the gradual establishment of the Siegfried Line forts opposite the Maginot Line defences came a need to equip them with modern weapons. With hindsight it would appear that the German General Staff attitudes were still active even when orders for weapons for the new forts were issued to German industry, for the results were hardly very impressive. The largest piece developed was a 15 cm version of the standard field howitzer and most of the rest were lighter weapons. But this lack of fire power did not prevent the developed weapons from being very complex and expensive items.

At the lower end of the calibre scale was a Krupp product, the *5 cm KK (Kasematt-Kanone)* intended for use against tanks. For its time (1936) the use of a 5 cm gun as an anti-tank weapon was ahead of current practice, but the great rival of Krupp, Rheinmetall, produced a design of casemate fitted with a turntable to take a *3.7 cm Pak 35/36* without its wheels and this also went into production during 1937. Not content with this a special version of casemate capable of taking a *5 cm Pak 38* was designed during 1939 by Rheinmetall although it did not go into production. A similar casemate was produced in some numbers to take the *7.5 cm FK 16nA*, again by Rheinmetall, who were able to get most of the fortress weapon contracts.

While the last three were adaptations of existing weapons there were more specialised developments in the 10.5 cm calibre range. These comprised four different versions of *10.5 cm KK* which varied not only in barrel length but also in whether they were to be fitted in turrets or casemates. Not many were made. The largest piece was a *15 cm sHT* but very few of these were delivered. To back up these weapons a number of specialised machine gun installations were produced, usually for the *7.92 mm MG 34* but some were also adapted to take the *7.92 mm sMG 08* and other weapons. Most of these were simply heavy machine gun mountings, with limited traverse, firing through armoured embrasures and protected by armoured shutters.

Conventional mortars also figured in German fortress weapon plans. Rheinmetall went ahead with a special fortress installation of the standard *8 cm GrW 34* during 1940 but that scheme never materialised. A more successful venture by the same company was the *leHT*, an adaptation of the *10.5 cm NbW 40* for use under an armoured turret. Even more adventurous was the *automatischer 5 cm Granatwerfer M.19*, which was produced in two versions. Both were intended for use against troops attacking along lines 'blind' to sectors of fortifications, typical examples of which would be ravines or beaches beneath headlands. The *M.19* could fire a rapid sequence of mortar bombs at quite close range and with some accuracy but it was a complex and expensive weapon. Development started in 1934 and the bulk of the production run by Rheinmetall was concentrated on a version intended for use in an armoured pill-box. The second version got only as far as two development prototypes, but it was a fully automatic model that required no crew to man it as all loading and aiming were carried out by remote control. However, most *M.19*s were produced too late to be used in their intended role and suffered the inevitable fate of being incorporated into the Atlantic coast in France and the Channel Islands, although the numbers involved were not large.

Almost as soon as production of fortress weapons began the early war years saw the by-passing of the defence systems they were supposed to counter. The Sudetenland defences came under German control without a fight during 1938 and in 1940 the Maginot Line was taken virtually without a struggle by simply going round the fixed line of defences. As a result the production of fortress weapons in Germany came to an abrupt halt. When the 'Atlantic Wall' began to grow there was no great problem in supplying weapons to arm the various strongpoints. All that had to be done was remove the existing armament from the Sudeten and Maginot Line installations. The Czech Sudeten defences would have proved to be formidable obstacles indeed had the Wehrmacht ever had to attack them but they fell into German hands intact. Much of their armament found its way to the 'Atlantic Wall', the most numerous item being the 4 cm kanon vz. 36, a 47 mm anti-tank gun specially designed for use in fortress with a co-axial 7.92 mm machine gun. The barrel of this gun moved in a special ball fitting mounted in a well-designed armoured embrasure set in concrete. Large numbers of these guns were incorporated into the beach defences of the 'Atlantic Wall', so many in fact, that it is tempting to surmise that the type was kept in production by Škoda at Pilsen to meet the German demands. Apparently most of the heavier Czech guns were left where they were as they were probably more trouble to dismount than they were worth, and the same seems to have been true of the few weapons in the so-called 'Siegfried Line'.

But the Maginot Line offered a richer selection, and in a very short time the Atlantic coast defences were bristling with a number of French weapons. Among the most important of these was the fortress version of the old Schneider Canon de 105 mle 1913, the L 13 S. Another French weapon much used by the Germans was the Mortier de 50 mle 1935 which was a breech-loading mortar designed specifically for the Maginot Line. The mortar barrel was mounted in a traversing steel plate and range alterations were made by adjusting a gas bleed-off valve. Observations were made from a separate position. The German engineers simply took these mortars from their installations and set them up as coast defence weapons in open pits, complete with their mounting plates; a large number were so used in the Channel Islands. Small numbers of 81 mm mortars were removed from their fortress installations and issued to field formations, but much of the Maginot Line armament was so specialised (typical examples were the numbers of 135 mm 'lances-bombes' designed specially for the Maginot Line forts) that it was deemed best to leave it where it was.

One form of fortress weapon that was used in some numbers by the Germans was the *'Panzerstellung'* or *'Panzerturm'*. The use of old tank turrets in fortifications was the result of the German policy of converting captured tanks to accommodate artillery or anti-tank guns in place of their outmoded main armament. Very

few captured tanks could take the new heavier weapons in their turrets and consequently the turrets were removed and superstructures were built up on the tank hulls to protect the new armament. As a result of this policy the German armouries had a stockpile of tank turrets complete with their main armament. Most of these turrets were then built into the Atlantic coast defences. Some, such as the elderly Renault FT17 turrets, carried only a single machine gun and were thus of little use other than as armoured observation posts. Most of the turrets so used came from captured French tanks but use was also made of British tanks abandoned at Dunkirk: they were frequently stripped and dug in complete with their turrets as a form of 'beach strongpoint' along the Atlantic and North Sea coasts.

In typical German fashion the use of an extemporised expedient soon became part of an official policy and it was not long before German industry was called upon to supply tank turrets for fixed defences. Many outdated PzKpfw I, II, III, and IV turrets were transported to the Atlantic coast and other selected locations, and in late 1943 a number of Panther tank turrets were specially built for the 'Gothic Line' defences in Italy. These turrets were simplified versions of the standard Panther part but featured the usual main and co-axial armament, and whereas the other turret types were mounted on turret rings set in concrete the Panther version incorporated an armoured steel box in two horizontal halves. The upper half mounted the turret and was the 'fighting' compartment, while the lower half provided very basic living accommodation for the crew of three to four men. Once emplaced and camouflaged these Panther turrets were almost impossible to locate or hit and were so effective that attacks on the 'Gothic Line' had to be made by trying to bypass the areas where these turrets were emplaced.

As the war ended there were plans to utilise available Tiger and Königstiger turrets in fixed defences, but none appear to have been emplaced.

Twin 15 cm guns in a revolving armoured turret installed on the French coast

5 cm Maschinengranatwerfer M 19

German designation 5 cm M 19
Calibre 50 mm
Traverse 360°
Elevation +48° to +85°
Muzzle velocity 50–91 m/sec
Bomb weight 0.9 kg
Range 50–750 m
Type of feed 6 bomb clips
Ammunition stowage 3944 bombs
Weight of weapon (without mount): 220 kg
Manufacturer Rheinmetall-Borsig AG, Düsseldorf

Remarks: Development started 1934 but by the time limited production commenced during 1940 their original purpose was no longer in being and small number built were installed in Atlantic coast defences near Boulogne and in the Channel Islands. A remotely-controlled and fully automatic version was designed in conjunction with M 19 but only two prototypes were completed.

Loading the 5 cm M 19

5 cm Kasematt-Kanone

German designation 5 cm KK
Calibre 50 mm
Muzzle velocity 830 m/sec
Shot weight 2.06 kg
Manufacturer Friedr. Krupp AG, Essen

Remarks: Accepted for service and in production 1936. No other data available.

leichte 10.5 cm Turm-Haubitze

German designation l HT
Design designation 10.5 cm leichte Haubitze im Turm (l HT 73 P9)
Calibre 105 mm
Weight in action 1262 kg
Traverse 360°
Elevation +48° to +85°
Muzzle velocity 271 m/sec
Projectile weight 8.3 or 9.6 kg
Maximum range 6000 m
Rate of fire 6–7 rpm
Manufacturer Rheinmetall-Borsig AG, Düsseldorf

Remarks: Development started 1935, in production by 1939.

mittlere 10.5 cm Kasematt-Kanone and kurze 10.5 cm Kasematt-Kanone

German designation m 10 cm KK; kz 10 cm KK
Design designation m 10.5 cm K im Kasematte (30 und 31 P8, 405 und 406 P9)
Calibre 105 mm
Length of piece (L/31): 3255 mm
Weight of piece 4313 kg
Traverse 60°
Elevation −5° to +10°
Muzzle velocity 540–616 m/sec
Shell weight 15.1 kg
Maximum range 6995 m
Manufacturer Rheinmetall-Borsig AG

Remarks: Development started 1934, accepted for service in 1936; in production by 1939. If necessary this gun could be drawn back into casement for extra protection. A co-axial MG 34 was also fitted.

Version designated kz 10 cm KK was similar but with a shorter barrel.

mittlere 10.5 cm Turm-Kanone

German designation m 10 cm KT
Design designation m 10.5 cm K im Turm (43 P8, 92, 93 und 94 P9)
Calibre 105 mm
Length of piece (L/31): 3255 mm
Weight of piece 4313 kg
Weight installed (approx): 718,000 kg
Traverse 360°
Elevation −10° to +10°
Muzzle velocity 540–616 m/sec
Shell weight 15.1 kg
Maximum range 6995 m
Manufacturer Rheinmetall-Borsig AG, Düsseldorf

Remarks: Development started 1934. Accepted for service in 1936 as counterpart to m 10 cm KK; in production 1939.

lange 10.5 cm Kasematt-Kanone

German designation lg 10 cm KK
Design designation lg 10.5 cm K L/52 im Kasematte (Pak) 766 P3
Calibre 105 mm
Length of piece (L/52): 5460 mm
Traverse Up to 90°
Elevation −5° to +25°
Muzzle velocity 835 m/sec
Shell weight 15.1 kg
Maximum range 16,000 m
Manufacturer Rheinmetall-Borsig AG, Düsseldorf

Remarks: 1940 development of m 10 cm KK. Prototype only.

---

lange 10.5 cm Turm-Kanone

German designation lg 10 cm KT
Design designation lg 10.5 cm K L/52 in Turm (Pak) 765 P3
Calibre 105 mm
Length of piece (L/52): 5460 mm
Weight installed (maximum armour): 773,000 kg (approx); (standard armour): 440,000 kg (approx)
Traverse 360°
Elevation −10° to +45°
Muzzle velocity 835 m/sec
Shell weight 15.1 kg
Shot weight 14 kg
Maximum range 20,000 m
Manufacturer Rheinmetall-Borsig AG, Düsseldorf

Remarks: 1939 development of m 10 cm KT. Prototype only.

---

schwere 15 cm Turm-Haubitze

German designation sHT
Design designation 15 cm sH L/35 in Turm (33 P8)
Calibre 149.1 mm
Length of piece (L/35): 5218 mm
Weight of piece (approx): 18,610 kg
Weight installed (approx): 747,000 kg
Traverse 360°
Elevation −3° to +45°
Muzzle velocity 622–635 m/sec
Shell weight 43.5 kg
Maximum range 15,500 m
Manufacturer Rheinmetall-Borsig AG, Düsseldorf

Remarks: Development started 1935 and in limited production by 1939.

As an alternative lg 10.5 cm K L/52 could be fitted to same installation.

Schwere 15 cm Turmhaubitze *of the so-called 'Siegfried Line'*

---

15 cm Turm-Kanone

German designation 15 cm KT
Design designation 15 cm K 18 L/55 im Turm (464 P2)
Calibre 149.1 mm
Length of piece (L/55): 8200 mm
Length of barrel 6432 mm
Weight installed (approx): 740,000 kg
Traverse 360°
Elevation −3° to +45°
Muzzle velocity 890 m/sec
Shell weight 45 kg
Maximum range 24,500 m
Manufacturer Rheinmetall-Borsig AG, Düsseldorf

Remarks: 1938 development of sHT to take 15 cm K 18 barrel. Prototype finished in 1940 but not accepted for service.

4.7 cm Panzerabwehrkanone-Kasematt 36(t)

German designation 4.7 cm Pak K 36(t)
Original designation 40 mm kanon vz. 36
Length of piece (L/43.4): 2040 mm
Weight of weapon 1600 kg
Weight installed 2600 kg
Traverse 45°
Elevation −18° to +12°
Muzzle velocity (HE): 600 m/sec; (AP): 775 m/sec; (AP 40): 1080 m/sec
Shell weight 1.5 kg
Shot weight 1.65 kg; (AP 40): 0.83 kg
Maximum range 6000 m; (operational): 2900 m
Armour penetration (30°): 52 mm at 100 m; 35 mm at 1500 m; (AP 40)(30°): 100 mm at 100 m; 58 mm at 500 m
Manufacturer Škoda, Pilsen

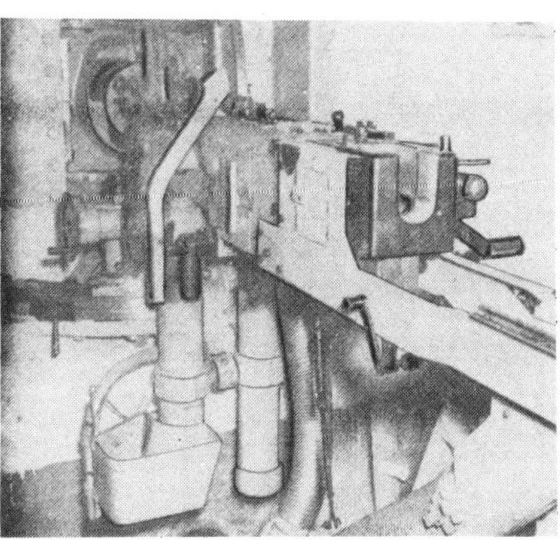

Remarks: Originally built for Czech Sudeten defences, these guns were removed in large numbers together with their embrasures, and shipped to Atlantic coast. A co-axial MG 37(t) was usually fitted.

4.7 cm Pak K 36(t). Interior view showing the ball mount for the gun barrel

5 cm Festungsgranatwerfer 210(f)

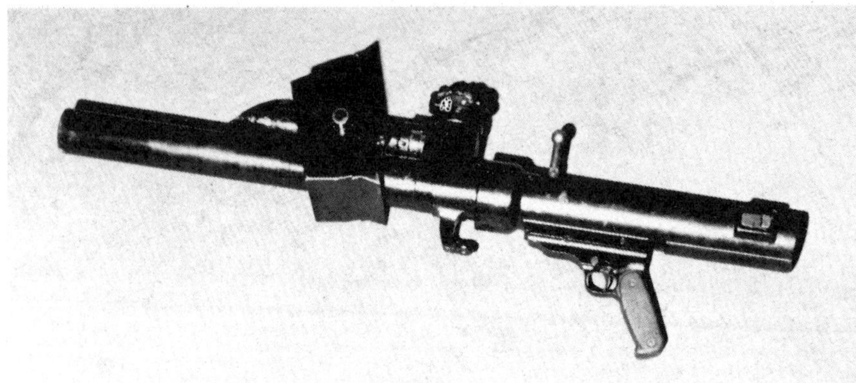

5 cm FestGrW 210(f) in action with German troops in an open concrete fortification ➤

German designation 5 cm FestGrW 210(f)
Original designation Mortier de 50 mle 1935
Calibre 50 mm
Weight of mortar 11 kg
Traverse Up to 360°
Elevation 45° fixed
Muzzle velocity 120 m/sec
Bomb weight 0.975 kg
Maximum range 1075 m

Remarks: Originally incorporated in Maginot Line defences, numbers of these mortars were taken to Atlantic coast defences along with their original embrasures. When used as coast defence weapons these mortars were usually placed on simple pivot mountings in pits and range alterations were made by bleeding off propellant gases through a variable aperture.

10.5 cm Kanone 331(f)

10.5 cm K 331(f) in a casemate on Jersey. This version is fitted with a curved armoured shield to protect the crew when the gun is traversed

10.5 cm K 331(f) in a casemate on the Atlantic coast

German designation 10.5 cm K 331(f)
Original designation Canon de 105 L mle 1936 Schneider
Calibre 105 mm
Length of piece (L/37.6): 3905 mm
Length of barrel 3802 mm
Weight installed 2300 kg
Weight of piece 1105 kg
Traverse 120°
Elevation (approx): −10° to +18°
Muzzle velocity 550 m/sec
Shell weight 15.74 kg
Maximum range 12,000 m
Manufacturer Schneider et Cie, Le Creusot

Remarks: Originally Maginot Line fortress guns. Extensively placed along the Atlantic Wall for beach defence.

Four different types of Panzerturm:

French Hotchkiss H 38 tank turret with 37 mm gun

French Renault FT 18 tank turret with 7.5 mm machine gun

French AMC 35 tank turret with 47 mm gun

German PzKpfw V Panther turret installed as part of the 'Gothic Line' defences in Italy

Infantry guns

Like so many weapons used between 1939 and 1945 the infantry gun was a result of the specialised nature of WW I. For centuries artillery had been used in direct support of massed infantry with the guns situated amongst, or in front of, the formations they were supposed to support, and right up to 1914, with the exception of a few tentative forays into the use of indirect fire during the Boer War, the same policy was adopted. By the end of 1914 those days were over: and the guns moved to the rear and their fire was indirect. As WW I continued the guns in the rear were used in larger and larger numbers in elaborate fire plans, and direct support of the infantry in the front line grew more and more remote. Not surprisingly the soldier in the trenches felt the need for some form of local heavy fire support and the trench mortar and the bomb thrower grew into prominence, until at the end of 1916 the first infantry guns came into being. The first of these were small handy weapons with simple carriages and sights capable of direct or elementary indirect fire. One of the first was the French Canon d'Infanterie de 37 mle 1916 TRP, a half-scale version of the famous French '75' mounted on a simple frame carriage. Škoda produced a miniature artillery piece in their 3.7 cm pěchotní dělo vz. 15 which could be used for trench warfare, but by 1918 the quantity of specialised infantry guns was still small and numbers of mountain guns were used in the trenches as a temporary expedient, an expedient which would be frequently repeated between 1939 and 1945.

After 1918 most armies tended to move away from the idea of a specialised infantry gun as many considered that the mortar could fulfil most of the tasks demanded by the infantry. But during the 1920s and 1930s the German Army did not subscribe to this concept, and during the 1920s German armament designers were busy working on designs for specialised infantry artillery. Rheinmetall were among the first to delve into this area and as early as 1927 proposed an unconventional light gun designed specifically for use by infantry formations. The design was unconventional in that when the breech opening lever was operated the breech block stayed still and the barrel, which was mounted in a slipper, was raised to present the breech for loading. Despite this odd feature the Rheinmetall design was accepted for service and production began in 1929. It was designated *7.5 cm leIG 18* and the first production models were made at Rheinmetall's Düsseldorf plant but later production was switched elsewhere. These other plants also manufactured a later variant of the *leIG 18*, the *7.5 cm leGebIG 18*, an interim mountain gun design which used the *leIG 18* barrel on a new lightened carriage. The design of this mountain warfare variant was started in 1933 and the first guns were issued in 1937. By 1937 the original version of the *leIG 18* with spoked wheels for horse traction was giving way to a new model with metal disc wheels and pneumatic tyres more in keeping with the gradual mechanisation of the German forces.

7.5 cm leIG 18 with pneumatic tyres in action with Waffen-SS troops on the Eastern Front

There was one further variant of the *leIG 18* produced during 1939, a much modified version intended for airborne use. Once again the basic barrel remained almost unchanged but the carriage was completely new. Small disc wheels were fitted and the whole weapon could be easily broken down for stowing in four containers. Eight guns were made and issued for troop trials, but the advent of the recoilless gun for use in the same role precluded any production.

In addition to the sizeable production contract for the *leIG 18*, Rheinmetall also evolved various new designs of infantry gun. One that was built in small series was the *7.5 cm IG L/13*, which appears to have been a purely commercial venture using a conventional gun design on a carriage with split pole trails. A small number were issued for troop trials but no production order was made. Another Rheinmetall design was the *2.8 cm WBA L/52*, which was built in 1938 to an OKH request. The gun was mounted on a tripod and the idea was to provide a really light gun for carriage by only two men, but this venture came to nothing. Other firms are believed to have been working on similar designs but none appear to have reached the hardware stage.

Around the same time that Rheinmetall began work on the *leIG 18* they also initiated the design of a heavier infantry howitzer. Design studies were made of howitzers with calibres of 10.5, 15 and 21 cm to be mounted on carriages with split or box trails. By 1935 the Army had made their choice and production of the *15 cm sIG 33* began. The *sIG 33* was a conventional design, but a rather heavy one, intended for horse traction; later versions had rubber-tyred wheels and air brakes to suit them for towing behind vehicles. Production began at Düsseldorf but was later

7.5 cm leichtes Infanteriegeschütz 18 *(leIG 18) with wooden wheels*

15 cm sIG 33 in action on the Eastern Front. This model has solid rubber-tyred wheels

switched to the AEG-Fabriken at Berlin-Henningsdorf and Böhm. Waffenfabrik at Strakonitz. The first examples were issued during 1938; these were the Model A of all-steel construction. During 1938 the first examples of the Model B were made which tried to reduce the substantial weight of the Model A by incorporating some light metal alloys in the carriage. This version went into small-scale production but a further version proposed in 1939, the Model C with all light-alloy carriage, was not accepted. By that time the use of light alloys was restricted as far as possible to Luftwaffe production and only a single example of the Model C was made.

In service the *sIG 33* proved to be a reliable and powerful weapon but its bulk and weight often restricted its mobility and consequently during 1940 the first attempts were made to mount the *sIG 33* on a self-propelled carriage. Even so the *sIG 33* remained in service until 1945 alongside the *7.5 cm leIG 18* in the support companies of German infantry regiments.

By the end of 1940 it was felt by OKH that a more powerful 7.5 cm infantry gun was needed and this time Krupp became involved in the infantry gun field. They designed and produced the prototype of the *7.5 cm IG 42*, a conventional light gun with a split-trail carriage and a prominent muzzle brake. But the *IG 42* did not get into production, for by the time the prototype was ready the need for a new gun was no longer a high priority. The fire power of the *leIG 18* had been increased by the use of anti-tank hollow-charge projectiles and mortars were becoming increasingly popular. But by 1944 the need for a new gun was once more acute and the *IG 42* was resurrected. The original Krupp design was redesigned *7.5 cm IG 42 aA* to distinguish it from the new *IG 42* which was the old *IG 42* barrel on a new carriage. The carriage selected was the one designed for the *8 cm PAW 600* adapted to take the 7.5 cm barrel. Not many appear to have been made. Another carriage that was adapted to take the *IG 42* barrel was that of the old *3.7 cm Pak 35/36* which by 1944 was obsolete and available in some quantity, with additional numbers provided by captured Soviet 37 mm obr. 1930 anti-tank guns which were copies of the *Pak 35/36* design. The *IG 42/Pak 35/36* combination was originally designated the *7.5 cm Pak 37*, but this was later changed to *7.5 cm IG 37*. Production of this variant began in 1944 and was still in progress when the war ended.

Still in the experimental stage when the war ended was a version of the *7.5 cm IG 42* with a smooth bore barrel intended for firing fin stabilised projectiles. A single example was found at Hillersleben in 1945.

As the war continued after 1941 the infantry gradually lowered their emphasis on infantry guns and began to increase the use of heavy mortars in their place. The reasons for this were simple enough as mortars were lighter, cheaper, and just as effective as the infantry gun, but when the war ended the infantry gun was still in service with many infantry regiments. By 1945 the standard German infantry guns had been supplemented by other designs, some of them mountain guns, and some light field guns such as the French '75'. One gun that was widely used by the German troops was the Soviet 76.2 mm Infantry Gun obr. 1927, the 76–27. The Soviets used infantry guns in much the same way as did the Germans so the Soviet gun was easily assimilated into German use on a grand scale. In German use the Soviet weapon was given the cumbersome title of *7.62 cm Infantriekanonehaubitze 290(r)*, or *7.62 cm IKH 290(r)*. Other foreign guns that were used by the Germans in the infantry support role were the Italian Cannone da 47/35 modello 35, a light gun that served the Italian Army as an anti-tank gun and pack gun as well as an infantry gun; a number of these were used by German troops based in Italy. A few second-line units based in France shortly after 1940 were issued with old French 37 mm infantry guns – the same guns that were among the very first to be designed for infantry use during WW I. In this case the wheel had turned full circle, but these elderly guns were used for only a very short while until more suitable equipment was available.

7.5 cm leichtes Infanteriegeschütz 18

7.5 cm leIG 18 with pneumatic tyres in action with Waffen-SS troops on the Eastern Front

7.5 cm leichtes Infanteriegeschütz 18 *(leIG 18) with wooden wheels*

German designation 7.5 cm leIG 18
Calibre 75 mm
Length of piece (L/11.8): 885 mm
Length of barrel 783 mm
Length of rifling 674 mm
Weight travelling (tractor): 515 kg; (horse): 405 kg
Weight in action (tractor): 570 kg; (horse): 400 kg
Traverse 11°
Elevation −10° to +75°30′
Muzzle velocity 221 m/sec
Shell weight 6 kg or 5.45 kg
Maximum range (normal charge): 3550 m; (super charge): 4600 m
Rate of fire 8–12 rpm
Barrel life 10,000–12,000 rounds
Original designers Rheinmetall-Borsig AG, Düsseldorf
Manufacturers Böhm. Waffenfabrik, Strakonitz; Habämfa, Ammendorf/Halle

Remarks: Development started during 1927; first issued for service 1932. Produced in two basic versions – for horse (spoked wheels) or mechanical traction (pneumatic tyres). Became one of most widely used guns in Wehrmacht service.

7.5 cm leichtes Gebirgsinfanteriegeschütz 18

German designation 7.5 cm leGebIG 18
Calibre 75 mm
Length of piece (L/11.8): 885 mm
Length of barrel 783 mm
Length of rifling 674 mm
Weight travelling 410 kg
Weight in action 440 kg
Traverse 35°
Elevation −10° to +73°30′
Muzzle velocity 221 m/sec
Shell weight 5.45 kg
Maximum range (normal charge only): 3550 m
Rate of fire 8–12 rpm
Barrel life 10,000–12,000 rounds
Original designers Rheinmetall-Borsig AG, Düsseldorf
Manufacturers Böhm. Waffenfabrik, Strakonitz; Habämfa, Ammendorf/Halle

Remarks: Basic leIG 18 modified for mountain service. Development started during 1935 and production commenced in 1937. Intended as temporary equipment until service debut of GebG 36 but remained in service until end of WW II. Could be broken down in 10 loads for manual pack transport or 6 loads for animal transport. A sled was an optional firing carriage.

7.5 cm leichtes Gebirgs-Infanteriegeschütz 18 *(leGebIG 18) showing the lateral traverse to the left*

7.5 cm leichtes Infanteriegeschütz 18 Fallschirmjäger

German designation 7.5 cm leIG 18 F
Calibre 75 mm
Length of piece (L/11.8): 885 mm
Length of barrel 783 mm
Length of rifling 674 mm
Weight in action 325 kg
Traverse 35°
Elevation −10° to +73°30′
Muzzle velocity 221 m/sec
Shell weight 5.45 kg
Maximum range 3550 m
Rate of fire 8–12 rpm
Manufacturer Rheinmetall-Borsig AG, Düsseldorf

Remarks: 1939 variant of 7.5 cm leIG 18 for use by airborne formations. Designed to be broken down into four container loads with each container weighing about 140 kg. Eight produced during 1939 and issued for troop trials but discontinued in favour of recoilless guns.

7.5 cm Infanteriegeschütz L/13

German designation 7.5 cm IG L/13
Calibre 75 mm
Length of piece (L/13): 975 mm
Weight in action 375 kg
Traverse 50°
Elevation −5° to +43°
Muzzle velocity (light shell): 305 m/sec; (heavy shell): 225 m/sec
Shell weight (light): 4.5 kg; (heavy): 6.5 kg
Maximum range (light shell): 5100 m; (heavy shell): 3800 m
Rate of fire Up to 20 rpm
Manufacturer Rheinmetall-Borsig AG, Düsseldorf

Remarks: Rheinmetall commercial project produced during mid-1930s. Basically progressive development of earlier leIG 18 with conventional barrel chambered for different ammunition and split telescopic trails. A small batch procured by German Army for trials but no production ensured.

7.5 cm Infanteriegeschütz 42

German designation 7.5 cm IG 42
Calibre 75 mm
Length of piece with m/b (L/24): 1815 mm
Length of rifling 1340 mm
Weight travelling 595 kg
Weight in action 590 kg
Traverse 60°
Elevation −6° to +32°
Muzzle velocity 280 m/sec
Shell weight 5.45 kg
Maximum range 4600 m
Manufacturer (barrel): Friedr. Krupp AG, Essen; (carriage): Rheinmetall-Borsig AG, Düsseldorf

Remarks: First IG 42 was Krupp prototype based on 1940 specification – not accepted for service. Designation revived 1944 for new barrel mounted on 8 cm PAW 600 carriage – relatively few issued for service. Original Krupp gun then became IG 42 aA – used for trials with smooth-bore barrel and finned projectiles.

7.5 cm IG 42

7.5 cm IG 42 aA (original Krupp design)

7.5 cm IG 42 glatt, *the basic IG 42 aA fitted with a new smooth-bore barrel and improved muzzle brake to test various fin-stabilized hollow-charge shells*

7.5 cm Infanteriegeschütz 37 or 7.5 cm Panzerabwehrkanone 37

Original designation 7.5 cm Pak 37
German designation 7.5 cm IG 37
Calibre 75 mm
Length of piece with m/b (L/24): 1815 mm
Length of rifling 1340 mm
Weight in action 510 kg
Traverse 58°
Elevation −10° to +40°
Muzzle velocity 280 m/sec
Shell weight 5.45 kg
Maximum range 5150 m

Remarks: Stop-gap anti-tank/infantry gun. Late 1944 expedient combining cut-down captured French 75 mm barrels with obsolete 3.7 cm Pak 35/36 or captured Soviet Pak 158(r) (37 mm obr. 1930) carriages.

15 cm schweres Infanteriegeschütz 33

German designation 15 cm sIG 33
Calibre 149.1 mm
Length of piece (L/11.4): 1700 mm
Length of rifling 1346 mm
Weight travelling (tractor): 1825 kg; (horse): 1700 kg
Weight in action (tractor): 1800 kg; (horse): 1680 kg
Traverse 11°
Elevation −4° to +75°
Muzzle velocity (HE): 240 m/sec; (hollow charge): 280 m/sec
Shell weight (HE): 38 kg; (hollow charge): 24.6 kg
Maximum range 4700 m
Rate of fire 2–3 rpm
Barrel life 10,000–15,000 rounds
Original designers Rheinmetall-Borsig AG, Düsseldorf
Manufacturers AEG-Fabriken, Berlin-Henningsdorf; Böhm. Waffenfabrik, Strakonitz

Remarks: Most powerful of all German infantry guns. Development started in 1927, approved for service 1933, in production from 1936 until 1945. Could fire HE, smoke and hollow-charge shells; even more effective in combat when mounted on SP carriages. Standard carriages produced in three versions: Modell A(1936) – all steel; Modell B(1938) – mixed steel/light alloy carriage. Modell C(1939) – all light alloy carriage, not accepted for service. sIG 33 developed into 15 cm Sturm-haubitze 43 for close support AFVs.

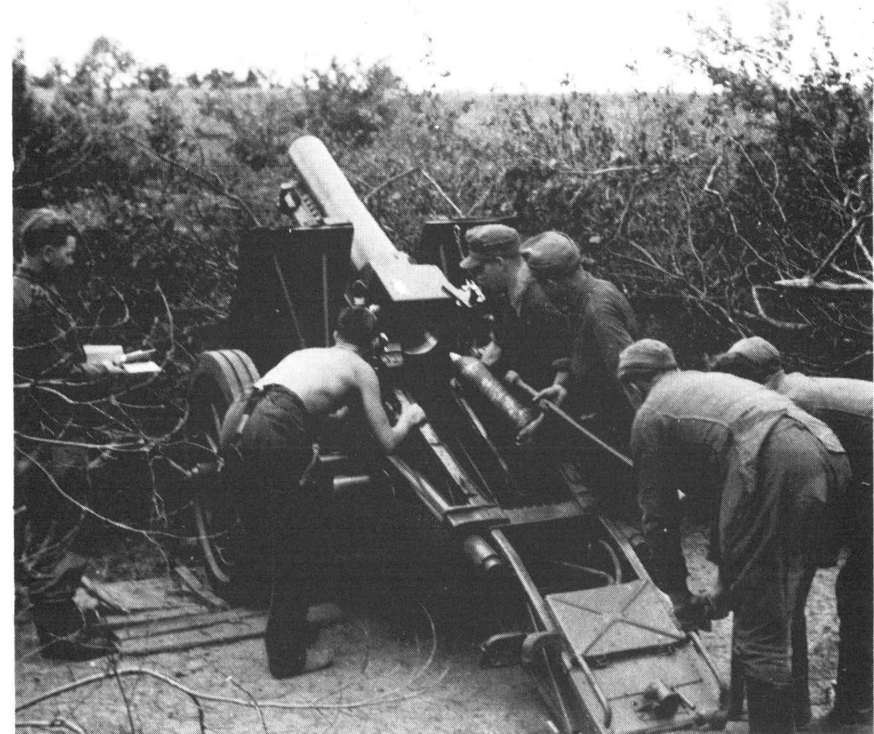

15 cm sIG 33 in action on the Eastern Front. This model has solid rubber-tyred wheels

15 cm sIG 33 with 15 cm Stielgranate 42

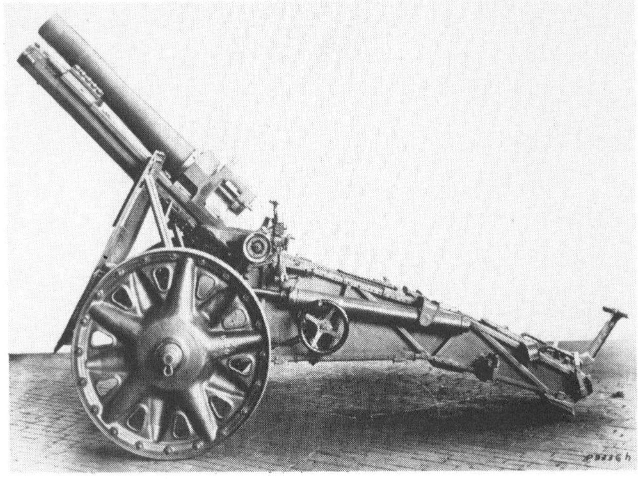

15 cm schweres Infanteriegeschütz 33 *(sIG 33)*

WBA L/52

Design designation WBA L/52
Calibre 28 mm
Length of piece (L/52): 1456 mm
Weight in action 140 kg
Traverse 360°
Elevation −15° to +20°
Muzzle velocity 853 m/sec
Projectile weight 0.381 kg
Manufacturer Rheinmetall-Borsig AG, Sömmerda

Remarks: 1938 project for light infantry gun that remained in prototype form only. No other data available.

3.7 cm Infanteriegeschütz 152(f)

German designation 3.7 cm IG 152(f)
Original designation Canon d'Infanterie de 37 mle 1916 TRP
Calibre 37 mm
Length of piece (L/22): 814 mm
Weight on wheels 160.5 kg
Weight in action 108 kg
Traverse 35°
Elevation −8° to +17°
Muzzle velocity 367 m/sec
Shell weight 0.555 kg
Maximum range 2400 m
Rate of fire 10–15 rpm

Remarks: French light trench support gun of WW I vintage. Numbers still in service 1939–40. Some captured guns issued to German garrison troops in France 1940–41.

7.6 cm Infanteriegeschütz 260(b)

German designation 7.6 cm IG 260(b)
Original designation Canon de 76 FRC
Calibre 76 mm
Length of piece (L/9.2): 699 mm
Length of barrel 593 mm
Weight travelling 275 kg
Weight in action 243 kg
Traverse 40°
Elevation −6° to +80°
Muzzle velocity 160 m/sec
Shell weight 4.64 kg
Maximum range 2200 m
Rate of fire 4 rpm
Manufacturer Fonderie Royale des Canons, Liège

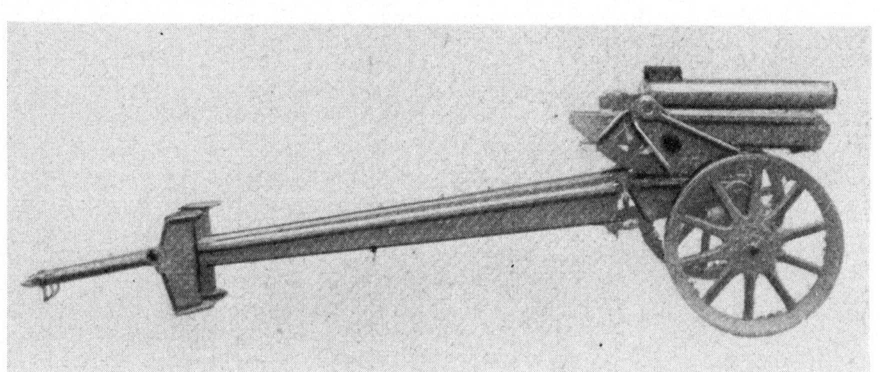

Remarks: Alternative 47 mm barrel could be fitted on same carriage. Total of 198 in Belgian service in September 1939. After 1940 only small number impressed in German service and used mainly by local garrison troops until 1941.

7.62 cm Infanteriekanonehaubitze 290(r)

German designation 7.62 cm IKH 290(r)
Original designation *76 mm Polkovaya Pushka obr. 1927 g* (76-27)
Calibre 76.2 mm
Length of piece (L/16.5): 1250 mm
Length of barrel 1163 mm
Length of rifling 765 mm
Weight travelling with limber 1595 kg
Weight in action 780 kg
Weight of piece 229 kg
Traverse 6°
Elevation −6° to +25°
Muzzle velocity 387 m/sec
Shell weight 6.4 kg
Maximum range 8550 m
Rate of fire 14 rpm

Remarks: Simple and sturdy regimental support weapon of original design. Produced in very large numbers. Many captured by German troops and widely used after 1941. Its popularity is shown by revised versions to take German gun sights and other modifications; special ammunition for these guns was also produced in Germany. Used on all fronts.

Mountain artillery

Mountain guns became a specialised form of artillery during the latter half of the 19th century, especially among the various states of Central Europe. Any nation with a mountain border to defend obtained some mountain guns to support their highly-trained mountain troop formations, and these guns themselves grew so specialised in form that they became a separate arm of the gun designer's art. Mountain guns had to be compact as they were often used in very restricted areas. They had to be light and easily broken down into loads for transport by mules or even men, and they had to be robust to withstand the hard service they would be exposed to in their environment. To add to an already difficult specification they also had to be capable of firing at extreme angles of elevation and depression, a condition imposed by the mountainous terrain in which they would be used.

One firm that devoted a great deal of attention to the mountain gun was the Austro-Hungarian concern of Škoda, based at Pilsen. Over a long period this firm produced a series of mountain gun designs which culminated in their 75 mm Model 15, one of the most successful mountain gun designs ever produced. This gun was adopted for German service during WW I and was still in use in 1933. By that time it was becoming obsolete and new designs were considered. Work on new mountain guns had already been started by Krupp and Rheinmetall in 1926 when both firms were asked by the Army Command to build prototypes of a 7.5 cm gun and a 10.5 cm howitzer for mountain use. By the time wooden mock-ups were made Krupp had decided to withdraw from the competition while the Rheinmetall designs fell short of the requirements and were not accepted. They were the *7.5 cm GebK L/21* and the *10.5 cm GebH L/15*, both of which used the same design of box trail.

It soon became apparent that it would be some time before new equipment would be available for issue to the mountain artillery units so an interim solution was sought. In the meantime the old Škoda M 15s were kept in service as the *7.5 cm GebK 15*, and a small quantity of Bofors 75 mm mountain howitzers were bought direct from Sweden and designated *7.5 cm GebH 34*. A more long-lasting expedient was to be the *7.5 cm leGebIG 18*, a version of the standard infantry gun with a lightened carriage. As a mountain gun this weapon had some limitations but it was kept in service until the end of WW II.

By 1935 Rheinmetall decided that they could draw on the experience gained from building and testing their unsuccessful prototypes and proceeded with a new design at OKH request. The eventual result was the *7.5 cm GebG 36* which went into production at Düsseldorf in 1938. Unlike the earlier design, the *GebG 36* used a split trail and could be broken down into eight loads for pack transport. In service it proved to be a very successful mountain gun and it was kept in production until at least 1944.

To complement the *GebG 36* OKH requested a new 10.5 cm howitzer. This time the Austrian firm of Gebr. Böhler at Kapfenberg were invited to submit a design in competition with Rheinmetall and their proposal was awarded the production contract while Rheinmetall's entry, the *10.5 cm GebH L/30*, did not get past the prototype stage. The Böhler proposal, initially known as the *Gerät 77*, was originally designed in 1936 but did not get into production until 1940. For a mountain weapon it was rather large but it had several novel features, one of which was the use of a firing pedestal under the axle. When the split trails were opened the wheels were 'toed-in' and raised off the ground so that the gun rested on the trail legs and the forward pedestal. A more unusual feature for a mountain howitzer was that the piece could be dismantled not into pack loads but into four wheeled loads that were towed by SdKfz 2 *kleines Kettenrad* tractors. A version with a strengthened axle was projected for use by airborne troops, but this did not get very far as recoilless guns were considered more suitable for that role.

Not content with the success of the *7.5 cm GebG 36*, in October 1940 OKH requested a redesign of the gun with view to improving carriage stability, for at maximum charge the gun could not be fired at low angles of elevation. Rheinmetall and Gebr. Böhler prepared projects under the designation *Gerät 99*, but the Rheinmetall submission was rejected and once again Böhler were given the contract. That was in 1942, but in the following year the programme was terminated after only four examples had been produced under the designation *7.5 cm GebG 43*, or *'Gebhard'*. The reason for this termination was no reflection on the design but rather a matter of production priorities, for by 1943 there was no urgent need for a new mountain gun. Some development work was carried out with the four prototypes, and at least three different types of muzzle brake, all of them rather large and efficient, were fitted for experiments.

In 1944 the firm of Böhler were involved in the design of a new type of mountain howitzer, the *15 cm GebH*. The choice of such a large calibre for mountain warfare was made possible by the application of the high-low pressure system as used in the *8 cm PAW 600* which would make the weapon light and suitable for its role. The war ended before the design reached the hardware stage.

After 1940 captured mountain guns were incorporated into the German mountain units, but even before that date Austria and Czechoslovakia had already contributed their mountain gun parks to the German armoury. Both nations used large numbers of Škoda guns including many M.15s and 100 mm Model 16 and 16/19s. Some of this booty was passed on to Italy and even Turkey, but most was impressed by the German troops. More Škoda guns came from Poland and Yugoslavia, the latter contributing the 75 mm Model 1928, an updated version of the M.15.

From Belgium came a small quantity of the Canon de 75 mle 1934 which was easily assimilated into the German armoury as it was the same Bofors Model 1934 already in German service. The war booty in France included a wide range of mountain artillery, most of which was taken over for German use. Among the total were numbers of elderly Canon de 65 M mle 1906 guns, the Canon de 75 mle 1919 and its counterpart the Canon Court de 105 M mle 1919. Additional guns of this type came from Yugoslavia as well. But the most valuable part of this booty comprised numbers of Canon de 75 M mle 1928, a sound modern design.

Mountain troops in Italy received small numbers of Obice da 75/18 M34 howitzers and were sometimes issued with additional Škoda M.15s from Italian sources. From the Soviet Union came large numbers of 76.2 mm obr. 1936 mountain guns. Originally these were a Škoda design but the type was licence-built in the Soviet Union in large quantities and the German troops found them good serviceable guns. Another Soviet mountain gun that had a varied design history was the Model 1909 which was the brainchild of a Greek inventor, Colonel Danglis. It was built by Schneider in France and sold to the old Tsarist Army. It was still in widespread use in 1941, and substantial numbers were impressed into German service.

From British sources came small numbers of 3.7 in howitzers, but although these were originally designed as mountain howitzers the Germans appear to have issued them as infantry guns in the Balkans. But a measure of the Germans' anxiety to use as many weapons as they could acquire can perhaps be best illustrated by their recorded use of the *7.5 cm Krupp Modell 1912* mountain gun. Only a single battery was taken into German service prior to WW I but at least one of these guns was still in service during WW II.

7.5 cm Gebirgsgeschütz 36

7.5 cm GebG 36 on skis

German designation 7.5 cm GebG 36
Calibre 75 mm
Length of piece (L/19.5): 1450 mm
Length of rifling 972 mm
Weight travelling (8 loads): 715 kg
Weight in action 750 kg
Traverse 40°
Elevation −2° to +70°
Muzzle velocity 475 m/sec
Shell weight 5.74 or 5.83 kg
Maximum range 9250 m
Rate of fire 6 rpm
Barrel life 6000–8000 rounds
Original designers Rheinmetall-Borsig AG, Düsseldorf
Manufacturer R. Wolf, Magdeburg-Bückau

Remarks: Development started during 1935, in production 1938; output continued until at least 1944 by which time production lines were established at Magdeburg-Bückau.

7.5 cm GebG 36 being transported by pack animals in eight loads or horse drawn

10.5 cm Gebirgshaubitze 40

German designation 10.5 cm GebH 40
Calibre 105 mm
Length of piece (L/30): 3150 mm
Length of barrel 2870 mm
Length of rifling 2407 mm
Weight travelling (4 loads): 2600 kg (approx)
Weight in action 1656 kg
Traverse 50°
Elevation −5° to +70°
Muzzle velocity 570 m/sec
Shell weight 14.81 kg
Maximum range 12,625 m
Rate of fire 4–6 rpm
Barrel life 8000–10,000 rounds
Manufacturer Gebr. Böhler, Kapfenberg

Remarks: An advanced design with unusual toed-in wheeled carriage. Development started in 1938, first issued during 1942. Intended to be broken down in four sub-assemblies towed by SdKfz 2 k.Kettenkrad motorcycle half-tracks.

10.5 cm Gebirgshaubitze 40 *(GebH 40) showing full traverse to the right*

10.5 cm GebH 40 at maximum elevation

7.5 cm Gebirgskanone Modell 1912

German designation 7.5 cm GebK M 1912
Calibre 75 mm
Length of piece (L/14): 1050 mm
Weight 525 kg
Traverse 5°
Elevation −10° to +30°
Muzzle velocity 325 m/sec
Shell weight 5.3 kg
Maximum range 5900 m
Manufacturer Friedr. Krupp AG, Essen

Remarks: Obsolescent mountain gun. Pictorial evidence shows only one example used for training after 1939. Dismantled into five loads for transport.

7.5 cm Gebirgskanone L/21

German designation 7.5 cm GebK L/21
Calibre 75 mm
Length of piece (L/21): 1575 mm
Weight travelling (8 loads): 879 kg
Weight in action 756 kg
Traverse 8°
Elevation (long carraige): −10° to +72°; (short carriage): −7° to +45°; (angled carriage): −30° to +52°
Muzzle velocity 431 m/sec
Shell weight 6.6 kg
Maximum range 9015 m
Manufacturer Rheinmetall-Borsig AG, Düsseldorf

Remarks: Development started 1926. Prototype used for troop trials but no contract resulted. Krupp submitted a design to same specification but withdrew at wooden mock-up stage.

7.5 cm GebK L/21 at maximum elevation

7.5 cm Gebirgsgeschütz 43 or Gerät 99

German designation 7.5 cm GebG 43
Design designation Gerät 99, 'Gebhard'
Calibre 75 mm
Length of piece with m/b 1630 mm
Length of piece less m/b (L/18.5): 1390 mm
Length of rifling 973 mm
Weight complete 582 kg
Weight of piece 110 kg
Traverse 40°
Elevation −5° to +70°
Muzzle velocity 480 m/sec
Shell weight 5.74 or 5.83 kg
Maximum range 9500 m
Manufacturer Gebr. Böhler, Kapfenberg

Remarks: Development started 1940 but terminated in 1942 by which time only four examples had been built. These four were subsequently used for extensive trials.

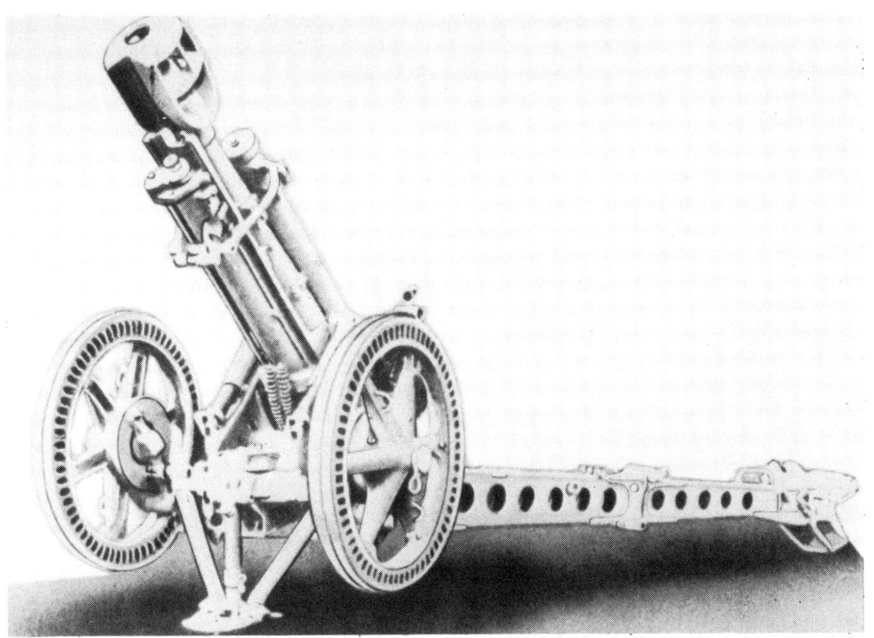

7.5 cm GebG 43 elevated on its frontal support

10.5 cm Gebirgshaubitze L/15

German designation 10.5 cm GebH L/15
Calibre 105 mm
Length of piece (L/15): 1575 mm
Weight travelling (9 loads): 989 kg
Weight in action 841 kg
Traverse 8°
Elevation (long carriage): −10° to +72°; (short carriage): −7° to +75°; (angled carriage): −30° to +52°
Muzzle velocity 295 m/sec
Shell weight 15.6 kg
Maximum range 7010 m
Manufacturer Rheinmetall-Borsig AG, Düsseldorf

Remarks: Development started 1926 in parallel with 7.5 cm GebK L/21. Troop trials with prototype did not result in any production contracts.

10.5 cm Gebirgshaubitze L/30

German designation 10.5 cm GebH L/30
Design designation Gerät 77
Calibre 105 mm
Length of piece (L/30): 3150 mm
Weight travelling (4 loads): 2488 kg (approx)
Weight in action 1725 kg
Traverse 50°
Elevation −5° to +70°
Muzzle velocity 570 m/sec
Shell weight 14.81 kg
Maximum range 11,980 m
Manufacturer Rheinmetall-Borsig AG, Düsseldorf

Remarks: Development started 1938 and two prototypes completed during 1942. Design passed over in favour of competing Böhler development which became 10.5 cm GebH 40.

10.5 cm GebH L/30 at maximum elevation

6.5 cm Gebirgskanone 216(i)

German designation 6.5 cm GebK 216(i)
Original designation Canone da 65/17
Calibre 65 mm
Length of piece (L/17.7): 1150 mm
Length of rifling 905 mm
Weight in action 556 kg
Weight of piece 100 kg
Elevation −7°30′ to +20°
Traverse 8°
Muzzle velocity 348 m/sec
Shell weight (HE): 4.24 kg
Shot weight (AP): 4.23 kg
Maximum range 6500 m
Rate of fire 4 rpm
Manufacturer Vickers-Terni, Turin

Remarks: First produced in 1913 as mountain gun and used as such during WW I. By 1940 most of remaining 700 guns relegated to infantry support role. Small number used by German troops in North Africa and later Italy.

6.5 cm GebK 216(i) in action with the Afrika Korps in Libya

6.5 cm Gebirgskanone 221(f)

German designation 6.5 cm GebK 221(f)
Original designation Canon de 65 M mle 1900
Calibre 65 mm
Length of piece (L/20.5): 1334 mm
Weight in action 400 kg
Traverse 6°
Elevation −9°30′ to +35°
Muzzle velocity 330 m/sec
Shell weight 4.4 kg
Maximum range 6500 m
Rate of fire Max 18 rpm
Manufacturer Schneider-Ducrest

Remarks: Unusual in employing soft-recoil system. In French Army of 1939 used mainly as infantry support gun. Only small numbers taken into German service.

7.5 cm Gebirgshaubitze 34 or 7.5 cm Gebirgskanone 228(b)

German designations 7.5 cm GebH 34; 7.5 cm GebK 228(b)
Original designations Bofors 75 mm Model 1934; (b): Canon de 75 mle 1934
Calibre 75 mm
Length of piece (L/24): 1800 mm
Length of barrel 1583 mm
Length of rifling 1296.7 mm
Weight travelling 928 kg
Weight in action 928 kg
Weight of piece 265 kg
Traverse 7°54′
Elevation (short carriage): −4° to +56°; (long carriage): −10° to +50°
Muzzle velocity 455 m/sec
Shell weight 6.59 kg
Maximum range 9300 m
Manufacturer AB Bofors, Bofors, Sweden

Remarks: Bofors design purchased for prolonged trials and training in 1934. Kept in service after 1939 but numbers involved were small (about 12). The official Belgian designation appears not to have been used but the few pieces involved were taken into German use.

7.5 cm Gebirgskanone 237(f) or 283(j)

German designations 7.5 cm GebK 237(f); 7.5 cm GebK 283(j)
Original designations (f): Canon de 75 M mle 1919 Schneider; (j): 75 mm M 19
Calibre 75 mm
Length of piece (L/18.6): 1398 mm
Length of barrel 1063 mm
Weight travelling 721 kg
Weight in action 675 kg
Weight of piece 217 kg
Traverse 10°
Elevation −10° to +40°
Muzzle velocity 400 m/sec
Shell weight 6.5 kg
Maximum range 9000 m
Manufacturer Schneider et Cie, Le Creusot

Remarks: Sturdy and efficient design, developed to 1914 specifications but appeared too late for WW I. For transport broken down in 7 pack loads. Also served in Greek and Polish armies. Substantial numbers of captured guns impressed into German service.

7.5 cm Gebirgskanone 238(f)

German designation 7.5 cm GebK 238(f)
Original designation Canon de 75 M mle 1928
Calibre 75 mm
Length of piece (L/18.6): 1397 mm
Length of barrel 1060 mm
Weight travelling 721 kg
Weight in action 660 kg
Weight of piece 218 kg
Traverse 10°
Elevation −10° to +40°
Muzzle velocity 375 m/sec
Shell weight 7.25 kg
Maximum range 9000 m
Manufacturer Schneider et Cie, Le Creusot

Remarks: Designed as replacement for mle 1919; fired heavier and more powerful shell. Between wars small numbers also sold to Poland (German designation unknown). All captured guns impressed into German service.

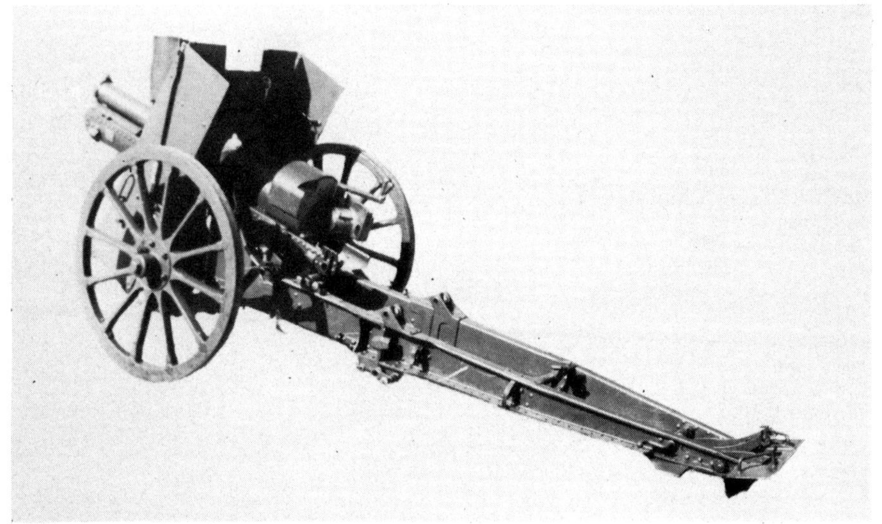

7.5 cm Gebirgskanone 247(n)

German designation 7.5 cm GebK 247(n)
Original designation 7.5 cm GebK M.11
Calibre 75 mm
Length of piece (L/17): 1275 mm
Weight travelling 843 kg
Weight in action 509 kg
Traverse 6°
Elevation −5° to +36°
Muzzle velocity 315 m/sec
Shell weight 6.5 kg
Maximum range 6900 m
Manufacturer Erhardt, Düsseldorf

Remarks: Actually light howitzer. Designed for export in 1910 and some 36 sold in Norway. Guns could be broken down in six loads for transport. Captured remaining guns issued to German mountain troops in Norway.

7.5 cm Gebirgshaubitze 254(i)

German designation 7.5 cm GebH 254(i)
Original designation Obice da 75/18 M34
Calibre 75 mm
Length of piece (L/20.7): 1557 mm
Length of barrel 1374.6 mm
Length of rifling 1133.5 mm
Weight travelling 820 kg
Weight in action 780 kg
Weight of piece 172 kg
Traverse 48°
Elevation −10° to +65°
Muzzle velocity 425 m/sec
Shell weight 6.4 kg
Maximum range 9560 m
Rate of fire 6–8 rpm
Manufacturer Ansaldo, Pozzuoli

Remarks: Evolved to replace older Italian maintain guns. Essentially Obice da 75/18 M35 field howitzer barrel mounted on a special mountain carriage. Could be broken down in eight loads for transport. Of good performance and very popular with Italian and German mountain troops. Also used as field gun.

7.5 cm Gebirgskanone 15 or 259(i)

7.5 cm GebK 15 with the gun shield removed

German designation 7.5 cm Gebk 15 oder 259(i)
Original designation (i): Obice da 75/13
Calibre 75 mm
Length of piece (L/15.4) 1155 mm
Length of barrel 990 mm
Length of refling 802.5 mm
Weight travelling 613 kg
Weight in action 613 kg
Weight of piece 106 kg
Traverse 7°
Elevation −10° to +50°
Muzzle velocity 349 m/sec
Shell weight 6.35 kg
Maximum range 8250 m
Rate of fire 6–8 rpm
Manufacturer Škoda, Pilsen

Remarks: One of the most widely used, and one of the best European mountain guns. Introduced 1915; after 1918 in service with Austrian, Bulgarian, Czech, Hungarian, Romanian and Turkish armies. Italian guns were weapons captured during WW I. In German service intended as substitute until introduction of GebG 36 but remained in use until end of WW II.

7.5 cm Gebirgskanone 28 or 285(j)

German designation 7.5 cm GebK 28 oder 285(j)
Original designation (j): 75 mm M 28
Calibre 75 mm
Length of piece (L/18): 1345 mm
Weight travelling 716 kg
Weight in action 700 kg
Traverse 7°
Elevation −8° to +50°
Muzzle velocity 425 m/sec
Shell weight 6.3 kg
Maximum range 8700 m
Rate of fire 4 rpm
Manufacturer Škoda, Pilsen

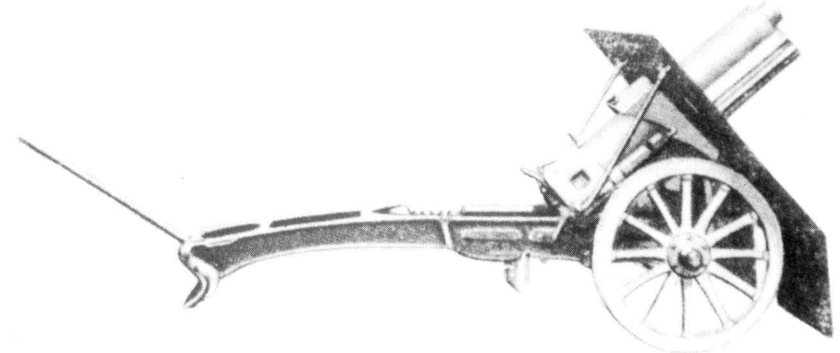

Remarks: Revised and modernised version of M.15, produced for export. Unusual in having provision also to take 90 mm instead of standard 75 mm barrel. Full German service designation allocated to captured guns was 7.5 cm GebK 28 (in Einheitslafette mit 9 cm GebH). Many also issued to pro-German Croat forces during WW II.

7.62 cm Gebirgskanone 293(r)

German designation 7.62 cm GebK 293(r)
Original designation (r): *76.2 mm Gornaya Pushka obr.1909 g* (76-09)
Calibre 76.2 mm
Length of piece (L/16.5): 1258 mm
Length of barrel 1165 mm
Length of rifling 963 mm
Weight travelling (7 loads): 1225 kg
Weight in action 627 kg
Weight of piece 208.9 kg
Traverse 4°50′
Elevation −6° to +28°
Muzzle velocity 387 m/sec
Shell weight 6.23 kg
Maximum range 8550 m
Rate of fire 10–12 rpm
Original manufacturer Schneider et Cie, Le Creusot
Later modifications Various Soviet State arsenals

Remarks: Export version of mountain gun designed by Greek Col. Danglis; originally also known as Schneider-Danglis 06/09. Former Tsarist Army guns updated during 1930s; for transport could be broken down in 7 loads. Most captured guns showed minor variations.

9.4 cm Gebirgshaubitze 301(e)

German designation 9.4 cm GebH 301(e)
Original designation QF 3.7-inch Howitzer Mk I on Carriage 3.7-inch Howitzer Mk IVP
Calibre 93.9 mm
Length of piece (L/12.6): 1188.7 mm
Length of barrel 1112.5 mm
Length of rifling 906.4 mm
Weight travelling 870 kg
Weight in action 830 kg
Weight of piece 253 kg
Traverse 40°
Elevation −5° to +40°
Muzzle velocity 294 m/sec
Shell weight 9.08 kg
Maximum range 5490 m
Rate of fire 8–10 rpm

Remarks: Modernised WW I design known in British Army as 'Pack Howitzer'. Although officially replaced by other weapons served with British and Dominion troops throughout WW II. Only small number captured and used by German troops until available ammunition expended.

7.62 cm Gebirgskanone 307(r)

German designation 7.62 GebK 307(r)
Original designation *76.2 mm Gornaya Pushka obr. 1938 g* (76-38)
Calilbre 76.2 mm
Length of piece (L/21.4): 1630 mm
Length of barrel 1430 mm
Length of rifling 1122.1 mm
Weight travelling (tractor): 1410 kg; (horse): 1450 kg
Weight in action (short carriage: 722 kg; (long carriage): 785 kg
Weight of piece 245 kg
Traverse 10°
Elevation (short carriage): −8° to +70°; (long carriage): −8° to +65°
Muzzle velocity 495 m/sec
Shell weight 6.23 kg
Maximum range 10,100 m
Original designers Škoda, Pilsen
Manufacturers Various Soviet State arsenals

Remarks: Czech 1936 design adopted by Red Army in 1938 and licence-built in Soviet Union. In transit broken down into three basic loads or 10 pack transport loads. Captured guns used by German mountain troops in Caucasus and Balkans.

7.62 cm GebK 307(r) being wheeled out of an underground bunker in France

10 cm Gebirgshaubitze 16, (ö), (t) and 316/1(i)

German designation 10 cm GebH 16, 16(ö), 16(t) und 316/1(i)
Original designations (t): 10 cm horska houfnica vz. 16; (i): Obice da 100/17 modello 16
Calibre 100 mm
Length of piece (L/19): 1930 mm
Length of barrel 1705 mm
Length of rifling 1500 mm
Weight travelling 2150 kg
Weight in action 1235 kg
Weight of piece 392 kg
Traverse 5°30′
Elevation −8° to +70°
Muzzle velocity 406 m/sec
Shell weight 13.375 kg
Maximum range 9280 m
Rate of fire 6–8 rpm
Manufacturer Škoda, Pilsen

Remarks: Rather large and heavy weapon for mountain warfare. For transport could be broken down in only three loads. Despite these negative points used widely during WW II. Italian howitzers taken into German service after September 1943.

10 cm GebH 16, also known as GebH 16(ö) or GebH 16(t)

10 cm Gebirgshaubitze(t) or 16/19(t)

German designation 10 cm GebH(t) oder 16/19(t)
Original designation (t): 10 cm horska houfnice vz. 16/19
Calibre 100 mm
Length of piece (L/24): 2400 mm
Weight in action 1350 kg
Traverse 5°30′
Elevation −7°30′ to +70°
Muzzle velocity 395 m/sec
Shell weight 16 kg
Maximum range 9800 m
Rate of fire 6–8 rpm
Manufacturer Škoda, Pilsen

Remarls: Developed from M.16, differing mainly in longer barrel.

10.5 cm GebH, a modernised version of the GebH 16/19. It could be broken down into three loads carried in two-wheeled carts. Each cart was drawn by two horse or mules in tandem

Trail cart

The tube load of the 10.5 cm GebH

Top carriage cart

10.5 cm leichte Gebirgshaubitze 322(f), 323(f) and 329(j)

German designation 10.5 cm leGebH 322(f), 323(f) und 329 (j)
Original designation 322(f): Canon Court de 105 M Mle 1919 Schneider; 323(f): Canon de 105 M mle 1928; 329(j): 105 mm M 19
Calibre 105 mm
Length of piece (L/12.4): 1304 mm
Length of barrel 988 mm
Weight travelling 806 kg
Weight in action 750 kg
Weight of piece 236.5 kg
Traverse 9°
Elevation 0° to +40°
Muzzle velocity 350 m/sec
Shell weight 12 kg
Maximum range 7850 m
Manufacturer Schneider et Cie, Le Creusot

Remarks: Entered French service same time as 75 mm mle 1919. Some sold to Spain and Yugoslavia between wars. Could be broken down in 8 parts; barrel dismantled in two loads. 105M (Montagne = Mountain) mle 1928 differed only in minor details.

10.5 cm GebH 322(f) with its gun trail folded. The Schneider folding trail feature was also copied on Soviet mountain guns

Mortars

When it first appeared on the battlefields of WW I the infantry mortar seemed to offer the ideal combination of light weight, ease of use and firepower that the hard-pressed infantry had been seeking for centuries. It did have, and still has, some tactical limitations, but it was just the weapon to provide the close and long range defensive or offensive fire needed to suit the character of the close-quarter trench warfare prevalent during that conflict. The modern mortar was a British invention that, for once, was exploited to its full extent. It was designed by Mr Wilfred Stokes, later Sir Wilfred Stokes, KBE, and the first model was made during 1915. Thereafter the Allies produced a range of mortars in a variety of calibres, but the German Army was rather slow in taking up the idea. They tended to rely of various kinds of *Minenwerfer* which were virtually scaled-down howitzers, with all their complexities and high cost, to say nothing of their relative lack of mobility. Post-war tactical analyses by the German General Staff highlighted the importance of the mortar as a weapon of the future, and as soon as possible after 1933 they issued specifications for an 81 mm mortar for use by infantry battalions.

The production contract was won by Rheinmetall in 1934 with their *8 cm GrW 34*. The design of this mortar was begun in 1932 at OKH's request and it remained in service until the end of WW II. The design of the *GrW 34* was entirely conventional. A bipod supported the smooth-bore barrel, which had a calibre of 81.4 mm. The firing pin at the end of the barrel could be shielded for safety and when fired all the recoil forces were absorbed by a large baseplate embedded in the ground. A simple sight was attached to the left-hand side of the barrel and fine elevation and traverse controls were fitted to the bipod. The bombs fired were finned projectiles with their propellant charges secured round the tails; this propellant was detonated when the bomb hit the fixed firing pin at the bottom of the barrel. The bomb was then fired at a high angle of elevation and fell almost vertically onto its target – one of the most important tactical advantages of the mortar.

Six prototypes of the *GrW 34* were made by Rheinmetall before the type was adopted in 1934. Very few were actually manufactured by Rheinmetall for production lines were set up elsewhere. In service the *GrW 34* proved a very serviceable weapon and earned a high reputation from friend and foe, particularly on account of its high rate of fire and accuracy, but it must be stated that the *GrW 34* had few advantages over its opponents' mortars and its success was due in no small measure to the crews that manned it. Later versions of the *GrW 34* were the *8 cm GrW 67* which later became the *8 cm GrW 34/1*, a version produced for self-propelled mountings. A subsequent improvement was the *8 cm GrW 73* but this variant did not advance beyond the trials stage. A wide range of ammunition was developed for the *GrW 34* including the *Wurfgranate 39* 'bouncing bomb' which rebounded into the air after hitting the ground nose first. At a height of between 20 and 50 feet the bomb detonated and scattered fragments over a wide area but in service the bomb proved unreliable and was withdrawn about 1942, but the *Wurfgranate 39* was encountered in action until the remaining stocks were used up.

While the *8 cm GrW 34* was intended as a battalion support weapon the infantry still required a mortar for use at company level or below. As a result of an OKH request the *5 cm leGrW 36* was evolved during 1936 but proved less than successful. Despite its calibre the *leGrW 36* was a complex weapon with the barrel permanently mounted on a large baseplate. The barrel elevation and traverse controls were rather complicated and up to 1938 an optical sight was provided, but this was so much trouble to use that it was replaced by a simple white line painted on the barrel. The performance of the *leGrW 36* was not particularly outstanding and after early 1942 it was withdrawn from use and replaced by larger calibre mortars.

Another German mortar that was not a particularly successful venture was the *kz 8 cm GrW 42*, or *'Stummelwerfer'*. Originally intended for use by airborne troops and other special purpose formations the *GrW 42* was a shorter and lighter version of the *GrW 34*. With the gradual decline of the German airborne arm after the costly assault on Crete in 1941 (Operation 'Mercury') the *GrW 42* became a more general issue item and many ended up with infantry companies.

At the same time as the infantry mortars were being planned for conventional use, a larger calibre mortar was planned for use by the *'Nebeltruppen'* chemical units which were normally called upon to provide smoke screens but, like in all other armies, retaining the ability to wage chemical warfare if the need arose. The requirement formulated by OKH during 1934 called for a 10.5 cm mortar capable of firing smoke bombs and the result was yet another Rheinmetall design, the *10 cm NbW 35*. However, Rheinmetall made only 38 examples before production was split up among a number of other firms, and almost as soon as the *NbW 35* entered service OKH requested a mortar with increased range. That was during 1937 but it was not until May 1940 that troop trials took place using three examples of each of two trial weapons. One was the *10.5 cm NbW 51* and the other the *10.5 cm NbW 52*, both designed by Rheinmetall engineers. As a result of these trials the best of both mortars was incorporated into the *10 cm NbW 40* which was a much more complex weapon than the *NbW 35*. The range was extended but at the cost of a great increase in weight and complexity, for the *NbW 40* featured breech loading, a recoil mechanism and an integral wheeled carriage. Perhaps the best comparison between these two smoke mortars can be seen in their costs – the *NbW 35* cost RM1500 and the *NbW 40* cost RM14,000. Not surprisingly very few *NbW 40*s were made, and another factor which did not help their long-term prospects was that by the end of 1940 the *Nebeltruppen* were beginning to re-equip with rocket equipments. As a result after 1941 the *NbW 35* and *NbW 40* were used in conventional mortars firing useful HE bombs.

When Germany invaded the Soviet Union in 1941 the booty included large numbers of 120-PM 38 mortars, a 120 mm weapon that was later to be judged as the best mortar design to emerge from WW II. The German troops made use of substantial numbers of these heavy support weapons and were so impressed by the overall fire power, mobility and simplicity of the design that it was decided to copy the Soviet design direct as the *12 cm GrW 42*. Most of the production of this mortar was carried out by Erste Brünner Masch.-Fabrik at Brünn at a cost of RM1200 each. Variants were the *12 cm GrW 42/1* and *42/2* intended for use on self-propelled carriages. With the *12 cm GrW 42* the German Army started a switch towards heavier calibre mortars and the *12cm GrW 42* was produced in such numbers that it took the place of infantry guns in some formations.

This change-over to heavier calibres of mortars took on a new slant with the emergence of the *21 cm GrW 69*. This mortar was designed by Director Vamberski, one of the technical heads of the Skoda Werke at Pilsen. For various reasons Vamberski had reached the conclusion that the heavy mortar was the weapon of the future which would replace the field gun and howitzer on the grounds of weight and cost. He also deduced that a 21 cm mortar could be made at a fraction of the cost of a 21 cm howitzer and the resulting weapon would still be capable of carrying out most of the howitzer's fire tasks. To prove his point the Skoda plant built, on their own initiative, a 22 cm mortar, the *B 14*, and offered it to the German forces. They ordered a 21 cm version, and two further prototypes, the *B 19* and *G 69*, were built before the latter was accepted for production. All this took place in late 1944 and early 1945 and production was only just getting under way when the war ended. Only a few examples reached the front-line troops, who nicknamed the mortar *'Elefant'*, not only for its size

but no doubt also for the amount of work needed to prepare the mortar for firing. Each fire position had to be roughly levelled to a diameter of about 6 metres and a hole dug to accommodate the large baseplate. The area of ground that took the track upon which the wheels of the carriage rested for traversing had to be accurately levelled as there was no provision for levelling on the carriage. To add to the complexity of this mortar it was breech-loaded and had a heavy recoil mechanism.

Two other specialised developments must be mentioned, both spigot mortars intended for engineer units. The lighter of the two was the *20 cm le Ladungswerfer*, and this version was produced and used in greater numbers than the heavier mortar, the *38 cm sch Ladungswerfer*. Both fired large bombs which fitted over a heavy spigot; these were intended for the demolition of concrete strongpoints, but only the 20 cm version was actually used in combat as the 38 cm mortar was a large and massive weapon that required careful and prolonged emplacement. For the assault engineer role the *20 cm le Ladungswerfer* fired a variety of projectiles including HE and smoke bombs, and also the *'Harpunengeschosse'* which carried a rope to enable charges to be pulled across minefields to clear them.

In the field of more conventional mortars the success of the *12 cm GrW 42* copy of the Soviet weapon was such that the German military decided to go one further and increase the calibre to 15 cm. Two firms submitted designs, Škoda and the Gustloff Werke at Suhl, but the prototypes of what would have been the *15 cm GrW 43* did not advance beyond the prototype stage. The *12 cm GrW 42* also featured in a series of trials that led to the *12 cm Granatwerferfünfling 43*, a combination of five barrels with two barrels laid over three, and a loading arrangement which held the bombs over the barrel until they were released either together or in a ripple. Škoda were also involved in producing a similar weapon with their *12 cm Mehrrohrwerfer*, but neither weapon reached service. However, the idea of an automatic loading device did not end with the 12 cm project, for Škoda went on to develop a rotary loader mounted on the muzzle of 5 cm and 8 cm mortars. Both types of mortar involved were pre-war Škoda models and were used only for trials.

By the end of the war Škoda were involved in all kinds of mortar developments, one of which was the *5 cm leGrW 40/S* intended for street fighting and thus capable of firing horizontally, with recoil forces being absorbed by heavy springs round the barrel. The *15 cm Minenwerfer 30/260/ohne Rücklauf* was an offshoot of the *15 cm GrW 43* programme, and of the three made two were sent for trials at Kummersdorf. Perhaps the most unusual of the Škoda projects was the *Gerät 170*, a 10.5 cm mortar which propelled its bombs by compressed air, in itself not a new idea but one which had definite limitations as it relied upon there being access to an air supply. This difficulty could no doubt be solved with a self-propelled mounting but only two prototypes were built.

As mentioned above the ideas of Director Vamberski were directed towards replacing conventional heavy artillery by more economical and easier-to-manufacture heavy mortars and starting in January 1945 he gave his ideas full rein by designing and producing a 30.5 cm mortar in the remarkable time of only twenty days. The finished product closely resembled a conventional howitzer and was breech-loaded, but the barrel was smooth and finned projectiles were fired. A self-propelled version mounted on a tank was also planned. A yet more ambitious proposal was for a 42 cm mortar but early development work soon revealed that the weapon would be far too heavy and the project was abandoned. This weapon too would have had a self-propelled version using a lengthened Tiger tank chassis.

Škoda were not the only firm in the Third Reich involved in the development of super-heavy mortars. The Krupp concern also did some design work on 30.5 and 42 cm mortars, but no weapons were actually built.

When it comes to considering the types of captured mortars in German service one is almost overwhelmed by the range of weapons involved for the German troops seem to have used every type and calibre of mortar they could lay their hands on. If ammunition was available any mortar type was considered suitable for German use, although most captured mortars were issued to second-line and garrison units to free the standard German mortars for front line use. One factor that was of great assistance to the Germans was that between the wars the French Stokes-Brandt concern became the virtual design leader of infantry mortars in Europe and produced mortars not only for the French Army but for most of Europe. Their standard calibre was 81.4 mm, which was exactly the same as the calibre chosen by the German Army, so interchangeability of ammunition was a simple matter. The same did not apply to Soviet mortars, which had a calibre of 82 mm but so much ammunition was captured that this was not a major problem. As related above the Soviet 12 cm mortar so impressed the Germans that they copied the design direct with virtually no changes.

8 cm GrW 34 in France, June 1940. Although of conventional design, it was a very sturdy and popular weapon and remained in frontline service until the end of WW II

5 cm leichter Granatwerfer 36

5 cm leGrW 36 in action

5 cm leGrW 36 on a concreted pivot installation on Guernsey

German designation 5 cm leGrW 36
Calibre 50 mm
Length of barrel (L/9.3): 465 mm
Length of bore 350 mm
Weight in action 14 kg
Traverse 33°45′
Elevation +42° to +90°
Muzzle velocity 75 m/sec
Bomb weight 0.9 kg
Maximum range 520 m
Rate of fire 15–25 rpm
Barrel life 20,000–25,000 rounds
Original designers Rheinmetall/Borsig AG

Remarks: Development started 1934, adopted for service 1936. Until 1938 used rather complicated telescopic sight. By 1941 seen as too complex for intended role, firing too light a bomb and production terminated. Gradually withdrawn from front-line service after 1942 but available mortars remained in use until 1945 with second-line and garrison units.

---

8 cm Granatwerfer 34

German designation 8 cm GrW 34
Calibre 81.4 mm
Length of barrel (L/14.1): 1143 mm
Length of bore 1033 mm
Weight travelling (3 loads): 64 kg
Weight in action (steel barrel): 62 kg; (alloy barrel): 57 kg
Traverse 10° to 23°
Elevation +45° to +90°
Muzzle velocity 174 m/sec
Bomb weight 3.5 kg
Maximum range 2400 m
Rate of fire 15–25 rpm
Barrel life 16,000–20,000 rounds
Original designers Rheinmetall-Borsig AG
Manufacturers (by 1943): Gellnow, Oberdorla/Thüringia; Ruhrstahl AG, Hattingen; Güttler, Brieg n. Breslau; Haas und Sohn, Neuhoffnungshütte n. Wetzlar.

Remarks: Development started 1923/33, adopted for service 1934. Thereafter in production and use until end of WW II. Sturdy and accurate weapon, fitted with RA 35 dial sight. Carried in action as 3-man load, but also in various SP carriages. 8 cm GrW 34/1, originally known as GrW 67, was a variant for use on self-propelled mountings.

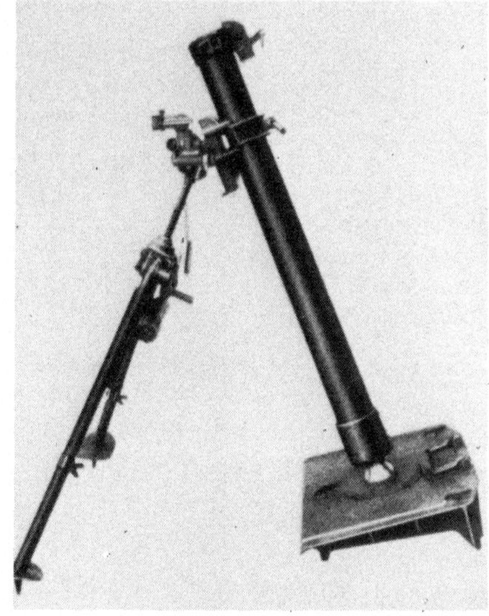

8 cm GrW 34 in France, 1940

kurzer 8 cm Granatwerfer 42

German designation kz 8 cm GrW 42
Calibre 81.4 mm
Length of barrel (L/9.2): 747 mm
Length of bore 650 mm
Weight travelling (3 loads): 30 kg (approx)
Weight in action 26.5 kg
Traverse 14° to 34°
Elevation +40° to +90°
Muzzle velocity Not recorded
Bomb weight 3.5 kg
Maximum range 1100 m
Rate of fire 15–25 rpm

Remarks: Development started 1940, first issued late 1941. Originally intended for airborne troops but subsequently adopted by Army and Waffen-SS and gradually replaced 5 cm leGrW 34. In action some were provided with lanyard-operated loading/firing mechanism for remotely-controlled use. Was generally known as 'Stummelwerfer'.

kz 8 cm GrW 42 in action on the Eastern Front

10 cm Nebelwerfer 35

German designation 10 cm NbW 35
Calibre 105 mm
Length of barrel (L/13): 1344 mm
Length of bore 1207 mm
Weight travelling (3 loads): 11 kg
Weight in action 105 kg
Traverse 28°
Elevation +45° to +90°
Muzzle velocity 193 m/sec
Bomb weight 7.38 kg
Maximum range 3025 m
Rate of fire 10–15 rpm
Barrel life 15,000–18,000 rounds
Original designer Rheinmetall-Borsig AG.

Remarks: Virtually an enlarged 8 cm GrW 34. Development started 1934, first produced 1939, output terminated 1941. Initially issued only to 'Nebeltruppen' artificial fog/smoke units, later also to other formations as heavy HE mortar. In Nebeltruppen service replaced by 15 cm rocket equipments.

10 cm NbW 35. The mortar, ammunition and crew were carried by the 3-ton Sd Kfz 11/4 half-track; the ammunition was stowed in racks on the sides of the vehicle

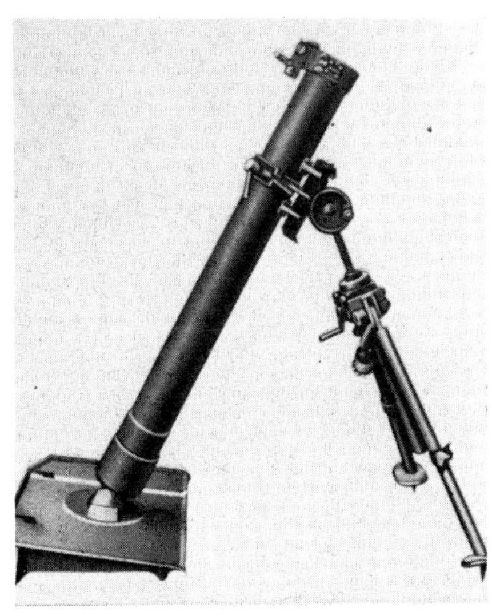

10.5 cm Nebelwerfer 51 L/12 and 52 L/12

German designations 10.5 cm NbW 51 L/12 und 52 L/12
Calibre 105 mm
Length of barrel (L/12): 1260 mm
Weight in action 651 kg
Traverse (approx): 14°
Elevation +45° to +85°
Muzzle velocity 271 m/sec
Bomb weight 9 kg
Maximum range 6000 m
Rate of fire 8 rpm
Manufacturer Rheinmetall-Borsig AG

Remarks: Development of these two almost identical mortars commenced in 1937. Three examples of each model issued for troop trials in May 1940 but adjudged not suitable for service. Trials led to development of 10 cm NbW 40.

10.5 cm NbW 51 L/12 prototype

10.5 cm NbW 52 L/12 prototype

10 cm Nebelwerfer 40

German designation 10 cm NbW 40
Calibre 105 mm
Length of barrel (L/17.7): 1858 mm
Length of bore 1720 mm
Weight travelling 892 kg
Weight in action 800 kg
Traverse 14°
Elevation +45° to +84°
Muzzle velocity 310 m/sec
Bomb weight (HE): 8.65 kg; (smoke): 8.9 kg
Maximum range 6350 m
Rate of fire 8–10 rpm
Barrel life 3000–5000 rounds
Original designers Rheinmetall- Borsig AG

Remarks: Evolved from NbW 51 and NbW 52. First issued to Nebeltruppen during late 1940, shortly before re-equipped with 15 cm rocket equipments. Only limited number manufactured; unit cost was RM14,000.

10 cm NbW 40 detachment of a Luftwaffe Field Division preparing for action

12 cm Granatwerfer 42 and 378(r)

12 cm GrW 42, with the two-wheeled carriage used for towing

German designation 12 cm GrW 42 und 378(r)
Original designation (r): *120 mm Polkovoy Minomyot obr. 1938 g* (120 PM-38)
Calibre 120 mm
Length of barrel (L/15.5): 1865 mm
Length of bore 1536 mm
Weight travelling 560 kg
Weight in action 285 kg
Traverse 8° to 17°
Elevation +45° to +84°
Muzzle velocity 283 m/sec
Bomb weight 15.6 kg
Maximum range 6050 m
Rate of fire (maximum): 8–10 rpm; (normal): 6 rpm
Barrel life 3000 rounds
Manufacturer (r): Various Soviet State arsenals; (GrW) 42): Erste Brünner Masch.-Fabr., Brünn.

Remarks: An almost direct copy of Soviet regimental mortar. In action could use both Soviet and German ammunition. A powerful and very popular weapon with German front-line troops. In some units replaced standard infantry guns.

12 cm NbW 42 on the Eastern Front

21 cm Granatwerfer 69

German designation 21 cm GrW 69 Elefant
Calibre 210.9 mm
Length of barrel (L/14.2): 3000 mm
Length of bore 2400 mm
Weight travelling 2800 kg
Weight in action 2800 kg
Traverse 60°
Elevation +45° to +75°
Muzzle velocity (light bomb): 285 m/sec; (heavy bomb): 247 m/sec
Bomb weight (light): 85 kg; (heavy): 110 kg
Maximum range (light bomb): 6300 m; (heavy bomb): 5190 m
Manufacturer Škoda Werke, Pilsen

Remarks: Originally a Škoda 22 cm design altered to 21 cm at OKH request. Two prototype designations were B 19 and G 69. Production commenced late in 1944; total some 200 delivered.

21 cm GrW 69 Elefant. *Traversing was effected by manhandling the equipment around the circular section of a single rail track*

20 cm leichter Ladungswerfer

German designation 20 cm leLdgW
Spigot diameter 90 mm
Length of spigot 540 mm
Weight in action 93 kg
Bomb weight 21.27 kg
Maximum range 700 m
Manufacturer Rheinmetall-Borsig AG

Remarks: Specialised spigot mortar evolved for assault engineers to demolish obstacles and strongpoints. Fired HE, smoke and special 'Harpunengeschosse' which carried ropes with hooks across minefields. Used operationally during 1940 Western campaigns and in North Africa; gradually diverted to second-line engineer units after 1942.

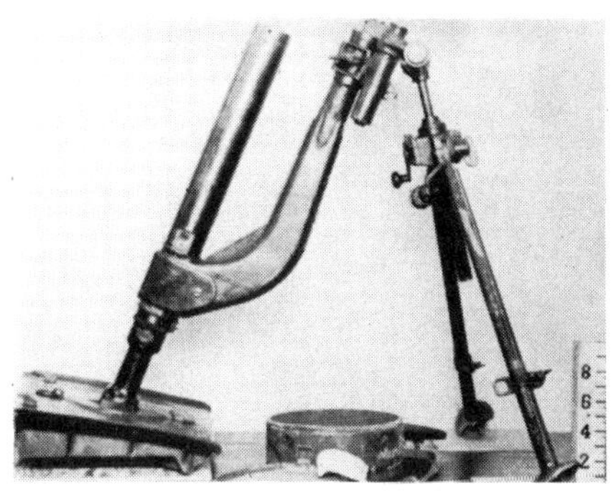

20 cm leichter Ladungswerfer *(le LdgW). This weapon fired thin-walled HE, smoke and the so-called 'Harpun' (inert) bombs used to protect lines for drawing bundles of explosive charges over minefields*

38 cm schwerer Ladungswerfer

German designation 38 cm sLdgW
Spigot diameter 169 mm
Length of spigot 1680 mm
Weight in action (approx): 1600 kg
Traverse 360°
Elevation +37° to +85°
Muzzle velocity 107 m/sec
Bomb weight 149 kg
Maximum range 1000 m
Manufacturer Rheinmetall-Borsig AG

Remarks: Specialised spigot mortar for assault engineers, evolved to 1938 OKH specifications. Only limited number built and issued as weapon proved too heavy for its role and needed careful emplacing.

38 cm schwerer *LdgW. Like the 20 cm le LdgW it was an electrically fired weapon. The spigot mortar was mounted on a traversing bracket upon a drum-shaped platform staked to the ground*

5 cm leichter Granatwerfer 40 Skoda

German designation 5 cm leGrW 40 S
Design designation Skoda B 13
Calibre 50.9 mm
Weight in action 28 kg
Traverse 20°
Elevation 0° to +90°
Muzzle velocity 75 m/sec
Bomb weight 0.99 kg
Maximum range 550 m
Manufacturer Skoda-Werke, Pilsen

Remarks: Specialised mortar for use in street fighting. Projected in 1943 after Stalingrad, and three prototypes built and tested. Proved too heavy, with a complex recoil mechanism, and development discontinued.

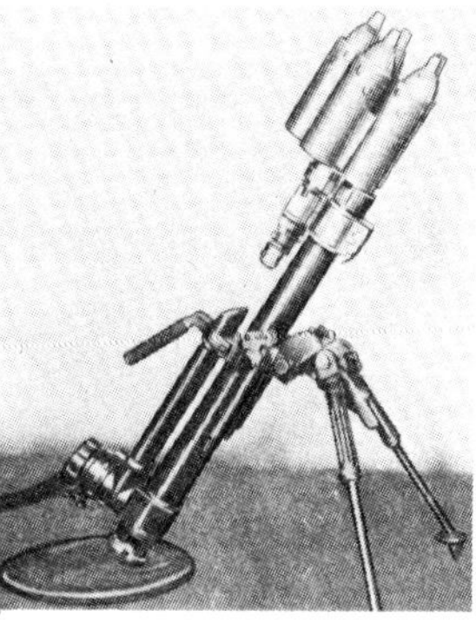

Gerät 170

German designation Gerät 170; 10 cm GrW/Druckluftwerfer
Calibre 105 mm
Weight in action 450 kg
Traverse 45°
Elevation 0° to +75°
Firing pressure of air 200 kg/cm²
Bomb weight 9 kg
Manufacturer Skoda-Werke, Pilsen

Remarks: Compressed air-operated mortar. Only two prototypes completed during 1943-early 1945, one of which was tested on Kummersdorf firing ranges.

10.5 cm GrW (Druckluftwerfer)

15 cm Granatwerfer 43, 15 cm Minenwerfer 30/600

German designation 15 cm GrW 43; 15 cm MW 30/600/ohne Rücklauf
Project designation Skoda B 17
Calibre 150 mm
Weight travelling 900 kg
Weight in action 633 kg
Traverse 20°
Elevation +40° to +80°
Muzzle velocity 260 m/sec
Bomb weight 30 kg
Maximum range 5000 m
Manufacturer Skoda Werke, Pilsen

Remarks: Recoilless mortar evolved by Skoda engineers. Three prototypes built and tested during 1943 but weapon considered unsuitable for front-line service. Both prototypes subsequently used for Skoda for trials with various recoil mechanisms. Gustloff-Werke of Suhl built a very similar prototype weapon.

22 cm schwere Granatwerfer B 14

German designation 22 cm sGrW B 14
Calibre 220 mm
Length of barrel (L/13.6): 3000 mm
Weight in action 1680 kg
Traverse 20°
Elevation +40° to +75°
Muzzle velocity 155 m/sec
Bomb weight 177 kg
Maximum range 2000 m
Manufacturer Skoda Werke, Pilsen

Remarks: Prototype built in 1944; led to 21 cm GrW 69.

30.5 cm schwerer Granatwerfer

German designation 30.5 cm sGrW
Design designation Skoda B 20
Calibre 305 mm
Length of barrel 5100 mm
Weight travelling (2 loads): 17,800 kg
Weight in action 9300 kg
Traverse 360°
Elevation +40° to +75°
Muzzle velocity 430 m/sec
Bomb weight 160 kg
Maximum range 10,000 m
Manufacturer Skoda Werke, Pilsen

Remarks: Design work commenced in January 1945 and was completed in 20 days. First prototype ready for test firing early in April 1945. A self-propelled version was also projected.

42 cm schwerer Granatwerfer

German designation 42 cm sGrW
Calibre 420 mm
Length of barrel (L/12.2): 5150 mm
Weight travelling 21,600 kg
Weight in action 16,000 kg
Traverse 360°
Elevation +40° to +75°
Muzzle velocity 385 m/sec
Bomb weight 400 kg
Maximum range 10,000 m
Manufacturer Škoda Werke, Pilsen

Remarks: Project design begun early in 1945, was terminated by end of WW II, but temporarily resurrected by Czechs, for trials only, in 1946. A self-propelled version was also projected.

4.5 cm Granatwerfer 176(i)

German designation 4.5 cm GrW 176 (i)
Original designation Mortaio da 45/5 modello 35
Calibre 45 mm
Length of barrel (L/5.4): 260 mm
Weight in action 15.5 kg
Traverse 20°
Elevation +10° to +90°
Muzzle velocity 83 m/sec
Bomb weight 0.48 or 0.465 kg
Maximum range 536 m
Rate of fire 8–10 rpm
Manufacturer O.T.O., Turin

Remarks: Light, accurate but complicated Italian weapon, with many novel features. Generally known as 'Brixia'. Some used by German troops in North Africa and Italy.

4.6 cm Granatwerfer 36(p)

German designation 4.6 cm GrW 36(p)
Original designation 46 mm granatnik wz. 36
Calibre 46 mm
Length overall 648 mm
Length of barrel 396 mm
Weight 12.6 kg
Muzzle velocity 95 m/sec
Bomb weight 0.76 kg
Maximum range 800 m
Rate of fire 15 rpm

Remarks: An indigenous Polish design evolved during 1932–34. In service from 1937. Only limited numbers used by German troops.

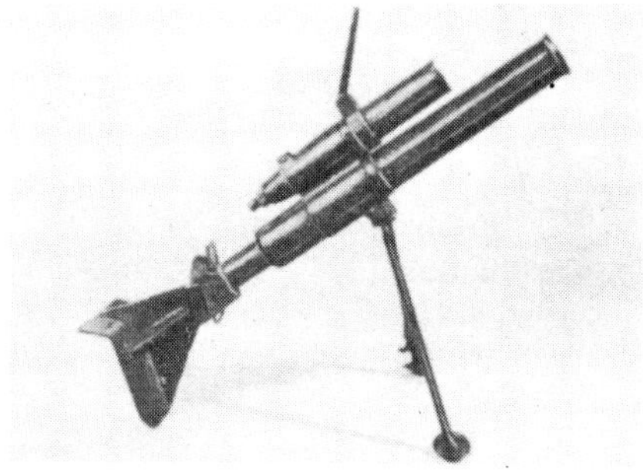

5 cm Granatwerfer 201(b)

German designation 5 cm GrW 201(b)
Original designation Lances grenades de 50 mm DBT
Calibre 50 mm
Length of barrel 200 mm
Length of bore 190 mm
Weight carried 8.8 kg
Weight in action 7.7 kg
Traverse 360°
Elevation +30° to +50°
Muzzle velocity 75 m/sec
Bomb weight 0.6 kg
Maximum range 585 m
Rate of fire 12–15 rpm

Remarks: More a grenade launcher than mortar, of rather complex construction. Very few captured weapons issued to German troops, and then only to local occupation forces.

5 cm Granatwerfer 203(f)

German designation 5 cm GrW 203 (f)
Original designation Lances Grenades de 50 mm mle 37
Calibre 50 mm
Length of barrel 415 mm
Length of bore 280 mm
Weight in action 3.65 kg
Traverse 8°
Elevation 45° fixed
Muzzle velocity 70 m/sec
Bomb weight 0.435 kg
Maximum range 460 m
Rate of fire 15–20 rpm
Manufacturer Stokes-Brandt, Paris

Remarks: Introduced in French service in 1939. Small and light, but short-range weapon only. Very few issued to German troops.

5 cm GrW 203(f) was introduced in the French army in 1939 to replace rifle grenades used at platoon level

5 cm Granatwerfer 205/1(r)

German designation 5 cm GrW 205/1(r)
Original designation *50 mm Rotny Minomyot obr. 1938 g* (50-RM 38)
Calibre 50 mm
Length of barrel (L/15.6): 780 mm
Length of bore 555 mm
Weight carried (3 parts): 16.2 kg
Weight in action 15.35 kg
Traverse 7° to 16°
Elevation Fixed at 45°, 75° and 82°
Muzzle velocity (maximum): 96 m/sec
Bomb weight 0.85 kg
Maximum range (45° elevation): 800 m; (75° elevation): 400 m; (82° elevation): 100 m
Rate of fire Up to 30 rpm
Manufacturer Soviet State arsenals

Remarks: Evolved during 1934–37, accepted in Red Army service as company mortar in 1938. Rather complex for large-scale production and only limited numbers delivered before replaced by 50-RM 39. Range variations made by gas escaping through variable port in barrel base. Only small numbers captured and impressed into German service.

5 cm Granatwerfer 205/3(r)

German designation 5 cm GrW 205/3(r)
Original designation *50 mm Rotny Minomyot obr. 1940 g* (50-RM 40)
Calibre 50 mm
Length of barrel (L/12.6): 630 mm
Length of bore 525 mm
Weight carried (3 parts): 12 kg
Weight in action 9.3 kg
Traverse 6° to 15°
Elevation Fixed at 45°, 75° and 82°
Muzzle velocity 80 m/sec
Bomb weight 0.9 kg
Maximum range (45° elevation): 800 m; (75° elevation): 400 m; (82° elevation): 100 m
Rate of fire Up to 30 rpm
Manufacturers Various Soviet State arsenals

Remarks: Produced and used by Red Army troops as company mortar in very large numbers. Design introduced simplified bipod incorporating a novel method of cross-levelling. In German service used mainly on Eastern Front and against partisans.

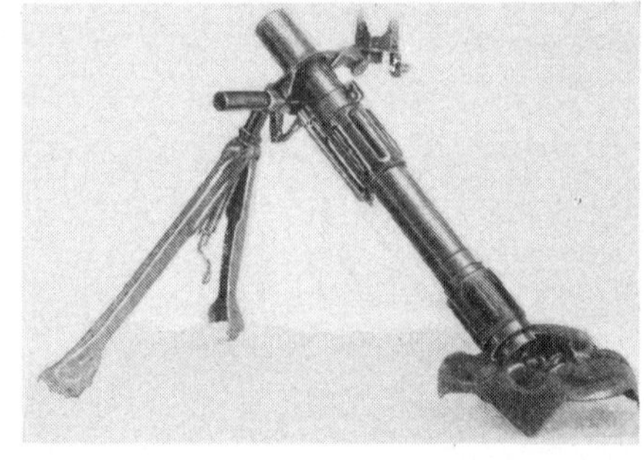

6 cm Granatwerfer 225(f)

German designation 6 cm GrW 225(f)
Original designation Mortier de 60 mm mle 1935
Calibre 60.7 mm
Length of barrel 724 mm
Length of bore 655 mm
Weight in action 17.8 kg
Traverse 5° to 12°
Elevation +45° to +85°
Muzzle velocity 158 m/sec
Bomb weight (light): 1.3 kg; (heavy): 2.2 kg
Maximum range (light bomb): 1700 m; (heavy bomb): 950 m
Rate of fire Up to 30 rpm
Manufacturer Stokes-Brandt, Paris

Remarks: Evolved by Edgar Brandt bureau. Entered French Army service 1937, and copied in USA as 60 mm M1 (and later M2 and M19). Total of 4940 with French troops in 1940. Captured examples issued mainly to German garrison units.

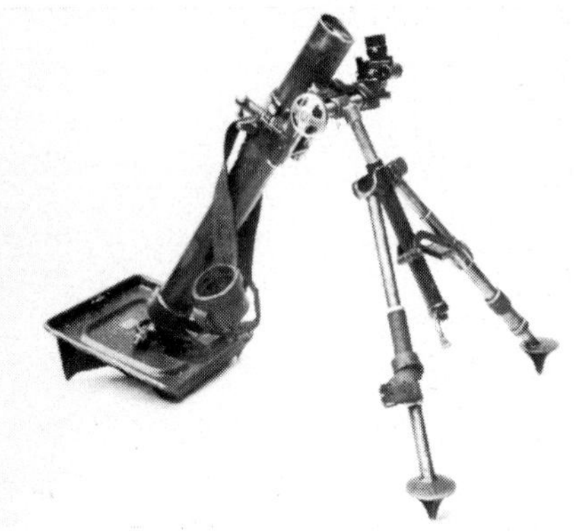

production examples were not produced until August 1944, by which time there was no longer any need for their specialised street-fighting role.

Another specialised use of the rocket was in the anti-tank role. The German forces made considerable use of the hollow-charge warhead on almost every type of artillery weapon, but results soon showed that the hollow charge would only be really effective if it was delivered at a low velocity, and the rocket seemed to offer many advantages as a delivery system. As a result the HWA initiated the development of a small 8.8 cm rocket that eventually emerged as the *RP Gr 4312* (or *8.8 cm R-Pz Gr 43*). Activated by a percussion igniter the *RP Gr 4312* was originally intended for firing from what appeared to be a miniature artillery piece known as the *8.8 cm RW 43*, or '*Puppchen*'. When the *RW 43* was fired all the recoil forces were absorbed by the mass of the barrel and carriage, and the rocket was discharged to a range of about 500 metres. Early in 1943 the design of the 'Puppchen' was approved for production, but almost as soon as it went into service a new form of rocket projector was captured from American troops in Tunisia, the Launcher M1, usually known as the 'Bazooka', which fired 2.36-in rockets. As soon as the first examples fell into German hands they were sent back to Germany where the rockets and their launcher were closely examined and their potential was quickly recognised. While the rocket design was quite complex the launcher system was little other than a long tube open at both ends, fitted with a simple electrical firing system. In a very short time the 8.8 cm rockets were adapted to take the new electrical igniters and a new launcher, the *Raketen Panzerbüchse 43 (8.8 cm RP 43)* was quickly developed and put into production. Once in service the *RP 43* proved an immediate success, for it gave the front-line soldier a simple weapon able to destroy almost any Allied tank, but a price had to be paid for this ability. For one thing the *RP Gr 4322* rocket shell had a relatively short range of about 150 metres. Another drawback was that the backblast of the rocket firing produced a dangerous sheet of flame for some distance to the rear of the launcher and this flaming exhaust also extended from the rocket immediately after firing so the gunner had to wear protective clothing and a gas mask. Despite these disadvantages the *RP 43* was a considerable success and the launcher design was soon adapted to take a protective shield as the *8.8 cm RP 54*. This combination of a long tube and shield made the *RP 54* rather clumsy to use but eventually an improved rocket, the *8.8 cm RP Gr 4992,* was introduced, the propellant of which was burnt up in a very short space of time and entirely within the tube. Such was the measure of this improvement that a shorter tube was possible and the *RP 54/1* was produced. With the *RP 54* and *54/1* in service the *RP 43* was relegated to second-line use, and the *RP 54* and *54/1* flowed in huge numbers from the production lines situated all over Germany.

By late 1944 the Allies had virtual air supremacy over most of Europe and the tactical fighter-bombers of the Allied air forces made movement and action almost impossible during daylight hours for the vast bulk of the German ground forces. In their frustration the front-line soldiers demanded some form of personal anti-aircraft weapon which would give them at least a partial return to their mobile offensive tactics, and the German weapon designers responded with a novel solution to the situation. Their answer was the *Luftfaust,* a shoulder-fired rocket projector firing salvos of very effective 2 cm warheads against low-flying aircraft. Overall the design of the *Luftfaust* demanded very little industrial potential for almost every part was intended for manufacture in small and simply-equipped workshops by unskilled labour. The *Luftfaust* used eight launcher tubes clustered around a central cylinder and all the parts were made either of steel tubes or simple steel stampings; a simple electrical firing system was added along with rudimentary shoulder rests and fore grips. The rockets were in a pre-loaded clip of nine shells and, once loaded, the clip locked the rockets ready to fire. Aiming was via simple blade sights and the rockets were fired in two salvos – first four and then five. Each *Luftfaust* unit comprised a box containing a launcher and five clips of rockets ready to use. When the war ended production of the *Luftfaust* was under way in numerous small workshops but none were issued to the fighting troops as by the time they were ready the complete supply and distribution system had broken down.

The most advanced of all the German solid fuel rockets was the *RSpr 4831* or *Rheinbote*. This was another Rheinmetall project and in retrospect must be regarded as one of the most remarkable of all the many German rocket designs. It was a four-stage ballistic rocket missile launched from a modified *A4* transporter without any guidance, and such were the power and efficiency of the rockets used that a maximum speed of Mach 5.5 was reached by the final stage, a speed not exceeded by any other rocket until years later. Rheinmetall initiated the design in response to an HWA request for a tactical rocket missile capable of carrying a 40 kg warhead to a range of 160 km. Initial work centred on a two-stage rocket but this was replaced by a three-stage rocket under the project designation *RhZ 61/9*. On *Rheinbote* a fourth stage was added for take-off, and after successful trials the rocket went into small-scale production. However, for all its technical ingenuity *Rheinbote* was almost useless as a military weapon because its warhead, which contained only 40 kg of explosive, would have little other than a morale effect on its target. This shortcoming was discovered during trials held at Blizna in occupied Poland when a rocket missile fell short and a shallow crater just over one metre across was the only damage caused. General Dornberger recommended that the trials should be suspended, but he was overruled by Hitler himself who ordered that a rocket battery should be set up to use *Rheinbote*, and this battery eventually went into action at Zwolle in Holland in November 1944. Over 200 *Rheinbote* rockets were fired at Antwerp, but their impact was minimal as the small warheads made little impression at a time when that important supply port was under bombardment by *A4s* and *V1s*.

On a much smaller scale was another very advanced German rocket, one of the very first to employ a wire guidance system. This was the *X-7 Rotkäppchen* anti-tank missile, the result of an OKH request made during 1944. Ruhrstahl, AEG and Rheinmetall all made design submissions but only Ruhrstahl actually produced a prototype known as the *X-7*. Overall the design of the *X-7* was simple and cheap but it worked very well during early development and troop trials. The warhead carried proved effective against even the heavy IS-2 tank but the guidance system was never fully developed and the planned full-scale production did not commence before the war ended. Another firm, BMW, also developed an anti-tank guided missile early in 1945, but it was not given a designation even though primary trials were carried out at Dachau as the war drew to a conclusion.

After 1942 the attention of German weapon designers was increasingly diverted towards anti-aircraft defence as growing numbers of Allied bombers threatened to disrupt German communications and industrial potential. Needless to say the rocket featured as a possible counter to this threat and many and various were the rocket designs that flowed from the design and drawing offices of many German concerns. Once again Rheinmetall were early in the field and in 1939 built a small number of an experimental rocket known as *Hecht* which was used as an experimental vehicle only. By 1941 it had been superseded by *Feuerlilie*, a ground-to-air missile evolved in two versions (*F 25* and *F 55*) under the auspices of the *Luftfahrtforschungsanstalt* (*LFA*) centred at Volkenrode. During 1942 all further development work was handed over entirely to Rheinmetall. This firm carried on various trials with a number of variants of the basic design but the *Feuerlilie* project was suspended in January 1945 without a production version being finalised.

The work carried out by Rheinmetall with *Rheinbote* convinced the HWA that the firm could produce a guided AA rocket

and a development contract was issued in November 1942. The initial contract was for a missile designated *Rheintochter R-I* which had several very advanced design and guidance features. Unfortunately many of these novel features were too advanced, so that by July 1944 only 34 trial missiles had been fired with varying success. By December 1944 the success rate was no better and the project was abandoned in favour of a more advanced design, the *R-III*, which used a liquid fuel rocket motor for the main stage in place of the earlier solid fuel rocket. A later version still was designated *R-IIIf* but both were suspended by the end of December 1944. All versions of *Rheintochter* were launched from a modified *8.8 cm Flak 41* carriage and guidance was line-of-sight, altered by radio control.

Rheintochter was not the only German attempt to evolve missile for the Henschel Flugzeugwerke AG based at Schönefeld near Berlin contributed the Hs 297 *Schmetterling*. Henschel were among the leaders of German guided missile design and since 1939 had been busy producing a long string of air-to-air guided missile designs which eventually culminated in their Hs 296, but that project did not proceed far before it was overtaken by the Hs 297 in 1941. In 1943 OKL issued a development contract and the designation then became the 8-117 or Hs 117. Guidance of the swept-wing missile was again line-of-sight with course alterations made by radio control. For take-off from a much-modified *3.7 cm Flak 18* carriage two solid fuel booster rockets were fitted, but once airborne a liquid fuel rocket was used for propulsion. Flight trials began at Karlshagen in May 1944 and by December 1944 production was ordered with rates rising to a planned 3,000 a month by November 1945. By early 1945 such rates were obviously unrealistic and none ever saw active service, but as the war ended the first Luftwaffe units were in training to accept and use the missile. The 8-117C was to have been a radar-guided version and some trials were carried out by the Luftwaffe's Flak *Lehr- und Versuchsabteilung 700* formed in September 1944.

A late starter in the guided AA missile programme was the Messerschmitt *Enzian,* work on which began in late 1944. The overall design was based on the Me163 rocket interceptor and the construction was mainly of wood. Carrying a very heavy warhead, *Enzian* was guided to the locality of its target by radio control and the warhead was then detonated. Launching was from a modified *8.8 cm Flak 36* carriage. The *Enzian E1, E2* and *E3* were all trial missiles, and it was intended that the *E4* would be the production version. However, the project was reviewed in January 1945 and as it was not considered sufficiently advanced the *Enzian* series was cancelled.

While all the above missiles were the results of development work carried out by larger German industrial concerns, the State had its own official research station concerned with rockets and rocket guidance at Peenemünde on the Baltic Sea coast. This station operated under the cover name of the Elektromechanische Werke, or EWM, and branches were situated at Anklam, Kummersdorf, Bodensee and Bleicherode. This concern will be mentioned again when the A4 is considered, but it was also involved from about 1942 onwards with the design and development of guided AA missiles. By early 1943 the design of a Flak rocket with supersonic performance based on the aerodynamic shape of the A4 missile was adopted and designated *Wasserfall*, and from then onwards a great deal of effort and facilities was diverted towards making it a viable weapon. The task was not easy, for *Wasserfall* embodied many very advanced design features, not the least of which was the adoption of a liquid fuel rocket engine which had to be capable of standing in the open fully fuelled until the moment of launching. As the liquids involved were highly volatile this involved a great deal of design work using a variety of metals for fuel tank construction. More innovations were introduced with the guidance system, which used two separate radar sets, one on the ground and one in the rocket. On the ground two cathode ray tubes emitted visual blips which had to be kept close together by an operator using a joystick control and when the missile was in range of its target the warhead was detonated; as a back-up a proximity fuse was fitted. For its destructive effect *Wasserfall* relied on a massive 145 kg warhead and as an insurance against the rocket missing its target a 90 kg self-destruct charge was carried. Guidance on the rocket was effected by a combination of graphite rudders in the rocket exhaust after take-off, and four short wings on the rocket body for most of the flight. No launcher ramp was needed as the rocket was designed to take off from its tail fins vertically. As the *Wasserfall* development continued through 1943 and 1944 more and more delays were caused by the priority given to the A4 programme and many components used by *Wasserfall* which were common to the A4 were often diverted so that by the end of the war only about 35 rockets had actually been launched, but some *Wasserfall* components were tested in modified *A4*s. When the war ended pilot production of the projected *Wasserfall* C-2 8/45 was in the planning stage with an output of 900 a month planned for March 1946. The factory was to have been a new underground plant near Bleicherode, but it was never built.

The cost of developing *Wasserfall* must have been enormous and in the end the Germans had little to show for it.

The cost and complexity of *Wasserfall* did have a profound effect on one of the EMW designers working on the project, namely one Dipl. Eng. Schleufeln. He was so dissatisfied with the progress and viability of *Wasserfall* that he started design work on a simple and small unguided *10 cm Flak* rocket on his own initiative during mid-1944 and in September 1944 he submitted his design to the RLM. It was accepted the following month and Schleufeln was put in charge of the project, given the cover-name *Taifun*. *Taifun* used a simple liquid fuel rocket motor which gave the missile a high speed. A contact fuse was fitted to the warhead in the nose and launching was to have been from batteries of frames each holding 30 rockets mounted on converted *8.8 cm Flak* carriages (the *Dobgerät*). Many features of the *Taifun* rocket design were ingenious and novel but so simple was the overall design that it offered every promise of becoming a successful weapon. As a result great hopes were placed on it and ambitious production plans were made. Pilot production began at Peenemünde in January 1945 but the approaching Soviet armies made evacuation from the Baltic coast necessary. It was intended that development and production would continue at the Mittelwerke at Neuhausen but the war ended before arrangements were completed. It is not known if any production *Taifuns* were fired but it would seem unlikely even though about 600 were made before production ceased. Had production proceeded as planned it was hoped that about 400 batteries each with 12 launchers would have been in service by September 1945.

From the above it can be seen that work on rocket development comprised a considerable part of German weapon evolution. Many programmes that did not reach service status were carried out by nearly all the major German weapon manufacturers and a short listing of some of these follows.

Škoda of Pilsen were involved in a low priority programme to design and produce a *10.5 cm Flak* rocket. Various launchers were built, one of which involved the use of multiple rails mounted on an *8.8 cm Flak 36* carriage; a similar launcher was intended for use on a mobile mounting fitted on a Panther tank chassis. Also built was a turret-like mounting using similar launching rails for shipboard installation.

Rheinmetall were involved in the development of two 21 cm rockets intended to carry cables to altitude, the intention being to foul the wings or propellers of enemy aircraft. These two rockets were the *Drahtseil-Rakete (RSK) 1000* and *2000*. The *RSK 1000* got as far as troop trials but like so many other similar projects the idea proved unsuccessful.

Mention has already been made of the establishment of a rocket research station at Kummersdorf-West during 1930 for research into solid-fuel artillery rockets, but the station was also set up to investigate the possibilities of using liquid-fuel rockets for long-

range bombardment of military targets. During the 1920s a great deal of research was carried out on an amateur and largely unfunded basis and eventually this work attracted the attention of the Army, whose artillery experts realised the potential value of rockets as heavy long-range bombardment weapons, and in 1929 allotted funds for further official research. A Reichswehr artillery officer, *Hauptmann* Walter Dornberger, was assigned to the task of centralising and co-ordinating the rocket research taking place all over Germany, with a view to possible military weapons being evolved.

Hauptmann Dornberger set to his task with a will and remained in charge of all German rocket research until 1945, eventually reaching the rank of Major-General. He gathered around him a nucleus of engineers and scientists such as Walter Riedel, Wernher von Braun, Heinrich Grünow and Walter Thiel, who were able to pool their knowledge and capabilities in such a successful manner that in a few years they placed Germany in the vanguard of rocket research, and as a team they had no equal anywhere in the world. Although much of the early work by this team was directed towards a series of fairly simple solid-fuel artillery rockets, from the start work was also carried out on the design of experimental liquid-fuel rockets.

In 1932 the rocket team moved to the old ammunition depot at Kummersdorf-West, near Berlin, and there they began work on actually constructing a liquid fuel research rocket. The motor for this rocket used liquid oxygen and alcohol and by 1933 the first rocket, designed *Aggregat 1 (A1)*, was completed. Unfortunately it was a failure due to the bulk and weight of the stabilisation mechanism which rendered the design nose-heavy, so a revised design, the *A2*, was built with the stabilising mechanism housed in the centre section of the rocket and in this form successful test firings of two rockets were made from Borkum Island in the Baltic during December 1934.

The use of Borkum was dictated by the restrictions on range imposed by the built-up areas close to Kummersdorf-West. There the site was suitable for the relatively short range solid-fuel rockets but the ranges contemplated for liquid-fuel rockets were too great for the Kummersdorf facilities. A new range was needed, so during December 1935 the region around Peenemünde was surveyed and by April 1936 the area had been purchased jointly by the Army and the Luftwaffe. Peenemünde was situated off the Baltic coast and was almost ideal for its purpose as it encompassed the islands of Rügen and Greifswalder Oie which could be used as range stations. To the north-west stretched an area of open sea eminently suitable for the main firing range. By agreement the Army occupied the eastern part of the island and there they set up the *Heeresversuchsanstalt Peenemünde (HVP)*. The Luftwaffe took over the western half of the island and set up their *Erprobungsstelle Karlshagen*. From the very beginning no expense was spared in setting up the very extensive facilities needed for such a research station and millions of Reichsmarks were spent in providing laboratories, test stands, liquid oxygen manufacturing plant, range stations, and all the many other items needed for such establishment.

During 1937 tests began of a new research rocket, the *A3*. Much of the early development work was carried out at Kummersdorf-West but the actual testing was among the first to be carried out at the HVP. The *A3* was a significant step forward in rocket design for it was the first to use molybdenum or graphite rudders situated in the rocket exhaust to provide accurate steering and stabilisation during the critical take-off stage, which was carried out at relatively low speed. Like the earlier designs, the *A3* used a liquid oxygen and alcohol motor which took time to build up to its full thrust, and stabilisation during the first part of flight was a crucial problem. The rudders were a considerable advance, but the method of controlling them was still too rudimentary and thus the *A3* was unable to be launched in windy conditions. But the motor developed a thrust of about 1500 kg and the rocket weighed 750 kg. The first *A3* was fired on 6 December 1937 and despite the many stabilisation problems a military rocket was seen to be a distinct and viable military weapon.

To follow on from the *A3* a military version carrying a complex guidance system and a 1000 kg warhead was proposed. This was the *A4*. But there were still very many problems associated with the rocket stabilisation system to be overcome before a military rocket could be considered. Thus the *A4* project was temporarily shelved in favour of the *A5*, which was to serve as the development model for the *A4*. To discover more about the aerodynamics of rocket bodies, more money was spent in expanding the Peenemünde facilities. A new 40 × 40 cm wind tunnel of ingenious design capable of providing air velocities up to Mach 4.5 was installed in the new research division premises known as the *Forschungslabor* (the wind tunnel was not fully operational until 1943) and other such facilities were added, but much of the work on the *A5* involved air drops from an He-111 bomber. The motor of the *A5* was the same as that used on the *A3* and the *A5* had no warhead, but it proved to be a valuable research tool and gradually the stabilisation problems were solved until the stage was reached where the *A4* could be revived.

For the *A4* a larger version of the liquid oxygen/alcohol motor developed by Dr Walther Thiel was used. In appearance the *A4* closely followed the aerodynamic shape of the *A5* and was thus well streamlined with a graceful and well-proportioned outline that belied its aggressive employment. Four tail fins provided the stabilisation and guidance at high speeds, but once again it was the graphite exhaust rudders that enabled the rocket to be launched vertically from a small and simple launching table. A great deal of research went into the exact nature and form of the rocket motor itself before the final version was decided upon, but it was not until 21 March 1940 that the first static motor firings were begun. At around the same time the construction of a pilot factory was also begun at Peenemünde for the mass production of *A4*s, but in early 1940 the *A4* had a low development priority as many of the German war leaders regarded the war as virtually won, and the Peenemünde research was seen as a form of technical novelty and German achievement, despite the huge sums of money spent on setting up the establishment. To add to the problems of the programme directors the first prototype *A4* exploded under static tests on 18 March 1942.

By mid-1942 many of the early problems besetting the *A4* seemed well on the way to being fully solved, and on 13 June 1942 the first test firing of an *A4* took place. Unfortunately the launch was a failure and so were the next two attempts. On 3 October 1942 the first really successful launch of an *A4* was made and it flew to a range of 190 km.

One of the reasons for the slow development rate was the low priority given to the project, which resulted in a poor supply of manpower and raw materials. All that was changed on 22 December 1942 when General Dornberger, with the support of Reichsminister Speer, was at last able to obtain Hitler's permission to commence mass production plans.

Even before the *A4* was approved for production the Luftwaffe had been watching its gradual emergence with some trepidation. Political as well as the normal inter-service rivalry coloured their judgement when they realised that the *A4* would be able to bombard London at a time when it was evident that conventional Luftwaffe bombers would be unable to make much impact. Therefore they too began to search for some form of long-range unmanned missile and by March 1942 they decided upon the Fieseler Fi 103.

The Fi 103 was not a rocket as it used a pulse-jet engine for its propulsion. Overall the Fi 103 was a much simpler and cheaper weapon than the *A4*, and the disparity in costs was a major factor in the inter-service bickering that went on between the Luftwaffe and the Army. The Fi 103 emerged as an unmanned airframe with an 850 kg warhead (the *A4* was destined to carry a 975 kg warhead). The engine design dated back to 1929 when a Dr Paul Schmidt carried out unofficial trials at Munich-Wiesenfeld. Dr Schmidt's engine was eventually sponsored

A4 (V2) long range rocket missile being removed from its underground storage tunnel

by a joint Luftwaffe/Army fund, but in 1934 the Army withdrew and the Luftwaffe then took over the programme. By 1939 the Argus-Motoren-Gesellschaft became involved and gradually Dr Schmidt faded into the background. Argus went on to produce a motor giving about 120 kg of thrust and tested it slung beneath the wings of a Gotha Go 145 trainer during April 1942. The motor seemed to be the most suitable for use on a pilotless flying bomb but there were still many unknowns when the Luftwaffe decided to order the flying bomb as their reaction to the *A4* programme. That was on 19 June 1942. Airframe design was placed in the hands of the Gerhard Fieseler Werke GmbH at Kassel-Bettenhausen and they eventually produced the Fi 103, later to be known by the cover name of *Flakzielgerat 76 (FZG 76)* or the code name of '*Kirschkern*' (Cherrystone). H. Walter KG were made responsible for the catapult design and Siemens produced the autopilot which was originally an Askania design. The Argus pulse-jet was finalised as the Argus 109-014 and early trials were carried out on Dornier Do 17 and Junkers Ju 88 aircraft. During December 1942 the first air-launch of an unpowered airframe had been made and the first trial catapult launch took place at Peenemünde on 24 December 1942. By that time production plans were already under way with a view to an output of 2,000 a month which would rise to 3,000 a month by December 1943 when the Fi 103 was expected to become operational.

With the emergence of a Luftwaffe project that looked simpler and cheaper than their own *A4* the Army began to become alarmed for the long-term prospects for their own programme. During early 1943 they began to inject more energy towards the *A4* by the setting-up of a 'Special A4 Committee' under the guidance of *Direktor* Gerhard Degenkolb, an industrialist of considerable ability and dynamism who soon made himself felt by the pushing-through of approval for mass production on 15 January 1943. The initial plan was ambitious and called for a monthly output of 900 rockets. Already the Peenemünde pilot plant was turning out development rockets but other production was planned at the Zeppelin factory at Friedrichshafen and the Henschel Rax-Werke at Wiener Neustadt. The all-important combustion chambers were to be mass-produced at the Linke-Hoffman plant in Breslau, but what was to appear as the major bottleneck in production was the assembly of the fuel pump turbines which would take place at the Heinkel plant at Jenbach.

As soon as mass production was approved the initial military plans were laid. From the start it was obvious that the United Kingdom would be the target, so massive launching bunkers were planned for Northern France. After a considerable amount of survey work the site for the first bunker was found in the Bois d'Eperlecques near Watten. Work soon began on the excavation and construction of this site and 120,000 cubic metres of concrete were allocated for its completion. Soon after a second site near Wizernes was earmarked. Both sites attracted the attention of the Allied air forces and they inflicted their first unwitting damage to the *A4* when the US 8th Air Force bombed the Zeppelin factory on 22 June 1943 and destroyed many of the tools and jigs assembled for the first production *A4*s. Soon after the Henschel Rax-Werke were also attacked, and later the launching bunkers were destroyed. But the *A4* programme kept going with the formation of an Army training unit at Peenemünde known as the *Heimat-Artilleriepark 11 (HAP 11)*, and the building of new factories for the production of the chemicals needed for the rocket motors. For the new rocket the German chemical industries were considerably stretched, for to supply the needs of the planned 900 rockets a month some 13,000 tons of liquid oxygen were needed – the normal monthly output to supply the rest of German industry was 25,000 tons and all of that was already allocated. Other requirements that had to be met every month for the *A4* were 4,000 tons of 99% pure alcohol, 2,000 tons of methyl-alcohol, 500 tons of hydrogen peroxide and 1,500 tons of high explosive (60% amatol, 40% aluminium and other metals). All this was a considerable burden on German industry, but during May 1943 Hitler himself gave the *A4* programme the production rating of DE12, the highest production priority in the Third Reich, and from then on the *A4* absorbed men and materials that had been set aside for many other projects, including the *Wasserfall* development, the Me 262 fighter programme and eventually the *FZG 76*.

The inroads into the *FZG 76* allocations raised a considerable political storm and eventually the flying bomb was exempted from the ever-increasing demands of the *A4*, but despite the early promise the Luftwaffe project soon ran into protracted technical problems. Nearly every one of the early prototype airframes failed to function correctly.

Despite the many problems new production plans were drawn up increasing the output to 3,000 a month. These flying bombs would require 300 tons of hydrogen peroxide a

month and 2,000 tons of low-octane fuel. For the warheads 4,500 tons of high explosive were needed. The December 1943 output was raised to 5,000 a month and all seemed set for introduction into service during that month.

The development of the *A4* and the *FZG 76* received a severe setback on the night of 17–18 August 1943 when the Royal Air Force carried out a concentrated bombing raid on Peenemünde (Operation Hydra). The raid did considerable damage to the Peenemünde facilities and many members of the research team were killed, but the main disruption came from the need to move many of the Peenemünde operations to decentralised locations. More important at the time was the removal of the major test range facilities to Blizna in Poland. Other parts of the Army programme at Peenemünde were scattered all over Germany, with only the *Wasserfall* and other smaller missile programmes remaining.

The *FZG 76* was less affected by the bombing raid, but as a precaution the main test site was moved to Zempin not far from Peenemünde, and the launching ramps were erected at Brüster-ort. On 1 July 1943 the Luftwaffe set up a trials unit under the title of *Lehr- und Erprobungskommando Wachtel* under the command of one *Oberstlt*. Wachtel and eventually this unit, intended to work out the firing drills and service operation of the *FZG 76*, became part of Flak-Regiment 155(W). Work on the development of the *FZG 76* progressed to the extent that many of the early design problems were resolved and corrected. In France the construction work began to built 96 launching sites aimed at London and by September 1943 40,000 workers, many of them French, were involved. In addition to the launching sites, two massive bunkers and numerous other storage buildings were projected but the activity was soon noted or reported across the Channel and the Allied air forces began a prolonged bombing programme to put the new sites out of action.

Meanwhile production of the *FZG 76* was not running smoothly. The Fieseler-Werke at Kassel was bombed on the night of 22 October 1943 and the line was moved to Rothwestern which was not equipped to turn out the numbers needed. On 16 November 1943 the first initial series airframe was launched without much of its control gear installed. Soon after, the first production versions were delivered and launched, but once again troubles began because the standards of the mass-produced bombs were so low that at the end of November 1943 it was decided to scrap some 2,000 examples from the Volkswagen lines and start again with tighter control standards.

With the bombing of the Zeppelin and Henschel works it was decided to plan new production facilities in an underground factory beyond Allied bomber range. At about the same time the *A4* project was reorganised on a more normal commercial basis by the formation of the Mittelwerke GmbH with its main office at Berlin-Charlottenburg. This organisation was intended to run the new underground factory which was eventually situated under the Kohnstein mountain at Nordhausen. At the same time the Demag plant at Berlin-Falkensee became incorporated into the *A4* programme. The Henschel and Zeppelin plants were combined into the Südwerke and an Ostwerke was projected at Riga in German-occupied Latvia but did not materialise.

The Nordhausen factory was soon manned by slave and convict labourers who worked under appalling conditions to expand the rudimentary storage tunnels already there into one of the most advanced assembly plants in Europe. As well as *A4* production the Nordhausen tunnels were used for assembling jet engines and eventually numbers of the *FZG 76* and the *Taifun* AA rocket missile. Nordhausen alone was a good example of how the *A4* was able to absorb huge amounts of money and labour, but by the time the first production rockets were delivered the *A4* had changed from being the 'ultimate weapon' into a weapon of pure malice and revenge. The *A4* had become *Vergeltungswaffe 2* (*V2*) and the *FZG 76* was the *V1*. The Allied bomber raids that took place during 1943 grew in scope and size until they culminated in the mass raids on Hamburg during July 1943. That series of raids drove home to the German people the power of the Allies, but instead of regarding the raids as a portent Hitler began to think only in terms of retaliation, and the flying bomb and the *A4* rocket were the ideal vehicles.

By early 1944 the first production examples of the *A4* were ready and a series of trial launchings began at Blizna with the newly-formed *Versuchsbatterie 444* at nearby Köslin. The first trial firings by this unit had actually taken place during November 1943 but those had revealed a problem that had not been foreseen: most of the rockets exploded before they hit the ground. When the first production examples were launched more problems came to light as a result of the lower mass-production standards used. Until March 1944 some 57 rockets were launched from Blizna but only 26 actually took off. Of those only four hit the target area and the rest exploded as they re-entered the atmosphere. Despite great efforts and improved production standards the problem seemed insoluble and once more emphasis turned towards the *FZG 76*. Production of the *A4* dropped drastically and by July 1944 totalled only 86, compared with 437 in May.

By early 1944 the *FZG 76* was thus in a state closer to actual service use than the *A4* and the tempo of the programme increased. Despite the slow production rate, plans were made for a massive flying bomb bombardment of London to start during May. New types of concrete launching ramps were designed and built which could be installed fairly easily and quickly. All the 96 original sites had been discovered by the Allies and nearly all had been rendered useless from the air and it was not intended to instal the new ramps until a few days before the offensive would start. *Flak-Regiment 155(W)* was placed under the command of a new formation, 65 Armeekorps, based at Saint-Germain and formed to supervise the use of both the *V1* and the *V2*, by which terms the new weapons were then known.

In this atmosphere of conspiracy the target date for the beginning of the attack on London slipped by. On 6 June 1944 the Allies landed in Normandy and the need for the *V1* became even more desperate. A hasty attempt to open some sort of attack on London was scheduled for the night of 12–13 June, but in the event the attack was a failure. Only ten *V1*s were launched and of those only six became airborne – the other four crashed soon after take-off. Of the six only four made a landfall and the first *V1* came down near Gravesend at 0418 hours. The *V1*, in Churchillian terms 'the malignant robot', had become a weapon of war.

The relative failure of the 13 June attack was a distinct contrast to the projected barrage of hundreds of rockets to be sent over in a massive stream, but on the night of 15 June another more concerted attack began. In 12 hours 244 *V1*s were launched and this time 199 reached England. Once again the aiming point was 'Ziel 42' – a point 1,000 metres east of Waterloo station. From then on the numbers of launchings and arrivals grew until by 29 June the 2,000 mark was reached. By then the *V1* had killed 1,769 people in and around London and so much damage was being done to morale and property that during the week of 20–27 June some 40% of all Allied air activity was directed against the *V1*. By the beginning of July plans were put into effect to air-launch the *V1* from the Heinkel He 111s of III/KG 3 and on 7 July the first air launching was made against Southampton.

Once airborne the *V1* was controlled by an auto-pilot which kept the speed and course of the missile constant and it was this that caused the undoing of the weapon, for it made an almost ideal target for anti-aircraft guns. By 19 July the defences of London were rearranged along the approach paths of the *V1*, and by the use of advanced tracking radar sets and proximity fuses the defenders were able to bring down more and more *V1*s over open country. Even so, some *V1*s still got through and it was only the gradual land advances in France that overran the launching sites and prevented further launchings. By the end of August

Three operational A4 (V2) rocket missiles on their simple launching pads with protective circular covers on their nose fuses; these also served as partial stabilisers

the Germans were out of France and Flak-Regiment 155(W) redeployed for the new target of Brussels, soon changed for Antwerp.

Antwerp, at that time the only Allied supply port, became the focus of the greatest use of the V weapons but London was by no means forgotten. Some airborne launchings continued and as late as March 1945 *V1*s were launched from points in Western Holland.

By July 1944 the gradual Allied advances in France revived the interest in the *V2* and by early August 1944 the re-entry problem had at last been solved Under the control of the 65 Armeekorps new rocket batteries moved into position. Near the Hague were two batteries of *Art. Abt. 485 (mot.)*, a mobile detachment. Two batteries of *Art. Abt. 836 (mot.)* along with original *Batterie 444* were situated near Liège. A new launching platform that did not require fixed installations had been developed for the *V2* and the units using the rocket were thus flexible and mobile.

The first attempt to fire the *V2* was made on 6 September 1944, and the target was Paris. *Batterie 444* was one of the units involved and the other was SS *Mörser-Batterie 500*, but the two rockets launched failed. On 8 September 1944 *Art. Abt. 485 (mot.)* fired two rockets at London and the first *V2* fell on Chiswick at 1843 hours. Thereafter the *V2* attacks continued and by 18 September 25 had reached England. By the end of 1944 the *V2* had been fired at thirteen cities. The unwilling recipients were as follows: Antwerp 924, London 447, Norwich 43, Liège 27, Lille 25, Paris 19, Tourcoing 19, Maastricht 19, Hasselt 13, Tournai 9, Arras 6, Cambrai 4, Mons 3, Diest 2, Ipswich 1.

Nordhausen continued to manufacture the *V2*s right until the war ended, but by early 1945 the V-weapon programme was entirely under SS control. By February the output had reached a total of 2,275 *V2*s, but by that time the main problem was delivering them to the launcher units due to the Allied air supremacy over the Reich. The field units continued to remain in action as long as possible, but by early March 1945 the Allied advances into Holland forced them to withdraw into the Reich. Their retreat was marked by a final heavy attack on London and on 27 March 1945 the last *V2* fell on the capital. A total of 1,054 *V2*s actually fell on the United Kingdom but many more that were launched failed for one reason or another – a total of 61 were observed to fall into the sea just offshore.

The last *V1* fell on London on 29 March 1945. It was the 2,419th to reach its target since the offensive had begun, but a total of 10,492 had actually been fired. The much vaunted V-weapons had failed in their task to a massive extent, for since 1935 an ever-increasing amount of facilities and manpower had been devoted towards producing weapons that despite all their ingenuity and technical brilliance could only carry warheads weighing about a ton. Such a warhead, whatever its delivery system, was not worth all the effort, money and suffering that it took to produce. For reasons that could only apply in the frantic, fevered political state of the Third Reich, the V-weapons began to dominate the thinking of German military and political leadership and diverted them to a goal which, far from ensuring victory, hastened their eventual defeat.

Even as the war entered its final stages German scientists were develping more and better war rockets. A winged *A4*, the *A4b*, was designed with a range that would enable all of the United Kingdom to be reached. The *A4b* was actually a winged *A4* research rocket intended as a pilot study for the eventual *A9* which was to have been the production version. The most ambitious project was mooted as early as 1940, the *A9/A10*, which was to have been an *A10* booster carrying an *A4* or an *A9* with a range of 4,800 km. Although this futuristic idea remained on the drawing-board a proposal was made for an *A4* inside a special launching tube to be towed across the Atlantic by a submarine – a similar proposal was made for the *V1* which would have been launched from a ramp on a submarine.

Almost until the advancing Soviet forces reached Peenemünde, experimental work was being carried out. The last *A4* – a production test missile from Mittelwerke – was fired there on 14 February 1945 and soon after the establishment was moved, under SS control, to Bleicherode near Nordhausen, but by March 1945 the Reich was in such a state of chaos that little further research work was done. In one way or another the design teams departed to the USA and the Soviet Union soon after the end of the war and in both countries they were actively encouraged to carry on the work they had started so many years before.

A Fieseler Fi 103 robot bomb – better known as V1 – shortly after launch

Brief summary of rockets in 'Aggregat' series

A1
First German experimental liquid-fuel rocket, designed and built under Army auspices during 1931–33. Powered by rocket motor using liquid oxygen and 75 per cent alcohol developing 300 kg thrust for 16 secs. Length of rocket 1395 mm; body diameter 304 mm. First test launch attempted in 1933 proved unsuccessful; rocket was nose-heavy and did not take off.

A2
Essentially same as A1, of same dimensions and with same power unit, but stabilising controls moved from nose to centre section of rocket. Designed and constructed at Kummersdorf during 1934, two examples successfully test-launched from Borkum near Peenemünde early in December 1934.

A3
First German liquid-fuel rocket with gyroscopic controls. Basic layout and rocket motor established pattern for all subsequent liquid-fuel rockets. Length 7600 mm, body diameter 750 mm, weight approx 750 kg. Rocket motor delivered 1500 kg thrust for 45 secs. A series of A3 rockets launched at Peenemünde between 4 and 11 December 1937, of which only four proved successful.

A4
Became the V2.

A4b
Originally known as A9. Winged A4 used as development vehicle for projected A9 long-range rocket missile. Work on this test rocket held up between October 1942 and June 1944 to complete A4 development. First A4b fired on 27 December 1944 reached 78,000 m altitude but shed its wings on re-entry. Winged test rocket A4b G3 was flown successfully on 24 January 1945, penetrated the sound barrier, and reached a terminal velocity of 1200 m/sec. No more test launches could be made due to evacuation of Peenemünde in February 1945.

A5
Designed and used as development vehicle for A4 missile. Basically modified A3 with control vanes in gas exhaust stream and improved guidance equipment. Used same rocket motor as A3. First launched in June 1938; first fully controlled test flights in October 1939.

A6
Design study only.

A7
Proposed winged version of A5. Not built.

A8
Design study only.

A9
Winged version of advanced A4. Proposed in 1942; designation later changed to A4b.

A9/A10
Proposed intercontinental rocket weapon. Combination of an advanced A4 or A9 and a new A10 booster stage. First proposed in July 1940 but postponed until more advanced development of A4 rocket. Destined never to be built. A9/A10 would have stood 26 m high with a maximum body diameter of 4.75 m. The A10 booster was to have had take-off thrust of 183,000 kg, while the A4 or A9 rocket motor would have provided 30,500 kg thrust. The planned range was 4800 km with a 925 kg warhead, reaching an altitude of 350 km and a terminal speed of 7680 km. This would have brought New York and Washington within target range.

A11–A15
Proposed variations of A9/A10 theme to increase range to max 5600 km.

Appendix 1

Annual production totals for field rockets and launchers, 1941 to 1945 (January totals only).

Launchers

	1941	1942	1943	1944	1945
15 cm NbW 41	650	970	1188	2336	139
21 cm NbW 42	648	970	100	835	73
30 cm NbW 42*			380	544	30

* includes 30 cm RkW 56

Rockets (in thousands)

	1941	1942	1943	1944	1945
15 cm	418	1208	1096	1985	120
21 cm		9	120	258	12
28/32 cm	125	169	143	140	4
30 cm		24	106	145	6

Appendix 2

Artillery rocket data

Rocket group	15 cm	21 cm	28 cm	30 cm	32 cm
Width at widest point (mm)	158	214	280	300	337
Weight complete (kg)	34.15	112.6	86–83	127	79
Weight of warhead filling (kg)	1.31	9.5	50	45	46
Weight of propellant (kg)	5.9–6	18	6.5	15.1	6.5
Length overall (mm)	931	1260	1260	1249	1290
Initial velocity (m/sec)	340	320	145	230	145
Range (m)	6900	7850	1925	4550	2200
Types of warhead	HE Smoke	HE	HE	HE	Incendiary
Projectors	15 cm Do-Gerät 15 cm NbW 41 30 cm RWerfer 56	21 cm NbW 42	28/32 cm NbW 41 sWG 40 sWG 41 sWuR 40	30 cm NbW 42 30 cm RWerfer 56	28/32 cm NbW 41 sWG 40 sWG 41 sWuR 40

2 cm Luftfaust

German designation 2 cm Luftfaust
Calibre of launcher tubes 22 mm
Length of launcher tubes 1308 mm
Weight of launcher 6.6 kg
Weight of one rocket clip (9 rockets): 2.5 kg
Initial velocity 280–310 m/sec
Rocket weight 0.11 kg
Maximum effective range 500 m
Maximum range 2000 m
Designers Hugo Schneider AG (HASAG), Leipzig

Remarks: Revolutionary one-man anti-aircraft weapon, designed for ease of operation and manufacture. Intended for use against low-flying aircraft, Luftfaust projectiles comprised specially adapted 2 cm cannon shells. In action all nine rocket shells would be fired in two salvos of four and five 0.2 secs apart. Ordered into large-scale production in many small workshops early in 1945, but none reached service.

The 2 cm Luftfaust *in firing position. The container suspended from the demonstrator's shoulder carried spare 2 cm rocket rounds* ➤

7.3 cm Föhn-Gerät

German designation 7.3 cm Föhn-Gerät
Calibre 73 mm
Length of launcher rails 787 mm
Number of launchers 35
Traverse 360°
Elevation −10° to +90°
Initial velocity 380 m/sec
Weight of rocket 2.74 kg and 3.75 kg
Design agency Waffenprüfstelle der Luftwaffe, Tarnewitz
Manufacturer Not known

Remarks: Ground adaptation of Luftwaffe aircraft rocket missile for use against low-level aircraft, first delivered to service late in 1944. Produced in two versions – as static and as mobile weapon, mounted on 3.7 cm Flak 18 carriage.

7.3 cm fortress rocket projector

7.3 cm Föhn-Gerät *static equipment*

7.3 cm Propagandawerfer 41

German designation 7.3 cm PgW 41
Calibre of launching rail 73 mm
Length of launcher rail 749 mm
Weight of launcher 12.26 kg
Traverse (approx): 20°
Elevation 45° fixed
Rocket weight 3.24 kg

Remarks: Specialised and expensive single-rocket launcher developed for Propagandatruppen. Fired 7.3 cm PgGr 41 containing approx 0.5 kg of leaflets wrapped around a spring. Small explosive charge blew off sides of shell and released leaflets after set time. Introduced in service during 1941. Only small numbers produced.

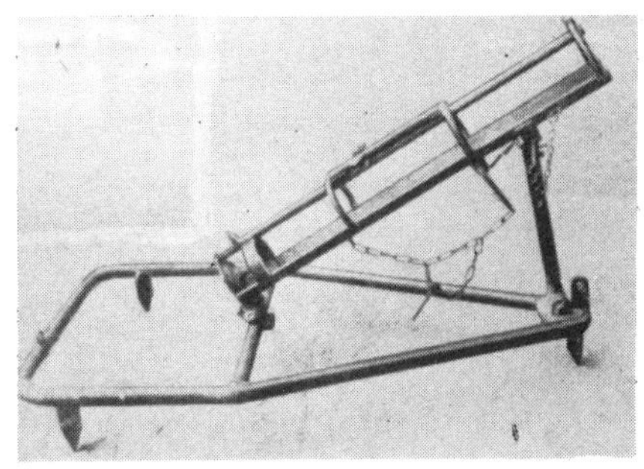

8 cm Raketen-Vielfachwerfer

German designation 8 cm R-Vielfachwerfer, 'Himmlerogel'
Number of launching rails 24
Number of rockets in one salvo 48
Weight of vehicle in action 6853 kg
Traverse 360°
Elevation 0° to +37°
Initial velocity 290 m/sec
Rocket weight 6.9 kg
Maximum range 5300 m

Remarks: German copy of Soviet RS-82 rocket and M-8 launcher rails ('Katyusha') produced exclusively for Waffen-SS. Mounted mostly on ex-French Army SOMUA armoured half-tracks. Rocket shells launched by this weapon differed from majority of German field rockets by being fin-stabilised.

8 cm Raketen-Vielfachwerfer *('Himmlerorgel')* ➤

8.8 cm Raketenwerfer 43

8.8 cm Raketen-Werfer 43 *('Puppchen'). This weapon fired hollow-charge anti-tank rocket projectiles similar to those used by the* Raketen-Panzerbüchse 43. *The equipment could be dismantled for easier transport into seven loads*

8.8 cm R-Werfer 43 with its wheels removed

Rear view of the 'Puppchen' showing the rocket being inserted into the breech

German designation 8.8 cm R-Werfer 43 'Puppchen' ('Dolly')
Calibre 88 mm
Length of barrel (L/18): 1600 mm
Weight travelling 146 kg
Weight in action 100 kg
Traverse 60°
Elevation −18° to +15°
Initial velocity 150 m/sec
Maximum range 700 m; (anti-tank): 230 m
Rocket weight (RGr 4312): 2.66 kg
Rate of fire 10 rpm
Barrel life 1000 rounds
Manufacturer Westfallische-Anhaltische Sprengstoff AG, Reinsdorf

Remarks: Introduced into service 1943 but not produced in large numbers because later RP 43, RP 54 and 54/1 proved to be cheaper and easier to use.

8.8 cm Raketenpanzerbüchse 43

German designation 8.8 cm RPzB 43 Panzerschreck or 'Ofenrohr'
Calibre 88 mm
Length of barrel 1638 mm
Weight in action loaded 9.5 kg
Initial velocity 100–110 m/sec
Rocket weight 3.25 kg
Maximum range 150 m
Range of fire 4–5 rpm
Barrel life 1000 rounds
Manufacturer HASAG, Meuselwitz, and others

Remarks: An enlarged German copy of US Army 2.36-in Rocket Launcher M1 ('Bazooka'). First issued late in 1943 and produced in large numbers. Its short range and need for special protective clothing for both crew members proved important tactical limitations. On availability of later equipment remaining RPzB 43 launchers relegated to second-line and Volkssturm units.

8.8 cm RPzB 43. The gunner wears a gas mask for protection against the flashback

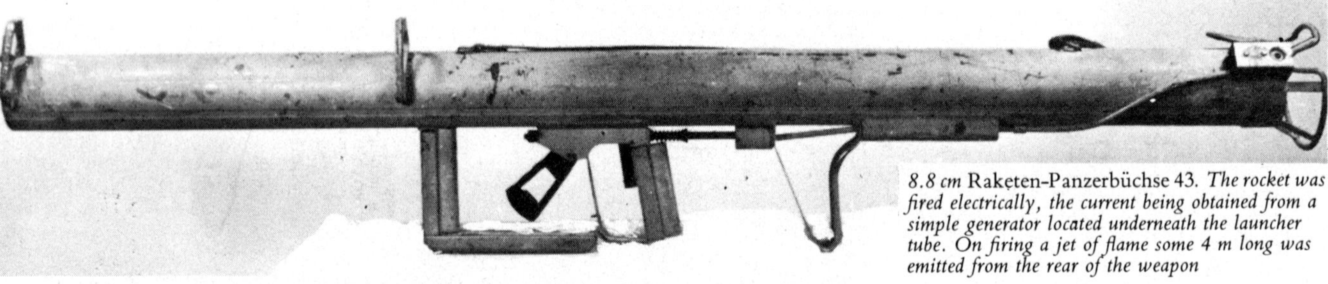

8.8 cm Raketen-Panzerbüchse 43. *The rocket was fired electrically, the current being obtained from a simple generator located underneath the launcher tube. On firing a jet of flame some 4 m long was emitted from the rear of the weapon*

8.8 cm Raketenpanzerbüchse 54

8.8 cm RPzB 54, showing the second member of this team attaching the circuit wire to the rocket projectile in the launcher tube

8.8 cm RPzB 43. The second crewman is preparing the rocket nose fuse

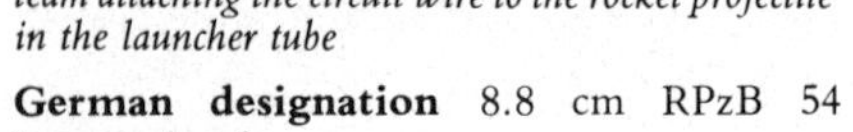

German designation 8.8 cm RPzB 54 Panzerschreck
Calibre 88 mm
Length of barrel 1640 mm
Weight (with shield): 11 kg; (without shield) 9.3 kg
Initial velocity 100–110 m/sec
Rocket weight 3.25 kg
Maximum range 150 m
Rate of fire 4–5 rpm
Barrel life 1000 rounds
Armour penetration 160 mm at 60°

Manufacturers HASAG, Meuselwitz; Enziger-Union-Werke, Pfeddersheim bei Worms; Schriker & Co., Vach bei Nürnberg; Kronprinz, Solingen-Ohligs; Jäckel, Freistadt-Oberschloss; Gebr. Scheffler, Berlin

Remarks: Evolved from RPzB 43 and fitted with protective shield. Introduced in service during 1944 and remained in production until early 1945. Could fire RPzBGr 4322 and 4992 hollow-charge armour-piercing missiles.

8.8 cm Raketenpanzerbüchse 54/1

German designation 8.8 cm RPzB 54/1 Panzerschreck
Calibre 88 mm
Length of barrel 1350 mm
Weight (with shield): 9.5 kg; (without shield): 7.8 kg
Initial velocity 100–110 m/sec
Rocket weight 3.25 kg
Maximum range 180 m
Rate of fire 5 rpm
Barrel life 1000 rounds
Armour penetration 160 mm at 60°
Manufacturers *See* RPzB 54

Remarks: Issued during 1944. Used a shorter barrel to fire RPzGr 4992 hollow-charge armour-piercing missiles.

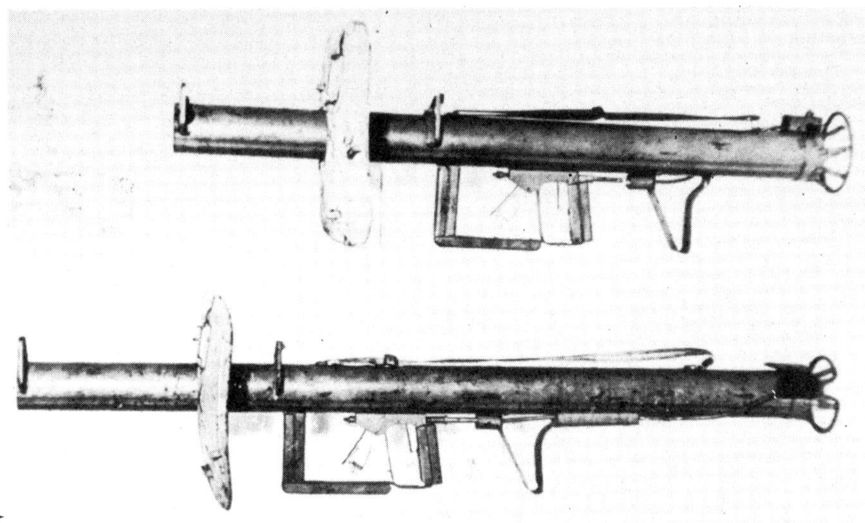

Comparison of the 8.8 cm RPzB 54 and RPzB 54/1 ➤

10.5 cm Raketenwerfer

German designation 10.5 cm RW
Calibre of rocket 105 mm
Length of launcher rails 3500 mm
Weight in action (approx): 7000 kg
Traverse 360°
Elevation −3° to +85°
Initial velocity 700 m/sec
Weight of rocket 19 kg
Number of launcher rails 16
Manufacturer Škoda-Werke, Pilsen

Remarks: Experimental anti-aircraft rocket equipment mounted on 8.8 cm Flak 36 carriage. Only one equipment completed by 1945 but a ship-board mounting had been built and another version for mounting on a Panther tank chassis was in design stage.

10.5 cm Raketenwerfer *on transport limber*

10.5 cm R-Werfer in firing position

11 cm Rauchspurgerät

German designation 11 cm R-Gerät
Calibre of interior launching rail 114 mm
Length of launcher rails 3200 mm
Traverse (approx): 30°
Elevation (approx): 15° to +40°
Rocket weight (approx): 4 kg
Maximum range 4500 m

Remarks: Smoke-laying rocket equipment. Issued to 2/Art. Abt. Königsbruck during late 1934 and used for trials and experiments only.

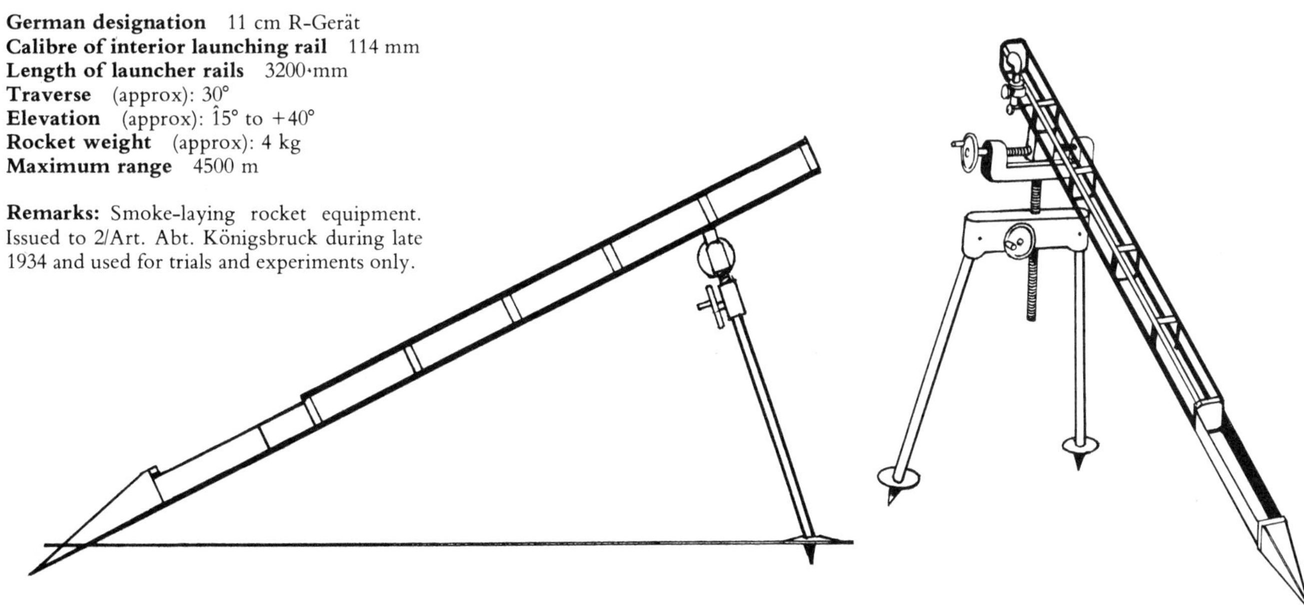

15 cm Do-Gerät

German designation 15 cm Do-Gerät
Calibre of launcher rail 158.5 mm
Length of launcher rail (approx): 2140 mm
Weight travelling (3 loads): 19 kg
Weight in action (loaded): 53.15 kg
Traverse (approx): 6°
Elevation (approx): î0° to +20°
Initial velocity 340 m/sec
Rocket weight 34.15 kg
Maximum range (approx): 5000 m

Remarks: Developed for airborne troops. First issued 1941, but only limited numbers used in action.

15 cm Do-Gerät *with parachute attached*

15 cm Nebelwerfer 41

German designation 15 cm NbW 41
Calibre of launcher barrels 158.5 +0.4 mm
Length of barrels 1300 mm
Weight travelling 590 kg
Weight empty 510 kg
Weight in action (loaded): 770 kg
Traverse 24°
Elevation −5° to +45°
Initial velocity 340 m/sec
Rocket weight (HE): 34.15 kg; (smoke): 35.48 kg
Maximum range 6900 m
Rate of fire 6 rockets in 10 secs; 3 salvos of six rockets in 5 mins
Manufacturers Frame-Werke, Heinichen/-Saxony; Sächs. Textil-Masch.-Fabrik, Chemnitz

Remarks: Six-barrelled launcher on standard 3.7 cm Pak 35/36 carriaage, with only minor modifications. First issued to Werfer-Abteilungen (formerly Nebeltruppen) during late 1941 and remained standard German ground rocket missile launcher. Fired mostly 15 cm Wurfgranate 41 Spreng HE missile powered by 7 WASAG R61 (Diglycol-Dinitrate) sticks ignited by ERZ.39 initiator.

Loading 15 cm rocket shells into the top launcher tubes of the NbW 41 ►

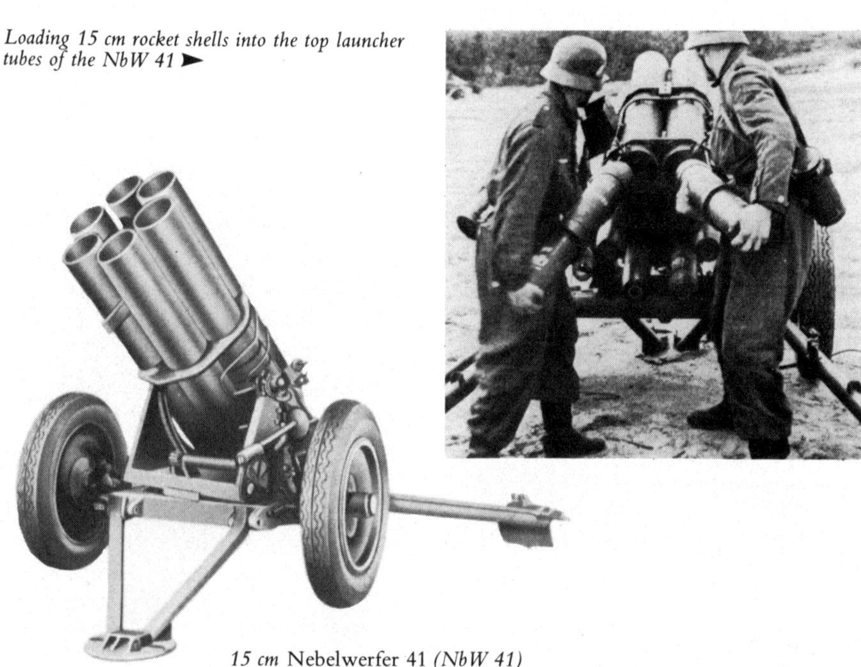

15 cm Nebelwerfer 41 *(NbW 41)*

15 cm Scheinsignalrakete

German designation 15 cm SSR
Calibre of rocket 150 mm
Weight of loaded crate 53 kg
Weight of rocket 42 or 43 kg
Maximum ceiling (approx): 2000 m

Remarks: Large flare rocket designed to act as illumination for night fighters, marking of navigation points and decoys for 'Pathfinder'-led RAF bombers. Various types and colours of flare could be fired direct from carrying crates. First used operationally late in 1943.

15.2 cm Kz.1000 (Kp)*

German designation 15.2 cm Kz.1000 (Kp)
Body diameter 152.4 mm
Length 1478 mm
Weight 73.7 kg
Weight of warhead (approx): 1 kg
Main parachute diameter 3356 mm
Pilot parachute diameter 152 mm
Length of cable (approx): 900 m
Manufacturer Friedr. Krupp AG, Essen

Remarks: Little survives about this AA rocket, no precise data regarding its performance has been found, and its exact status is uncertain. It was intended to be fired ground-to-air ahead of approaching enemy aircraft. A pilot parachute would then pull out the main parachute and release a cable which suspended rocket nose section with its warhead. Theoretically, an aircraft fouling this cable would pull warhead to impact explosion. An unrealistic and wishful device, probably as unsuccessful as similar developments elsewhere.

* Full designation not known.

15 cm Raketengranate 19/40

German designation 15 cm RGr 19/40
Calibre 149.1 mm
Length of projectile 790 mm
Initial velocity 525 m/sec
Maximum velocity 761 m/sec
Weight 45 kg
Maximum range 20,000 m
Manufacturer Rheinmetall-Borsig AG, Düsseldorf

Remarks: Rocket-assisted projectile intended for firing from 15 cm sFH 18. Rocket charge timed to ignite when shell reached top of its trajectory. Troop trials showed projectile to be too unreliable and inaccurate for service use.

21 cm Nebelwerfer 42

German designation 21 cm NbW 42
Calibre of launcher barrels 214.5 + 0.4 mm
Length of barrels 1300 mm
Weight travelling 605 kg
Weight empty 550 kg
Weight in action (loaded): 1100 kg
Traverse 24°
Elevation −5° to +45°
Initial velocity 320 m/sec
Rocket weight 112.6 kg
Maximum range 7850 m
Rate of fire 5 rockets in 8 secs; 3 salvos of five rockets in 5 mins
Manufacturer Masch. Fabr., Donauwörth

Remarks: Essentially 15 cm NbW 41 with larger barrels; same carriage. Fired only HE ammunition. Troop trials took place early in 1942, in full service 1943.

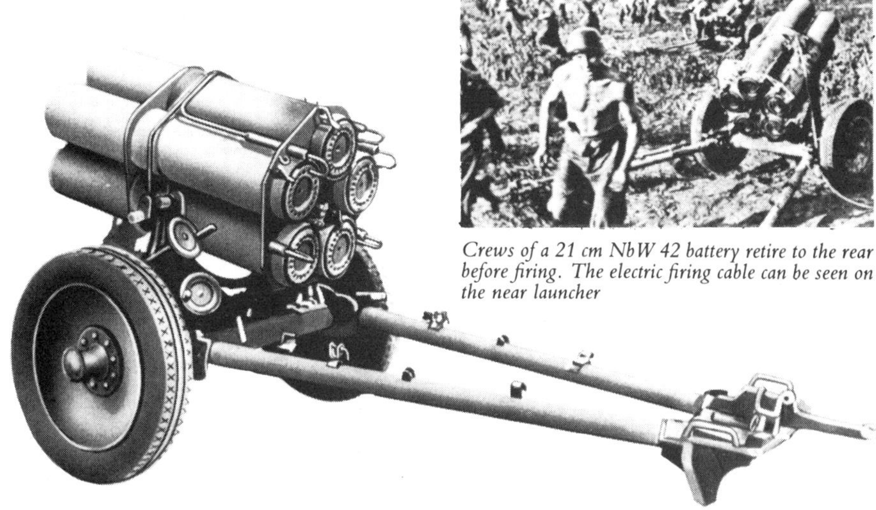

Crews of a 21 cm NbW 42 battery retire to the rear before firing. The electric firing cable can be seen on the near launcher

28/32 cm Nebelwerfer 41

German designation 28/32 cm NbW 41
Weight travelling (loaded): 1630 kg; (empty): 1130 kg
Weight in action (loaded 28 cm): 1630 kg; (32 cm): 1600 kg
Traverse 22°30′
Elevation +13°30′ to +45°
Initial velocity 145 m/sec
Rocket weight (28 cm (HE)): 82 kg; (32 cm (Inc)): 79 kg
Maximum range (28 cm): 1925 m; (32 cm): 2200 m
Rate of fire 6 rockets in 10 secs; 2 salvos of six rockets in 5 mins
Manufacturers Masch.-Fabrik, Donauwörth; Frame-Werke, Heinichen/Saxony; Sächs. Textil-u. Masch. Fabrik, Chemnitz; Skoda-Werke, Pilsen

Remarks: First mobile German field rocket launcher. Frames contoured for 32 cm missiles, with liner rails for 28 cm rounds. Entered service in 1941 but not used in large numbers due to limited range.

28/32 cm NbW 41 equipment being loaded with 32 cm incendiary rocket shells

◄ *Loading the 28/32 cm NbW '1 w th 28 cm rocket shells. Liner rails have been placed into the six frames to accommodate the smaller projectiles*

30 cm Nebelwerfer 42

German designation 30 cm NbW 42
Calibre of launcher rails (front): 301 + 2 mm; (rear): 217 + 2 mm
Weight travelling 1100 kg
Weight empty 1100 kg
Weight in action (loaded): 1860 kg
Traverse 22°30′
Elevation +13°30′ to +45°
Initial velocity 230 m/sec
Rocket weight 127 kg
Maximum range 4550 m
Rate of fire 6 rockets in 10 seconds; 2 salvos of six rockets in 5 mins
Manufacturer Masch. Fabrik, Donauwörth

Remarks: Used same carriage as 28/32 cm NbW 41. Only a limited number were made after introduction in 1943.

Battery of 30 co NbW 42s preparing foraction in the Russian winter

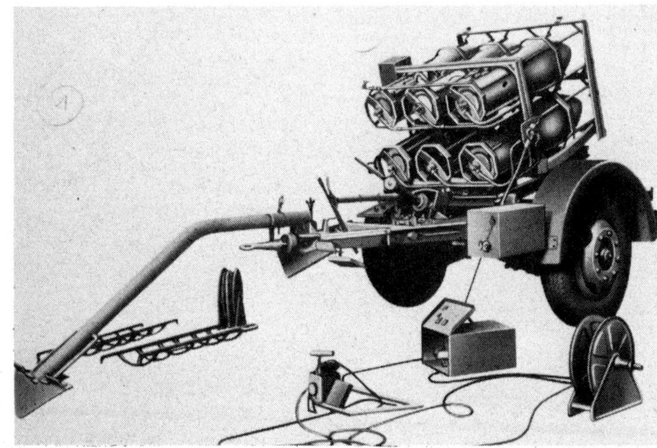

30 cm NbW 42, showing the electric firing equpment

30 cm Raketenwerfer 56

German designation 30 cm R-Werfer 56
Calibre of launcher rails (30 cm): 303 mm; (15 cm): 159 mm
Weight travelling 1004 kg
Weight in action (loaded 30 cm): 1735 kg; (15 cm): 1175 kg
Traverse 22°30′
Elevation −3° to +45°
Initial velocity (30 cm): 230 m/sec; (15 cm): 340 m/sec
Rocket weight (30 cm HE): 127 kg; (15 cm HE): 34.15 kg
Maximum range (30 cm): 4550 m; (15 cm): 6900 m
Rate of fire 6 rockets in 10 secs; 2 salvos of six rockets in 5 mins
Manufacturer Maschinen-Fabr., Donauwörth

Remarks: Introduced in service during 1944. Comprised 30 cm NbW 42 launching rails mounted on 5 cm Pak 38 carriage, but could also fire 15 cm rocket ammunition via special liners. These were stacked on launcher when not in use.

30 cm R-Werfer 56, showing connecting the electric firing cable

30 cm Raketenwerfer 56. *This equipment also fired two different calibres of rocket shells, the 30 and 15 cm. When firing the smaller projectiles liner rails were inserted into the frames*

schweres Wurfgerät 40

German designation sWG 40
Weight of frame alone 52 kg
Weight of each rocket crate 30 kg
Weight loaded (32 cm): 488 kg; (28 cm): 500 kg
Traverse 0°
Elevation +10° to +45°
Initial velocity 145 m/sec
Rocket weight (32 cm): 79 kg; (28 cm): 82 kg;
Maximum range (32 cm): 2200 m; (28 cm): 1925 m
Rate of fire 4 rockets in 6 secs
Manufacturer J. Gast, Berlin-Lichtenberg (not confirmed)

Remarks: Simplest German field rocket launcher comprising wooden frames to fire max of four 28/32 cm missiles directly from their carrying crates (*Packkisten*). First operational on Eastern Front in June 1941. Often referred to as 'Stuka zu Fuss'.

schweres Wurfgerät 41

The 28/30 cm rocket shells could also be fired individually from their carrying crates. Eastern Front, winter 1943

Several batteries of sWG 41 launchers emplaced before Sevastopol in summer 1942

German designation sWG 41
Weight of frame alone 110 kg
Weight of each rocket crage (28/32 cm): 30 kg; (30 cm): 20 kg
Weight loaded (28 cm): 558 kg; (32 cm): 548 kg; (30 cm): 738 kg
Traverse 0°
Elevation +10° to +45°
Initial velocity (28/32 cm): 145 m/sec; (30 cm): 230 m/sec
Rocket weight: (28 cm): 82 kg; (32 cm): 79 kg; (30 cm): 127 kg
Maximum range (28 cm): 1925 m; (32 cm): 2200 m
Rate of fire 4 rockets in 6 secs
Manufacturer J. Gast, Berlin-Lichtenberg

28 cm Wurfkörper Spreng *HE rocket shell in its steel carrying crate*

Schweres Wurfgerät 41

Remarks: Differed from sWG 40 by being constructed of tubular steel framework. Could launch 28/32 cm and 30 cm field rockets straight from their carrying crates. Often referred to as 'Stuka zu Fuss'; sometimes also as 'Heulende Kuh' ('Howling cow').

schwerer Wurfrahmen 40

German designation sWuR 40; 'Stuka-zu-Fuss'.
Traverse 0°
Elevation +14° to +50°
Number of rockets carried 6
Initial velocity 145 m/sec; (30 cm): 230 m/sec
Rocket weight (28 cm): 82 kg; (32 cm): 79 kg; (30 cm): 127 kg
Maximum range (28 cm): 1925 m; (32 cm): 2200 m; (30 cm): 4550 m
Rate of fire 6 rockets in 10 secs
Manufacturer J. Gast, Berlin-Lichtenberg

Remarks: Most successful mobile launcher of 28/32 cm field rocket missiles first produced late in 1940. Comprised metal framework fitted over and alongside an SdKfz 251 armoured half-track, or (from 1944) other armoured cross-country vehicles. Rocket missiles loaded on frames and fired still in their carrying crates. Could also launch 30 cm field rockets.

Four schwerer Wurfrahmen 40 *mounted on an SkKfz 250 in action in Warsaw, 1944* ►

28 cm Raketengranate 4331

German designation 28 cm RGr 4331
Calibre 283 mm
Initial velocity 1130 m/sec
Weight of projectile 248 kg
Weight of warhead 14 kg
Maximum range 86,500 m
Manufacturer Rheinmetall-Borsig AG (not confirmed)

Remarks: Pre-rifled extended range projectile fired from 28 cm K 5(E). Rocket motor timed to ignite 19 secs after projectile left gun muzzle to boost speed when initial muzzle velocity began to fall. Used in action 1944–45.

38 cm Raketenwerfer 61

German designation 38 cm RW 61 'Sturmtiger'
Calibre of launcher and rocket 380 mm
Length of launcher 2060 mm
Length of liner 1886 mm
Traverse 60°
Elevation 0° to +70°
Initial velocity 300 m/sec
Weight of rocket 345 kg
Maximum range 5650 m
Rocket designer Rheinmetall-Borsig AG
Mounting manufacturer Alkett, Berlin-Spandau

Remarks: Designed as heavy assault/support weapon for street fighting based on experiences at Stalingrad. Prototype completed in summer 1943 comprised a specially-designed superstructure mounted on a PzKw VI Tiger 1 chassis. Designated 'Sturmtiger', only 10 of these massive and most unusual heavy rocket launchers were made during late 1944. Each carried 12 38 cm rocket missiles on internal racks. Although a powerful weapon, very small number of available RW 61 launchers had little impact on progress of any campaign.

38 cm R-Werfer 61 on a converted PzKpfw VI Tiger I chassis

Naval rockets used for coast defence

8.6 cm Raketen

German designation 8.6 cm R
Calibre 86 mm
Length of rocket (L/4.5): 387 mm; (L/4.8): 413 mm; (L/5.5): 475 mm

Weight of rocket (L/4.8): 8.15 kg
Maximum ceiling 2440 m
Manufacturer Friedr. Krupp AG (not confirmed)

Remarks: Produced in several versions including HE, smoke and signal flares. HE versions fired from open rails and intended as a defence against dive bombers.

21 cm RaketenLeuchtgeschoss

German designation 21 cm RLg
Calibre 210 mm
Length of rocket 1175 mm
Velocity 550 m/sec

Weight 60 kg
Maximum range 9000 m
Manufacturer Rheinmetall-Borsig AG

Remarks: Used for target illumination at night.

Raketen-Spreng-Granate 35 cm

German designation RSpgr 35 cm
Calibre 350 mm
Length of rocket 1225 mm

Weight of rocket 150 kg
Range 2000 m
Manufacturer Rheinmetall-Borsig AG

Remarks: Intended for use by naval landing units for demolition of strong-points. Tested but not accepted for service.

Raketen-Tauchgranate 1.5 and 3

German designation RTg 1.5 und 3
Calibre of rocket 380 mm
Length of rocket 1400 mm
Velocity (1.5): 115 m/sec; (3): 180 m/sec
Weight of rocket (1.5): 305 kg; (3): 296 kg
Maximum range (1.5): 1350 m; (3): 3000 m
Manufacturer Rheinmetall-Borsig AG, Düsseldorf

Remarks: Depth-charge-carrying rockets used in small numbers for coast defence. Developed into 38 cm RW 61.

X-7 Rotkäppchen

German designation X-7 Rotkäppchen
Body diameter 150 mm
Wing span 600 mm
Length 950 mm
Weight 9 kg
Weight of warhead 2.5 kg
Power unit WASAG 109-506 solid fuel rocket
Initial velocity 100 m/sec
Rocket burning time 2.6 secs
Maximum speed 360 km/h
Maximum range 1200 m
Manufacturer Ruhrstahl, Brackwede

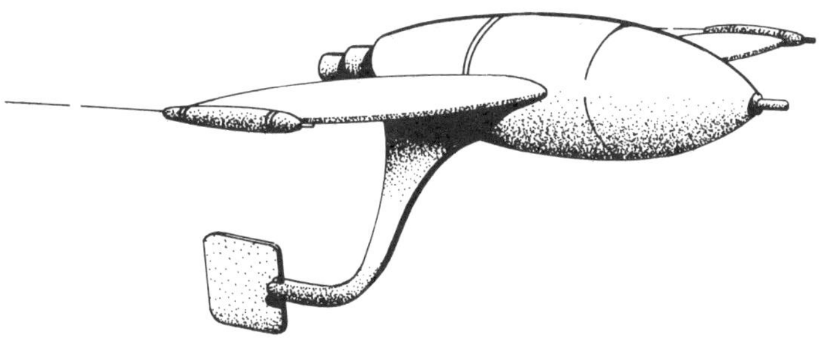

Remarks: Proposed wire-controlled anti-tank missile evolved from experimental X-4 air-to-air rocket missile. Development commenced in summer 1944 and troop trials took place on Eastern Front early in January 1945.

Taifun

German designation (code name): Taifun F
Body diameter 100 mm
Length 1930 mm
Weight 20.3 kg
Weight of warhead 0.5 kg
Thrust 840 kg
Maximum speed 3600 km/hr
Maximum operational ceiling 15,000 m
Rocket burning time 2.5 secs
Time to 10,000 m 14 secs
Design development Elektromechanische Werke (EMW) Peenemünde
Manufacturer Mittelwerke GmbH, Nordhausen

Remarks: Unguided spin-stabilised AA missile, intended to be launched in salvos of 30 from converted 8.8 cm Flak 18, 36 or 37 carriages. Development started at Peenemünde in September 1944 and progressed via at least four experimental rockets; Taifun F was planned production version. First series of 10,000 ordered from Mittelwerke GmbH/Nordhausen in January 1945; 600 completed before end of war. Earliest date Taifun estimated cleared for operational service was May 1945, but some reportedly used in action.

Rheinbote

German designation RSpgr 4831 Rheinbote
Total length 11400 mm
Total weight 1715 kg
Diameter of final stage 190 mm
Weight of final stage 200 kg
Weight of warhead 40 kg
Number of stages 3 + booster
Thrust of booster 38,000 kg
Thrust (1st stage): 5600 kg; (2nd stage): 5600 kg; (3rd stage): 3400 kg
Initial velocity 230 m/sec
Maximum speed 5900 km/hr
Maximum range (65° launch): 218 km
Maximum trajectory height 78,000 m
Manufacturer Rheinmetall-Borsig AG, Marienfelde

Remarks: Development of this multi-stage ramp-launched ballistic rocket began in 1942. An experimental rocket, RhZ 61/9, was built and tested successfully during 1943 and 1944. Production version, Rheinbote, was used against Antwerp in November 1944. It was fired from modified A4 transporters or modified 8.8 cm Flak 41 carriages. Rheinbote's maximum speed of Mach 5.55 was higher than that of A4.

'Rheinbote' on a modified A4 transporter

Feuerlilie F 25

German designation Feuerlilie F 25 (code name)
Body diameter 250 mm
Length 2100 mm
Wing span 1150 mm
Initial velocity 200 m/sec
Maximum speed 840 km/hr

Weight 120 kg
Original design centre Luftfahrtforschungsanstalt, Volkenrode and Braunschweig
Manufacturer Rheinmetall-Borsig AG, Leba

Remarks Experimental research rocket built to assess various forms of rocket guidance, except for propulsion unit similar to earlier Hecht. Initial design work started during 1942; development cancelled early in January 1945. Featured 40° wing speed, was ramp-launched at 60° inclination.

Feuerlilie F 55

German designation Feuerlilie F 55 (code name)
Body diameter 550 mm
Length 4800 mm
Tail span 2500 mm
Initial velocity 400 m/sec
Maximum speed 1500 km/hr

Weight 470 kg
Maximum ceiling 5000 m
Original design centre Luftfahrtforschungsanstalt, Volkenrode and Braunschweig
Manufacturer Rheinmetall-Borsig AG, Leba

Remarks: Experimental research rocket, essentially an enlarged version of F 25. Of tailless configuration, wing sweep 50°, ramp-launched at 70° inclination. Initially with R-I solid fuel (WASAG R61) rocket motor; later versions with liquid fuel rocket unit. Development cancelled early in January 1945.

Rheintochter R-I

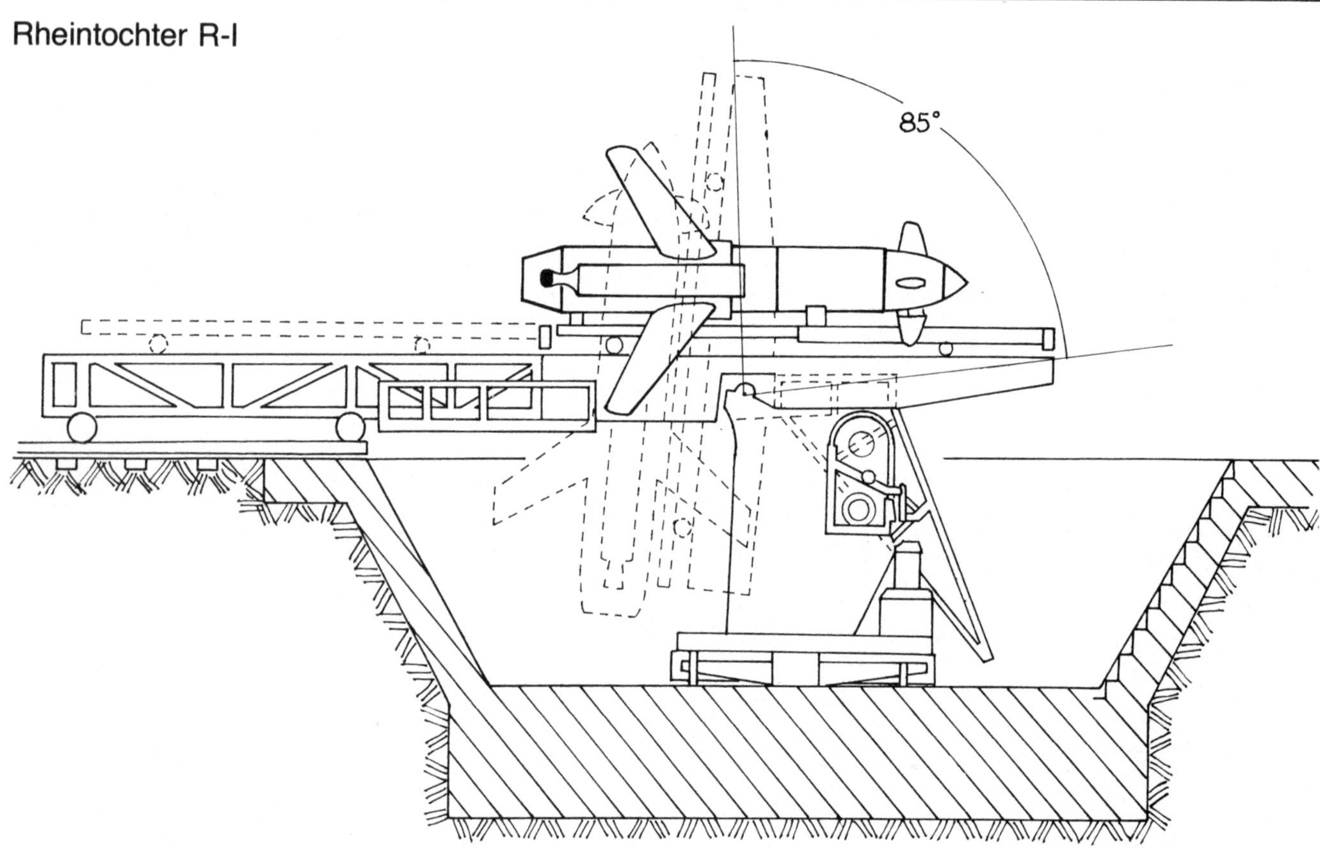

German designation Rheintochter R-I (code name)
Body diameter 540 mm
Length (with booster): 6300 mm; (less booster): 4000 mm
Fin span 2220 mm
Thrust of booster 75,000 kg for 0.6 sec
Thrust of main motor 4000 kg for 10 secs
Weight when fired 1750 kg
Weight at target 750 kg
Weight of warhead 100–150 kg
Maximum speed 1288 km/hr
Maximum ceiling 6000 m
Maximum range 40,000 m
Traverse of launcher 360°
Elevation of launcher 0° to +85°
Manufacturer Rheinmetall-Borsig AG

Remarks: Initially intended as a radio-controlled AA missile, later used as experimental research rocket for trials with various guidance systems for planned R-III production version. Featured 2-stage solid fuel rocket motor specially built by Rheinmetall. Development started in November 1942; total of 82 launched, of which 22 with full guidance equipment. Cancelled early in January 1945.

'Rheintochter' R-I ready for test firing, June 1944

Rheintochter R-IIIf

German designation Rheintochter R-IIIf (code name)
Body diameter 540 mm
Length (approx): 5000 mm
Fin span 2200 mm
Thrust of booster 28,000 kg for 0.9 sec
Thrust of main motor 1800–2200 kg for 43 secs
Weight when fired 1500 kg
Weight at target 685 kg
Weight of warhead 160 kg
Maximum speed 1075 km/hr
Maximum ceiling 15,000 m
Maximum range 40,000 m
Traverse of launcher 360°
Elevation of launcher 0° to +85°
Manufacturer Rheinmetall-Borsig AG

Remarks: Radio-controlled 2-stage AA missile, essentially a smaller fuel rocket development of Rheintochter R-I. HWA development contract awarded to Rheinmetall on 23 November 1942 but only a few missiles had been completed and test-launched when programme cancelled in January 1945. Was intended to be equipped with 'Kehl-Strassburg' or 'Kogge-Brigg' radio command systems.

Schmetterling

German designation Henschel Hs 117 Schmetterling (8-117); (originally Hs 297)
Body diameter 350 mm
Length 4280 mm
Wing span 2000 mm
Weight when fired 420 kg
Weight of warhead 23 kg
Power unit BMW 109-558 or Walther 109-729 liquid fuel rocket
Boosters 2 × Schmiddling 109-553 solid fuel
Maximum ceiling 10,000 m
Maximum range 32,000 m
Manufacturer Henschel Flugzeugwerke AG, Berlin-Schönefeld

Remarks: Radio-controlled subsonic AA missile with liquid-fuel rocket motor. Designed by Dr Henrici in 1941 as Hs 297 but initial proposal ignored by RLM. Ordered in experimental production in August 1943 under a new designation Hs 117 code-named 'Schmetterling'. Selected as standard AA missile in autumn 1944 and ordered in large-scale production, with deliveries scheduled to commence in March 1945. About 150 built for various tests. Of these only about 25 were fully completed and launched during troop trials beginning late in 1944. Initial version had line-of-sight radio guidance; later Schmetterling Hs 117C was intended to have radar guidance. In service would have been launched from modified 3.7 cm Flak 38 carriages.

'Schmetterling' being moved into its launching ramp at the Karlshagen test centre, February 1945

'Schmetterling' on its launching ramp during tests at Karlshagen in February 1945 ►

Enzian

German designation (production model): Enzian E-4
Body diameter 880 mm
Length 3900 mm
Wing span 4000 mm
Weight when fired 1800 kg
Weight of warhead 300 kg
Power unit Walther 109-739 or Konrad VfK 613A
Booster 4 × Schmiddling 109-553
Maximum speed 975 km/hr
Maximum ceiling 12,500 m
Operational ceiling 2500 m
Operational range at 2500 m 24,500 m
Range at 12,500 m 9900 m
Designers Messerschmitt AG
Manufacturer, motors DVK, Berlin and Dresden
Manufacturer, airframes Holzbau-Kissing KG, Sonthofen/Allgäu

Remarks: Radio-controlled subsonic AA missile designed by Dr Walther Konrad early in 1944 as a relatively cheap and easy to produce weapon against US day bombers. Experimental Enzian E-1, E-2 and E-3 missiles flight-tested at Peenemünde, E-4 planned as production version, with E-5 as ultimate supersonic development. Total of 38 missiles test-fired when project considered too underdeveloped for quick operational use and cancelled in January 1945. An 8.8 cm Flak 36 carriage was used to launch Enzian missiles from rails 6800 mm long.

'Enzian' experimental AA missile ready for test firing, June 1944

Vergeltungswaffe 1

Fi 103 (V1) shortly after launching

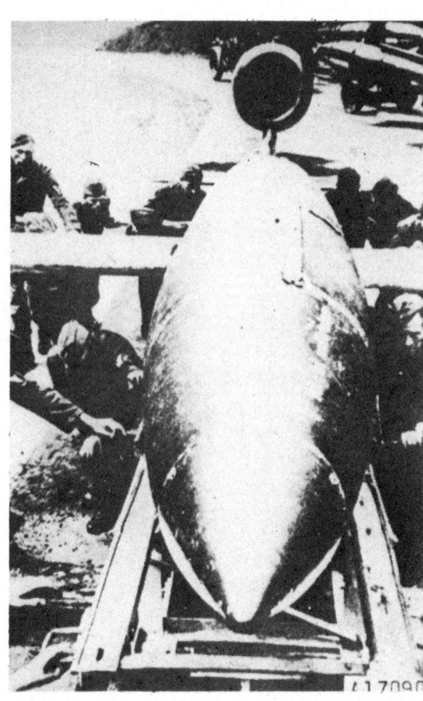

German designations 8-103; Fieseler Fi 103; 'Kirschkern' (code name); FZG 76 (cover designation); V1 (propaganda designation)
Body diameter 838 mm
Length overall 7900 mm
Length of body 7550 mm
Wing span (tapered): 4870 mm; (straight): 5300 mm
Weight when fired 2180 kg
Weight of warhead 850 kg
Thrust of motor 350 kg
Operation life of motor 20 mins
Initial launch velocity 105 m/sec
Maximum speed 645 km/hr
Operational ceiling 3000 m
Operational range 240 m
Designers (airframe): Gerhard Fieseler Werke GmbH, Kassel
Manufacturers (airframe): Volkswagen Werke, Fallersleben and Schönbeck; Mittelwerke GmbH, Nordhausen
Manufacturers (power unit): Argus-Motoren-Gesellschaft, Berlin; (guidance): Siemens-Werke (various locations); (catapult): H. Walter KG

Remarks: For detailed development see Section introduction.

Fi 103 (1)

Vergeltungswaffe 2

German designation A4, later V2
Body diameter 1650 mm
Length 14036 mm
Weight when fired 12840 kg
Weight of warhead 975 kg
Thrust of motor 25,000 kg for 68 secs
Maximum velocity 5580 km/hr
Impact velocity 2900 km/hr
Operational range 305 km
Maximum ceiling 97,000 m
Design centre Elektromechanische Werke, Peenemünde
Main production centre Mittelwerke GmbH, Nordhausen

Remarks: For detailed development history see Section introduction.

A4 rocket missile on its Meilerwagen *transporter/launcher postwar*

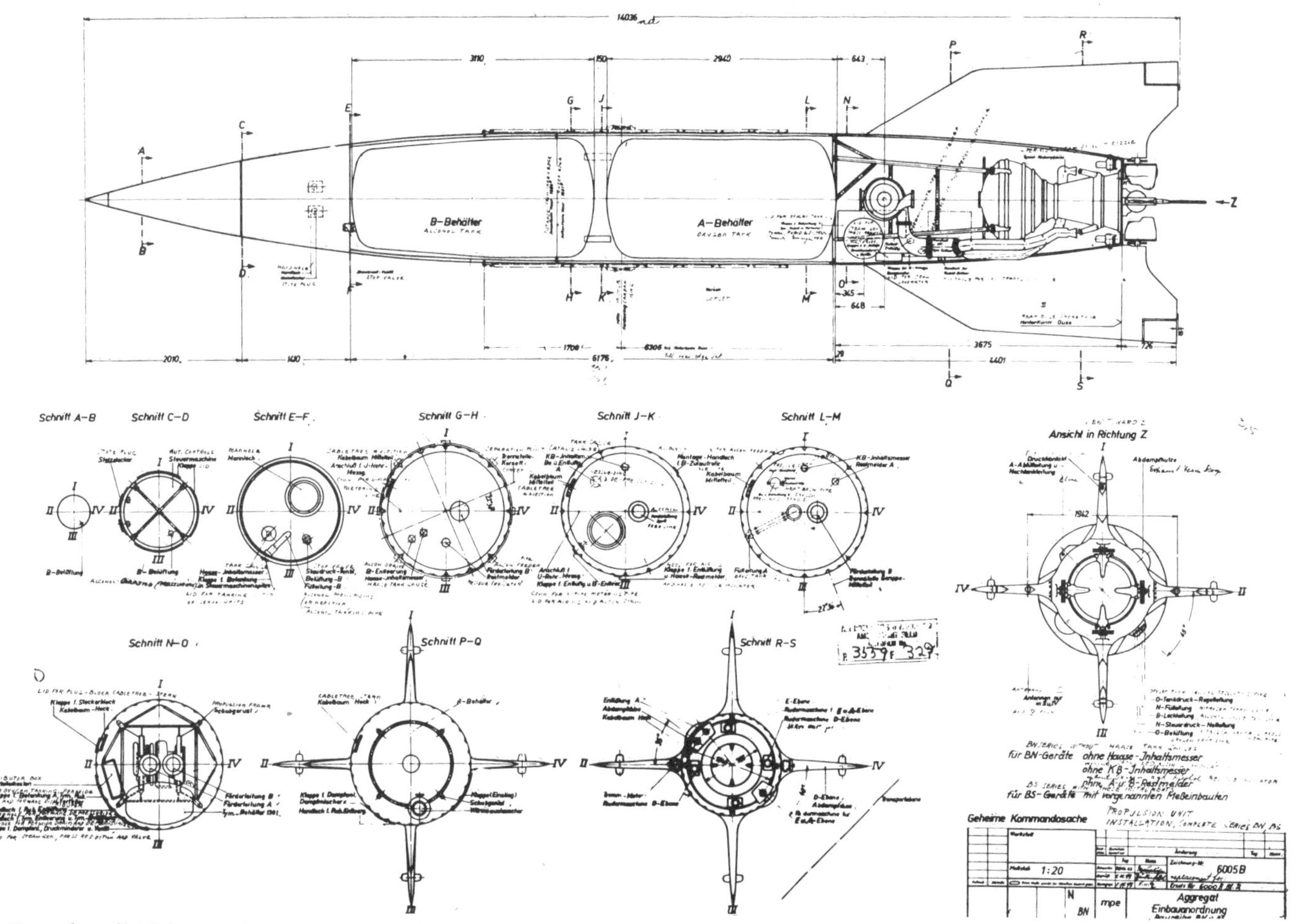

Copy of an official drawing showing the A4 propulsion unit installation – the fuel tanks, turbo-pump system and rocket motor with gyroscopically-controlled graphite vane stabilisers inside the jet nozzle

A standard A4 rocket missile lifted on its simple mobile launching stand during the postwar trials

A4 (V2) undergoing fuelling and final adjustments of control apparatus prior to launching

The second experimental winged A4b shortly before its test launch. All experimental Peenemünde missles were painted black/white to facilitate accurate observation of their rotating movements in flight

Wasserfall

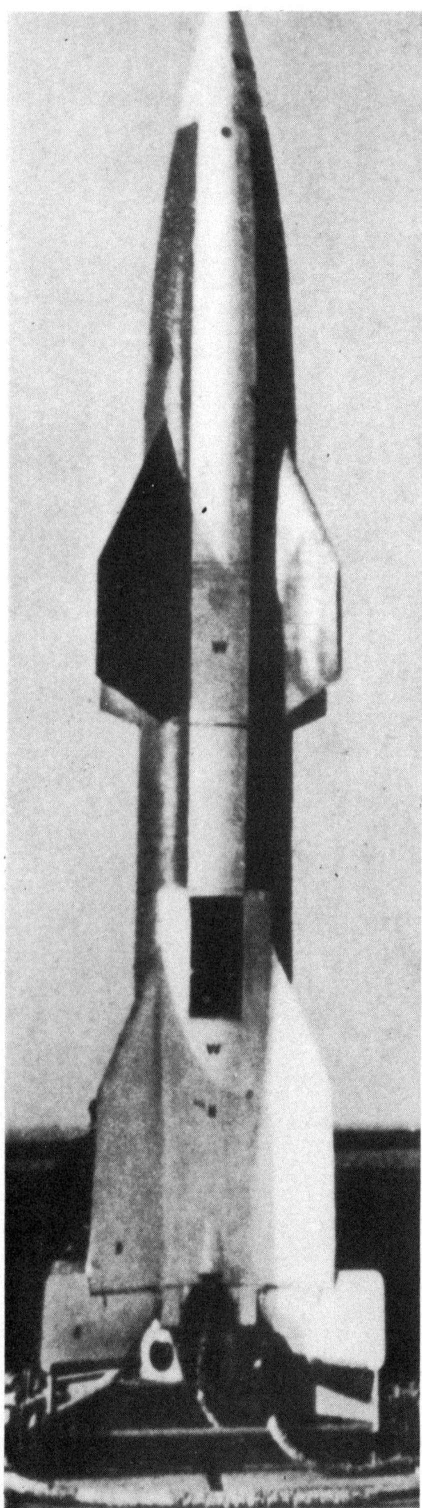

German designation Wasserfall C-2 8/45 (production version)
Body diameter 880 mm
Length 7835 mm
Wing span 1890 mm
Tail fin span 2510 mm
Weight when fired 3850 kg
Weight of warhead 235 kg
Power unit Peenemünde P IX liquid fuel rocket
Thrust of motor 8000 kg for 40 secs
Maximum speed 1370 km/hr
Operational ceiling 17,700 m
Operational range 26,500 m
Design centre Elektromechanische Werke, Peenemünde

Remarks: Radio/radar guided AA missile. Design evolved by Dr Wernher von Braun in 1942 and based on A4 rocket missile. Development work commenced in 1943, with first successful test-launch on 29 February 1944. Total of 35 test missiles launched before evacuation of Peenemünde on 17 February 1945; development not completed before end of WW II. Notable as most complex and expensive German AA missile demanding most protracted development – but also as most advanced.

An experimental Wasserfall *development missile immediately after lift-off*

Flamethrowers

Fire has been an inseparable element of warfare since time immemorial, but the flamethrower added yet another element of horror to WW I when the German Army introduced the *Flammenwerfer* against French troops during the Argonne fighting of October 1914. Despite gradual development and modifications the early German flamethrowers already had all the design elements of later models based on two portable tanks, one containing an inflammable liquid and the other a compressed gas. When a valve was opened on the projector connection the gas forced the liquid out along the tube where it was ignited and the resultant flame was ejected to a range of about 35 metres. The early *Flammenwerfer* equipments were large bulky items that required up to three men to move and use them but in 1935 a much improved model was entering service with the German Army.

This new flamethrower was the *Flammenwerfer 35* and it remained in production until 1941. Although intended to be a one-man load it weighed some 35.8 kg, rather too heavy for one man to handle in action, and was consequently often used by two men. The fuel was in a large tank carried on a soldier's back and a smaller cylinder contained pressurised nitrogen. A single trigger valve on the projector both released the fuel and ignited it. But the weight problem limited the use of the *Flammenwerfer 35* and a new and lighter model was introduced in 1940 as the *Flammenwerfer klein verbessert 40*. On this smaller model the fuel and nitrogen were arranged in two 'lifebuoy' rings with the outer ring containing the fuel. Weight was reduced to about 21.3 kg but this saving had to be paid for by a subsequent loss of about one third in fuel capacity.

The next development was the *Flammenwerfer 41* with two horizontal tanks of which the lower and larger one contained the fuel. The same projector as before was used but a new hydrogen ignition system was introduced with a long thin cylinder to contain the hydrogen mounted over the projector. Combat experience on the Eastern Front during 1941 and 1942 showed that the hydrogen ignition system became unreliable under conditions of extreme cold leading to the introduction of a new cartridge ignition system. This was combined with a new projector which had a magazine containing a feed system for ten incendiary cartridges. Every time the trigger was pressed a cartridge was loaded, fired and ejected and the cartridge ignited the fuel jet. The new system, known as the *Flammenwerfer mit Strahlpatrone 41,* became the standard manpack flamethrower of the bulk of the German forces until the end of the war.

The flamethrowers mentioned above were intended for use by specialised engineer assault troops but other branches of the German forces also requested some form of flamethrower equipment that would be light and easy to use. Chief among the list of requests was one from the Luftwaffe airborne troops and for them the *Einstossflammenwerfer 46* was produced during 1944. The equipment consisted of a cylindrical tube about 597 mm long and with a diameter of about 70 mm. A trigger mechanism was fitted at the forward end and when fired an explosive cartridge provided the gas pressure to project a single burst of flame to a distance of about 27 metres for half a second. The projector was then discarded. Despite its airborne origins many were issued to infantry assault units as well.

In addition heavier and more specialised flamethrowers were also developed. One of these was a larger version of the *Flammenwerfer 35*, introduced into service at about the same time. Known as *mittlerer Flammenwerfer* it was mounted on a trolley pulled by two soldiers using a double harness, but was soon regarded as obsolete and withdrawn from field use. Many were subsequently installed in static positions in beach defences.

Another static flamethrower was the *Abwehrflammenwerfer 42* copied from a 1941 Soviet design. Large numbers of these flamethrowers were buried with only the nozzles showing at intervals of between 10 to 25 metres. A 22.7 litre fuel tank was pressurised by a slow-burning explosive charge whenever they were operated from a remote station or by tripwires. These powerful static flamethrowers were first encountered along the so-called 'Stalin Line'. Most were installed to cover road blocks, likely landing points, harbour walls, etc., and wire entanglements were usually used to hide the nozzles.

The *Flammenwerfer Anhänger* was a pump operated mobile equipment mounted on a two-wheeled trailer and intended for towing behind a vehicle. The projector was protected by an armoured shield and mounted on top of the tank which contained 180 litres of creosote oil. This projector tube could be hand-traversed through 90° and elevated up to 40°. This equipment was produced only in limited numbers.

Intended as a one-man load, the Flammenwerfer 35 *was rather too heavy and was often used by two men in action*

Flammenwerfer 35 *had a range of 25–30 metres with a duration of fire of 10 seconds. The fuel container held 11.8 litres of oil*

Flammenwerfer 35 *in action against a concrete bunker*

Flammenwerfer klein verbessert 40. *This model superseded the 1935 pattern, the advantages being that it weighed only 21.3 kg and was a better fit on the operator's back. The fuel content was less than the 1935 version, but the range remained the same*

Rear view of the 1940 pattern flamethrower showing the fuel and compressed gas containers in concentric rings

A paratrooper carrying a Flammenwerfer 41 goes into action

Flammenwerfer 41 *consisted of two cylindrical containers in the horizontal position, the lower one for fuel and the upper for nitrogen* ➤

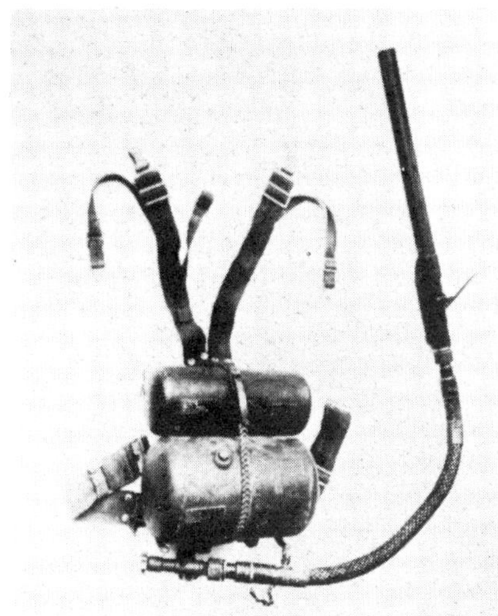

Flammenwerfer mit Strahlrohrpatrone 41 *was a development of the 1941 model with a new flame gun incorporating a flash cartridge ignition system. This equipment weighed 18 kg and had a range of 25 to 35 metres*

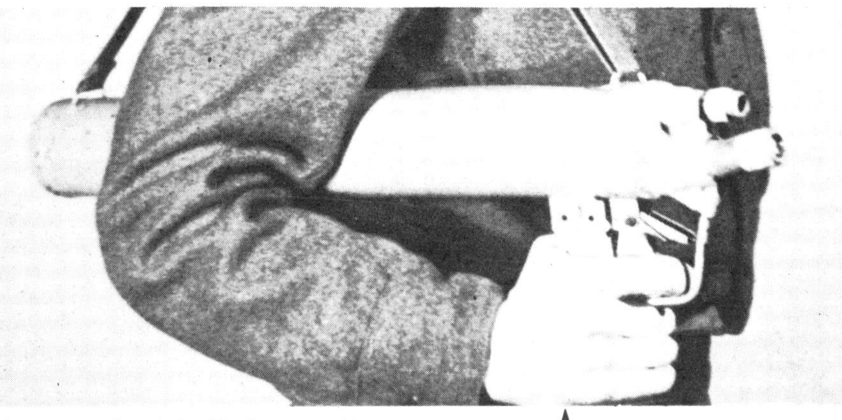

The single-shot flamethrower in carrying position

A medium flamethrower ready for action

Mittlerer Flammenwerfer *was fundamentally an enlarged trolley-borne version of the 1935 pattern portable. This equipment weighed 102 kg, had a fuel capacity of 30 litres, a range of 23–37 metres and gave a continuous flame jet of about 25 seconds*

Abwehrflammenwerfer 42 *was a static one-shot weapon having a range of some 50 metres and a duration of 5–10 seconds*

Close up of an AFmW 42 buried in the ground

Flammenwerfer-Anhänger *trailer. Flame-throwing was effected by a petrol-driven pump. Total duration of the flame jet was 24 seconds over a maximum range of 40 metres*

Hand grenades

Hand grenades were first introduced into warfare during the mid-15th century but at first their use was usually restricted to specialised assault units formed from specialised troops who became known as grenadiers. The spread and scope of warfare during the late 19th and early 20th centuries did away with early differentiations and gradually the hand grenade became part and parcel of the foot soldier's armoury. WW I produced a rapid spread of their use and from then on hand grenades were as much a personal weapon as the rifle.

By 1939 the standard German hand grenades were of two basic types – the stick grenade or *Stielgranate,* and the smaller egg grenade or *Eiergranate.* Both types used a thin metal casing relying on blast effect rather than fragmentation of the casing for their role. The *Stielgranaten* in service in 1939 were of two almost similar models, the *Stielgranate 24* and *39*, differing mainly in the length. Both used the same type of action where before throwing the grenade was primed by pulling on a length of cord running through the wooden handle; the grenade detonated after a 4.5 second delay. A later variation of this type of grenade was the *Stielgranate 43*, which used a solid wooden handle as the friction primer, with the detonator fitted to the top of the grenade head. Not only did this simplify manufacture but it also enabled the grenade to be thrown without the stick.

All three models could have their anti-personnel effect enhanced by the addition of a serrated fragmentation sleeve clipped over the head of the grenade. For more specialised purposes six grenades could be secured round a seventh; in this configuration they were known as a *Geballte Ladung.* Such seven-grenade clutches were used as demolition charges and for anti-tank work, when they were usually laid under tank tracks. Individually the grenade heads could be used as small anti-personnel mines or booby traps in conjunction with the pressure igniter *DZ 35.* Another German practice was to replace the delay detonator with an instant detonator. Grenades so fitted were often left around after a withdrawal in the hope that the enemy would pick them up for their own use, when any attempt to pull the priming cord would result in an immediate detonation.

Stick grenades were also used for smoke production; the *Nebelgranate 39* was almost identical to the normal Stielgranate and was used in the same way. The *Nebelgranate 41* was another smoke grenade, but it lacked the throwing handle and used a friction igniter in a special adaptor.

The egg grenades were smaller than the stick grenades, one of the earlier models being the *Eierhandgranate 39*, an egg-shaped grenade using a thin steel case. A friction igniter was used and to arm the grenade a cover was unscrewed and pulled. Later some slight alterations were made to produce a second version but no change was made to the designation. The *Nebel Eierhandgranate 42* was the smoke version.

For anti-tank use the Germans devised an unusual grenade in the *Panzerwurfmine 1(L).* This used a hollow-charge head which had four folding canvas vanes to guide it towards its target when it was thrown. For throwing the vanes were held by the end, unfolding after being thrown. Also used against tanks and such tarkets as pillboxes were the *Blendkörper 1H* and *2U*, small glass flasks which broke on impact with their target and the filling then vaporised on exposure to the air. The resultant smoke was a definite offensive hazard to the tank or pillbox occupants.

One facet of the hand grenade scene in Germany prior to the end of the war was the emergence of the *Nipolit* grenade. Originally based on experimental work carried out by the Westfalische Anhaltische Sprengstoff Aktiengesellschaft (WASAG) the *Nipolit* grenade was manufactured from time-expired explosive propellants and had the considerable advantage that it could be moulded or machined direct from the original substance – no metal casings or handles were needed. The only additional requirement was a standard igniter. The *Nipolit* grenade could thus be used as a purely blast grenade, but a metal fragmentation sleeve could be added for a more offensive role. Prototypes were made as stick grenades (with the handles machined direct from the *Nipolit*) and as 250 and 500 gram egg grenades, but they were not used in action and the war ended before this promising idea could be carried further.

Although not strictly a grenade the *Nebelkerze 39* deserves mention here as it was a smoke canister or candle. Pulling an igniter on the lid produced smoke which continued to emanate for from four to seven minutes.

As well as using all the various German-designed and -made grenades mentioned above large numbers of captured grenades were also issued to all arms of the German forces. These comprised French, Dutch, Soviet and Danish grenades. Thus for example the German forces involved in the fighting at Arnhem in 1944 used captured Dutch grenades while captured British No. 75 Hawkins grenades were dug-in by the Germans as anti-tank mines in beach defences.

Stielhandgranate 24

German designation StiGr 24
Diameter 70 mm
Length 356 mm
Weight 595 g
Type of filling TNT
Delay 4–5 secs

Stielhandgranate 24 *(StiGr 24)*

Übungsgranate 24 *drill grenades used for training* ➤

Stielhandgranate 39

German designation StiGr 39
Diameter 70 mm
Length 406 mm
Weight 624 g
Type of filling TNT
Delay 4–5 secs

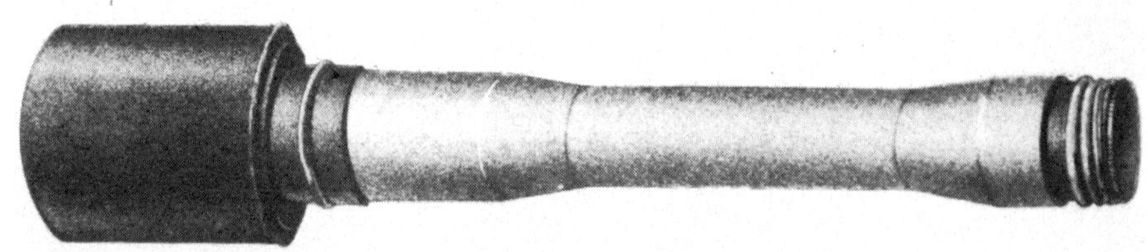

Stielhandgranate 43

German designation StiGr 43
Diameter 70 mm
Length 356 mm
Weight 624 g
Type of filling TNT
Delay 4.4 secs

Stielhandgranate 43 *(StiGr 43) featured a clipped-on framentation sleeve* ➤

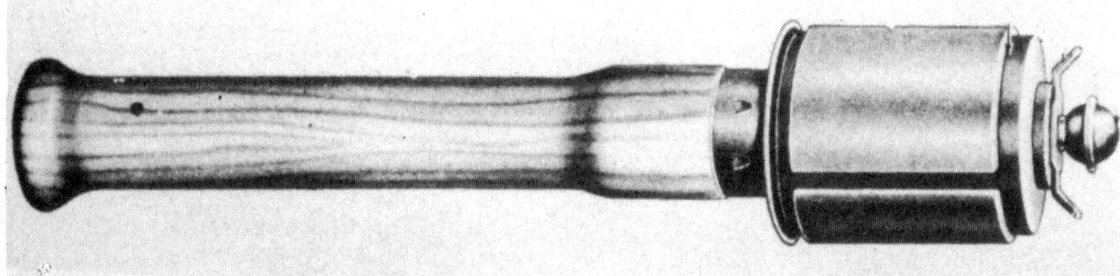

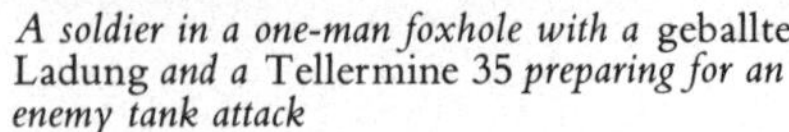

A soldier in a one-man foxhole with a geballte Ladung *and a* Tellermine 35 *preparing for an enemy tank attack*

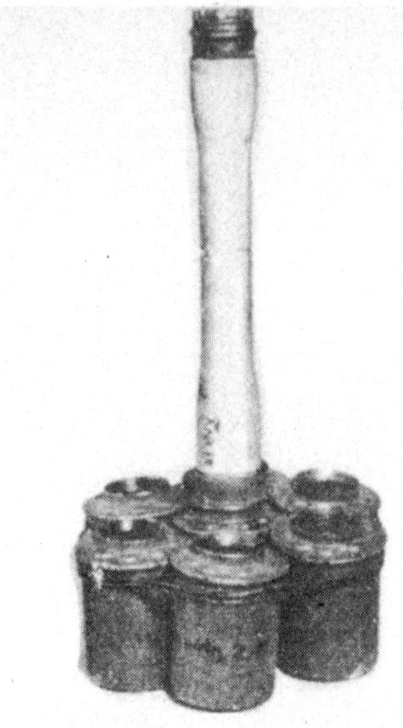

Geballte Ladung *was a general term used to describe various combinations of concentrated charges. In this case it is made up of six standard handgrenade heads secured to a complete grenade*

These geballte Ladungen *are being made up of six Dutch handgrenades clustered round a StiGr 39*

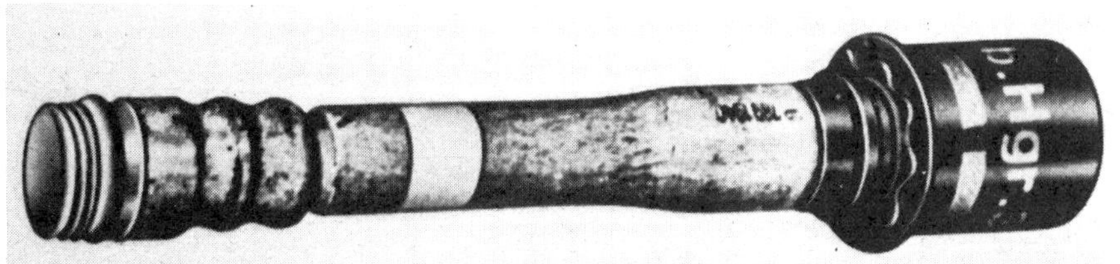

Nebelgranate 39 *(NbGr 39) artificial smoke grenade*

Eierhandgranate 39

German designation EihGr 39
Diameter 50 mm
Length 76 mm
Weight 340 g
Type of filling TNT
Delay 4–5 secs

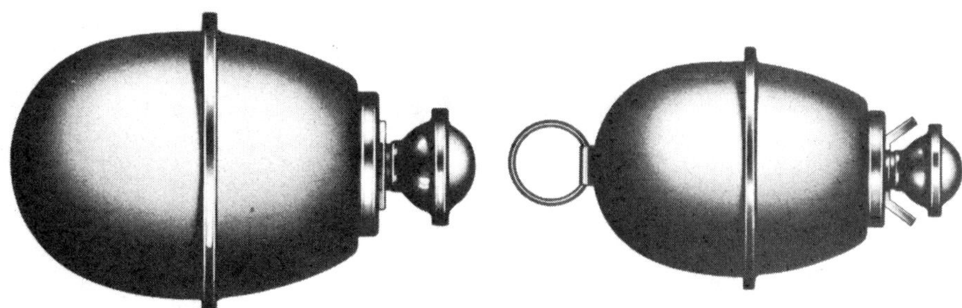

Eierhandgranate 39 *(EihGr 39)*

EihGr 39 (modified)

Nebeleierhandgranate 42 *(NbEihGr 42) artificial smoke grenade*➤

Panzerwurfmine(L)

German designation PzWuMi(L)
Diameter 114 mm
Length 533 mm
Weight 1.36 kg
Type of filling RDX/TNT

Remarks: Consisted of a metal body and a wooden handle to which four canvas fins were attached.

Throwing the PWM 1(L). Note the protective cap which holds down the springloaded fabric stabilizing fins still held in the demonstrator's hand ►

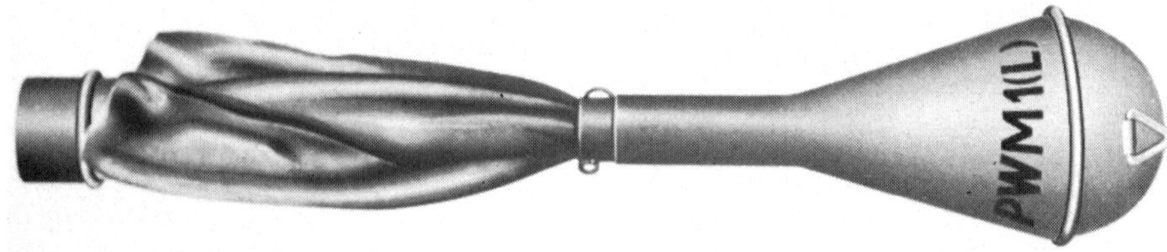

Blendkörper 1H

German designation BK 1
Diameter 63 mm
Length 152 mm
Weight 374 g
Type of filling Titanium Tetrachloride

A section of assault engineers with smoke canisters, demolition charges and wire cutters. Their armament consists solely of P 08 pistols and StiGr 39 handgrenades ►

Nebelkerze 39 *(NbK 39) artificial smoke canisters*

Nebelkerze 39

German designation Nb K 39
Diameter 89 mm
Length 146 mm
Weight 2.15 kg
Type of filling Zinc and Hexachlorethane

Blendkörper 1H *which produced acrid artificial smoke*

Landmines

During WW II all the combatants laid millions of landmines in all the theatres of war, not usually in an offensive role but in attempts to deny areas to an enemy or to slow down (or channel) enemy advances. Germany was one of the largest users, laying extensive coastal minefields and millions more landmines on the battlefields of North Africa and the Eastern Front.

Landmines were of two main types, anti-tank and anti-personnel. The anti-tank mine was a relative newcomer to warfare as the tank itself was a by-product of WW I. Anti-tank mines had to have sufficient destructive effect if they were to be of any use and were heavier than anti-personnel mines. Usually they were not designed to knock out tanks but just to blow off a track or wheel and thus disable the vehicle.

The most common of the various types of German anti-tank mines was the *Tellermine*. There were several models, the first of which was the *Tellermine 29* (*TMi 29*). By 1939 it had been largely replaced by the later *TMi 35*, but numbers were used in North Africa in 1942. The *TMi 35* remained the standard German anti-tank mine until 1943, but in that year a modified type came into use. It was the *TMi 35 Stahl,* designed to save on raw materials, but from 1943 onwards a new model, the *TMi 42*, came into use. The *TMi 42* introduced some new features such as a smaller pressure plate with a simpler form of main igniter. Like the earlier mines it had a socket for a subsidiary anti-lift igniter. The last of the *Tellerminen* was the *TMi 43 Pilz*, which had a mushroom-shaped head but it was used in smaller numbers than the other models.

In addition to the *Tellermine* the Germans produced a range of other anti-tank mine designs. One of the more specialised was the *leichte Panzermine* evolved for the Luftwaffe *Fallschirmjäger* and first used in action during the invasion of Crete. These mines were packed five to a special case and dropped by parachute in special containers (*Abwurfbehälter*) in batches of three or four cases.

Another German anti-tank mine was the *Riegelmine 43* (*RMi 43*) which was first encountered in North West Europe shortly after the Allied invasion of France in summer 1944. It was also known as the *Sprengriegelmine,* and its bar shape meant that fewer had to be laid to form an anti-tank barrier. In service it proved most effective.

The hollow-charge principle was used on the *Panzerstabmine 43* (*PzStabMi 43*) but relatively few of this model were produced. In service these mines were buried with the hollow-charge warheads pointing upwards towards the belly of the tank and were detonated when a rod connected to the igniter was tilted.

All the above mines were made of steel or other metals and could thus be traced by Allied mine detectors, so the Germans began to use mines made from non-metallic materials. The first of these was the *Holzmine VB*, soon replaced in service by the *Holzmine 42*, both of which had wooden bodies, but as some metal had to be used in their construction, they could still be detected. The *Panzerschellminen A und B* were quick-laying anti-tank mines also made of non-metallic substances but these also had metal carrying handles and detection was still possible. Allied mine detectors became more and more sensitive and consequently by 1944 the Germans were experimenting with mines that had bakelite, plastic, fibre composition or even clay bodies. The end result was the *Topfmine* (*ToMi*) which had a plastic body, and from October 1944 onwards it superseded all previous models apart from some mines which were made with clay bodies.

In most cases anti-tank mines were laid with anti-lift devices fitted, the most common of which were pull or tension igniters fitted to the base or sides. By 1944 several more sophisticated anti-lift devices were in service, one of which was the *Entlastungszünder 44* (*EZ 44*) which used a small clockwork device fitted to a small explosive charge.

To enhance the effectiveness of anti-tank mines several arrangements of mines connected together were used. A simple expedient was to lay a plank or strip of metal across several mines but a more formal arrangement was the *Rampensperre,* a length of wood on which five Tellermines were fixed. In this form the ramp obstacle could be quickly laid across a track or road. There were many other anti-tank expedients ranging from an 8.8 cm rocket grenade, buried in the ground and fired remotely upwards, to satchel charges buried under a road. Many of the coastal defences of the 'Atlantic Wall' had captured artillery shells or aircraft bombs buried nose upwards or in remotely-operated minefields. A refinement of this was the use in some areas with overhanging cliffs, of the gantry mine. Bombs or shells were suspended from gantries overhanging beaches and released when the enemy landed beneath.

At the beginning of the war the standard German anti-personnel mine was the *Schrapnellmine 35* (*SMi 35*), known to the Allies as the 'Shoe mine'. The *SMi 35* was activated by push or pull igniters fitted to tripwires and when fired a propellant charge in the base ejected a small cylinder which exploded one to two metres above the ground, scattering some 350 steel balls to a range of about 150 metres. In late 1944 a modified version, the *SMi 44*, came into service, while a version with a wooden body, the *Schützenmine 42* (which was used in three or four versions) had been introduced a year earlier as a rapid-laying mine.

The *Stockmine* operated on a similar principle to the *SMi 35* but the mine was fixed above ground on a wooden picket. The body comprised a concrete cylinder which held the charge and the shrapnel which was scattered when a tripwire was moved.

To try to avoid detection by mine detectors the Germans introduced the *Glasmine 43* (GLMi 43), which had a body made entirely of glass. Part of the mechanism was metal but the mine was very difficult to detect. Another mine made from glass was the *Eismine 42* or *Flascheneismine* which resembled a large milk bottle. It was originally designed for use on the Eastern Front and intended for laying under the ice of frozen rivers. When the enemy approached the mines were detonated by pressure or from a remotely controlled electrical detonator. During 1944 these specialised glass mines were used in North-West Europe as anti-personnel mines, either buried by themselves or enclosed in concrete.

The *Rollbombe* was a very simple device consisting of a concrete ball containing an explosive charge and shrapnel. In use the igniter was pulled and the bomb rolled down a slope towards the enemy where it exploded. Usually these *Rollbomben* were made by units in the field and many garrison troops improvised their own anti-personnel mines, usually known as *Behelfsminen.* Some of these became approved designs such as the *Behelfsmine W-1,* which was originally a French 50 mm mortar bomb. The tail fins were removed and the nose fuse replaced by a pressure igniter and detonator. When buried the *W-1* made an effective mine. Another cheap and effective design was the *Behelfsmine E5* which was a case with a push-on lid containing five French hand grenades. The centre grenade was fitted with a pressure igniter. Similar charges were produced by lashing together a number of stick grenade bodies with the handles removed and with a pressure igniter fitted to the centre grenade. The *Brettstückmine* was a simple device which used a standard 1 kg demolition charge secured to a wooden base. A wooden lid rested on a push igniter so that when pressed from above the charge was detonated. There were many variations on this theme.

In addition to the landmines produced in Germany the German forces also made use of vast numbers of mines captured from many nations.

Anti-tank mines

Tellermine 29

German designation TMi 29
Diameter 454 mm
Height 70 mm
Weight 6 kg
Type of filling TNT
Firing pressure 45–125 kg

Remarks: Manufactured in 1931 and earlier, these mines featured three sockets for individual ZDZ 29 push-pull igniters in their lids. Three additional sockets (two in sides, one in base) were provided for subsidiary igniters and were used as anti-lifting devices.

Tellermine 35

German designation TMi 35
Diameter 320 mm
Height 80 mm
Weight 8.7 kg
Type of filling TNT
Firing pressure 80–180 kg

Remarks: Standard German anti-tank mine until supplemented late in 1942 by TMi 42. It was detonated by pressure on its lid which fired central igniter. Additional pull igniters could be screwed into base and side and anchored to ground to form anti-lifting devices.

Tellermine 35(Stahl)

German designation TMi 35(Stahl)
Diameter 320 mm
Height 80 mm
Weight 9.75 kg
Type of filling TNT
Firing pressure 225–295 kg

Remarks: Modification of TMi 35 differing only in a new type of steel pressure plate. Occasionally laid double, with pull igniter connected to handle of upper mine.

Tellermine 35 Stahl. *This mine had a pull igniter screwed into the side socket as an anti-lifting device*

Tellermine 42

German designation TMi 42
Diameter 324 mm
Height 102 mm
Weight 7.8 kg
Type of filling Amatol
Firing pressure 340 kg

Remarks: Differed from earlier TMi 35 series principally in reduced size of pressure plate, simpler form of main igniter and changed position of sockets for subsidiary anti-lifting igniters.

An anti-tank obstacle formed of linked TMi 42s. This was pulled into the path of enemy armoured fighting vehicles by a soldier from a concealed position

Tellermine 42

Tellermine 43

German designation TMi 43 Pilz ('Mushroom')
Diameter 318 mm
Height 90 mm
Weight 7.8 kg
Type of filling Amatol
Firing pressure 320 kg

Remarks: An even more simplified version of basic design. Used same type of igniter as TMi 42 but had no sprung pressure plate as in earlier models, igniter being actuated by crushing of central sheet metal cap. Last of Tellermine series encountered during WW II.

TMi 43 Pilz. This view shows the pull igniter screwed into the bottom of the mine as an anti-lifting device. If the mine is lifted a wire secured to a peg in the ground will pull the safety pin and fire the igniter, which in turn will explode the mine

---

leichte Panzermine

German designation le PzMine
Diameter 260 mm
Height 57 mm
Weight 4 kg
Type of filling TNT
Firing pressure 4.5 kg

Remarks: Standard anti-tank mine used by airborne troops. Could also be used in anti-personnel role by unscrewing five hexagonal nuts on top of cover plate and laying mine on a hard surface.

Riegelmine 43

German designation R-Mine 43; R Mi 43
Length 800 mm
Width 95 mm
Height 90 mm
Weight 9.6 kg
Type of filling Amatol
Firing pressure 200 kg on ends, 400 kg on centre

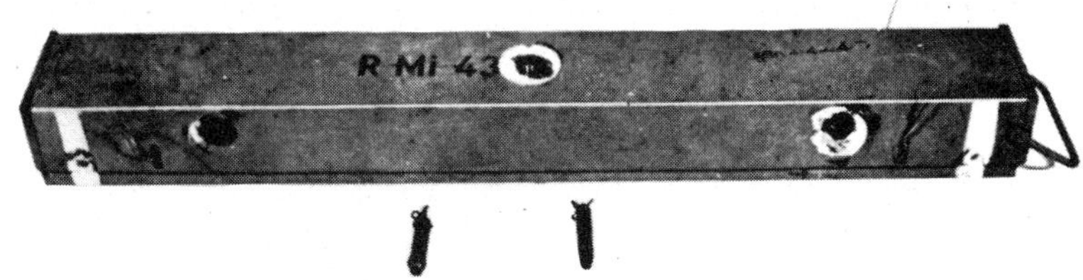

Remarks: High explosive bar mine used in open country, on roads and in minefields. Consisted of a rectangular encased charge supported when armed on two soft iron shear wires within a tray. It was covered by a lid which rested on explosive charge and acted as pressure plate. Used ZZ 35, ZuZZ 35(Mod) or ZZ 42 top igniters, often in combination with two ZZ 35 igniters fitted horizontally as anti-handling devices.

Panzerschnellminen

German designation PzSch-Mine
Length 527 mm
Width 330 mm
Height 127 mm
Weight 7.25 kg
Type of filling Picric acid

Remarks: Type 'A' was actuated by pressure on box lid, causing shearing of two 12.7 mm wooden dowels and pressing out igniter link pin. Type 'B' was actuated by pressure on box lid shearing 19.2 mm wooden dowels and exerting pressure on heads of two pressure igniters.

Panzerschellmine B

Panzerschellmine A

Holzmine 42

German designation Hz-Mine
Length 330 mm
Width 305 mm
Height 114 mm
Weight 8.2 kg
Type of filling Amatol
Firing pressure 90 kg

Remarks: Contained in a wooden box subdivided internally into four compartments. Main explosive charge contained in two side sections, central compartment had primer charge, while fourth compartment held operating mechanism. Often fitted with anti-lifting devices consisting of one or two pull igniters and 1 kg charge.

Holzmine VB *(VB Mi 1), the prototype version of* Holzmine 42. *It was issued in small numbers*

Holzmine 42

Topfmine

German designation ToMi A4531
Diameter 318 mm
Height 140 mm
Weight 10 kg
Type of filling Amatol
Firing pressure 150 kg

Remarks: An entirely non-metallic anti-tank mine with plastic body of black tar finish that resembled asphalt. Igniter assembly made of glass.

ToMi. This view shows the carrying handle and igniter assembly at the bottom of the mine ➤

leichte Panzerabwehrmine 407(f)

German designation le PzMi 407(f)
Length 240 mm
Width 140 mm
Height 114 mm
Weight 6.6 kg
Type of filling Picric acid
Firing pressure 190–225 kg

Remarks: Light French anti-tank mine used extensively by German troops in North Africa and Europe.

schwere Panzerabwehrmine 420(f)

German designation s PzMi 420(f)
Length 406 mm
Width 254 mm
Height 120 mm
Weight 12 kg
Type of filling Picric acid
Firing pressure 363 kg

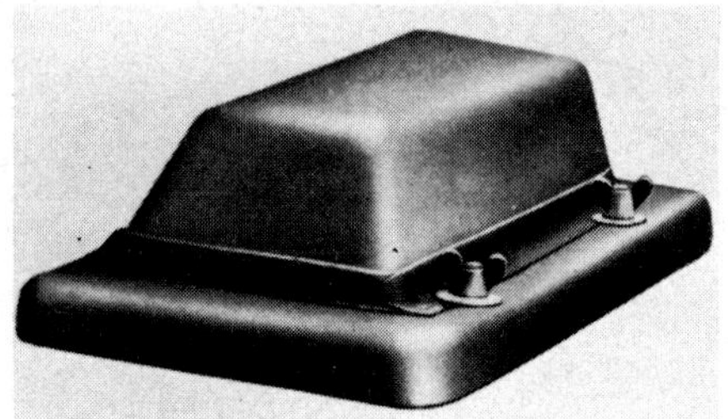

Remarks: French heavy anti-tank mine from captured stocks.

Panzerabwehrmine 406(b)

German designation PzMi 406(b)
Length 230 mm
Width 222 mm
Height 220 mm
Weight 10 kg
Type of filling TNT
Firing pressure 180 kg

Remarks: Belgian heavy anti-tank mine from captured stocks.

C V P I

German designation C V P I
Diameter 254 mm
Height 76 mm
Weight 3.6 kg
Type of filling TNT
Firing pressure 27 kg

Remarks: Hungarian anti-tank and anti-personnel mine, first used by German troops in Middle East in September 1942.

Panzerabwehrmine 410(r)

German designation PzMi 410(r)
Length 220 mm
Width 220 mm
Height 80 mm
Weight 5.1 kg
Type of filling 3.6 kg TNT
Firing pressure 200–700 kg

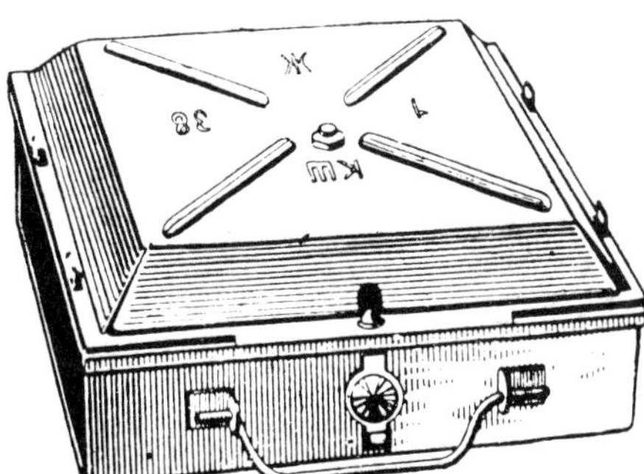

Remarks: Standard Soviet heavy anti-tank mine type TM/39 from captured stocks. Sheet metal casing.

Panzerabwehrmine 416(r)

German designation PzMi 416(r)
Length 215 mm
Width 215 mm
Height 100 m
Weight 4.2 kg
Type of filling Amatol 80/20 or 'Dynamon'
Firing pressure 200–700 kg

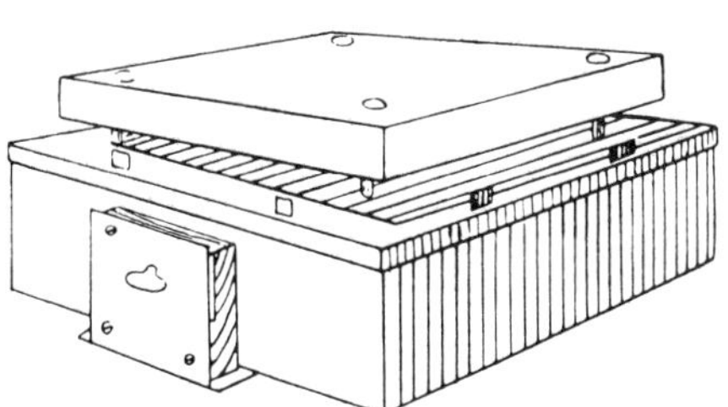

Remarks: Simplified Soviet anti-tank mine type T-IV from captured stocks. Some with wooden casing.

Panzerabwehrmine 404(e)

German designation PzMi 404(e)
Diameter 230 mm
Height 135 mm
Weight 5.4 kg
Type of filling TNT
Firing pressure 150–160 kg

Remarks: British anti-tank mine Mk IV from captured stocks.

Panzerabwehrmine 405(e)

German designation PzMi 405(e)
Diameter 230 mm
Height 102 mm
Weight 3.6 kg
Type of filling TNT
Firing pressure 150–160 kg

Remarks: British anti-tank mine Mk V from captured stocks.

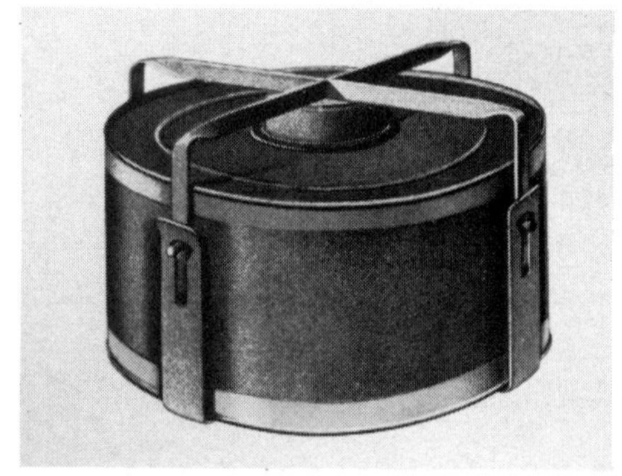

Anti-personnel mines

Schrapnellmine 35

German designation SMi 35
Diameter 102 mm
Height (less igniter): 127 mm
Weight 4 kg
Type of filling TNT
Firing pressure 6.8 kg

Remarks: Anti-personnel mine operated by direct pressure on igniter in head, or by a pull on one or more tripwires attached to pull igniters. Mine could also be fired electrically.

SMi 35 with a pressure igniter

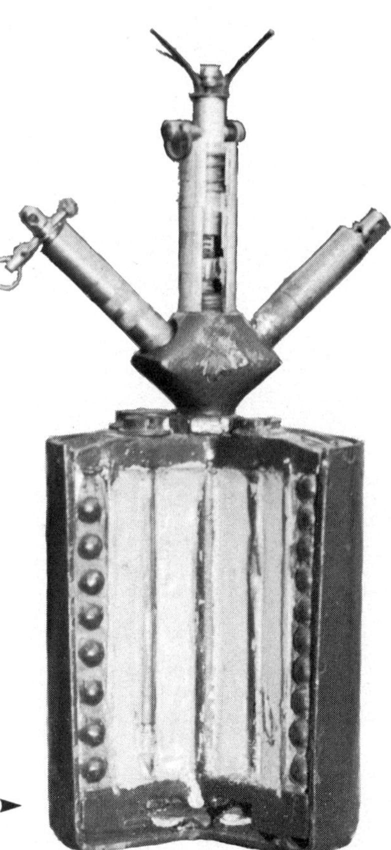

Schrapnellmine 35 *(SMi 35). This cut-away view shows the steel balls that were scattered over the area when the mine exploded. The SMi 35 is fitted with a pressure and two side-pull igniters* ►

Stock-Mine 43

German designation StoMi 43
Diameter 72 mm
Height 158 mm
Weight 2.1 kg
Type of filling TNT
Firing pressure 4–6 kg

Remarks: Consisted of a small hollow concrete cylinder made of weak cement mortar composition containing a shrapnel filling. Explosive charge, into which pull igniter and detonator were screwed, was set inside this cylinder. Mine was mounted on a wooden picket driven into ground.

Schützenmine 42

German designation SchüMi 42
Length 127 mm
Width 98 mm
Height 50 mm
Weight 0.5 kg
Type of filling TNT
Firing pressure 2.75–5 kg

Remarks: Consisted of light plywood box containing a 200 g demolition charge and pull igniter. Pressure on lid pushed out igniter pin and fired explosive charge.

Schützenmine 42, *more widely known as 'Schu-Mine'*

Glasmine 43

German designation Gl Mi 43
Diameter 152 mm
Height 120 mm
Type of filling Picric acid
Firing pressure 17–20 kg

Remarks: Made of glass and contained 200 g of TNT or picric acid. Thin glass lid served as a shear plate. Breaking of lid activated igniter.

Glasmine 43. *Like the 'Schu-Mine' it was most difficult to detect*

Eismine 42 or Flaschen-Eismine 42

German designation FlEsMi 42
Diameter 102 mm
Height 280 mm
Weight 2.38 kg
Type of filling Gelatine-Donarit
Firing pressure 4–6 kg

Remarks: An anti-personnel mine shaped like a wide-mouthed bottle, intended for use under ice. Contained 2.6 kg of explosive and was provided with a pull or pressure igniter. When used as an improvised mine completely encased in concrete to give a shrapnel effect on detonation.

Behelfs-Schützenmine A.200

German designation BehSchüMi A.200
Diameter 64–76 mm
Height 90 mm
Weight 354 g
Type of filling Picric acid
Firing pressure 4–6

Behelfs-Schützenmine A 200, *also known as 'Topfmine' (Pot mine), was a simple and cheap anti-personnel mine designed for mass-production by semi-skilled workers* ➤

Remarks: Known as 'Potmine' and first encountered in Normandy in June 1944. Thereafter used in considerable quantities throughout France. Consisted of a cylindrical mild steel body with a chemical igniter screwed in top. Two similar 'Potmines' were also in use, designated S.150 and A.202.

Demolition charges

The German forces used demolition charges for a variety of purposes. In an advance the charges were used for the destruction of bunkers, strong-points and similar obstacles. On retreat they were used to bring down bridges, buildings and similar structures. With demolition charges the German forces for once stuck rigidly to a standardisation of charges and all used standard screw sockets which would accept a wide range of detonators or igniters. The same screw socket was also used on German landmines, both anti-tank and anti-personnel. Charges were made up in five standard sizes – 100 grams, 200 grams, 1 kg, 3 kg and 10 kg. Charges of different weights could then be joined to provide a weight of explosive suitable for any task, including the making of improvised mines and booby traps.

A typical use of the standardised charge was when a number were combined to make up a pole charge, or *Stangenladung*. Numbers of charges were lashed to the end of a long pole and then inserted into the firing loophole or ventilation louvre of a bunker. Firing was then effected electrically or by a length of safety fuse. These pole charges were also used against tanks and other AFVs. To clear barbed wire obstacles a special charge, the *Rohrladung Stahl 3 kg*, was used – a long steel tube packed with explosives. Several of these tubes could be joined end-to-end by standard bayonet sockets for large wire obstacles. (Numbers of hand grenades on a length of wood could serve for the same purpose.) Hollow-charge explosives were also issued for the destruction of concrete or steel obstacles. Some of the smaller charges used magnets to hold them against steel; the larger charges varied in weight from 12.5 to 50 kg.

For more specialised purposes the German Army produced numbers of small remotely-controlled tracked vehicles which were used to carry demolition charges to heavily defended positions or minefields. The smallest of these vehicles was the *Goliath*, which was produced in two models – one was driven by an electric motor, the other by a petrol engine. The front compartment contained the explosive charge which weighed about 83 kg, and the vehicle was controlled remotely through a 2000-metre cable paid out from a drum in the rear of the vehicle and connected to the operator's control panel. *Goliath* was expendable and was destroyed by the charge when it was detonated.

In addition to the little *Goliath* a larger vehicle, the *Funklenkpanzer B IV* (SdKfz 301), was also used in small numbers, usually in conjunction with Tiger tank formations. There were several different versions of the *B IV*, all of them basically similar, and they were all taken by a driver to the area where they were to be used. Once in the general target area the driver took control by radio and the tracked *B IV* was then directed towards its target. When at the target the large 550 kg charge was then dropped from the front of the vehicle and in theory the vehicle was then driven to rear for re-use. In practice the *B IV* was usually destroyed by its own charge. A similar vehicle that did not go into series production was the NSU *Springer* (SdKfz 304). Like the *B IV* it was to have been radio controlled but it was to have been expendable and would have been destroyed by its charge.

An improvised Rohrlandung *made up with heads of stick handgrenades and a wooden plank*

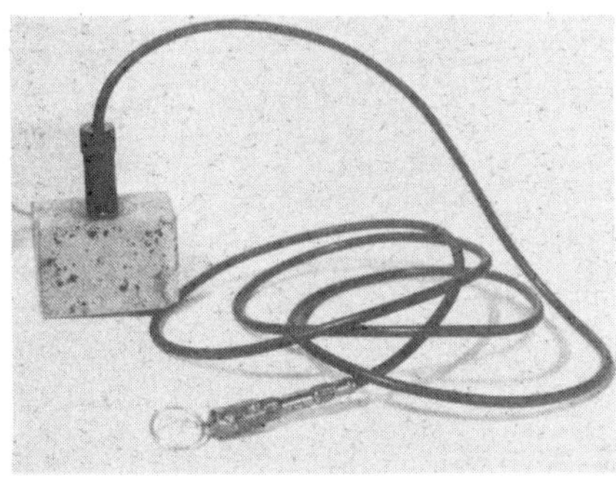

Standard German 200 g charge with detonator and holder inserted. Note the safety fuse with friction detonator

Brettstückmine. *An improvised mine made with 1 kg demolition charge secured to a wooden base and fitted with a pressure detonator. This type of mine was frequently used as a booby trap device*

Another type of Brettstückmine *made with 3 kg demolition charge and* Druckzünder 35 *pressure detonator*

Geballte Ladung 1 kg. *This explosive charge was packed in a watertight zinc container and provided with three threaded sockets for detonators*

Geballte Ladnung 3 kg *with a pull detonator in the top socket*

Geballte Ladung 10 kg. *This concentrated demolition charge had six standard detonator sockets embedded directly into the TNT*

An assault pioneer with a pole charge attacking an enemy pillbox

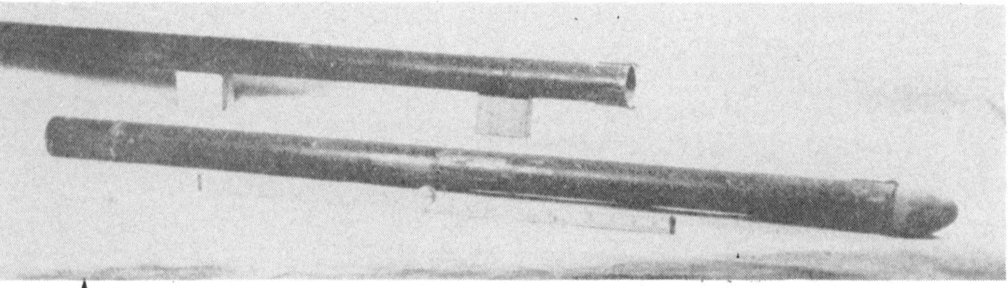

Rohrladung Stahl 3 kg *('Bangalore torpedo'). This demolition charge consisted of two sections of mild steel tube filled with 3 kg of TNT and a wooden nose cap. At the rear of the tube was a socket for a detonator and a safety fuse or delay detonator*

Haft-Hohlladung 3 kg *was a magnetic anti-tank assault weapon filled with hollow-charge explosive mixture. It was designed to be placed on enemy tanks and fired by means of a short delay detonator. Three strong magnets held it to the target*

Stangenladung *(Pole charge)*

An infantryman training to attach the 3 kg magnetic anti-tank charge to a tank

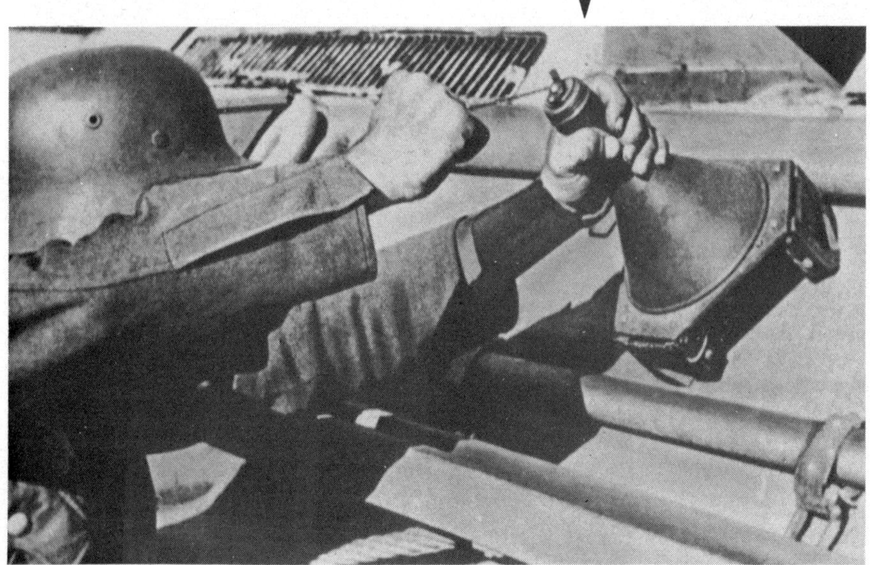

Schwerer Ladungsträger *(SdKfz 301). These demolition vehicles were produced in A, B and C versions, known collectively as B IV. All were fully tracked vehicles, large enough to carry a driver who controlled the machine during approach marches. In action the driver dismounted and the vehicle was controlled by radio*

'Goliath' Modell B *seen from above. The rear compartment contains the cable and drum, the central compartment – the power unit and control mechanism, and the front compartment – the explosive charge*

Leichter Ladungsträger 'Goliath' Modell B *(SdKfz 303) was powered by a 703 cc motorcycle engine and carried a heavier demolition charge*

Schwere Ladungsträger Modell A, *with explosive charge on the front cover and the driver's armoured shields raised. These were lowered when the vehicle was under radio control*

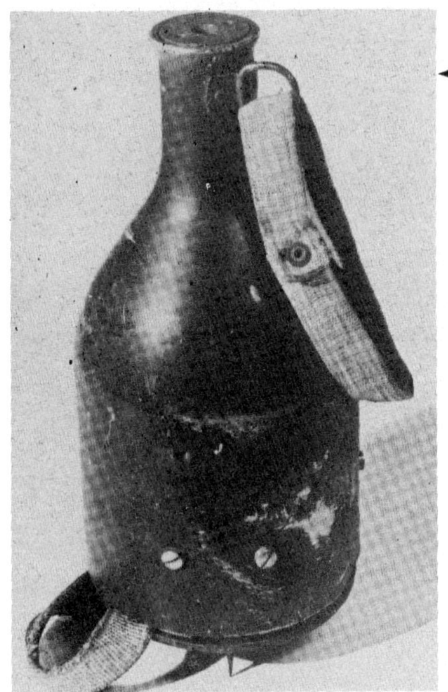

Panzerhandmine 3 *(PzHM 3). This assault weapon was similar in construction and use as the funnel-shaped 3 kg magnetic hollow-charge. The magnetic ring had three sharp steel spikes so that the weapon could be used with protective (anti-magnetic) skirting over the armour*

Abstandsladung H 15. *This hollow-charge weapon was used for attacking armour plate and reinforced concrete in static defences*

Leichter Ladungsträger 'Goliath' Modell A *(SdKfz 302) remotely-controlled demolition charge carrier was powered by electric motors. The operator's control box can be seen attached to the rear of the vehicle*

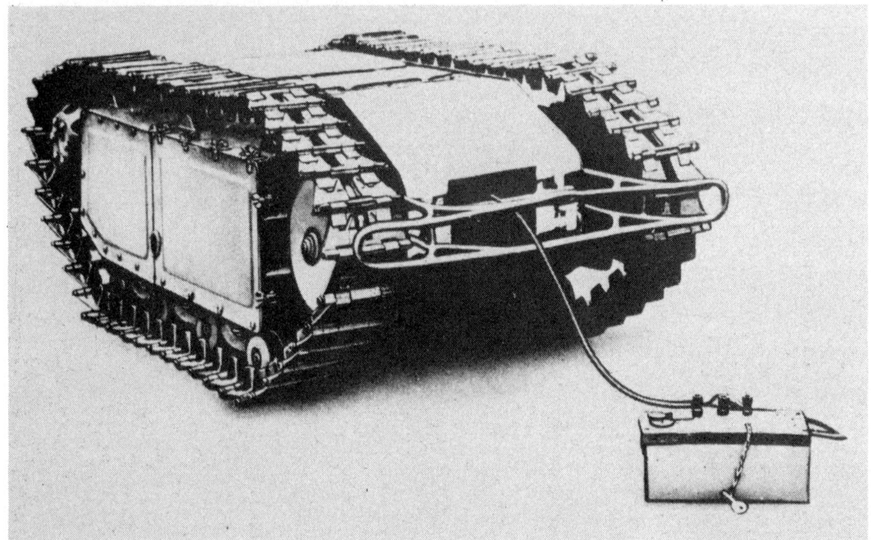

Attaching the control wires to the rear of the 'Goliath'

The 'Goliath' was brought to the point of release on a two-wheeled handcart

'Goliath' moving forward to blow up a disabled Soviet SU-85 assault gun

Glossary of German and Other Abbreviations Used in Text

Initial letter for country of origin

(a)	amerikanisch	American
(b)	belgisch	Belgian
(d)	dänisch	Danish
(e)	englisch	English, ie. British
(f)	französisch	French
(g)	griechisch	Greek
(h)	holländisch	Dutch
(i)	italienisch	Italian
(j)	jugoslawisch	Yugoslavian
(n)	norwegisch	Norwegian
(ö)	österreichisch	Austrian
(p)	polnisch	Polish
(r)	russisch	Russian; Soviet
(s)	schweizerisch	Swiss
(t)	tschechisch	Czech
(u)	ungarisch	Hungarian

Abbreviation	Meaning	Translation
A	*Aggregat*	Aggregate (experimental liquid-fuel rocket series)
AA	Anti-aircraft	
a/A	*alter Artillerie*	artillery of old pattern
abg	*abgeändert*	modified/converted model
Abt	*Abteilung*	detachment (of artillery etc.)
AFV	armoured fighting vehicle	
Anh	*Anhänger*	trailer
AP	armour-piercing	
AT	anti-tank	
Bett	*Bettung*	platform (railway gun); foundation, base (!xed gun)
BettGesch	*Bettungsgeschütz*	gun on platform mounting
BhMi	*Behelfsmine*	improvised mine
BhSkl	*Behelfssockellafette*	auxiliary pedestal mount
BK	*Blendkörper*	artificial smoke container
	Bordkanone	fixed aircraft cannon
C/	*Construktion*	model (German naval artillery only)
DrhL	*Dreheitslafette*	revolving gun carriage
E; (E)	*Eisenbahn*	railway
E-Flak	*Eisenbahn-Flak*	railway AA gun(s)
Ehl	*Einheitslafette*	dual-purpose gun mount/ carriage
Eihgr	*Eierhandgranate*	egg(shaped) hand grenade
EisbPzZ	*Eisenbahnpanzerzug*	armoured train
EisM; EsMi	*Eismine*	anti-personnel 'bottle' mine
EL	*Erdkampflafette*	ground mount
Fest	*Festung*	fortress
FH	*Feldhaubitze*	field howitzer
FK	*Feldkanone*	field gun
Flak	*Fliegerabwehrkanone*	anti-aircraft gun
FlaSL	*Fliegerabwehr-Sockellafette*	anti-aircraft pedestal mount
g	*goda* (Russian)	year, model (eg. 1939 g)
Geb	*Gebirgs-*	mountain
GebH	*Gebirgshaubitze*	mountain (pack) howitzer
Ger	*Gerät*	equipment
Gesch	*Geschütz*	artillery piece; gun
Gew	*Gewehr*	rifle
GewGr	*Gewehrgranate*	rifle grenade
GewGrPz	*Gewehrgranate Panzer*	anti-tank rifle grenade
GL Flak	*Generalluftzeugmeister, Amtsgruppe für Flak' Entwicklung*	
gl	*glatt*	smooth (bore)
GrW	*Granatwerfer*	trench mortar
H	*Haubitze*	howitzer
Haft; HaftHldg	*Hafthohlladung*	magnetic hollow anti-tank demolition charge
HE	high explosive	
Hgr; HdGr	*Handgranate*	hand grenade
HT	*Haubitze in Turm*	turret howitzer
HWA	*Heereswaffenamt*	Army Weapons Office
IG	*Infanteriegeschütz*	infantry support gun or howitzer
iHL	*in Haubitzen Lafette*	on howitzer carriage
iMrsLaf	*in Mörser Lafette*	on mortar carriage/mount
inc	incendiary	
K	*Kanone*	cannon; piece of ordnance
Kb; Kar	*Karabiner*	carbine
K(E)	*Kanone (Eisenbahn)*	railway gun
Kfz	*Kraftfahrzeug*	motor vehicle
KH	*Kanone-Haubitze*	gun-howitzer
KK	*Kasemattenkanone*	casement gun
KL	*Kasemattenlafette*	casemate gun mount
Kp	Krupp (arms manufacturer)	
KstBttr	*Küstenbatterie*	coastal battery
KstG	*Küstengeschütz*	coastal defence gun
KsL	*Küstenlafette*	coastal defence mount
L	*Lafette*	gun carriage; mount
L/	*Lang/*	length of barrel in calibres
LdgW	*Ladungswerfer*	spigot mortar
le	*leicht*	light
leFH; lFH	*leichte Feldhaubitze*	light field howitzer
LeG; LG	*Leichtgeschütz*	recoilless gun
leGebG	*leichtes Gebirgsgeschütz*	light mountain gun
leGrW	*leichter Granatwerfer*	light trench mortar
LM	*Lafette Marine*	naval pedestal mount
LPist	*Leuchtpistole*	signal/flare pistol
M	*Marine*	naval; Navy
	Mündungsbremse	(with) muzzle brake
m	*mit*	with
M-Schoss	*Minengeschoss*	thin-walled HE shell
m/b	muzzle brake	
MG	*Maschinengewehr*	machine gun
MK	*Maschinenkanone*	automatic cannon
MKb	*Maschinenkarabiner*	automatic carbine
MP; MPi	*Maschinenpistole*	sub-machine gun

Abbreviation	Meaning	Translation
MPL	*Marine Pivot Lafette*	naval pivot gun mount
Mrs	*Mörser*	short large-calibre howitzer; mortar
MV	muzzle velocity	
n/A	*neuer Artillerie*	new pattern artillery piece
Nb	*Nebel*	artificial smoke
NbK	*Nebelkerze*	artificial smoke candle
NbW	*Nebelwerfer*	artificial smoke projector; also rocket launcher
NSDAP	*National Sozialistische Deutsche Arbeiterpartei*	National Socialist German Workers Party; 'Nazi'
Workers Party;'Nazi'		party
obr.	*obrazets* (Russian)	pattern, model (year)
OKH	*Oberkommando des Heeres*	Army High Command
OKL	*Oberkommando der Luftwaffe*	Luftwaffe High Command
OKM	*Oberkommando der Marine*	Navy High Command
OKW	*Oberkommando der Wehrmacht*	Supreme Command of the Armed Forces
Pak	*Panzerabwehrkanone*	anti-tank gun
PivL	*Pivotlafette*	pivot mount; rotating gun mount
PPG	*Peenemünder Pfeilgeschoss*	Peenemünde dart-shaped missile
PzB	*Panzerbüchse*	anti-tank rifle
PzGr; Pzgr	*Panzergranate*	armour-piercing AT projectile; AP
R	*Rakete*	rocket
RfG	*Rückstossfreies Geschütz*	recoilless gun
RhB	*Rheinmetall-Borsig*	arms manufacturer
RPzB	*Raketen-Panzerbüchse*	anti-tank rocket launcher
RVfW	*Raketen-Vielfachwerfer*	multiple rocket launcher
RW	*Raketenwerfer*	rocket launcher

Abbreviation	Meaning	Translation
S	*Sonder*	special
s	*schwer*	heavy
SchüMi	*Schützenmine*	anti-personnel mine
sFH	*schwere Feldhaubitze*	heavy field howitzer
sGrW	*schweres Granatwerfer*	heavy trench mortar
SK	*Schiffskanone*	warship gun; naval gun
SK C/12	*Schiffskanone, Construktion 12*	naval gun 1912 pattern
sK	*schwere Kanone*	heavy cannon/gun
SLK	*Schelladekanone*	quick-firing/rapid fire gun
SkL; SockLaf	*Sockellafette*	pedestal mount for gun
SlGew	*Selbstladegewehr*	self-loading/automatic rifle
sMG	*schweres Machinengewehr*	heavy machine gun
S-Mi	*Schrapnellmine*	shrapnel anti-personnel mine
Sprgr	*Sprenggranate*	HE shell
sPzB	*schwere Panzerbüchse*	heavy anti-tank rifle
sWuR	*schwere Wurfrahmen*	heavy frame-type rocket launcher
TeMi; T-Mi	*Tellermine*	dish-shaped anti-tank mine
TbtsK	*Torpedobootskanone*	heavy torpedo boat gun
ToMi	*Topfmine*	pot-shaped land mine
u	*und*	and
UbtsK	*Unterseebootskanone*	submarine gun
umg	*umgeändert*	modified
V	*Versuchs*	experimental
vz	*vzor* (Czech)	pattern, model
W	*Werfer*	launcher (rockets etc.)
WaPrüf	*Waffenprüfstelle*	weapon test/approval centre
WG	*Wurfgerät*	heavy rocket launcher
Wgr	*Wurfgranate*	rocket shell/projectile
w/o	without	
wz	*wzór* (Polish)	pattern, model
Z; Zw	*Zwilling*	twin-barrel; twin mount

Index

Pistols

Revolvers

Flare Pistols

Bolt-action Rifles

Automatic Rifles

Sub-machine Guns

Light Machine Guns

Heavy Machine Guns

Anti-Tank Rifles

Anti-Tank Guns

Light Anti-Aircraft Guns

Medium Anti-Aircraft Guns

Heavy Anti-Aircraft Guns

Light Field Guns

Medium and Heavy Artillery

Railway Guns

Railway Anti-Aircraft Guns

Armoured Trains

Coastal Artillery

Fortress Weapons

Infantry Guns

Mountain Artillery

Mortars

Recoilless Weapons

Rockets

Coastal Artillery

Fortress Weapons

Infantry Guns

Mountain Artillery

Mortars

Recoilless Weapons

Rockets

Light Field Guns

Medium and Heavy Artillery

Railway Guns

Railway Anti-Aircraft Guns

Armoured Trains

Flamethrowers

Handgrenades

Landmines

Demolition charges